NATURAL ELECTROMAGNETIC PHENOMENA

BELOW 30 KC/S

Proceedings of a NATO Advanced Study Institute

NATURAL ELECTROMAGNETIC PHENOMENA

BELOW 30 KC/S

Proceedings of a NATO Advanced Study Institute
Held in Bad Homburg, Germany
July 22-August 2, 1963

Edited by

D. F. Bleil

U.S. Naval Ordnance Laboratory

SPRINGER SCIENCE+BUSINESS MEDIA, LLC
1964

ISBN 978-1-4899-6210-2   ISBN 978-1-4899-6425-0 (eBook)
DOI 10.1007/978-1-4899-6425-0

Library of Congress Catalog Card Number 64-25831

TABLE OF CONTENTS

TABLE OF CONTENTS
(continued)

PREFACE

A NATO Advanced Study Institute was conducted during the period of 22 July through 2 August 1963 in Bad Homburg, Germany. All Natural Electromagnetic Phenomena occurring in the environment of the earth with particular emphasis on those phenomena influenced by solar disturbances provided the subject matter of the Sessions. The meeting was organized and arranged through the joint efforts of the U. S. Naval Ordnance Laboratory, Silver Spring, Maryland, and the Office of Naval Research, largely through the London Office. The Institute was approved and financially supported by the NATO Science Committee.

The program was divided into sixteen one-half day sessions, each again divided approximately equally into a lecture period and a following discussion period, at which time the attendees contributed to the Session topic. In addition to the sixteen faculty members, there were in attendance throughout the Session thirty-six participants representing most of the NATO countries. A list of attendees may be found in the back of the book.

This book contains papers presented at the Institute, along with the lecturers' summary of the significant discussion pertaining to the paper.

The editorial work was done by D. F. Bleil, who is entirely responsible for all aspects of the book except for the technical accuracy of the papers. That responsibility belongs to the respective authors.

The Directors herewith express their appreciation to the NATO Science Committee for their encouragement and support without which this Institute could not have been held. In addition, we wish to offer thanks for the encouragement and material support provided by the many Offices within the U. S. Navy who, through their fine services, contributed immeasurably to the success of the Institute. Of these, the assistance given by the Office of Naval Research, Washington, D. C., and their London Office, the Navy Liaison Office, Frankfurt, Germany, and the U. S. Naval Ordnance Laboratory was exceptional. In particular, we gratefully acknowledge the enthusiastic and continued support of Captain J. Sloatman, Office of Naval Research, London, whose efforts and encouragement carried the planning of this Institute over many threatening minima.

To Professor C. Menneken goes the credit for the successful operation of the meeting. We are indebted to him for his insight, careful planning, good counsel, and untiring efforts before and during the meeting, all of which factors are directly responsible for the very smooth operating sessions we all enjoyed.

In addition, we offer our sincere thanks to Delores Williamson and Rebecca Larsen who served so ably as our conference secretarial staff, to Romelle Grabar, who as administrative assistant to one of us (D. F. Bleil) diligently pursued the continuum of correspondence attendant with the planning and terminal phase of the conference, and to Sylvia Kirkpatrick for her very fine job of retyping the entire proceedings.

In conclusion, we express our thanks to the Plenum Press for their guidance and prompt publication of the book.

D. F. BLEIL

I. ESTERMANN

INTRODUCTORY TALK

D. F. Bleil
Associate Technical Director for Research
U. S. Naval Ordnance Laboratory
White Oak, Silver Spring, Maryland

INTRODUCTION

Background

To Dr. William Gilbert goes the credit for the discovery of the magnetic field of the earth. In his book, written in 1600, he described many experiments including the magnetized sphere "terrella." The earth's magnetic field has since been a phenomenon of both scientific and commercial interest and our understanding of it has improved substantially over the years.

The instrumentation employed has been continuously improved with the result that vastly more knowledge exists about the earth's magnetic field and its temporal and spatial variations. But as it so often happens, the more one investigates a particular subject the more likely he is to uncover additional problems which must be investigated and geomagnetism is no exception. Today we are adding new information at a rate which is beginning to saturate our ability to assimilate and forge it into a coherent description of the phenomena involved. For example, the number of papers published on the subject of low frequency electromagnetic radiation as a function of the calendar year shows a sharp break in the curve corresponding to the beginning of the recent International Geophysical Year. The number of papers published since 1955 probably exceeds the total number published on the same subject in thirty or more years prior to that date. There are other indications of the rapid growth of interest in the field of the natural low frequency electromagnetic phenomena. It would be gratifying indeed if this new wealth of information were nicely integrated into a completely unified theory, but unfortunately this is not yet the case.

SOME UNSOLVED PROBLEMS

Instrumentation and Signal Processing

Today we measure the static magnetic field with high sensitivity (10^{-2} gamma) total-field or single-component magnetometers located almost everywhere on the surface of the earth, in rockets and satellites which reach out to many earth radii. The use of high-permeability cores in magnetic loops provides us with an extremely sensitive, low frequency electromagnetic measuring device. The time variations of the earth's magnetic field over a wide frequency range are measured and recorded automatically. Signal processing techniques employing high-speed computers make possible the accelerated reduction of the data and further, the use of correlation techniques, enables signals to be lifted out of the background noise. New electronic techniques are being continuously developed to make the life of the experimenter far easier and to help improve the quantity and quality of the data which he obtains. It is readily apparent that the researcher in geomagnetism today has both the equipments available to make his measurements and a wealth of supporting information with which to generate and test his theory. The obvious question is: how well does the theory account for the measurements we make today, and how well does it guide the future experiments? Although there is no doubt that the foundation of the present theory of the various geomagnetic field relations is sound, it is also not complete. One purpose of this Institute is to aid in the strengthening of the important structural members of the super-structure of this theory.

Conjugate Points

The magnetic field lines of the geocentric dipole pierce the air-earth interface at a pair of geographical points called the conjugate points or mirror points. The location of these points may be obtained by direct calculation. Charged particles rotating about the magnetic field lines with cyclotron frequency and oscillating back and forth along the line between mirror points should be observed at both points of a conjugate pair. The many simultaneous measurements made at the two ends of the magnetic field lines have resulted in variable degrees of success. Hook concluded that his coherent ionospheric absorption measurements were obtained at points located within $\pm$ 2° latitude of the calculated points. Yanagihara, investigating noise bursts and other micropulsations, found a "very loose conjugacy" at the terminal points. The manner in which the location of the conjugate points depends on the actual shape of the earth's magnetic field as a function of time and the influence of the ionic layers on these points requires description.

Time Variations of the Magnetic Field

Low frequency magnetic field variations fall in a band about seven decades wide. Although there is rather good agreement on the source of the excitation of these fluctuations, there doesn't appear to be agreement on the mechanism responsible for the measured frequencies nor on a description of the effects measured. The identification technique usually adopted for explaining the origin of the signal is to first classify it according to some frequency band, structure, and the time of occurrence of the signal. The frequency spectrum (Fig. 1) includes micropulsations (P_t, P_c, LP_c, PP, SIP, IPDP), gyromagnetic resonances, Schumann resonance, solar whistlers (Gold suggested that some low frequency rising or following signals, measured at PNL, were solar whistlers), dawn chorus, hiss, whistlers and sferics. (Of these, the sferics and whistlers are best understood.) Gallet and Helliwell suggest that "hiss" and the "dawn chorus" are noises excited in the exosphere by streams and bunches of high-speed ionized particles. The method of excitation of these signals is similar to that of the traveling wave tube. Ward calculated the transfer impedance of the magnetosphere and accounted for P_t's and magnetic bays. The P_t's travel as a hydromagnetic wave with a rise time of several minutes in keeping with the calculations given by Dessler. On the other hand, Madame Troitskaya's carefully timed measurements of the onset of SSC led her to state that it arose simultaneously all over the globe. Her results have been substantiated by Maeda and Ondoh for the Johnston Island measurements. Some arguments can be raised against most of the mechanisms proposed for the generation of the several effects mentioned above.

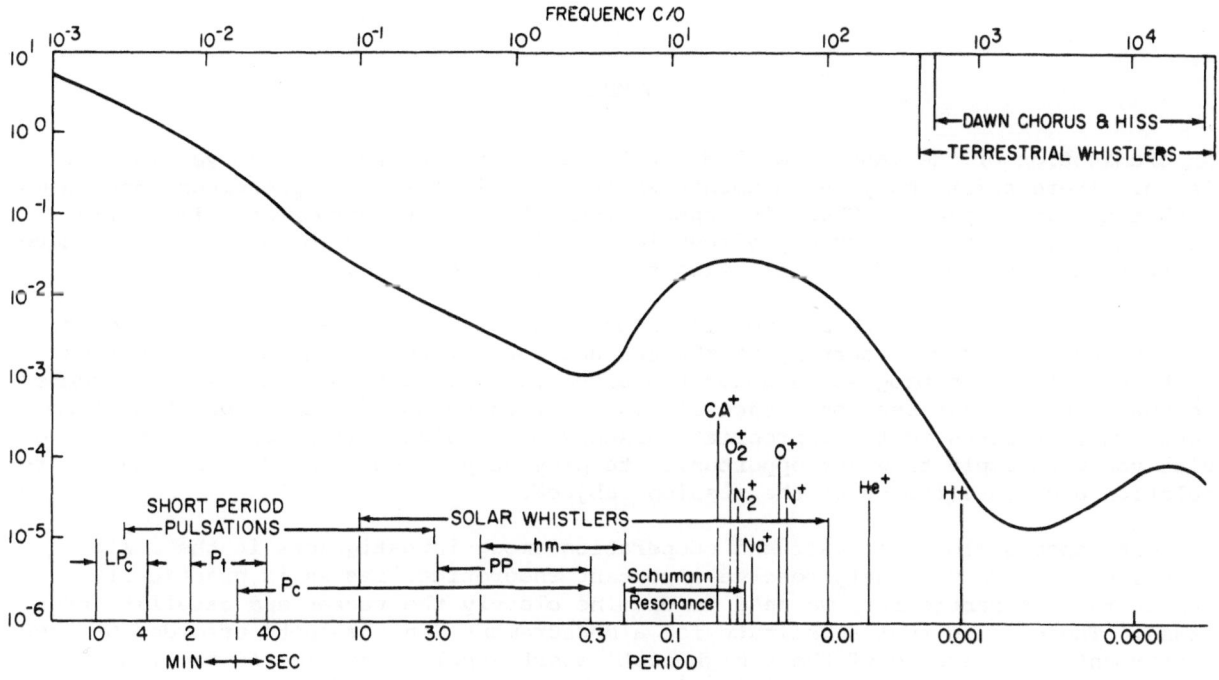

Fig. 1.   Geomagnetic Frequency Spectrum.

Trapped Particles

There are many calculations of trapped particles in the Van Allen Belt, but the relationship among these trapped particles, the aurora, upper ionosphere ion density and protonsphere has not been adequately developed. According to Wentzel, the magnetic moment of the proton does not remain constant in the region of the Van Allen Belt. He states that this effect will lower the mirror point which may influence the conjugate point measurements. Further, a change of the magnetic moment of the proton will change gamma ($\omega = \gamma H$, $\gamma = \frac{g}{h} \mu = \gamma = 2.67513 \cdot 10^4$ rad/gauss sec) and therefore change the Larmor precession frequency measured by Egeland.

Magnetosphere

Johnston points out that the rapidly-moving, highly-conducting solar plasma (solar wind) impinging upon the earth's magnetic field should deform it and confine it in extent. He proposed the tear drop model. He suggests that turbulence probably exists at the boundary. Dessler argues that it should not produce turbulence. However, Coleman and Sonett point out that from satellite measurements there is a 100 γ fluctuation at the interface, which unfortunately does not appear at the surface of the earth.

Calculations of the extent of bound-field confine it to several earth radii, whereas satellite measurements indicate that the field probably extends to 10 R_e. The shape and structure of the bound-field conceivably could influence the calculated velocity of the hydromagnetic wave.

3

SUMMARY

An examination of the many unsettled questions indicates that we have yet much to learn. There exist many measurements which deal directly with phenomena associated with the magnetosphere. Thus, it appears that this is an appropriate time for a meeting, with planned leisurely discussion sessions, in order to review what we have measured and determine how it all fits together, if possible.

In the planning of this Study Institute, we attempted to arrange the program, in a natural sequence, starting at the Sun and working toward the Earth's surface. Each session of the program is under the direction and control of the faculty member responsible for that session. Each of you is urged to contribute information from your own researches to the appropriate discussion session. It is our intent to provide you with ample time and opportunity to present your work and discuss its interrelation and significance to the session subject.

It appears closer contact and cooperation among investigators in the design experiments which will help resolve important theoretical issues is practically mandatory. In particular, we need to examine closely the rocket and satellite programs. There is no real substitute for a measurement "on-the-spot" afforded by these instruments but because of their high cost, short supply, and limited sampling ability we must make maximum utilization of them through carefully designed experiments. We hope that this meeting will lead to a better contact and cooperation among the experimenters and produce a stronger motivation of significant cooperative research.

SOME ASPECTS OF THE STRUCTURE
AND DYNAMICS OF THE TERRESTRIAL MAGNETOSPHERE

W. I. Axford
Center for Radiophysics and Space Research
Cornell University
Ithaca, New York

THE INTERPLANETARY MEDIUM

The suggestion that the sun continuously emits corpuscular radiation (the "solar wind") was made by Biermann (1951, 1952, 1957 and 1961) as an explanation of the fact that gaseous comet tails tend to trail directly away from the sun. It had been known for many years that magnetic storms must be the result of isolated bursts of corpuscular radiation, and the fact that some aurora and geomagnetic disturbance always occur at high latitudes may be considered as additional evidence that the radiation is continuous. Biermann's suggestion was taken up by Parker (1958, 1960a, 1960b and 1961), who examined the dynamics of the solar wind and showed that it may be considered as a hydrodynamic expansion of the solar corona and a direct consequence of supplying heat to maintain the corona at its observed temperature. There is certainly no possibility that the sun has a static corona as suggested by Chapman (1957 and 1959).

Early estimates of the density and speed of the solar wind were understandably rather variable. However, we now have direct evidence from space probes (Neugebauer and Snyder, 1962), which confirms the existence of the solar wind and shows its flux to be of the order of 10^8 to $10^9\,cm^{-2}\,sec^{-1}$ during relatively quiet periods, rising to $10^{10}\,cm^{-2}\,sec^{-1}$ during magnetic storms. The solar wind speeds range from 300 to 800 km sec^{-1} and the densities are 5 to 10 protons cm^{-3}; the motion is apparently supersonic with a Mach number of ~ 5. (That is, the bulk velocity of gas is roughly 5 times greater than the mean thermal velocity of the individual particles.) There is no direct evidence that the solar wind moves other than radially from the sun. The solar magnetic field is relatively weak, being $\sim 5\gamma$ at 1 a.u. during quiet periods, and $\sim 50\gamma$ during magnetic storms ($\gamma = 10^{-5}$ gauss) (Coleman et al, 1962). As a consequence of solar rotation and the radial motion of the solar wind, the solar magnetic field lines take the form of a spiral, and may be considered as rotating rigidly with the sun. The latter effect causes an anisotropy in the cosmic radiation which provides a means of determining the solar wind speed without having recourse to space probes (Ahluwalia and Dessler, 1963).

Sudden heating due to a flare on the sun must enhance the expansion of the corona, and result in the formation of an outward moving shock wave in the interplanetary medium (Parker, 1958, 1960a, 1960b, and 1961; Gold, 1955). The shock wave, together with the associated increase in solar wind speed and density, produces a magnetic storm as it interacts with the geomagnetic field. Localized enhancements of the solar wind may occur due to "hot spots" in the corona; these may persist for several solar rotations and are believed to cause magnetic storms which repeat over a 27-day period (Chapman and Bartels, 1950). The leading edge of such a region of enhanced solar wind must form a shock wave on interacting with the normal solar wind, and it is easy to see that the shape of the shock wave must be a spiral making a slightly smaller angle with the radial direction than the solar magnetic field lines. This shock wave ensures that the resulting magnetic storm has a sudden commencement, as do the storms associated with transient heating of the corona. Its curved shape causes the storms to occur some days after central meridian passage of the "hot spot" around the sun.

Cavity Formation.

The magnetohydrodynamic concept according to which magnetic field lines are "frozen" into a perfectly conducting fluid, makes it apparent that the geomagnetic field must produce a cavity in the solar wind - the whole of this cavity down to an altitude of about 100 km, forms the earth's magnetosphere. Clearly, it is an important prerequisite to an understanding of magnetospheric phenomena that we should know as much as possible about the shape and structure of the cavity. This is not easy, however, as cavitation problems in fluid dynamics are notoriously difficult. But we can make some theoretical predictions, and direct observations from space probes are now beginning to add considerably to our knowledge.

A few solutions to cavitation problems in magnetohydrodynamics are known: these concern incompressible flow of perfectly conducting inviscid fluid in two dimensions. The cavities produced as a result of uniform flow past a two-dimensional magnetic dipole (Kulikovskii, 1957), and a line current (Cole and Huth, 1959), are shown in Fig. 1. At the boundary of such cavities the normal pressure (fluid plus magnetic)

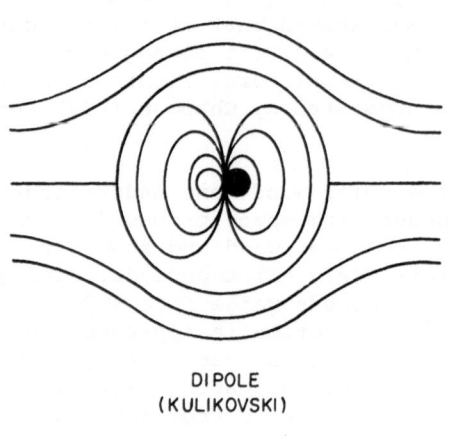

DIPOLE
(KULIKOVSKI)

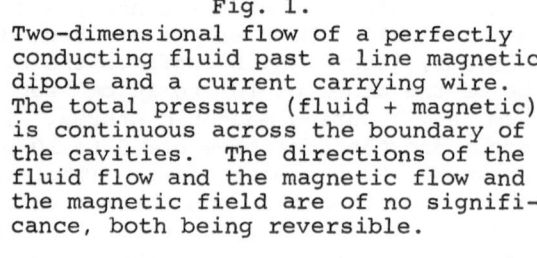

Fig. 1.
Two-dimensional flow of a perfectly conducting fluid past a line magnetic dipole and a current carrying wire. The total pressure (fluid + magnetic) is continuous across the boundary of the cavities. The directions of the fluid flow and the magnetic flow and the magnetic field are of no significance, both being reversible.

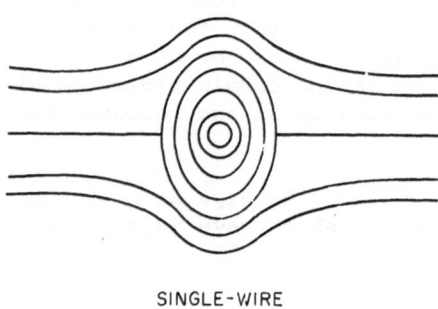

SINGLE-WIRE
(COLE AND HUTH)

must be continuous. It should be noted that in each case the pattern is symmetrical and the direction of flow therefore reversible - this is a consequence of the neglect of dissipative effects. However, in the case of the magnetosphere even if dissipative effects at the boundary are ignored, we still have to contend with the three-dimensional nature of the problem and the fact that the solar wind is highly supersonic.

Shock Waves.

It is necessary to consider the solar plasma as a continuum rather than a "free molecular" or "Newtonian" flow in its interaction with the magnetosphere since collective motions such as plasma oscillations and magnetohydrodynamic waves can occur with wavelengths much smaller than the typical length scale involved (that is, about 10^5 km). Thus, since the solar wind is supersonic, some form of shock wave must be produced on the upstream side of the magnetosphere (Axford, 1962a). This reduces the bulk velocity of the plasma to a low subsonic value near the forward stagnation point and permits the plasma to flow around the magnetosphere. The motion of the plasma eventually becomes supersonic once again on the flanks of the magnetosphere, although the free stream conditions may not be regained. It should be noted that the shock wave is an irreversible phenomenon and hence the magnetosphere can be expected to be nonsymmetrical even if all other dissipative processes are neglected.

This shock wave and also the shock wave which is believed to cause the sudden commencement of a magnetic storm (Gold, 1955) must be of the type known as "collision free." It is unfortunate this term has been allowed to come into common usage as it is rather misleading. In an ordinary non-ionized gas the collision frequency of molecules and the highest possible frequency of a sound wave are essentially the same. Thus, the steepening of a compressive pulse is halted when its width becomes comparable to the mean free path. In a plasma, however, one tends to equate the mean free path to the distance a particle must travel in quiescent plasma before its momentum is significantly altered, whereas sound waves do not become heavily attenuated until the product of their frequency and the electron-ion collision frequency becomes comparable to the product of the electron and ion gyrofrequencies. Thus, a compressive pulse in a plasma can steepen until its width is much less than the mean free path, although thermal equilibrium may not be restored until the plasma has moved through the latter distance. The processes underlying shock waves in a plasma are by no means understood; however, it is usually considered that non-linear interactions of waves produced by instabilities cause the required randomization of particle motions (Fishman et al, 1960). The shock structure and the region downstream should therefore appear to be highly turbulent and bursts of supra-thermal particles may occur.

Theoretical Discussion of the Structure of the Magnetosphere.

It is generally agreed that the magnetosphere must have a more or less tear drop shape as indicated in Fig. 2, the first to suggest this being Johnson (1960) and Piddington (1960). However, it is not always realized that such a magnetosphere is likely to have essentially the same topological features as the early model of Chapman and Ferraro (1931), which made use of an image dipole to distort the geomagnetic field into an infinite half-space. Details of the Chapman-Ferraro model are shown in Fig. 3. The main features to notice are that all field lines on the surface of the region of magnetic field run between two neutral points (N_n, N_s) which are in turn connected by a single field line to points (M_n, M_s) at high latitudes on the central (noon) meridian of the earth. Field lines which intercept the earth at higher latitudes than M_n, and M_s are shown shaded. These constitute the geomagnetic "tail." The low latitude field lines (unshaded region) form a donut around the earth and are completely enclosed by the field lines forming the tail. Obviously, a uniform solar wind must cause the magnetosphere to take up a shape more or less as sketched in the Fig. 2. However, this distortion is not likely to change the topology of the magnetic field from that of the Chapman-Ferraro model. In particular, all the features described above should be retained and thus, we have labeled and shaded the various diagrams in a similar manner to make this apparent. It is interesting to note that Hones (1963) has recently produced a more sophisticated version of the Chapman-Ferraro model using parallel unequal dipoles so that the field of the stronger dipole completely encloses that of the weaker, which is taken to be the geomagnetic field; not surprisingly, this model also retains all the topological features of the simple image dipole model.

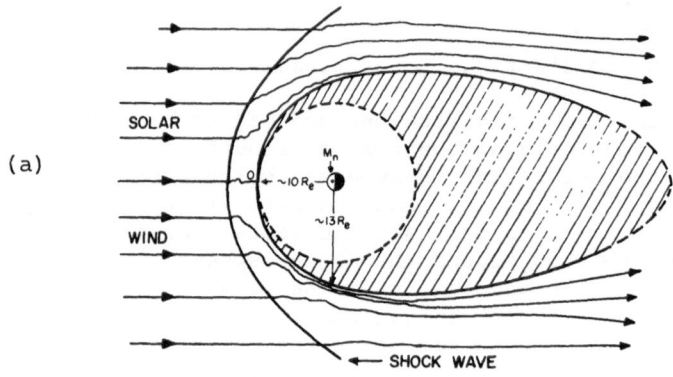

(a)

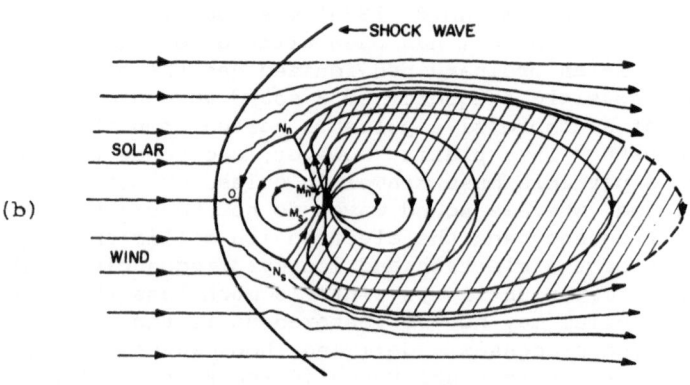

(b)

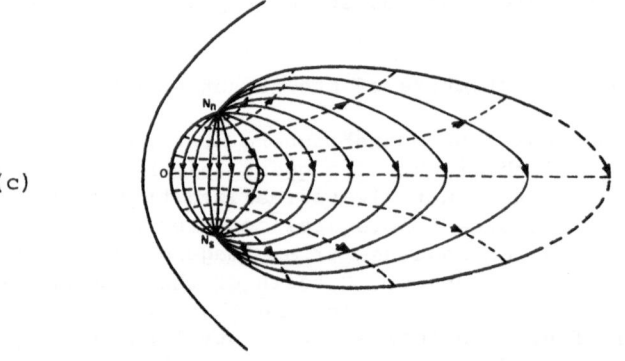

(c)

Fig. 2. (a) Equatorial section of the magnetosphere looking from above the north
             pole. The geomagnetic tail (shaded) completely envelopes the low
             latitude donut (unshaded). The downstream boundary of the tail is
             dotted to indicate our uncertainty with regard to its extent.

        (b) North-south section of the magnetosphere.

        (c) Elevation of the surface of the magnetosphere showing the surface
             field lines running between the neutral points (N_n, N_s). Flow
             lines of the eastward surface current are indicated by the dashed
             lines.

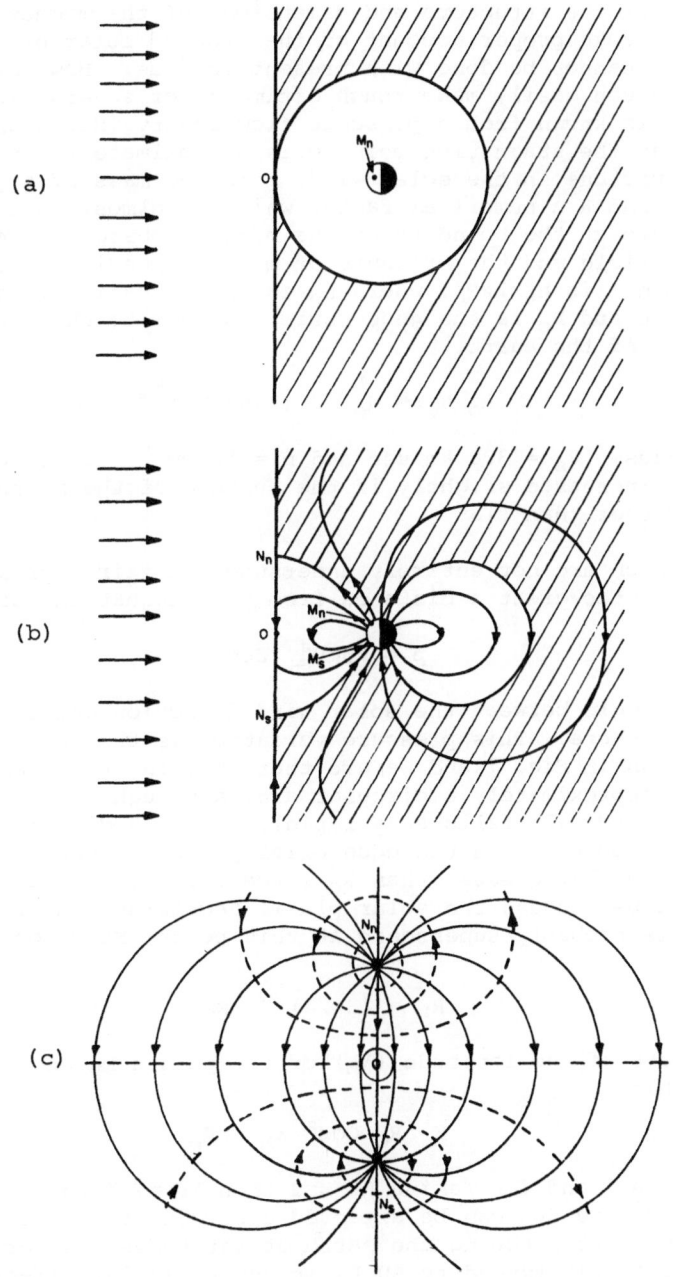

(a)

(b)

(c)

Fig. 3. Features of the Chapman-Ferraro model of the magnetosphere
corresponding to the features sketched in Fig. 2. The surface elevation (c) is shown as it would appear looking in
the sun-earth direction.

9

Exact details of the structure and dimensions of the magnetosphere cannot be given as these must take proper account of the flow of solar plasma around the magnetosphere and this cannot be done with present methods. However, this is not a great disadvantage as one can easily make rough estimates of several key dimensions of the magnetosphere. It is known from hypersonic flow theory (Hayes and Probstein, 1959) that the pressure at the stagnation point O is approximately $p = KnmV_s^2$, where n is the number density of protons in the solar wind, m is the mass of a proton, V_s the speed of the solar wind, and K a numerical factor which is almost unity. This must equal the magnetic pressure at the boundary of the magnetosphere (if we neglect the interplanetary magnetic field and the pressure of magnetospheric gas), which according to the image dipole model is $H_e^2/2\pi R_O^6$, where H_e is the magnetic field strength at the equator of the earth and R_O is the geocentric distance to the stagnation point. Thus, if R_e is the radius of the earth,

$$R_O/R_e \approx H_e^{1/3} / (2\pi KnmV_s^2)^{1/6} \tag{1}$$

Taking $H_e = 0.32$ gauss, $V_s = 300$ km/sec and $n = 10$ cm^{-3}, it is found that $R_O \simeq 10\ R_e$. The value of R_O is increased slightly if the surface of the magnetosphere is taken to be hemispherical rather than flat.

For reasons which are not entirely understood, a fair approximation to the solar wind pressure at not too great a distance from the stagnation point is given by

$$p = KnmV_s^2\ \cos^2\theta, \tag{2}$$

where $\cos\theta$ is the angle between the solar wind direction and the normal to the surface of the magnetosphere. This pressure variation is the same as for flow past a circular cavity shown in Fig. 1 and we are therefore led to expect that the surface field line $N_n ON_s$ is approximately a circular arc, although it should be remembered that the accuracy of Eq. (2) falls away rapidly as $\cos\theta$ diminishes. The geocentric distance (R_1) to the boundary in the equatorial plane perpendicular to the sun-earth direction must be somewhat greater than R_O since the pressure is decreased from the stagnation point value because the solar plasma accelerates as it passes around the magnetosphere and is probably supersonic at this point; Harrison (1962) suggests

$$R_1 \simeq (3\pi/4)^{1/3}\ R_O \tag{3}$$

A rough estimate of the latitude (λ_M) of the points M_n and M_s can be obtained from the relationship

$$K^1\ R_O\ \cos^2\lambda_M = R_e, \tag{4}$$

where K^1 lies between 1 and 2. Taking $K^1 = 2$ then $\lambda_M = 77°$ when $R_O = 10\ R_e$, and a variation between 70° and 80° can be expected according to the solar wind pressure. Since all field lines intersecting the earth at latitudes greater than λ_M is as small as 70° the tail must extend to 60 R_e in the antisolar direction, assuming the average width of the magnetosphere to be 20 R_e and the average field strength in the tail to be 30γ. This assumes that the component of the magnetic field perpendicular to the equatorial plane is 30γ, however, it is possible that the tail could be much longer if the field lines lie mostly parallel to the equatorial plane; if so, the tail could be as much as 20 R_e in radius if the field is 20γ and λ_M has the more modest value of 75°.

A number of workers have calculated the approximate shape of the magnetosphere assuming a Newtonian solar wind with specular reflection of particles incident on the surface (Beard, 1960; and Spreiter and Briggs, 1962). In this case Eq. (2) holds everywhere and K is taken to be 2. The shape of the magnetosphere appears to be rather insensitive to the physical nature of the exterior wind, and the results of such calculations (if done correctly) (Davis and Beard, 1962) are not substantially different from what has been described above, apart from the discrepancy in K. In particular, one does not expect any significant deviation from the topological features exhibited by the Chapman-Ferraro model.

The question of whether or not the magnetosphere is stable is of some significance. Analyses based on the assumption of specular reflection of solar wind particles suggest that the surface is unstable in the Helmholtz manner (Dungey, 1955, and Parker, 1957). However, the equivalent continuum solution (Axford, 1960, 1962b) suggests that the surface could be stable when the relative velocity is sufficiently low. Dessler (Van Allen, 1959) has argued that the surface is stable on the basis of the steadiness of low-latitude magnetograms during magnetically quiet periods when the solar wind is presumed to be relatively steady. However, there is no good reason for believing that the effects of any instability would be noticeable at low latitudes, while it is known that high-latitude magnetograms show continuous disturbance at all times. Since the relative velocity of the solar wind and the surface of the magnetosphere is greatest on the afternoon side of the magnetosphere (see Fig. 9), one would expect instability to occur in this region if anywhere. The instability would be observable only as a waviness of the surface of the magnetosphere since any small scale "spray" would probably be indistinguishable from the interplanetary medium. Rayleigh-Taylor instability due to the acceleration of the surface of the magnetosphere during a magnetic storm sudden commencement may also occur (Axford, 1962a). However, it would be short-lived and very difficult to observe directly.

Observations of the Shape of the Magnetosphere.

Space probe observations are now beginning to confirm the general picture of the magnetosphere which has been outlined here. There is unfortunately very little information about the state of the geomagnetic tail. However, a good deal is known about the magnetic field structure on the sunward side of the magnetosphere.

The earliest suggestions of the occurrence of a region of turbulent plasma lying between the surface of the magnetosphere and the standing shock wave produced by the solar wind came from Pioneer I (Sonett, Smith and Sims, 1960) and Pioneer IV (Van Allen, 1959). The Geiger counter record from Pioneer IV shows what is apparently trapped radiation out to about 60,000 km in the sunward direction, a region of fluctuating counting rate between 60,000 km and 90,000 km, and a steady count beyond 90,000 km (Fig. 4). We interpret the counts in the intermediate region as being due to bremsstrahlung from electrons with energies of the order of tens of kev. produced by random electric fields in the turbulent plasma behind the standing shock wave. The steady count beyond 90,000 km is apparently the cosmic ray background. The magnetometer on Pioneer I showed similar effects, although the record is not continuous - the portion of the record taken at 12.6 R_e shows the strong fluctuations characteristic of the turbulence behind the shock wave while the record at 14.8 R_e shows a much smaller and quieter field which one would associate with undisturbed solar wind (Fig. 5).

More recently, observations from Explorer XII have shown the surface of the magnetosphere to be quite well-defined, in terms of both the magnetic field (Cahill and Amazeen, 1963) and trapped electrons (Freeman, Van Allen and Cahill, 1963) (Fig. 6). A remarkable feature of the magnetic measurements is that the magnetic field tends to reverse its direction at the magnetospheric boundary - no satisfactory explanation of this effect has yet been proposed. These results have been extended by Explorer XIV (Frank, Van Allen and Macagno, 1963), the orbit of which makes much greater angles with the sun-earth line than Explorer XII. A summary of the electron flux measurements (for energies $\simeq$ 40 kev.) is given in Fig. 7. Note that the magnetosphere is indeed wider perpendicular to the sun-earth line in the equatorial plane than it is in the sun-earth direction. Also the electron flux appears to be low in the interior of the geomagnetic tail, and increases toward the surface. In the low-latitude region, the lines of constant flux are somewhat asymmetric with respect to day and night, and a rapid decrease of flux occurs on the midnight meridian at about 7 R_e, which is rather closer than the boundary of the magnetosphere on the noon meridian ($\sim$ 10 R_e).

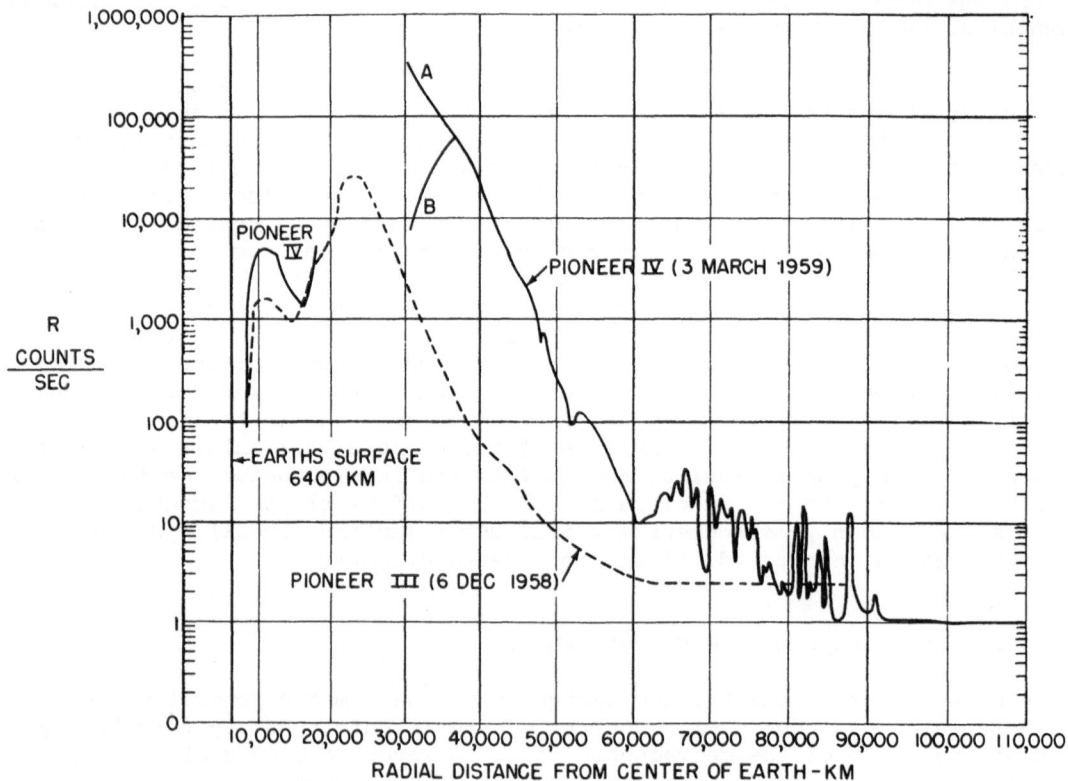

Fig. 4.   Counting rate of GM counters carried aboard Pioneers
III and IV (Van Allen, 1959).

This effect, which was observed earlier at low altitudes by Injun I (O'Brien, 1963),
may be due partly to the asymmetry of the magnetic field, and partly to the effect of
electric fields which drive the particles toward lower latitudes, as will be explained
in a later section.

The boundary of the magnetosphere was first observed by Explorer X, which passed
through it at a distance of about 40 R_e on the downstream flank of the magnetosphere,
well below the equatorial plane (Heppner et al, 1963; Bonetti et al, 1963).   Within
the boundary and magnetic field was observed to be relatively steady, while outside
it was weaker and highly disturbed, and a supersonic flow of plasma (presumably the
solar wind) always appeared.   The boundary moved across the trajectory of the satellite several times, possibly as a result of changes in the solar wind, or alternatively as a result of a large scale waviness of the boundary.   A striking feature of the
Explorer X observations is that the geomagnetic field at great distances from the
earth is almost radial, and that there was no sign of the magnetosphere closing on
itself.   The geomagnetic tail must therefore be very extensive, and there is a likeli-
hood of interesting effects taking place near the equatorial plane in the tail region,
which may almost be a magnetically null surface.

The results obtained from Russian space probes are on the whole disappointing.
Gringauz and his colleagues (Gringauz et al, 1960a, 1960b) have reported the existence
of a "third radiation belt" which may be connected with the turbulent region between
the magnetospheric boundary and the standing shock wave.   However, the concurrent
magnetic measurements did not make this clear.   There is one particularly interesting

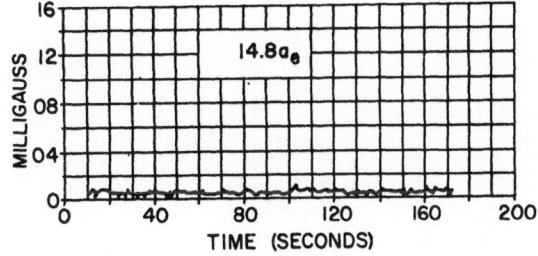

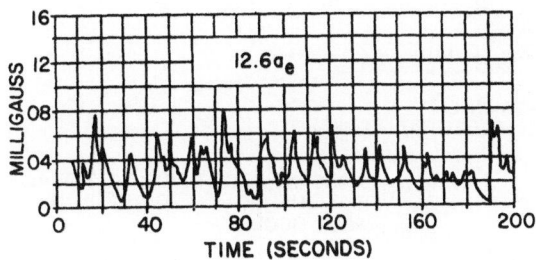

Fig. 5.
Magnetometer record from Pioneer I
(Sonett, Smith and Sims, 1960).

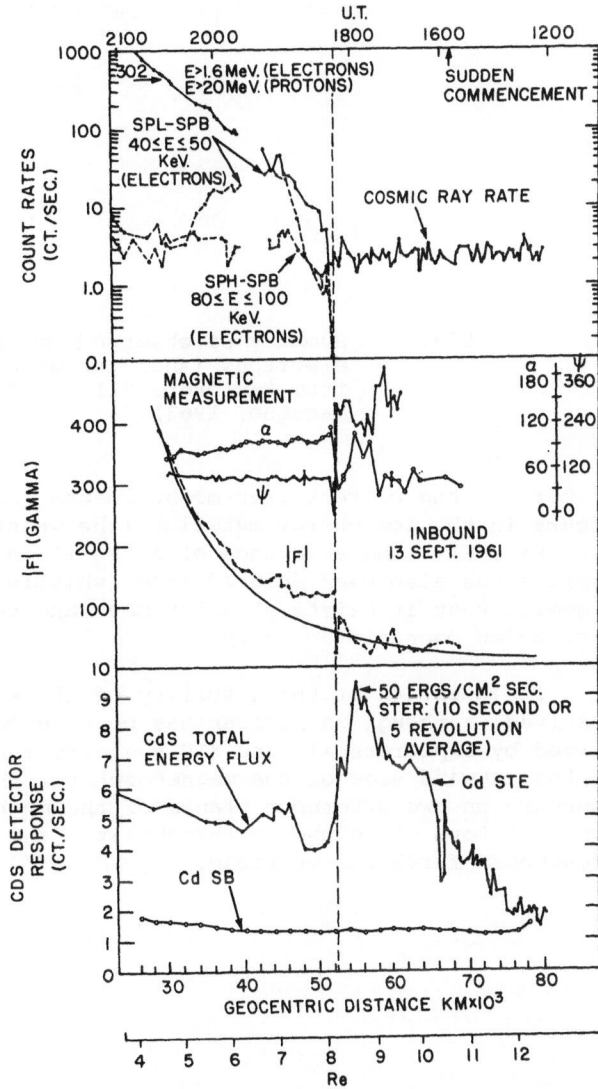

Fig. 6.
Details of the transition occurring
at the surface of the magnetosphere
as observed by Explorer XII (Freeman,
Van Allen and Cahill, 1963).

13

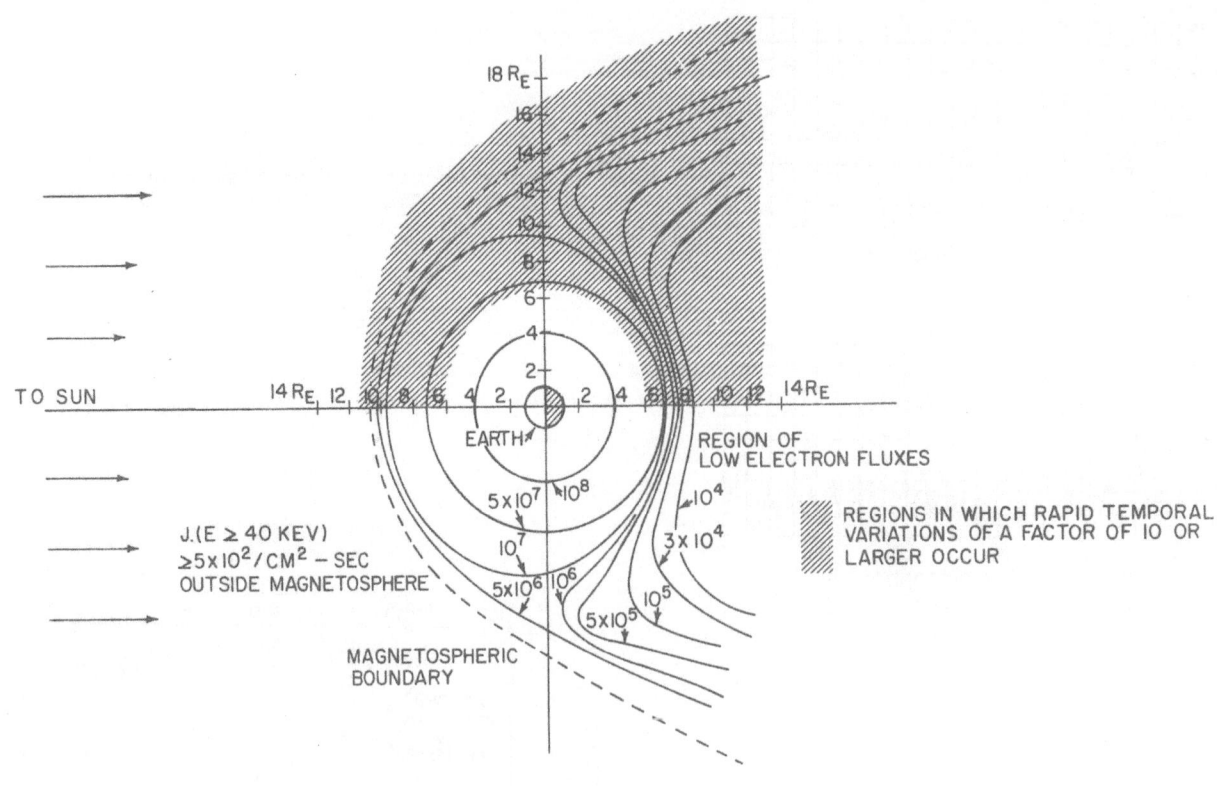

Fig. 7. Summary of observed omnidirectional intensities of
         electrons (energies greater than 40 kev) obtained
         from Explorers XII and XIV (Frank, Van Allen and
         Macagno, 1963).

feature of the current trap measurements (Gringauz, 1963), namely that an "edge"
occurs in the low energy material (the whistler medium) at a distance of about 4 R_e
(see Fig. 8). The existence of a "knee" in the magnetospheric ionization density
profile has also been deduced from whistler measurements by Carpenter (1963), who
suggests that it exists at all times, and tends to move inward toward the earth with
increasing magnetic activity.

The question of the stability of the surface of the magnetosphere is not entirely
resolved, although as Dessler has pointed out (1962), the well-defined "edge" observed by Explorers XII and XIV suggests that small scale instabilities do not occur
on the sunward side of the magnetosphere. However, the possible waviness of the
boundary on two afternoon flanks of the magnetosphere, observed by Explorer X may be
due to a form of Helmholtz instability, and one would expect this to be even more
pronounced further downstream.

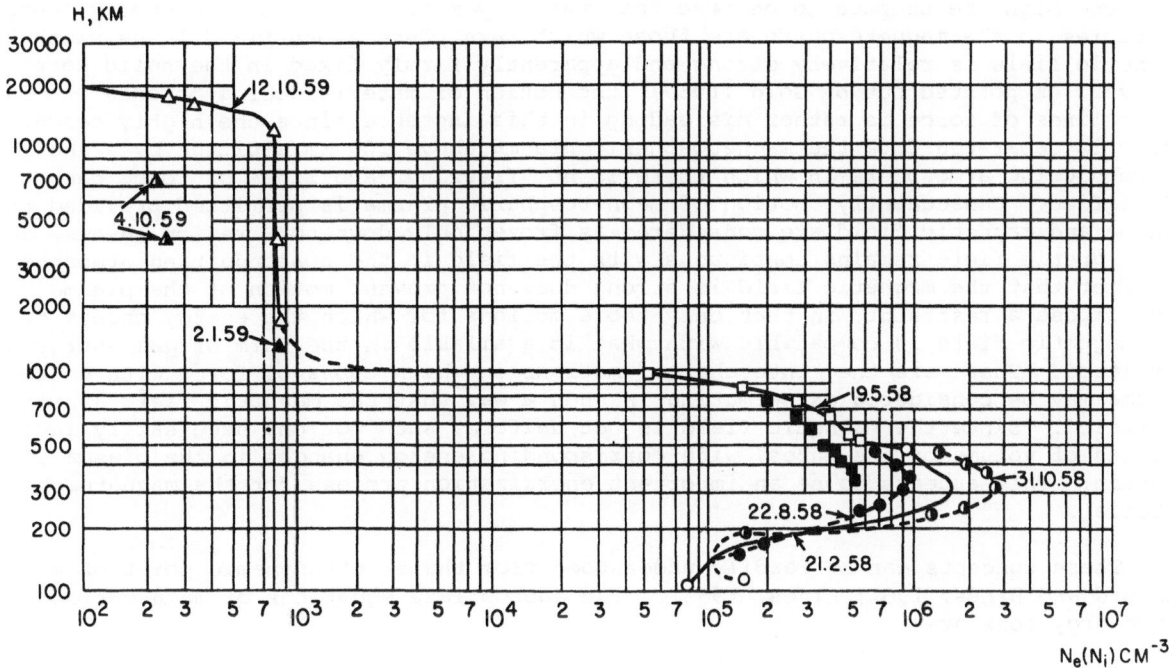

Fig. 8. Approximate altitude distribution of the charged
particle concentration in a period of nearly maximum
solar activity (Gringauz, 1963).

MOTIONS IN THE MAGNETOSPHERE

Theoretical Aspects.

It is a most significant feature of the magnetosphere that the energy density of
the gas it contains is usually everywhere much less than the energy density of the
magnetic field; thus, the magnetic field dominates the mechanics of the interior of
the magnetosphere. That is, the magnetosphere has a shape determined by the solar
wind and the attitude of the earth relative to the sun, and provides a sort of magnetic framework which is largely unaffected by internal processes. In fact, this is
not always exactly true as the gas pressure is believed to become significant occasionally due to the temporary trapping of large numbers of energetic particles during
the main phase of geomagnetic storms, producing the so-called "ring current" effect;
even in this case, however, the gas pressure probably never exceeds the magnetic
pressure.

One might be tempted to believe that the only motions open to low energy charged particles in the magnetosphere are those which take place along field lines since the magnetic field is relatively strong and apparently firmly fixed in the solid earth. However, as pointed out by Gold (1959), the notion of material being "frozen" to magnetic lines of force is rather misleading in this instance since the highly conducting earth is separated from the magnetospheric plasma above 100 km altitude by a shell of nonconducting atmosphere in which there is no "freezing" and lines of force cannot be identified. Consequently, motion of magnetospheric plasma is permitted provided the plasma and magnetic field are considered as frozen only down to E region levels, and the magnetic field remains continuous with the field in the nonconducting atmosphere. The fact that the magnetic field is strong does not prevent motion of the plasma but does impose a restraint, in that only those motions for which the energy change of the magnetic field is compatible with what is available in the form of gas energy are permitted. These are the "interchange" motions in which tubes of force with enclosed plasma may be considered as permuting in such a way that the magnetic field appears unaltered. Since the magnetic field is not uniform, such motions necessarily involve changes of volume of the tubes, with corresponding energy changes in the plasma they contain. This appears to be an important energization process for the magnetospheric material.

These concepts can be readily understood from the electrodynamic point of view (Axford and Hines, 1961; Hines, 1963). The appropriate equations of mean motion for low energy ions are

$$M \, D\underline{V}_{\perp} / Dt = e_i \, (\underline{E}_{\perp} + \underline{V}_{\perp} \times \underline{B}) - MK_n \, (\underline{V}_{\perp} - \underline{V}_{\perp} \, n) \qquad (5)$$

$$M \, D\underline{V}_{\parallel} / Dt = e_i \, \underline{E}_{\parallel} - MK_n \, (\underline{V}_{\parallel} - \underline{V}_{\parallel} \, n), \qquad (6)$$

where M is the mass of the individual ion, e_i is its charge, K_n the "frictional" frequency (related to the collision frequency) of the ions with neutral particles, $\underline{V}_n$ the mean velocity of neutral particles, $\underline{V}$ the mean velocity of the ions, $\underline{E}$ the electric field and $\underline{B}$ the magnetic induction. The subscripts $\perp$ and $\parallel$ refer to the directions perpendicular and parallel to the magnetic field, respectively. We have ignored collisions between charged particles, partial pressure gradients, viscous effects, and gravity, although each of these may be important in restricted regions. A similar set of equations hold for electrons. On considering the order of magnitude of the various terms, it can be readily seen that provided the frequencies of the mean motion are small compared with the gyro-frequency $e_i \, B/M$ the "inertia" term on the left of Eq. (5) can be neglected. Throughout most of the magnetosphere (namely above about 150 km altitude) the frictional frequency for ions is also small compared with the gyro-frequency, and this is true for electrons down to 80 km altitude. Thus, for all particles above 150 km the equation of motion reduces to

$$\underline{E}_{\perp} + \underline{V}_{\perp} \times \underline{B} = 0. \qquad (7)$$

In the "dynamo" region between altitudes of 90 to 140 km the frictional frequency for ions is large compared with the gyro-frequency and Eq. (5) may be written approximately as

$$e_i\underline{E}_{\perp} - MK_n \, (\underline{V}_{\perp} - \underline{V}_{\perp} \, n) = 0, \qquad (8)$$

while the electrons obey Eq. (7). Along the field lines, particles accelerate in a quasi-steady electric field until collisions become significant and the inertia term on the left of Eq. (6) may be neglected; thus

$$e_i \underline{E}_\parallel - MK_n \left(\underline{V}_\parallel - \underline{V}_{\parallel n} \right) = 0. \tag{9}$$

Since the frictional frequency is small throughout most of the magnetosphere, it is reasonable to reduce Eq. (9) further to $\underline{E}_\parallel \simeq 0$, although small departures from this exact condition may have interesting effects.

As a consequence of Eq. (7) all constituents above about 150 km altitude move approximately with mean velocity

$$\underline{V}_\perp = \underline{E}_\perp \times \underline{B} / B^2 \tag{10}$$

We describe this as the "motion" of lines of force with the plasma which identifies the lines being frozen to them, since the application of

$$\partial \underline{B}/\partial t = \mathrm{curl}\ \underline{E} \tag{11}$$

to Eq. (7) leads to a relation which is analogous to Kelvin's vorticity theorem in ordinary hydrodynamics. In quasi-steady conditions $\underline{E}$ is derivable from a potential such that

$$\mathrm{grad}_\perp \emptyset = \underline{V}_\perp \times \underline{B}; \tag{12}$$

hence, the lines of force and the streamlines of the motion must be equipotentials of the electric field.

Below 150 km altitude, the electrons can still be considered as frozen to the lines of force down to 90 km and they therefore move perpendicular to $\underline{E}_\perp$, according to Eq. (10). The heavy ions, however, can drift only slowly relative to the neutral gas in the direction of the electric field, as indicated by Eq. (8). Thus, there is a Hall current perpendicular to the electric field carried mainly by the electrons, and a much smaller direct (Pedersen) current parallel to the electric field carried mainly by the ions. We can, therefore, interpret ionospheric current systems as indications of the presence of large scale electric fields and these in turn imply motions elsewhere in the magnetosphere as given by Eq. (10). If there were no motion of the neutral atmosphere ($\underline{V}_n = 0$) it would be possible to interpret the current systems directly in terms of motion of the "feet" of lines of force in the magnetosphere since the predominance of the Hall current would require only the reversal of the sense of the current system to give the flow lines. However, when $\underline{V}_n = 0$, and particularly when $\underline{V}_n$ and $\underline{V}$ are comparable, this procedure may be very difficult as the motion of the neutral atmosphere at ionospheric levels is not very well known.

In this discussion we have neglected the effects of the movement of plasma on the neutral atmosphere. Although the density of neutral gas in the ionosphere is much larger than that of the ionized gas, given sufficient time the latter can force the former into motion. In fact, the time required is not long in comparison with most periods of interest here, being about 20 minutes in F region of the ionosphere and several hours in the E region.

Energization of Charged Particles Due to Motions in the Magnetosphere.

We have suggested above that changes in volume associated with magnetospheric motions in which tubes of magnetic flux are permuted, cause changes in the energy of the particles concerned. In fact, the energy changes take place in such a manner that the magnetic moment and "longitudinal" adiabatic invariant of the individual particles are conserved. It is the nonuniformity of the magnetic field which leads us to speak of "changes in volume" causing energy changes. Not surprisingly, if we consider the motion of individual particles it is this nonuniformity which causes the energy changes, since the motion of the particles is given by

$$\underline{V} = \underline{E} \times \underline{B}/B^2 + \underline{V}_d \tag{13}$$

17

if we ignore electric fields parallel to the magnetic field lines. $\underline{V}_d$ is the drift due to nonuniformity of the magnetic field - this is energy and charge dependent, and it is more or less in the longitudinal direction in the case of the geomagnetic field. All particles take part in the $\underline{E}$ x $\underline{B}$ / B^2 motion and this in itself does not produce any change in energy since the motion is along equipotentials of the electric field. However, the energy dependent drift $\underline{V}_d$ does in general cause the particles to move across these equipotentials and changes in energy occur accordingly.

Low energy particles follow the $\underline{E}$ x $\underline{B}$ / B^2 motion quite closely, and the longitudinal drift can be considered as a perturbation of this motion. For high energy particles the $\underline{V}_d$ drift is more significant than the electric field drift and hence their motion may be considered as more or less longitudinal with perturbations due to the electric field gives us a measure of the maximum amount of energy which can be given to a particle as a result of the magnetospheric motion - this is usually of the order of tens of kilovolts, and thus the maximum energy change for any particle due to this process is of the order of tens of kev. Particles with energy greater than about 100 kev. are therefore only moderately perturbed by the magnetospheric motion, but particles of energies of about 1 kev. will tend to follow magnetospheric motion quite closely. Further discussion of these processes can be found in papers by Hines (1963), Dungey (1963) and Sonnerup and Laird (1962).

The type of energy change we have described takes place reversibly, with the energy of the particle increasing as it is carried to lower geomagnetic latitudes. However, it seems fairly certain that in addition, irreversible energy changes take place due to fluctuations in the electromagnetic environment of the particle, and these lead on the average to a general increase of energy with time. Thus, in considering the effects of magnetospheric motions, it should be realized that although the motion might ideally cause 1 kev. particles to be carried into the interior of the magnetosphere and be energized to perhaps 20 kev. in doing so, irreversible processes can be expected to increase the energy to a much greater level if sufficient time is available and the motion does not carry the particle out too soon. Consequently, the net effect of such motions is irreversible and leads to the accumulation of energetic particles in the interior of the magnetosphere if they can escape precipitation into the atmosphere.

Rotational Motion of the Magnetosphere.

Due to viscous effects, the rotation of the earth is impressed upon the upper atmosphere. The electrons and ions of the ionosphere also take part in this motion and the electric field generated is transmitted throughout the magnetosphere since, as described above, the magnetic field lines must be equipotentials in quasi-steady conditions. The equipotentials in the equatorial plane of the magnetosphere corresponding to rotation are shown in Fig. 9. It should be noted that the low latitude donut co-rotates with the earth more or less rigidly, while the high latitude field lines which form the geomagnetic tail "twiddle" around so that the sense of rotation in the equatorial plane is reversed.

The total electric potential variation between the equator and the poles is 88 kilovolts of which only about 10 to 15 kilovolts are associated with the geomagnetic tail (depending on the value of λ_M). The gradient of this electric potential, however, becomes very large in the tail as the outer surface of the magnetosphere is approached. This can easily be seen from the following argument (Axford, 1963). The time taken to complete a circuit of any streamline or equipotential is 24 hours. Since the electric field and the magnetic induction are continuous in the interior of the magnetosphere, field lines with feet at latitude λ_M (which therefore lie exactly on the surface dividing the low latitude donut from the geomagnetic tail) take 24 hours to cover the portion of the streamline which lies inside the magnetosphere. The remaining portion which forms the surface of the magnetosphere should ideally be

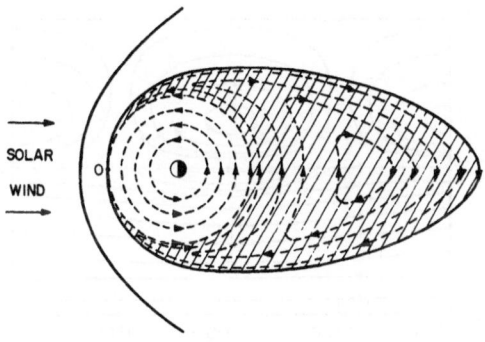

SOLAR
WIND

Fig. 9. Equatorial section of the magnetosphere looking from above the north
pole and showing the streamlines of the motion due to rotation of
the earth (or alternatively, the equipotentials of the corresponding
electric field). The geomagnetic tail is shaded as in Fig. 2.

completed at infinte speed, corresponding to a very large electric field or potential
gradient, although dissipative effects must prevent the singularity from actually occurring.

Motions Corresponding to the Current Systems S_q and L.

It is believed that the ionospheric electric current systems S_q and L are induced
by motions of the neutral atmosphere due to solar and lunar tidal effects respectively; the lunar tide is probably a gravitational phenomenon but the solar tide may be a
consequence of atmospheric heating (Hines, 1963). The S_q current system is shown in
Fig. 10 and the lunar-induced system is rather similar, but much weaker (Chapman and
Bartels, 1950). Diagrams of this type are deduced from the magnetic variations assuming that the currents are closed within the ionosphere and that currents which flow
along lines of force between conjugate points can be neglected. These assumptions are
inaccurate (Dougherty, 1963), but the general picture of the current system so found
is probably reasonably correct.

The S_q system produces the most significant motions that occur at low latitudes
and as it is possible that the motions of the neutral atmosphere are relatively weak
at these latitudes, the magnetospheric motion may be derivable with reasonable accuracy
from the current system. Thus, we see that there is motion in the magnetosphere from
east to west at the equator during the day with a movement of the feet of lines of
force toward higher latitudes in the morning and lower latitudes in the evening. When
combined with rotation (see Fig. 10), it can be seen that a closed cell is formed at
the equator on the sunward side of the magnetosphere. It seems likely that the
motion away from the equator before noon is partly the cause of the curious diurnal
variations of the equatorial F region ionization (Duncan, 1960). The lunar current
system L is much weaker than S_q and the corresponding magnetospheric motions are accordingly little more than a perturbation of the pattern sketched in the slide; however, the perturbation does appear to be noticeable at times in that L influences the
intensity of sporadic E ionization associated with the equatorial electrojet
(Bandyopodhyay and Montes, 1963).

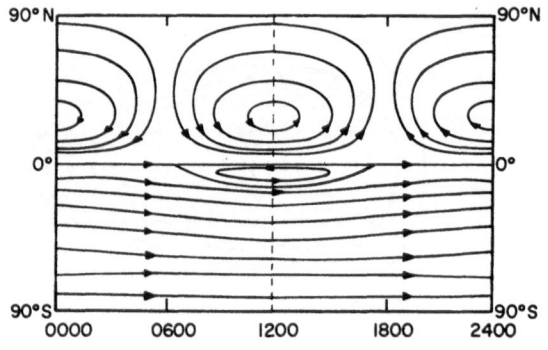

Fig. 10.
Sketches of the S_q current system in the latitude range 0-90°N. On revising
the arrows one obtains approximately the sense of motion of the feet of lines
of force in the ionosphere. This motion is combined with the motion due to
the rotation of the earth to give the net motion shown in the latitude range
0-90°S. Note the closed cell which is formed at low latitudes during the day.

Magnetospheric Motions Associated with Polar Current Systems.

The most intense magnetospheric currents at high latitudes are those forming the
D_S system which appears during magnetic storms (Chapman and Bartels, 1950). A somewhat idealized sketch of the pattern of currents is given in Fig. 11, where it will
be noticed that on reversing the arrows the sense of motion on the night side is in
accord with what is observed for ionization irregularities associated with the aurora
(Harang and Triom, 1960). The corresponding motion in the equatorial plane of the
magnetosphere is also sketched in Fig. 11, where it can be seen that the plasma moves
past the earth from roughly the antisolar direction toward the sun. The total variation in the electric potential appears to be of the order of 20 kilovolts (Axford,
1963).

There is some controversy concerning the origin of the D_S current system. We will
shortly consider the hypothesis that the motions and currents are driven by the solar
wind. However, some authors suggest a purely tidal mechanism (Cole, 1960) and others
suggest that the radiation belts are involved (Fejer, 1961; 1963). There seems nevertheless no doubt that this current system with its associated magnetospheric motions
contains the key to an understanding of the phenomenon of the aurora and probably to
the origin of much of the trapped radiation.

Whatever the mechanism of the D_S current system, it would be surprising if there
were not some vestige of it even during magnetically quiet periods. This has recently
been found by Nagata and Kokubun (1962), who give it the symbol S_q^P and show that it
has essentially the same pattern as the D_S system (Fig. 12).

Conjugate Point Variations.

Small differences of electric potential between conjugate points in the northern
and southern hemispheres tend to be annulled by currents flowing with little resistance along the lines of force. If, however, the potential difference is maintained
by some process such as an atmospheric tide, the current is continuous and it imparts
a twist to the field lines which is uniform along their length. Even if the angle of

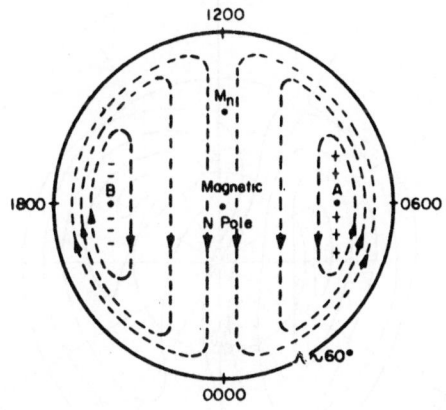

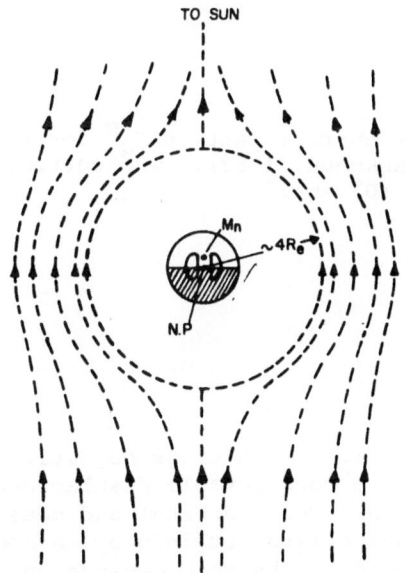

Fig. 11.

(a)   An idealized sketch of the ionospheric motion correspond-
      ing to D_S current system in the north polar cap above
      geomagnetic latitude $\Lambda \approx 60°$. The + and – signs indicate
      polarization charges associated with the electric field.

(b)   Motion in the equatorial plane of the magnetosphere
      corresponding to ionospheric motion in (a).

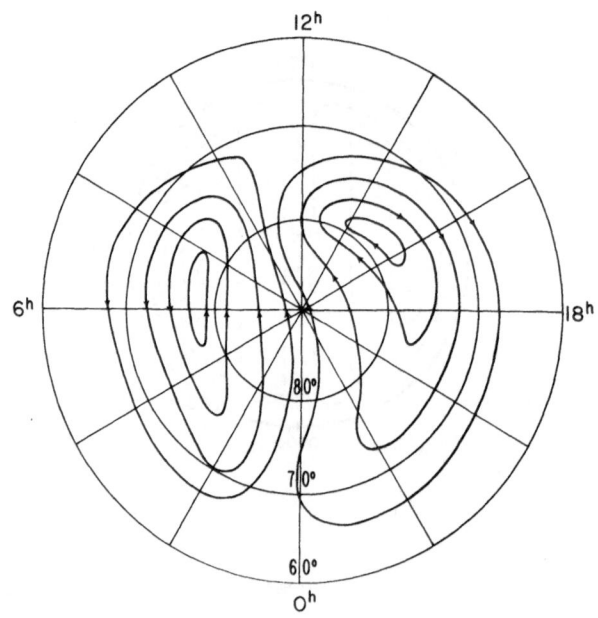

Fig. 12.   The quiet-day current pattern S_q^p for the south polar cap
(Nagata and Kokubun, 1962).   The current between adjacent
lines is 2×10^4 amp.

the twist is quite small the great length of geomagnetic field lines, at high latitudes especially, can result in a considerable displacement of the ends of the field
lines from their normal position.   Thus, diurnal and seasonal variations can be expected in the apparent conjugate points obtained by methods such as riometer techniques, since much of the absorption observed depends on the penetration of energetic
electrons into the atmosphere and these in turn move along lines of force in the
magnetosphere.

Information causing changes in the electrical potential on a magnetic line is
propagated in the form of hydromagnetic waves.   A transient change in the potential
in the ionosphere at one end of the field line can therefore lead to the formation
of a wave which may travel back and forth along the field line a number of times if
dissipation at the ends is not too great.   Some low frequency emissions can be expected
to be produced in this way from ionospheric variations.

The Theory of Axford and Hines.

Axford and Hines (1961) have suggested that the basic processes underlying high latitude disturbance phenomena (including the aurora) is the motion in the magnetosphere associated with the D_S current system. It is proposed that this motion, which is sketched for the equatorial plane of the magnetosphere in Fig. 13, is due to a viscous-like interaction between the solar wind and the surface of the magnetosphere.

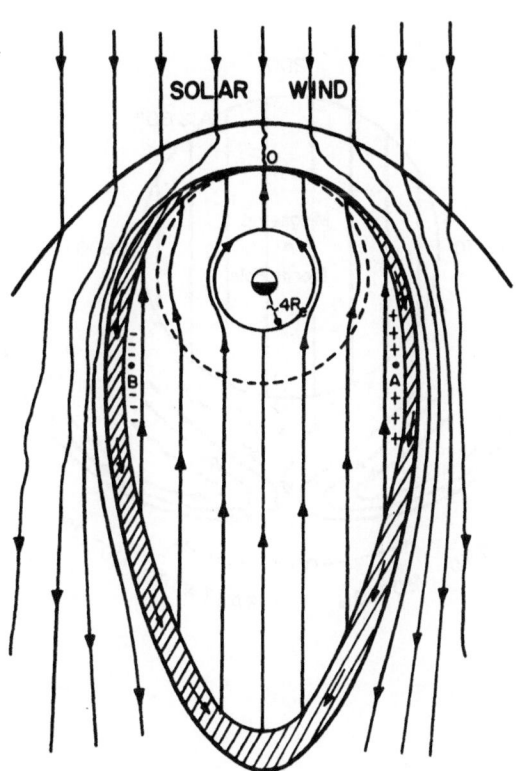

Fig. 13. The proposed circulatory motion in the equatorial plane of the magnetosphere looking from above the north pole. The points A and B and the polarization charges correspond to those in Fig. 11 (a). The hatched area indicates the regions in which field lines near the surface of the magnetosphere are dragged along by the solar wind on the exterior.

This interaction produces an internal circulation in the interior of the magnetosphere rather similar to that occurring in a falling rain drop. Note that the dragging of lines of force around the surface of the magnetosphere cannot be very noticeable at ionospheric levels due to the fact that, ideally anyway, the surface field lines are connected to only one point in each hemisphere; hence, only the internal part of the circulation is evident in the form of the D_S current system and, at quiet times, the S_q^p current system.

There are obvious merits to the suggestion. In the first place, the circulation provides a simple and effective accelerating mechanism which can energize captive solar wind particles from about 1 kev. up to 10 kev. (corresponding to conservation of magnetic moment between fields of 50γ at the magnetospheric surface and 500γ at a distance of 4 R_e). Secondly, the polar current systems and the pattern of alignment of auroral arcs at high latitudes (Fig. 14) are reproduced by the circulation at ionospheric levels, giving us some understanding of the nature of the current systems and

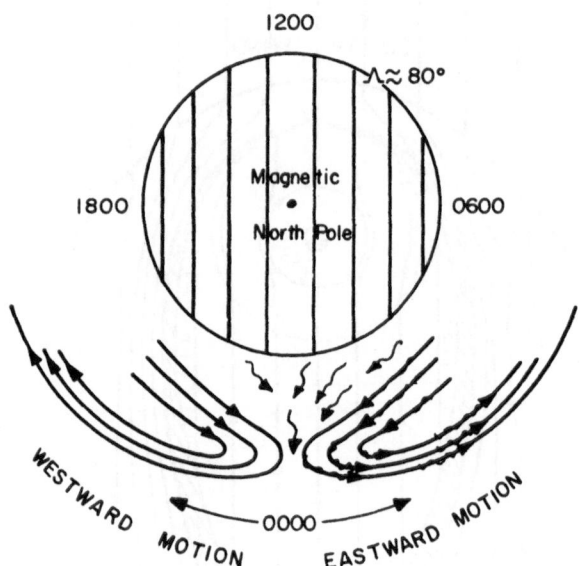

Fig. 14. Motion of auroral irregularities (Davis, 1960 and 1962b). Above geomagnetic latitude 80°, auroral arcs tend to be aligned in the sun-earth direction, as indicated. The lines in the post midnight region at lower latitudes are intended to represent the effects of breakup. The arrows indicate the sense of motion of irregularities.

their relationship to the motion of auroral irregularities. Thirdly, by combining the circulation with the motion associated with the rotation of the earth, it is possible to predict qualitatively much of the morphology of many high latitude disturbance phenomena. Finally, the circulation in combination with irreversible accelerating processes provide an input mechanism for the outer zone of geomagnetically trapped radiation.

In the view of Axford and Hines a magnetic storm can be considered as consisting of the sudden commencement due to the passage of a shock wave heralding enhanced solar wind density and velocity, an initial phase due to the increased solar wind pressure and a main phase caused by the trapping and energization to 20 kev. or so of solar wind particles, thus producing a "ring current" (Dessler and Parker, 1959). The decay of the main phase takes place as the solar wind returns to normal and energetic protons are lost from the ring current due to charge exchange with neutral hydrogen atoms (Dessler and Parker, 1959), or perhaps to de-energization associated with movement to outlying parts of the magnetosphere.

We will not discuss all the implications of the theory here, as these are to a large extent covered elsewhere (Axford and Hines, 1961, 1962; Hines, 1963a, 1963b; Chapman, 1956; Hartz, Montbriand and Vogan, 1963). Instead, we will confine our attention to the viscous interaction hypothesis itself and to the morphology of disturbance phenomena. However, it is pointed out in passing that a circulation of the type we are considering would contribute to the effect observed by O'Brien (1963) and others (Dessler, 1961), which implies that the high latitude boundary of the trapped electrons (energies $\geq$ 40 kev.) occurs further from the pole on the midnight magnetic meridian than it does on the noon magnetic meridian. The circulation is such that it carries all trapped particles to lower latitudes on the night side of the earth and to higher latitudes on the day side. This would therefore complement the mechanism of Reid and Rees (1961) which relies on the effect on the longitudinal adiabatic invariant of stronger fields at a given latitude on the day side of the magnetosphere than on the night side.

The Viscous Interaction Hypothesis.

Fejer and others (Fejer, 1961, 1963; Kern, 1961, 1962; Chamberlain, Kern and Vestine, 1960) have proposed alternative mechanisms to viscous interactions between the solar wind and the magnetosphere as an explanation of the D_S current system and the associated ionospheric motions. Whether or not such mechanisms contribute to the circulation, viscous interaction must in any case occur to some extent (Gold, 1962), and the following argument suggests that it is strong enough to produce effects which are consistent with observations.

There are two principal quantities associated with the various phenomena under discussion. These are the total electrical potential variation ($\emptyset$) corresponding to the magnetospheric motions, and the rate of dissipation of energy (Φ) resulting from such motions (Hartz, Montbriand and Vogan, 1963). The speed of auroral motions, the strength of the D_S current system, the position of the auroral zone, and the energy of primary auroral particles, all suggest that $\emptyset$ is of the order of 20 kilovolts during a typical magnetic storm. In the same conditions Φ is at very most 10^{19} ergs/sec according to estimates of energy dissipation due to auroral processes, joule heating of the atmosphere by the ionospheric currents, and stressing of the geomagnetic field by the accumulation of trapped particles which are later lost. As might be expected, Φ is several orders of magnitude less than the solar wind energy incident per second on the magnetosphere.

A series of simple arguments based on viscous boundary layer theory adapted from ordinary hydrodynamics and using very reasonable values for various relevant parameters shows that the thickness of the boundary layer associated with the observed value of $\emptyset$ is typically 500 km, the kinematic viscosity associated with the interaction is $\sim 10^{13} cm^2$/sec, the drag per unit surface area of the magnetosphere about 2×10^{-10} dynes/cm^2 and the energy dissipation rate about 10^{19} ergs/sec.

The fact that the viscous interaction hypothesis shows that the values of $\emptyset$ and Φ, deduced independently from observed quantities are compatible is in itself remarkable enough. In addition, it can be shown that the actual drag to be expected at the surface of the magnetosphere, based on observed characteristics of the solar wind, is consistent with the value estimated from the observed value of $\emptyset$. This drag is believed to be due to the highly turbulent state of the solar wind after it has passed through the standing shock on the sunward side of the magnetosphere. The energy density associated with longitudinal (sound) waves in the turbulence is $\sim 10^{-9}$ ergs/cm^3 (Sonett and Abrams, 1963). These sound waves can penetrate the surface of the magnetosphere and if there is a relative motion the residual radiation stress parallel to the surface causes a drag which a rough calculation suggests is of the order of 10^{-10} dynes/cm^2 - that is, within a factor 2 of the value required if $\emptyset$ = 20 kilovolts.

In Fig. 13 we have shown the circulation within the magnetosphere to be symmetrical about the sun-earth direction. In fact, this is not completely correct since the surface of the magnetosphere tends to move quite rapidly due to the rotational effect described earlier. Any viscous-like interaction between the solar wind and the surface of the magnetosphere depends on the relative velocity and this is clearly affected by the rotationally-induced motion in the magnetosphere as well as by the solar wind. The point of zero relative velocity on the surface of the magnetosphere is consequently well to the west of the sub-solar point, and the viscously-induced circulation is therefore likely to be roughly symmetrical about this direction, although it may be stronger on the afternoon side than on the morning side of the magnetosphere (Axford, 1963). This may be the explanation of the tendency of the polar current systems and the sudden commencement currents at high latitudes to be symmetrical about a direction which lies 40° to 50° to the west of the sun (Nagata and Kokubun, 1962; Chapman, 1956; Wilson and Sugiura, 1961).

The Morphology of High Latitude Disturbance Phenomena.

Relative to the earth, the magnetospheric circulation at ionospheric levels during periods of high magnetic activity, has essentially the pattern of the D_S current system with, of course, the sense reversed from that of the current. As can be seen from Fig. 11 and 14, this reproduces the observed motion of irregularities in the aurora (to the west before geomagnetic midnight and to the east after geomagnetic midnight). Furthermore, if there is any correspondence between the direction of the motion and the alignment of auroral arcs, as there is in the auroral zone, the pattern is consistent with the tendency of arcs in the polar caps to be aligned roughly in the sun-earth direction (Davis, 1960, 1962b).

The net motion in the magnetosphere at distances greater than about 4 R_e in the equatorial plane must be given approximately by superposition of the motions due to rotation of the earth and the circulation corresponding to the D_S current (which we have suggested is due to viscous drag of the solar wind), since the S_q and L current systems are weak at high latitudes. This superposition is indicated for the equatorial plane of the magnetosphere in Fig. 15. Note that the central streamline, which lies more or less in the sun-earth direction if rotation is not included, is deflected first to the afternoon side and then to the morning side as it passes the earth. This is of some significance since field lines which have been dragged round the surface of the magnetosphere, must move through the interior more or less along this streamline. Thus, a concentration of particles captured from the solar wind is to be expected here and we therefore suggest that it will be the locus of greatest activity at ionospheric levels. In contrast, there are areas, particularly on the afternoon side of the magnetosphere, in which activity must be relatively low, because the streamlines never approach the surface of the magnetosphere where the field lines could conceivably capture solar particles.

The corresponding net motion at ionospheric levels in the north polar cap is sketched in Fig. 16. It is noticeable that the central streamline, described above, makes a loop which runs from high latitudes, first to the afternoon side, passing through the magnetic midnight meridian at about auroral zone latitudes, and continuing along the auroral zone until the late morning when it moves north once again. One expects greatest activity along this loop, in the sense that at any given latitude the diurnal variation of activity should show a maximum wherever the loop is intersected. The activity should not be uniform along the loop since different degrees of acceleration of particles (due to compression) are involved according to the latitude concerned,

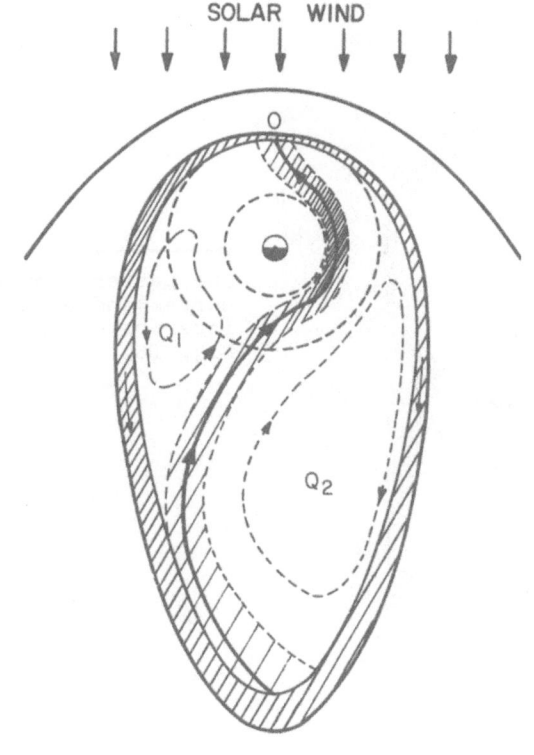

Fig. 15.
The net motion in the equatorial plane of the magnetosphere obtained
by superposition of the motion due to rotation (Fig. 9) and viscous
interaction with the solar wind (Fig. 13). Field lines in the areas
denoted Q_1 and Q_2 do not at any stage move very close to the surface
of the magnetosphere. The hatching in the area surrounding the central streamline is intended to indicate the degree of energization of
particles, maximum energization corresponding to the close hatching
in the morning hours.

and also the further particles progress along the loop, the more noticeable the effects of irreversible acceleration become. Thus, one expects a general increase of
the energy of precipitated particles along the loop, with energies of 1 to 10 kev.
occurring on the first section (resulting in sporadic E rather than absorption), and
energies of perhaps 100 kev. occurring in the morning hours (resulting in radio-wave
absorption at D-region levels). Regions of low activity are also predicted in the
interior of the loop, and more especially on the evening side at geomagnetic latitudes
of about 70°.

This description of the production of ionospheric disturbances is rather well
confirmed by observation. The quiet region on the evening side of the polar cap is
quite conspicuous for several phenomena. We have, for example, only to consider the
transition from quiet to disturbed aurora at about geomagnetic midnight (Fig. 14).
Plots of the diurnal maxima of sporadic E ionization (Hagg, Muldrew and Warren, 1959;
Thomas, 1960), geomagnetic agitation (Hope, 1961), spread F (Petrie, unpublished),
radio absorption (Hartz, Montbriand and Vogan, 1963), radio aurora (Forsyth, Green
and Mah, 1960; Green, unpublished), and visual aurora (Malville, 1959) are shown in
Fig. 17. These show up in the first part of the loop of activity quite clearly. A

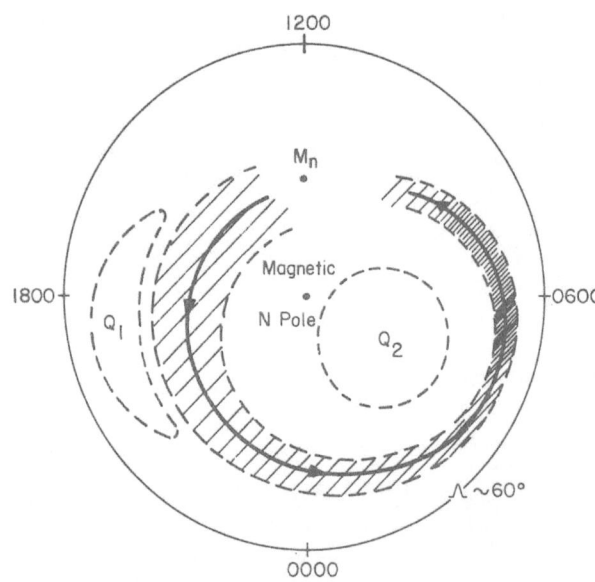

Fig. 16.
The pattern in the north polar cap
corresponding to that shown in the
interior of the magnetosphere in
Fig. 16.

Fig. 17.
Location of diurnal maxima of various
ionospheric disturbance phenomena,
namely geomagnetic agitation (I),
aurora (II), sporadic E (III), spread
F (o), and radio absorption (x).

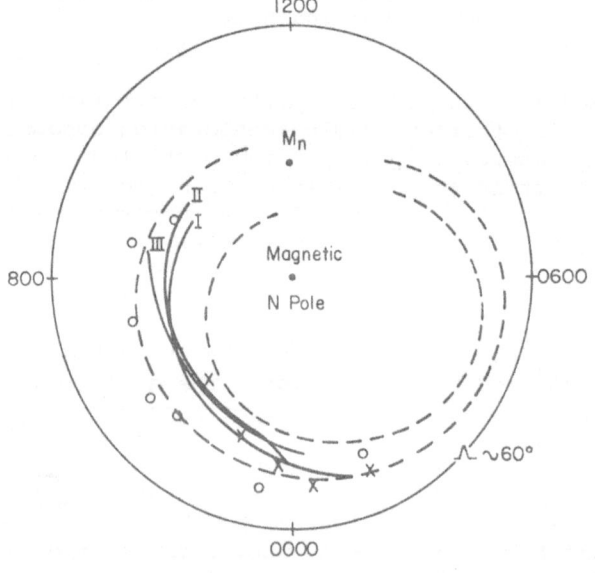

study of auroral absorption by Canadian workers (Hartz. Montbriand and Vogan, 1963; Hartz, 1963) is shown in Fig. 18. An examination of the contours will reveal the first part of the loop as being relatively weak as far as absorption is concerned, and that the absorption becomes extremely pronounced in the mid-morning hours, as one would expect since irreversible processes are required to produce the necessary energetic electrons in sufficient numbers. The quiet region on the evening side of the polar cap is very clearly indicated in this plot.

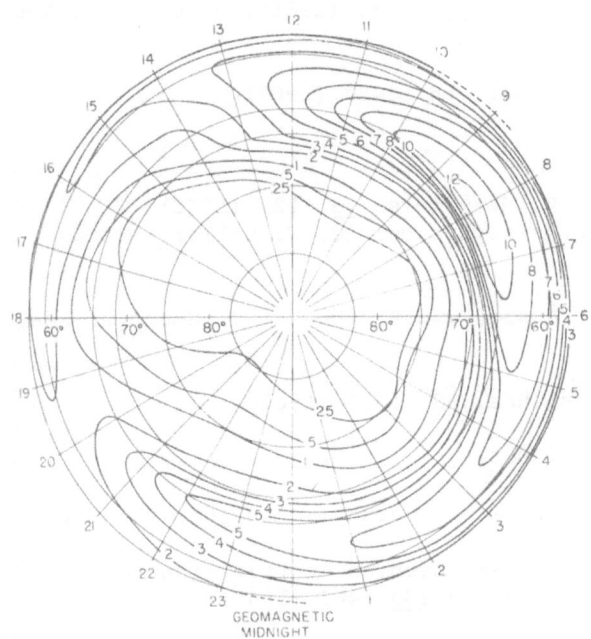

Fig. 18.
Contour diagram showing the percentage occurrence in time of auroral absorption of 1 db or more as a function of geomagnetic latitude and mean geomagnetic time (Hartz, Montbriand and Vogan, 1963).

If irreversible acceleration of electrons does in fact take place as we have described, then it can be considered a further prediction of the theory that any associated VLF emissions should have a diurnal maximum of occurrence at more or less the same time as does the auroral absorption. In looking for such a maximum, however, the screening effects of the absorption must be avoided, thus, the emissions should be observed at somewhat lower latitudes than the auroral zone. One would expect that direct observations of electron precipitation (as carried out by Injun I) (O'Brien, 1962) should also show a maximum in this region and that the main input of trapped electrons occurs at an equatorial distance of 4 to 6 R_e on the morning side of the earth.

ACKNOWLEDGEMENT

This work was supported under National Aeronautics and Space Administration Grant NsG - 382.

REFERENCES

Ahluwalia, H. S. and A. J. Dessler: Planetary and Space Science, 9, 195 (1963)

Axford, W. I.: Quart. J. Mech. App. Math. 13, 314 (1960); J. Geophys. Res. 67, 3791 (1962a); Canadian J. Phys. 40, 654 (1962b); J. Geophys. Res. (1963) (in preparation)

Axford, W. I., and C. O. Hines: Canadian J. Phys. 39, 1433 (1961); J. Geophys. Res. 67, 2057 (1962)

Bandyopodhyay, P. and H. Montes: J. Geophys. Res. 68, 2453 (1963)

Beard, D. B.: J. Geophys. Res. 65, 3559 (1960)

Biermann, L.: Z. Astrophys. 29, 274 (1951); Z. Naturforsh 7a, 127 (1952); Observatory 107, 109 (1957); Space Astrophysics, Ed. W. Liller, McGraw-Hill, New York, 150 (1961)

Bonetti, A., H. S. Bridge, A. J. Lazarus, B. Rossi, and F. Sherb: J. Geophys. Res. 68, 4017 (1963)

Cahill, L. J., and P. G. Amazeen: J. Geophys. Res. 68, 1835 (1963)

Carpenter D. L.: J. Geophys. Res. 68, 1675 (1963)

Chamberlain, J. W., J. W. Kern, and E. H. Vestine: J. Geophys. Res. 65, 2535 (1960)

Chapman, S.: Vistas in Astronomy 2, 912 (1956); Smithsonian Contribution to Astrophysics 2, 1 (1957); Proc. Roy. Soc. A 253, 462 (1959)

Chapman, S., and J. Bartels: Geomagnetism, II, Clarendon Press, Oxford (1950)

Chapman, S., and V.C.A. Ferraro: Terrest. Mag. and Atmosph. Elect. 36, 171 (1931)

Cole, J. D., and J. H. Huth: Phys. Fluids 2, 624 (1959)

Cole, K. D.: Aust. J. Phys. 13, 484 (1960)

Coleman, P. J., L. Davis, E. J. Smith, and C. P. Sonett: Science 138, 1099 (1962)

Davis, L., and D. B. Beard: J. Geophys. Res. 67, 4505 (1962)

Davis, T. N.: J. Geophys. Res. 65, 3497 (1960); 67, 59 (1962a); 67, 75 (1962b)

Dessler, A. J.: J. Geophys. Res. 66, 3587 (1961); 67, 4892 (1962)

Dessler, A. J. and E. N. Parker: J. Geophys. Res. 64, 2239 (1959)

Dougherty, J. P.: J. Geophys. Res. 68, 2383 (1963)

Duncan, R. A.: J. Atmospheric and Terrest. Phys. 18, 89 (1960)

Dungey, J. W.: Proc. Ionosphere Conf., Phys. Soc. London, 229 (1955); J. Geophys. Res. 68, 3540 (1963)

Fejer, J. A.: Canadian J. Phys. $\underline{39}$, 1409 (1961); J. Geophys. Res. $\underline{68}$, 2167 (1963)

Fishman, F. J., A. R. Kantrowitz, and H. E. Petschek: Rev. Mod. Phys. $\underline{32}$, 959 (1960)

Forsyth, P., F. Green, and W. Mah: Canadian J. Phys. $\underline{38}$, 770 (1960)

Frank, L. A., J. A. Van Allen, and E. Macagno: J. Geophys. Res. $\underline{68}$, 3543 (1963)

Freeman, J. W., J. A. Van Allen, and L. J. Cahill: J. Geophys. Res. $\underline{68}$, 2121 (1963)

Gold, T.: "Gas Dynamics of Cosmic Clouds," Ed. M. C. VanderHulst and J. M. Burgers, North Holland Publ. Co., Amsterdam, (1955); J. Geophys. Res. $\underline{64}$, 1219 (1959); Space Sci. Rev. $\underline{1}$, 100 (1962)

Green, F.: Upper Atmospheric Institute, University of Saskatchewan (unpublished report)

Gringauz, K. I.: Planetary and Space Sci. $\underline{11}$, 281 (1963)

Gringauz, K. I., V. G. Kurt, V. I. Moroz, and I. S. Shklovskii: Doklady Akad. Nauk. SSSR $\underline{132}$, 1062 (1960a); Astron. Zh. $\underline{37}$, 716 (1960b)

Hagg, E. L., D. Muldrew, and E. J. Warren: J. Atmospheric and Terrest. Phys. $\underline{14}$, 345 (1959)

Harang, L., and J. Troim: Planetary and Space Sci. $\underline{5}$, 33 (1960)

Harrison, E. R.: Geophys. J. Roy. Astro. Soc. $\underline{6}$, 679 (1962)

Hartz, T. R.: Radio Astronomical and Satellite Studies of the Atmosphere, Ed. J. Aarons, North-Holland Pub. Co., Amsterdam (1963)

Hartz, T. R., L. E. Montbriand, and E. L. Vogan: Canadian J. Phys. $\underline{41}$, 581 (1963)

Hayes, W. D., and R. F. Probstein: "Hypersonic Flow Theory," Academic Press, New York (1959)

Heppner, J. P., N. S. Ness, C. S. Scearce, and T. L. Skillman: J. Geophys. Res. $\underline{68}$, 1 (1963)

Hines, C. O.: Quart. J. Roy. Met. Soc. $\underline{89}$, 1 (1963a); Planetary and Space Sci. $\underline{10}$, 239 (1963b)

Hones, E. W.: J. Geophys. Res. $\underline{68}$, 1209 (1963)

Hope, E. R.: J. Geophys. Res. $\underline{66}$, 747 (1961)

Johnson, F. S.: J. Geophys. Res. $\underline{65}$, 3049 (1960)

Kern, J. W.: J. Geophys. Res. $\underline{66}$, 1290 (1961); $\underline{67}$, 2649 (1962)

Kulikovskii, A. G.: Doklady Akad. Nauk. SSSR $\underline{117}$, 199 (1957)

Malville, J. M.: J. Geophys. Res. $\underline{64}$, 1389 (1959)

Nagata, T., and S. Kobubun: Nature $\underline{195}$, 555 (1962)

Neugebauer, M., and C. W. Snyder: Science $\underline{138}$, 1095 (1962)

O'Brien, B. J.: J. Geophys. Res. $\underline{67}$, 1227 (1962); $\underline{68}$, 989 (1963)

Parker, E. N.:   Phys. Fluids 1, 171 (1957); Ap. J. 128, 664 (1958); 132, 175 (1960a); 132, 821 (1960b); 133, 1014 (1961)

Petrie, L. E.:   (unpublished)

Piddington, J. H.:   J. Geophys. Res. 65, 93 (1960)

Reid, G. C., and M. H. Rees: Planetary and Space Sci. 5, 99 (1961)

Sonett, C. P. and I. J. Abrams: J. Geophys. Res. 68, 1233 (1963)

Sonett, C. P., E. J. Smith, and A. R. Sims: Space Research, Ed. by H. K. Kallmann-Bijl, North-Holland Pub. Co., Amsterdam, 921 (1960)

Sonnerup, B.W.O., and M. J. Laird: Cornell University Report, CRSR 130 (1962)

Spreiter J. R., and B. R. Briggs: J. Geophys. Res. 67, 37 (1962)

Thomas, L.:   Some Ionospheric Results Obtained During the IGY, Ed. W. J. Beynon, Elsevier Pub. Co., Amsterdam, 172 (1961)

Van Allen, J. A.:   J. Geophys. Res. 64, 1683 (1959)

Wilson, C. R., and M. Sugiura: J. Geophys. Res. 66, 4097 (1961)

ON THE PENETRATION OF INTERPLANETARY PLASMA
INTO THE MAGNETOSPHERE

H. Alfvén, L. Danielsson, C-G. Fälthammer, and L. Lindberg
Royal Institute of Technology
Stockholm, Sweden

ABSTRACT

Theoretical and experimental evidence is presented to indicate that the usual magneto-spheric models may need essential modification. Theoretically, it is concluded that the low-density magnetospheric plasma, in contrast to an ideal magnetofluid, may support strong electric fields even along the magnetic field lines. This can profoundly influence the penetration of interplanetary plasma. Laboratory experiments show that a plasma can penetrate much closer to a terrella than what would correspond to the magnetosphere boundary in closed magnetosphere models.

INTRODUCTION

One of the fundamental problems in connection with aurorae and magnetic storms is the penetration of the interplanetary plasma into the earth's magnetosphere. This penetration seems difficult to account for on the basis of hydromagnetic models where the plasma is pictured as a perfectly conducting fluid with a "frozen-in" magnetic field. It is the purpose of the present paper to point out that <u>the low density plasma in the magnetosphere need not necessarily behave like a magnetofluid</u>.

Some properties of a low-density plasma in an inhomogeneous magnetic field are discussed in the second section of the paper and it is concluded that the magnetic field lines in the magnetosphere need not be equipotential lines. This is most important because electric fields across the magnetosphere can permit the interplanetary plasma to penetrate deep into the magnetosphere (Alfvén and Fälthammar, 1963). The depth of penetration is determined by the electric and magnetic fields and the temperature in the incoming plasma and not by the dynamic pressure as in hydromagnetic models.

An exact mathematical analysis of the motions in the magnetosphere is too complicated to be attempted at this stage. However, idealized models with simple geometry can be valuable for shedding light on certain aspects of the phenomena. This approach is used in the third section. Model experiments represent another approach which we believe is most important as a complement to the theoretical considerations. In the fourth section some recent experiments are discussed which may be of relevance for understanding some of the problems involved.

As the behavior of the plasma depends on whether the plasma magnetic field is parallel or antiparallel to the dipole field, there is a possibility that two categories of magnetic storms exist which is considered in the fourth section. This should be checked by observations.

Future observations by spacecraft-carried instruments may provide evidence to decide between the model discussed here and the hydromagnetic models. Possible experiments are suggested in the sixth section.

ELECTRIC FIELDS IN LOW-DENSITY MAGNETIC PLASMAS

By low-density magnetic plasma, we mean a plasma where the mean free path λ is much larger than the characteristic dimension l_c, which, in our application, is the separation of the magnetic mirrors confining the plasma.

In an electron-proton plasma the effective mean free path of an electron is given to the order of magnitude by

$$\lambda \approx 10^4 \ \frac{T_e^2}{n_e} \tag{1}$$

where T_e and n_e are the temperature and number density of electrons.

Let us consider the magnetosphere of the earth, in particular, a magnetic field line, which intersects the equatorial plane at, say five times the earth's radius, i.e. at $3 \cdot 10^9$ cm. The density in the equatorial plane is probably of the order 100 cm^{-3} or less. If we put $l_c = 3 \cdot 10^9$ we find that $\lambda \approx l_c$ if the temperature is about 10^4 degrees. If the temperature is higher, the magnetosphere should be counted as a low-density region. No reliable temperature measurements seem to exist and it can perhaps not be excluded that during undisturbed conditions the temperature may be low. However, during magnetic storms the magnetosphere is invaded by a presumably hot plasma (perhaps 10^6 degrees) ejected from the sun, and this plasma is heated still more by magnetic compression when it penetrates into the magnetosphere.

If we put the cross section of the magnetosphere equal to 10^{20} cm^2, an interplanetary plasma with a flux density of 10^8 protons cm^{-2} sec^{-1} each with an energy of 10^3 ev carries an energy of 10^{31} ev/sec to the magnetosphere. If the outer region of the magnetosphere has a volume of 10^{30} cm^3 and contains 100 particles/cm^3, the energy needed to heat it to 10^4 degrees ($\approx$ 1 ev) equals 10^{32} ev. Hence, the interplanetary wind carries energy enough to make the outer magnetosphere a "low-density region" in ten seconds. Therefore, <u>at least during magnetic storms the magnetosphere is likely to be a low-density region</u>.

As discussed in a different context elsewhere (Alfvén and Fälthammar, 1961), there is in a low-density plasma no relation of the simple form i = σ E between the current density and the electric field. Suppose that there is only an electric force parallel to <u>B</u>. Then an individual electron obeys the equation of motion

$$m_e \ \frac{dv_{\parallel}}{dt} \ = \ e_e E_{\parallel} \tag{2}$$

and the corresponding current density $i_{\parallel}$ is given by

$$\frac{di_{\parallel}}{dt} \ = \ \frac{n_e e_e^2 E_{\parallel}}{m_e} \tag{3}$$

34

This shows that in the low-density case the conductivity $\sigma_{\parallel}$ defined by $i_{\parallel}/E_{\parallel}$ has no meaning. Even if $E_{\parallel} = 0$ we may have $i_{\parallel} \neq 0$, and for $E_{\parallel} \neq 0$ the current at a given instant may be zero or even antiparallel to $E_{\parallel}$.

It is interesting to study the case when the magnetic field is inhomogeneous in such a way that the guiding centers of the particles oscillate along a field line between two mirror points at a distance l_c. Suppose that the plasma consists of electrons and one kind of positive ions, and that their magnetic moments are μ_e and μ_i. In the presence of an electric field $\underline{E}_{\parallel}$ parallel to $\underline{B}$ they are acted upon by the average forces

$$f_{e_{\parallel}} = - \mu_e \frac{dB}{ds} + e_e \cdot E_{\parallel} \tag{4}$$

and

$$f_{i_{\parallel}} = - \mu_i \frac{dB}{ds} + e_i \cdot E_{\parallel} \tag{5}$$

parallel to the magnetic field.

If under the influence of this force a particle oscillates with the velocity $v_{k_{\parallel}}$ between the points s_1 and s_2, it spends the time

$$dt = ds/v_{k_{\parallel}} \tag{6}$$

on the line element ds. Its half-period is

$$\tau_k = \int_{s_1}^{s_2} ds/v_{k_{\parallel}} \tag{7}$$

Hence, it gives rise to an average space charge

$$dq_k = \frac{e_k \, ds}{\tau_k \cdot v_{k_{\parallel}}} \tag{8}$$

on the line element ds. (The subscript k is used to distinguish between different kinds of particles.) If N_i positive ions and N_e electrons oscillate simultaneously between the same mirror points, the space charge at each line element is $N_i dq_i + N_e dq_e$ and if N is so large that we have a plasma (which must be quasi-neutral), we must have $N_i dq_i + N_e dq_e = 0$ and also $N_i e_i + N_e e_e = 0$. According to Eq. (8) this means that $\tau_i v_i = \tau_e v_e$, or, if we introduce $\alpha = \tau_i/\tau_e$,

$$v_{e_{\parallel}} = \alpha v_{i_{\parallel}} \tag{9}$$

Here α is a constant which can have any value (because electrons and ions need not have equal energies).

Eq. (9) and the equations of motion

$$m_e \ \frac{dv_{e\parallel}}{dt} = f_{e\parallel}$$

$$m_i \ \frac{dv_{i\parallel}}{dt} = f_{i\parallel}$$

give

$$f_{e\parallel} = m_e \ v_{e\parallel} \ \frac{dv_{e\parallel}}{ds} = \frac{\alpha^2 m_e}{m_i} \ m_i \ v_{i\parallel} \ \frac{dv_{i\parallel}}{ds} = \frac{\alpha^2 m_e}{m_i} \ f_{i\parallel}. \tag{10}$$

If we assume $e_i = -e_e = |e|$, we obtain by introducing Eqs. (4) and (5) into Eq. (10)

$$- |e| \ E_\parallel \ - \ \mu_e \frac{dB}{ds} \ = \ \frac{\alpha^2 m_e}{m_i} \ \left(|e| \ E_\parallel \ - \ \mu_i \frac{dB}{ds} \right) \tag{11}$$

or

$$E_\parallel \ = \ - \ K \frac{dB}{ds} \tag{12}$$

with

$$K \ = \ \frac{1}{|e|} \ \frac{\mu_e/m_e \ - \ \alpha^2 \mu_i/m_i}{1/m_e \ + \ \alpha^2/m_i} \tag{13}$$

Introducing Eq. (9) and $\mu = W_\perp/B$ we find the following expression for the
<u>invariant K</u>

$$K \ = \ \frac{W_{i\parallel} \ W_{e\perp} \ - \ W_{e\parallel} \ W_{i\perp}}{|e| \ B \ (W_{i\parallel} \ + \ W_{e\parallel})} \tag{14}$$

We have used the notations $W_\parallel = m \ v_\parallel^2/2$ and $W_\perp = m \ v_\perp^2/2$ with the indices e and i referring to electrons and positive ions.

Eq. (12) shows that in a low-density plasma the electric field parallel to the
magnetic field is zero only when the magnetic field is homogeneous ($\frac{dB}{ds} = 0$) or when
the relation

$$\frac{W_{i\parallel}}{W_{e\parallel}} \ = \ \frac{W_{i\perp}}{W_{e\perp}} \tag{15}$$

is satisfied. This relation means that the helices of the ions and of the electrons
have the same pitch angle. If Eq. (15) is satisfied, so that K = 0, both kinds of

36

particles oscillate with the same amplitude in the absence of an electric field. However, if K ≠ 0, so that $W_{i\parallel}/W_{i\perp} \neq W_{e\parallel}/W_{e\perp}$, electrons and ions would oscillate with different amplitude, if there were no electric field. As the quasi-neutrality of the plasma requires that $N_i dq_i + N_e dq_e \approx 0$ everywhere, particles of both kinds must oscillate with equal amplitude. This can only be achieved by setting up an electric field.

We integrate Eq. (12) between two points A and C where the magnetic field strengths are B_A and B_C. The voltage difference between C and A is

$$V = V_C - V_A = K (B_C - B_A) \tag{16}$$

We introduce

$$\gamma = B_C/B_A \tag{17}$$

Then Eq. (16) becomes

$$|e|V = (\gamma - 1) \left[\frac{W_{i\parallel} W_{e\perp} - W_{e\parallel} W_{i\perp}}{W_{i\parallel} + W_{e\parallel}} \right]_A \tag{18}$$

where the expression within the brackets refers to the point A.

Evaluation of the Voltage in a Simple Case.

We shall use this result on a simple model of the magnetosphere. We assume that a line of force of the geomagnetic dipole field intersects the ionosphere at C (Fig. 1). Outside the ionosphere, there is a low-density plasma, which is produced by injection of hot plasma in the equatorial plane (at A in Fig. 1) and evaporation of low-energy particles from the ionosphere. For the sake of simplicity, we assume that in the equatorial plane only particles of one sign, say electrons, are injected. At C, where the field strength is B_2 ($B_2 > B_1$), there is a source emitting ions with negligible energy. The emission takes place as soon as there is a voltage difference

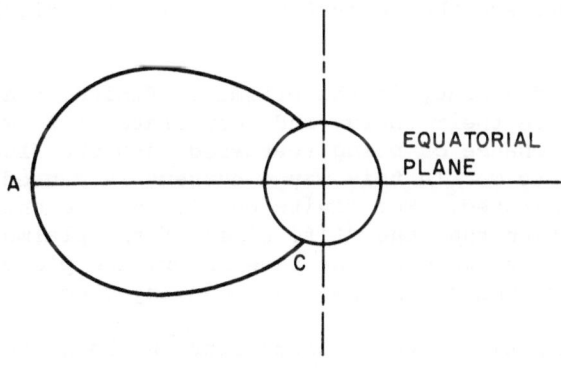

Fig. 1. The earth and a geomagnetic field line.

between the plasma at C and the source. In this way the voltage of the plasma at C is fixed by the voltage of the source. The injection of electrons at A lowers the voltage at this point, and as a consequence, positive particles are emitted from the source at C, until Eq. (18) is satisfied. Then we can calculate the voltage difference from Eq. (18).

As the positive ions have been accelerated by the voltage V before they reach A, we have $W_{i\parallel} = |e|V$, and $W_{i\perp} = 0$. Hence, we obtain from Eq. (18)

$$|e| \, V = (\gamma - 1) \, \frac{|e| \, V \, W_{e\perp}}{|e|V + W_{e\parallel}} \tag{19}$$

which besides the solution V = 0 also gives

$$|e| \, V = (\gamma - 1) \quad W_{e\perp} - W_{e\parallel} \tag{20}$$

If in a medium-density plasma (by which we mean a plasma where $\lambda \ll l_c$ but still $\lambda \gg \rho_L$, ρ_L being the Larmor radius) we have two similar sources on the same field line, we obtain a conduction current carrying electrons from the source A to C and ions from C to A. This current, which tends to annihilate any voltage difference between A and C, is proportional to V. As in cosmical medium-density plasmas, the conductivity usually is large, the current will often prevent any large voltage difference between A and C. Contrary to this, in a low-density plasma, the voltage difference V given by Eq. (20) is produced before any appreciable current can flow. If we inject electrons at A, they cannot reach C unless their energy is increased by V to such an extent as to displace the mirror point to C. Only if the voltage difference exceeds the value given by Eq. (20) can electrons reach C so that a current is produced. Further, as the ions emitted from C oscillate along the field line without collision, they cannot carry any average current when a stationary state is reached.

In our model a magnetic field line, which intersects the equatorial plane (field strength B_A) at 5 or 10 earth radii reaches the ionosphere at a point where the field strength (B_C) is so much larger that γ is 100 or 1000. This means that under our simplifying assumptions there may be a voltage difference between the equatorial plane and the ionosphere which exceeds the equivalent voltage of particles injected at A by a very large factor.

In case the collision frequency in the plasma is finite, the oscillating particles emitted from C will lose their energy and accumulate at A, whereas the particles emitted from A will have a chance of being scattered into the "loss cone" so that they move along the field lines to C. In this way a current is caused so that the voltage V given by Eq. (20) is eliminated. For finite collision frequencies the voltage V is large only under the condition that the "life time" of the plasma is shorter than the collision time. By "life time" we mean the time before the plasma has drifted away from the region we consider, and new plasma has been injected.

From the above considerations, we conclude that in the magnetosphere there may exist a voltage difference between the point where a field line intersects the equatorial plane and the point where it intersects the ionosphere. This voltage difference is not necessarily equal for different magnetic field lines, which means that there may well exist electric fields in the outer parts of the magnetosphere even if the ionosphere is considered as an equipotential surface.

Rigorous Analysis of the One-Dimensional Case.

The case of a low-density plasma situated in a straight narrow tube of force with magnetic mirrors at both ends has been studied in detail by Persson (1963). In contrast to the simple considerations given above, Persson's analysis takes into account the velocity distribution of the particles. From this analysis it is found that the electric field vanishes only if the pitch-angle distributions of electrons and ions are identical. When they differ appreciably, electric fields of the order of magnitude given by Eq. (12) prevail.

THEORETICAL CONCLUSIONS FROM SIMPLE MODELS

For an analysis of the outer regions of the magnetosphere, the strength and direction of the interplanetary magnetic field is essential. We shall confine our discussion to two simple cases.

(A) The interplanetary field is parallel to the magnetosphere field in the equatorial plane.

(B) The fields are antiparallel.

Case A.

Without any plasma the magnetic field consists of a dipole field superposed on a homogeneous field. The field strength in the equatorial plane is given by

$$B = B_o + a/R^3 \qquad (21)$$

The motion of a charged particle in the equatorial plane can be treated exactly. If there is a homogeneous electric field E in the equatorial plane, the closest approach to the dipole, of a particle with energy W is essentially

$$L = \left(\frac{\mu a}{|e|E}\right)^{\frac{1}{4}} \qquad (22)$$

where a is the magnetic moment of the dipole field, e is the charge of the particle, and μ is the adiabatic invariant of the particle. For a typical electron in a plasma with electron temperature T_e the adiabatic invariant is

$$\mu_e = \frac{kT_e}{B} . \qquad (23)$$

Notice that in the present case the plasma is stopped by diamagnetic repulsion, not by the magnetic pressure. For the actually observed values of magnetic fields and velocity and for reasonable values of the interplanetary electron temperature (so far observationally unknown) this means that the plasma can penetrate to depths where the auroral-zone magnetic field lines cross the equatorial plane.

The motion of the plasma will produce, as by charge separation, additional electric fields modifying the externally applied field. Using again a two-dimensional model, Karlson (1962) has analyzed this effect and concluded that the general character of the motion remains essentially unchanged.

The magnetic field given by Eq. (21) is modified considerably by currents produced by the plasma. For example, the front side of the magnetosphere is compressed considerably by the dynamic pressure of the plasma.

Also the topology of the field may be changed. In the undisturbed case, there is a sphere with the radius $(2a/B_0)^{1/3}$ on the surface of which the field is everywhere parallel to the spherical surface. This surface goes through the singular points $B = 0$, and can be considered as separating the magnetosphere from the interplanetary field.

The two-dimensional case treated by Karlson may give a first-order description of the motion of the interplanetary plasma near the equatorial plane. It has sometimes been argued that in the actual three-dimensional case there could not exist an electric field across the magnetosphere because it should be short-circuited along the magnetic field lines. That need not be true. As was found above in the second section, the low-density plasma may support a large potential difference along the strongly convergent field lines that end in the ionosphere. On the other hand, near the singular field line separating the magnetosphere from interplanetary space, there is no strong field-strength gradient parallel to $\underline{B}$. According to Eq. (12), this means that no appreciable electric fields should exist there. However, this exceptional region is small, and the incoming plasma may change the magnetic structure near the neutral point of the singular field line in such a way that an electric field can be maintained there too.

Case B.

Dungey (1961) has directed the attention to the importance of the antiparallel case.

In the undisturbed case the magnetic field in the equatorial plane is $B = a/r^3 - B_0$. It is zero in the equatorial plane along the circle $r_0 = (a/B_0)^{1/3}$. If similar to Case A, we treat the motion of charged particles in the equatorial plane, we find that no penetration through this circle is possible. On the other hand, field lines from the polar caps go to infinity, and as pointed out by Dungey, an interplanetary plasma may penetrate along these field lines to the ionosphere.

Plasma moving parallel to the equatorial plane but a few earth's radii to the north or to the south of it will experience an increasing magnetic field. The phenomena produced will be somewhat similar to the conditions in the equatorial plane in the parallel case. However, very little plasma will be captured near the equatorial plane, so we should not expect a ring current of the same strength as in the parallel case.

MODEL EXPERIMENTS

In order to study what happens when a magnetized plasma is shot toward a magnetic dipole field, a laboratory experiment has been made by Lindberg and Danielsson (1963).

It can be considered as a method developed by Birkeland and modified by Malmfors and later by Block (1958).

Apparatus.

The main parts of the device are shown in Fig. 2. A terrella of radius R = 1.2cm is situated in a volume filled with hydrogen at low pressure (0 - 10μ). The terrella gives a field of 10,000 gauss at its equator. The muzzle of a plasma gun with plane electrodes of height 30 cm and width 20 cm is placed 10 cm below the terrella. At the bottom of the gun there are three puff-valves, which are able to let in a cloud of hydrogen gas very fast. Some time after opening the valves, when the cloud has expanded a few centimeters upward between the electrodes, a discharge is fired

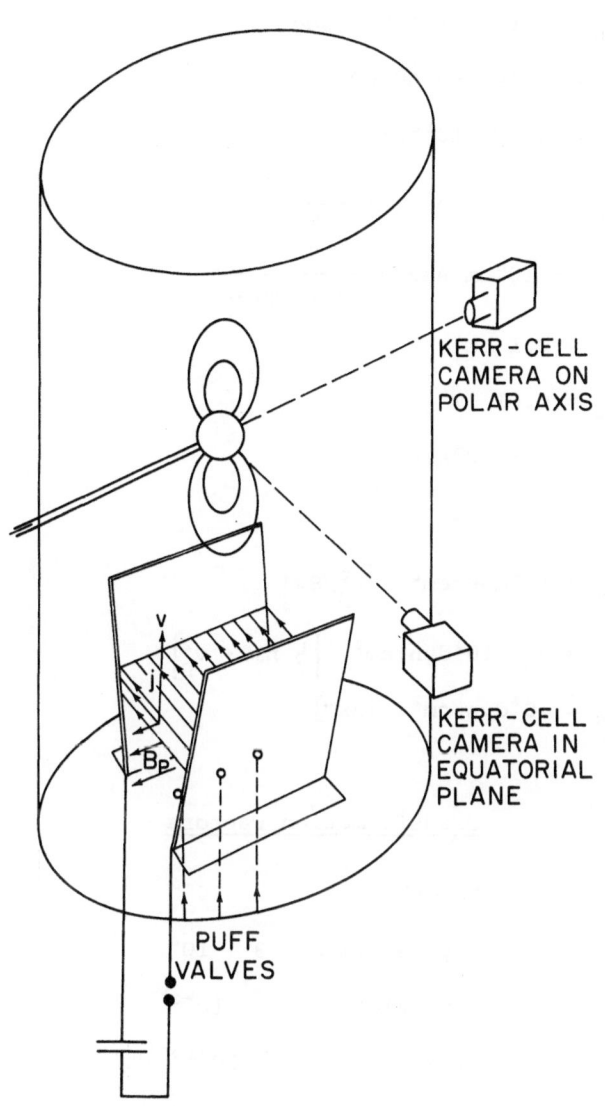

Fig. 2.
Schematic sketch of the apparatus in the Lindberg and Danielsson model experiment.

through the cloud. The high discharge current, 10 to 15 kA, ionizes the cloud into a plasma. The current gives rise to a magnetic field in the plasma and the $j \times B$ force accelerates it upward. In the vicinity of the terrella the velocity of the plasma is 2 to 5 cm/μs, the magnetic field in the plasma 300 to 700 gauss (originating from the discharge current), the depth and width of the plasma 10 x 20 cm (the current density about 100 A/cm^2). See further Table 1. The number density and temperature (and from these derived quantities) are only estimated.

TABLE 1
APPROXIMATE DATA

Plasma Quantities (undisturbed plasma)

Magnetic field	B_p = 300 to 700 gauss
Velocity of the plasma	v_p = 2 to 5 $\cdot$ 10^6 cm/sec
Temperature	T = 10,000 to 50,000 $^\circ$K
Number density	n = 10^{15} - 10^{16} cm^{-3}
Sound velocity	v_s = 10^6 cm/sec $\left[(\frac{\gamma \cdot kT}{m})^{\frac{1}{2}} \right]$
Magnetohydrodynamic velocity	v_{MH} = 10^6 cm/sec $\left[\frac{B_p}{(4\pi \cdot nm)^{\frac{1}{2}}} \right]$
Gyroradius of the ions	ρ_i = 0.1 cm
Gyroradius of the electrons	ρ_e = 0.01 cm
Gyrofrequency of the ions	ω_i = 5 $\cdot$ 10^6 sec^{-1}
Gyrofrequency of the electrons	ω_e = 10^{10} sec^{-1}
Magnetic pressure	p_B = 10^4 dyn/cm^2 $\left[B_p^2/8\pi \right]$
Dynamic pressure	p_d = 4 $\cdot$ 10^4 dyn/cm^2 $\left[\frac{1}{2} nm \cdot v_p^2 \right]$
Static pressure	p_s = 10^4 dyn/cm^2 $\left[nkT \right]$

Terrella Quantities

Terrella radius	R = 1.2 cm
Magnetic field at the surface in the equatorial plane	B_{eq}=10^4 gauss
Radius of the forbidden region	L $\approx$ 1.2 R
Radius of magnetosphere	R_s = 2 to 3 R

Transformation Factors

Length	10^{-9}
Dipole field	3 $\cdot$ 10^4
Potential	10^{-3}
Time	10^{-10}
Plasma field	10^7

Two different cases are studied:

(A) the plasma field is parallel to the dipole field in the equatorial plane,

(B) the fields antiparallel.

The diagnostic methods used are Kerr cell-photography and magnetic probe measurements.

Parallel Fields.

In the case when the equatorial field of the terrella is parallel to the field in the plasma, a dark region like Fig. 3 is observed between 11 and 14 μs after firing the gun. Compared with the dipole line, its border is considerably deformed as if "compressed" by the plasma. When it first appears, it extends to about 1.6 R in the equatorial plane, but expands to 1.8 R before it disappears. (All distances are measured from the center of the terrella.) On the backward or "night" side traces of a similar dark region are sometimes seen. The lower picture in Fig. 3 is taken simultaneously with the upper one, but from the pole axis. It shows a luminous ring around the pole at a certain latitude possibly corresponding to an "auroral zone."

Magnetic probes measuring the field components parallel to the terrella axis are located in the equatorial plane at 1.7 R in the directions 6^h 12^h 18^h and 24^h counted with the plasma gun (representing the sun) in the 12^h direction. When the plasma approaches the terrella, they all show an increase, followed by a smooth decrease, which can be interpreted as a temporary compression of the terrella field. On the day side, a probe, located at 3 R, shows a more steep front (Fig. 5a). The magnitude of the field change is usually 150 to 200 per cent of the undisturbed plasma field.

Antiparallel Fields.

When the equatorial field of the terrella is antiparallel to the field in the plasma, the dark zone gets quite a different shape (Fig. 4). The border is rather like the expected shape of the singular field lines, although somewhat compressed by the plasma. Also in this case the zone is visible between 11 and 14 μs after firing the gun, and its radius in the equatorial plane increases during this interval from 2.6 R to 3.3 R.

Magnetic probes in the equatorial plane show a compression followed by a decay and this occurs nearly simultaneously in all points. Of special interest are the two probes on the day side, at 1.7 R and 3.0 R, (Fig. 5b). The inner one shows just the compression mentioned, while for the outer one the compression is interrupted and followed by a peak of reversed field, indicating that the plasma and its field surround the probe for about 1 μs, before being reflected. The amplitude of this peak varies very much from shot to shot. The rapid reversal of the field shows that a strong current flows in a thin sheath separating the terrella field from the plasma field. We have here a phenomenon which is similar to the Cahill (1963) discontinuity.

The fact that zones with sharp boundaries are observed for as long as 3 μs may indicate that diffusion is not too serious during this time, which corresponds to a movement of the plasma about 15 cm (= 12 R).

Neither in the parallel nor in the antiparallel case is the plasma stopped at a limit which can be identified with the "border of the magnetosphere" in the closed magnetosphere models (see Fig. 6).

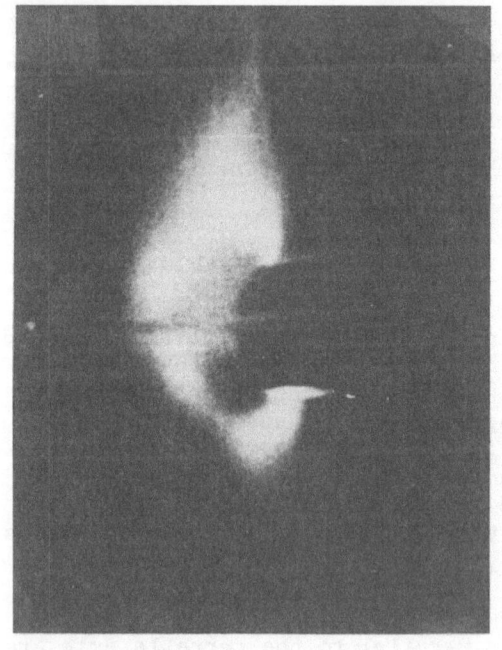

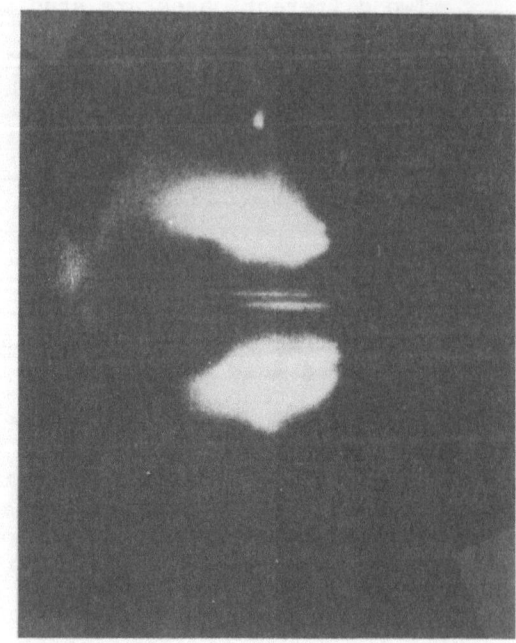

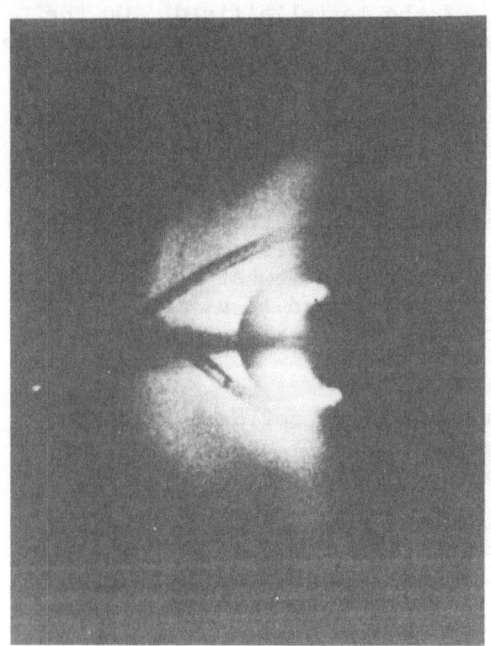

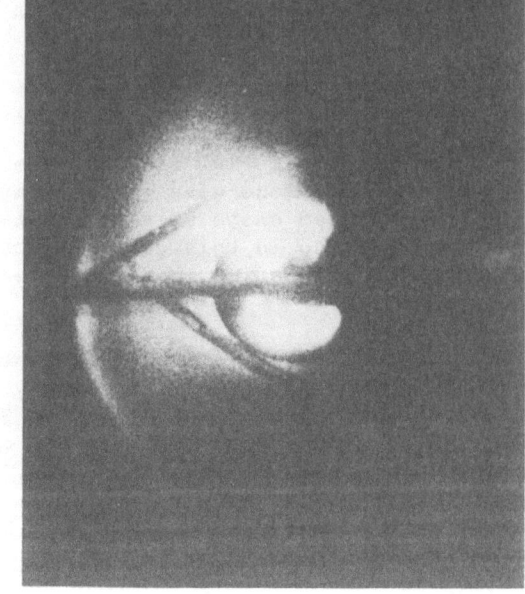

Fig. 3
Parallel fields.

Upper picture: seen from equator plane.
Lower picture: seen from polar axis.

Fig. 4
Antiparallel fields.

Upper picture: seen from equator plane.
Lower picture: seen from polar axis.

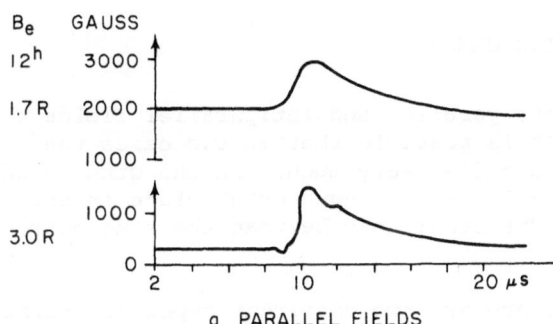

a PARALLEL FIELDS

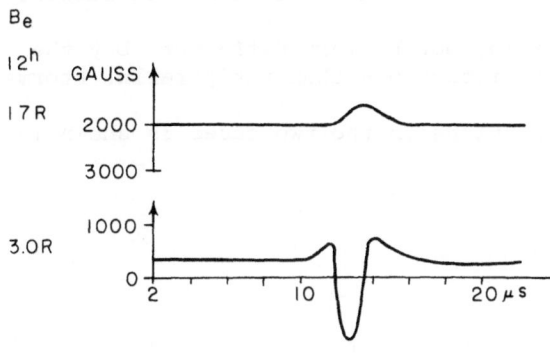

b ANTIPARALLEL FIELDS

Fig. 5.
Oscillograms showing the magnetic field
variation at two points in the direction
of the plasma source. (a) Parallel case
and (b) Antiparallel case.

Fig. 6.
Limits of the penetration of plasma in
the two experimental cases (full lines).
In these experiments the "magnetosphere
border" would correspond approximately
to the dashed line.

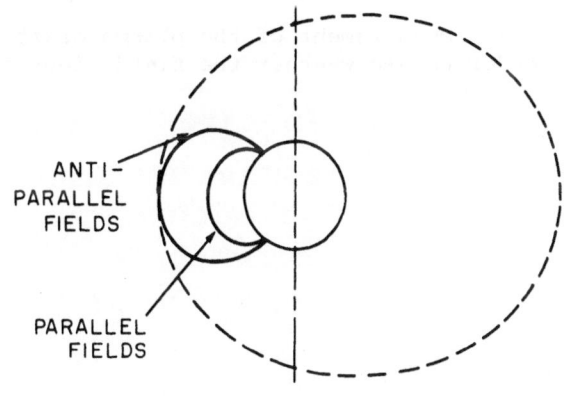

ANTI-
PARALLEL
FIELDS

PARALLEL
FIELDS

STORMS AND "ANTISTORMS"

It is of special interest to note that both with parallel and antiparallel fields the plasma hits the terrella at high latitudes. It is possible that in two cases the auroral phenomena and polar disturbances do not differ very much. On the other hand, we should expect that much more plasma is brought into the equatorial plane in the parallel case than in the antiparallel case. The result may be that the ring current is better developed in the former case.

Hence, one may tentatively suggest that there are two different types of storms, which we may call <u>parallel storms</u> and <u>antiparallel storms</u>.

A parallel storm is produced when a plasma with parallel magnetization hits the magnetosphere.

An antiparallel storm is produced by a plasma with antiparallel magnetization.

The polar phenomena (including the aurora) may not be very different, but the parallel storms should have a larger equatorial disturbance than antiparallel storms.

Possible magnetic field and current distributions in the two cases is shown in Fig. 7 and 8.

OBSERVATIONAL TESTS

A decision between the frozen-in model and the electric field model of the magnetosphere could be made in the following ways:

1. Determination of number density n and temperature T, and calculating the mean free path λ. The frozen-in model requires $\lambda \ll l_c$, the electric field model requires $\lambda \gg l_c$ (for each kind of particle).

2. Determination of the pitch angle distribution of both electrons and ions. Only if they are identical can the electric field be neglected.

3. Direct measurement of the electric field.

4. Measurement of the plasma drift $v_\perp$ at two points on the same line of force in order to see whether the field lines are frozen-in.

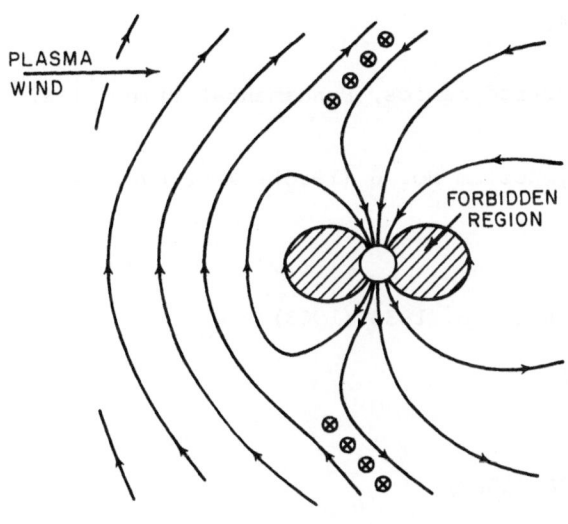

PLASMA WIND

FORBIDDEN REGION

Fig. 7
Possible configuration of magnetosphere field.

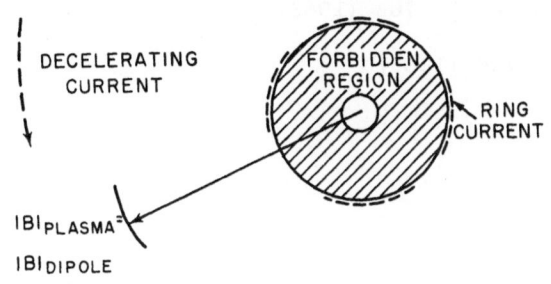

DECELERATING CURRENT

FORBIDDEN REGION

RING CURRENT

$|B|_{PLASMA}$

$|B|_{DIPOLE}$

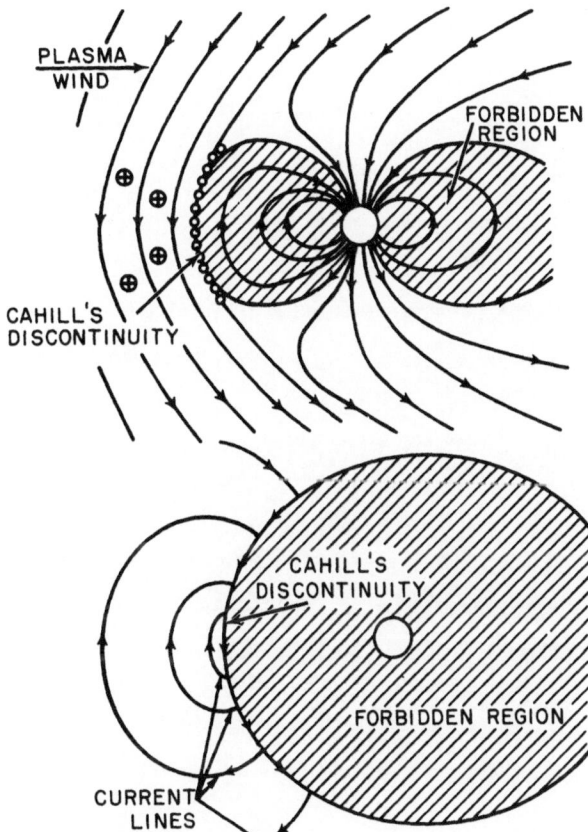

PLASMA WIND

FORBIDDEN REGION

CAHILL'S DISCONTINUITY

CAHILL'S DISCONTINUITY

CURRENT LINES

FORBIDDEN REGION

Fig. 8.
Possible configuration of magnetosphere field.

REFERENCES

Alfvén, H., and C.-G. Fälthammar: "Cosmical Electrodynamics, Fundamental Principles," Clarendon Press, Oxford (1963)

Block, L.: International Astronomical Union Symposium No. 6 (1958); Ark. Fys. 14, 153 (1958)

Bostick, W. H.: Phys. Fluids 5, 1305 (1962)

Cahill, L. J., and P. G. Amazeen: J. Geophys. Res. 68, 1835 (1963)

Dungey, J. W.: Phys. Rev. Letters 6, 47 (1961)

Helmer, J. C.: Phys. Fluids 6, 723 (1963)

Karlson, E.: Phys. Fluids 5, 476 (1962); 6, 708 (1963)

Kawashima, N., and H. Ishizuka: J. Phys. Soc. Japan 18, 763 (1963)

Lindberg, L., and L. Danielsson: Phys. Fluids 6, 736 (1963)

Persson, H.: Phys. 6, 1756 (1963)

GEOMAGNETIC STORMS

Masahisa Sugiura
National Aeronautics & Space Administration
Goddard Space Flight Center
Greenbelt, Maryland

INTRODUCTION

The earth's magnetic field undergoes transient variations of various types. Of these variations the most outstanding in amplitude and duration are magnetic storms.

One of the notable characteristics of magnetic storms is a simultaneous depression of the magnetic field in low and middle latitudes. Superimposed on this systematic depression of the magnetic field are both regular and irregular variations of varying magnitude and period that give greatly different appearances to the traces in the magnetograms at different locations of the world. Especially in the auroral zones and in the polar caps the variations are rapid, violent and of large amplitude, and appear completely irregular.

Nevertheless by separating the storm variations into different storm phases and resolving the variations into a few constituent parts the complex storm variations are found to have a systematic, if not simple, structure. Thus, we first review the several different phases that a magnetic storm undergoes in the course of its development and decay.

Most of magnetic storms begin abruptly with an increase in the horizontal component H in low and middle latitudes. This abrupt beginning is called the storm sudden commencement and abbreviated as ssc, or simply SC.

After the abrupt onset the field remains above the pre-storm level for a period of time, typically a few hours. This phase is called the initial phase.

This is followed by a large decrease in H that constitutes the main phase of the magnetic storm.

After H reaches a minimum, it starts to recover toward its normal level rather rapidly in the beginning and more slowly later. This decaying phase of the storm is called the recovery phase.

Each of these phases will be discussed below. It is noted that these phases can be identified clearly only in low and middle latitudes.

SUDDEN COMMENCEMENT

Fig. 1 shows a typical sudden commencement, observed at Tucson (geomagnetic latitude $40°.4$) on August 23, 1958, at 1840 (105th West M.T.). It is clearly an abrupt

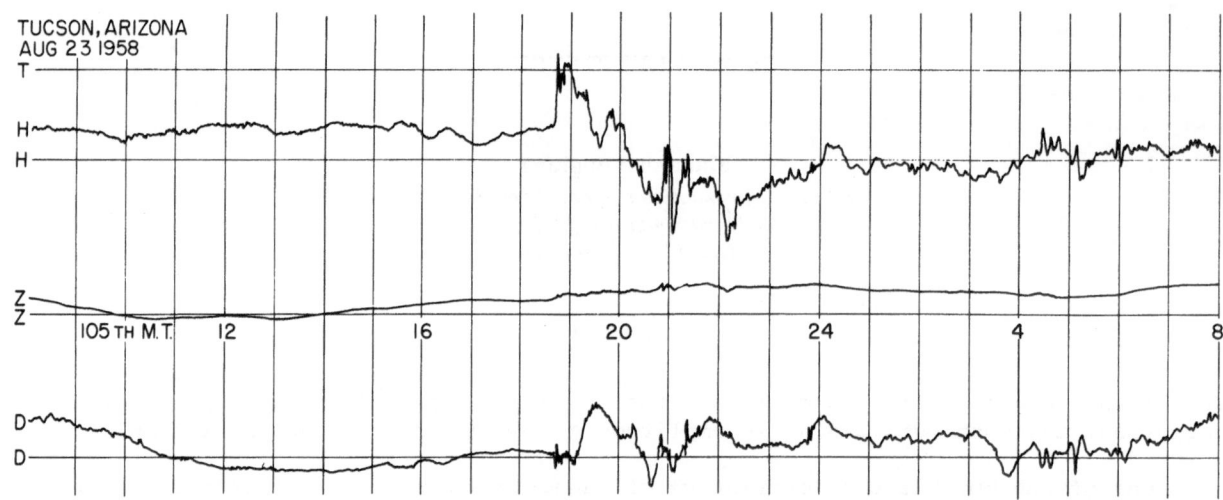

Fig. 1.   Magnetogram taken at Tucson, showing an ssc
on August 23, 1958, at 1840 105th M.T.
--U. S. Coast and Geodetic Survey.

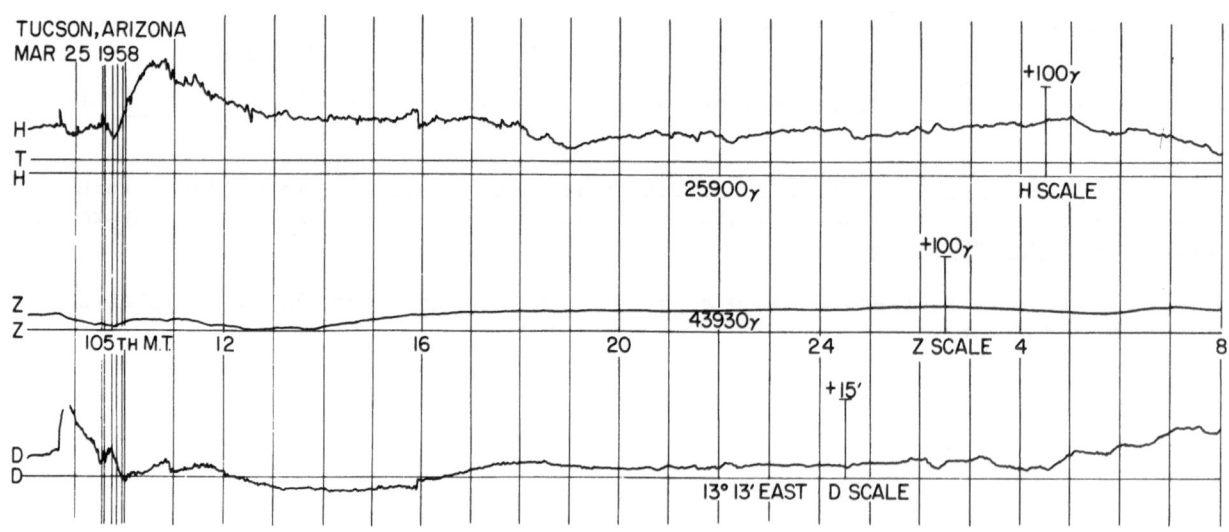

Fig. 2.   Magnetogram taken at Tucson, showing an ssc
on March 25, 1958, at 0840 105th M.T.
--U. S. Coast and Geodetic Survey.

increase in H, and the corresponding changes in declination D and in the vertical component Z are small. However, at Tucson an ssc is not always a simple increase in H, but it often takes a form of a pulse. An example is shown in Fig. 2; see the ssc at 0840 105th M.T. on March 25, 1958. In such a pulse-like ssc the trace for D also shows a pulse that is frequently of equal magnitude to that of H or even larger.

The occurrence of these pulse-like ssc's in H and D increases with increasing geomagnetic latitude. At the same time the amplitude of an ssc increases toward the auroral zone, reaching a maximum there and decreasing again toward the pole.

In the auroral zone the two components, H and D, of ssc's are of comparable magnitude, and an ssc coincides with the beginning of an oscillation lasting for several to a few tens of cycles.

Because of the distinct form of ssc's in H in low and moderate latitudes, ssc's were discussed and classified in the past according to their form in H alone. For instance, an ssc with a reversed impulse before the main positive change is marked ssc* (e.g., Newton, 1948). Akasofu and Chapman (1960) and Matsushita (1960) have proposed to indicate the shape of an ssc by indicating - or + or both signs, like SC(-+), etc.

Wilson and Sugiura (1961, 1963) and Wilson (1963) analyzed ssc's recognizing that the magnetic perturbation vector varying with time rather than the component in the magnetic meridian alone must be considered. We attempted to explain the observed characteristics of ssc's in terms of hydromagnetic waves.

Chapman and Ferraro (1931) and Ferraro (1952) attributed the magnetic field increase in the initial phase to the electric current induced in the front surface of the solar gas stream as it advances into the geomagnetic field. (See Chapman and Bartels (1940), Chapman (1960), and Ferraro (1960a) for a summary of the Chapman-Ferraro theory.) The effect of the current is to increase the magnetic field in front of the stream. The front surface is retarded due to the magnetic force acting on the current, and eventually stopped. The sharp increase in an ssc was explained by the fact that the above process is completed in a few minutes.

At the time Chapman and Ferraro proposed their theory, the existence of a plasma around the earth was not known, and they assumed that the space between the earth and the solar stream was vacuum.

We now know that the earth is surrounded by a plasma extending to great distances and that the steady flow of gas from the sun confines the earth's magnetic field in a finite region around the earth. The steady flow of solar gas was first suggested by Biermann(1951) from his study of comet tails, and was further explored by Parker (1958a, 1958b, 1960a, 1960b, 1961), who named the gas flow solar wind. The existence of steady solar wind has been confirmed by the measurements made by Mariner 2 (Neugebauer and Snyder, 1962).

The shape of the interface between the magnetosphere and solar wind has been investigated theoretically by a number of workers: (Zhigulev and Romishevskii, 1959; Ferraro, 1960b; Beard, 1960, 1962; Dungey, 1961; Hurley, 1961; Slutz, 1962; Beard and Jenkins, 1962; Spreiter and Briggs, 1962a, 1962b; Midgley and Davis, 1962, 1963; Mead, 1962, 1963).

The boundary of the magnetosphere has been observed by Pioneer 1 (Sonett et al, 1960), Pioneer 5 (Coleman et al, 1960), Explorer 10 (Heppner et al, 1963) and Explorer 12 (Cahill and Amazeen, 1963).

Intense magnetic storms are often preceded by a large solar flare by about one day. The beginning of a magnetic storm is then the mark of the arrival of a solar plasma ejected from the area of a solar flare.

Typically the steady solar wind is 300 to 400 km per second, and from the time lag between a solar flare and the onset of a storm the speed of the solar gas ejected by a flare is about 1000 km per second.

The boundary of the magnetosphere in steady solar wind appears to move considerably, but typically it is located at a geocentric distance of about 10 earth radii on the sunlit side. The boundary is at the position where the magnetic pressure balances the plasma pressure of the solar wind.

With an increased solar wind pressure due to the additional solar plasma ejected by the flare, the magnetospheric boundary is pushed inward until new balance is established.

Gold (1955), Singer (1957), and Axford (1962) attribute ssc to the arrival of the shock wave created by the supersonic plasma flow from the sun.

The impact of the solar plasma or the shock wave upon the magnetospheric boundary is transmitted through the magnetosphere toward the earth by hydromagnetic waves.

On the magnetic equatorial plane the magnetic field is perpendicular to that plane, and the longitudinal hydromagnetic wave generated by the compression of the magnetospheric boundary can be transmitted in the pure form of longitudinal hydromagnetic wave. Near the equatorial plane the situation is approximately the same as on the equatorial plane. Thus, in the equatorial and low latitudes the ssc hydromagnetic wave will be mainly in the longitudinal mode. Thus, the magnetic perturbation is linearly polarized and is paralleled to the permanent field. Examples are shown in Fig. 3 for Honolulu. In Fig. 3 the three diagrams on the left show the trace of the end point of the magnetic perturbation vector projected on to the horizontal plane; the numbers indicate time in minutes.

In higher latitudes, however, such a simple form of ssc is rarely observed if two orthogonal components in the horizontal plane are combined. Fig. 4 shows examples of ssc's not having a simple linearly polarized magnetic perturbation. The polarization of the perturbation is elliptical, in the clockwise or counterclockwise sense looking downward. It is noted that even at Honolulu ssc's sometimes show a tendency to be polarized elliptically.

A few remarks may be made on the projections of the perturbation vector onto the geomagnetic north and east directions, namely, the component changes in H and D. Of the eight cases shown in Fig. 4, ssc's marked No. 22 for College, No. 21 for Sitka, No. 23 for Honolulu, No. 2 for College, and No. 7 for Sitka show a preliminary negative impulse when projected on to the north-south direction; the negative impulse is small in the first three, and it is much larger in the last two examples.

The change in H shows an oscillatory character in No. 22 (College), No. 21 (Sitka) and No. 7 (Sitka): first a small negative pulse, then a large positive change, followed by a large negative change. In all the cases shown in Fig. 4, the change in D is oscillatory.

The direction of rotation when viewed downward in elliptically polarized ssc's is found to be a function of local time. In the northern hemisphere the polarization is clockwise from about 1000 to 2200 L.T., and counterclockwise in the remaining half day. In the southern hemisphere the direction is reversed. Fig. 5 illustrates these features of the ssc polarization.

These polarization rules are obeyed reasonably well, both statistically and in individual cases. Fig. 6 shows an example for the ssc at 0315 U.T., October 22, 1958 (Wilson and Sugiura, 1963).

The observed features described above can be interpreted in the following way.

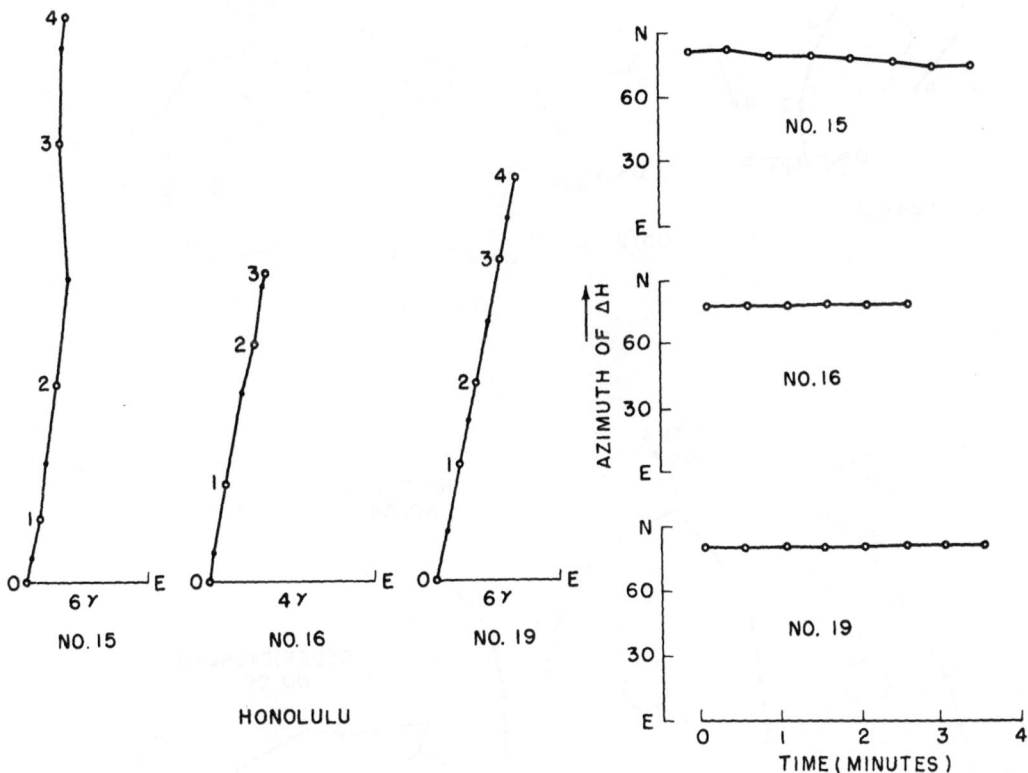

Fig. 3. Linearly polarized ssc's observed at Honolulu. Three diagrams on the left indicate the projection onto the horizontal plane of the trace of the end point of the magnetic perturbation vector as a function of time; the numbers are in minutes.
--Wilson and Sugiura (1961).

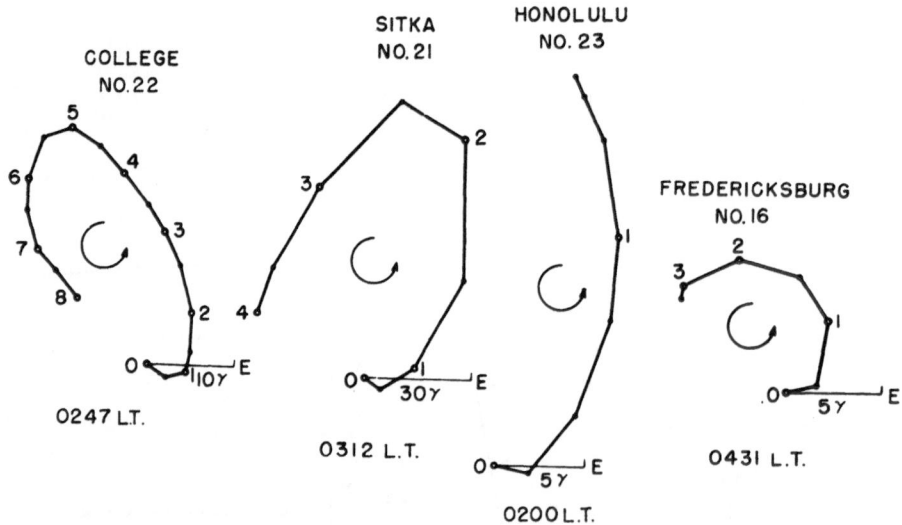

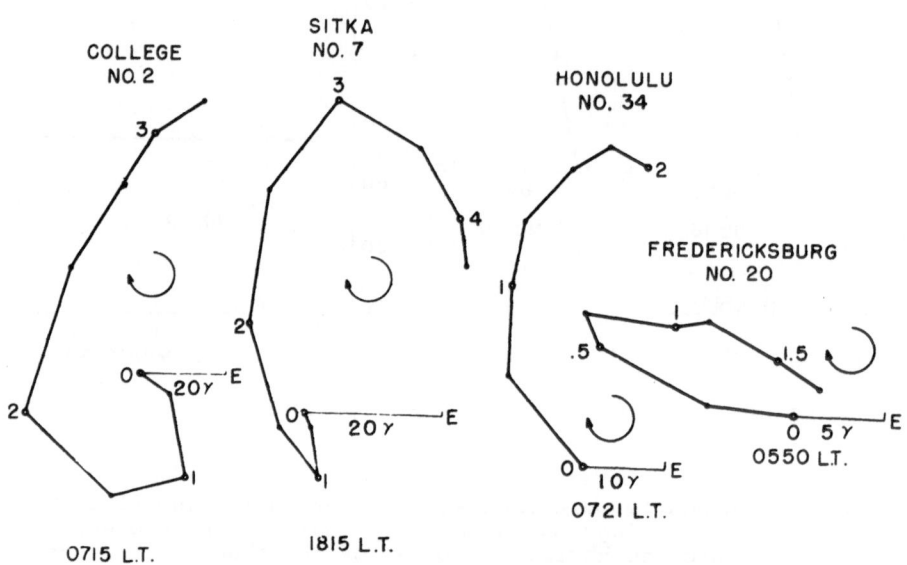

Fig. 4. Elliptically polarized ssc's. The traces are the
loci of the end point of the magnetic perturbation
vector projected onto the horizontal plane; the
numbers indicates are times in minutes.
                        --Wilson and Sugiura (1961).

54

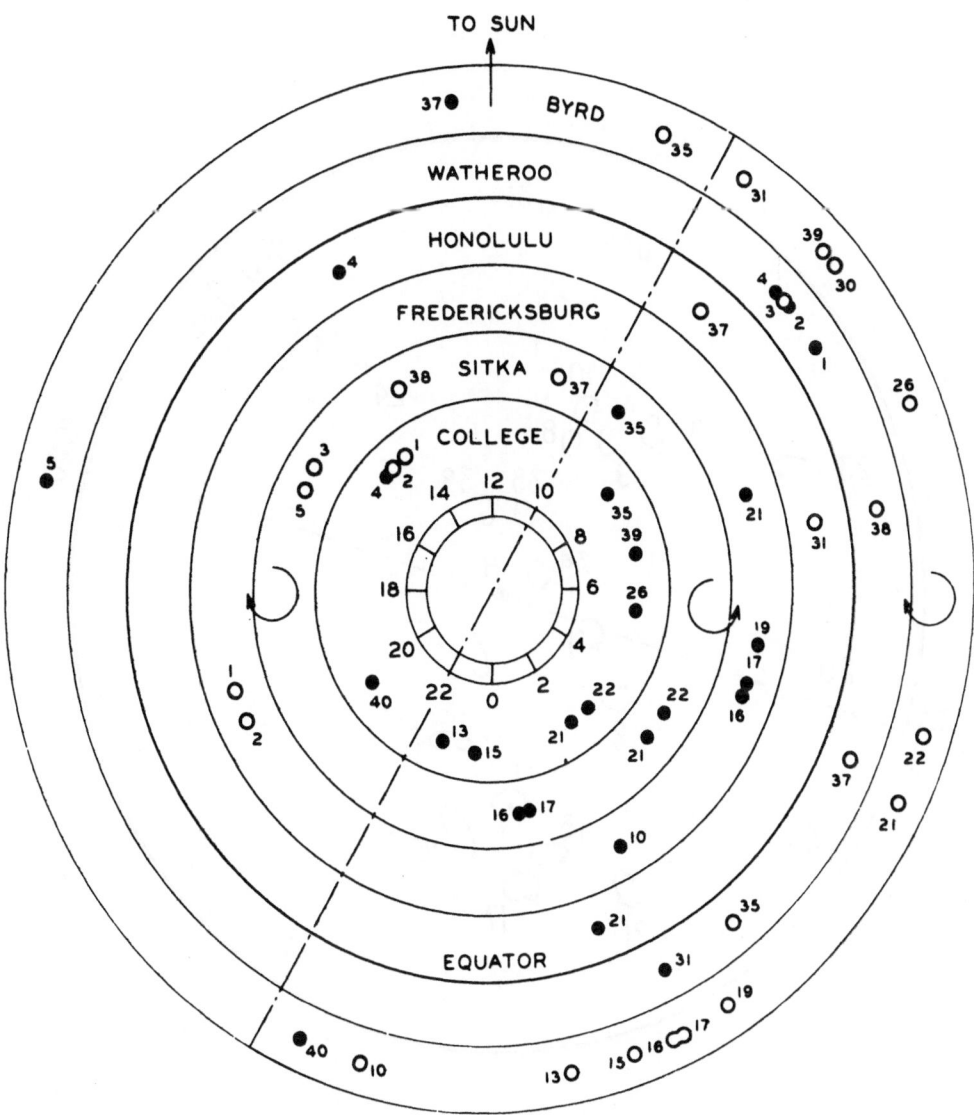

Fig. 5. Polarization rules for ssc's. Black circles represent counterclockwise rotation and open circles clockwise rotation (the sense of rotation when viewed downward). Note that the sense of rotation of the vector is opposite in the southern hemisphere to that in the northern hemisphere in each of the local time zones into which the earth is divided by the meridian plane through 1000 and 2200 hours. The equator is represented by the heavy circle between the belts for Honolulu and Watheroo.
                                    --Wilson and Sugiura (1961).

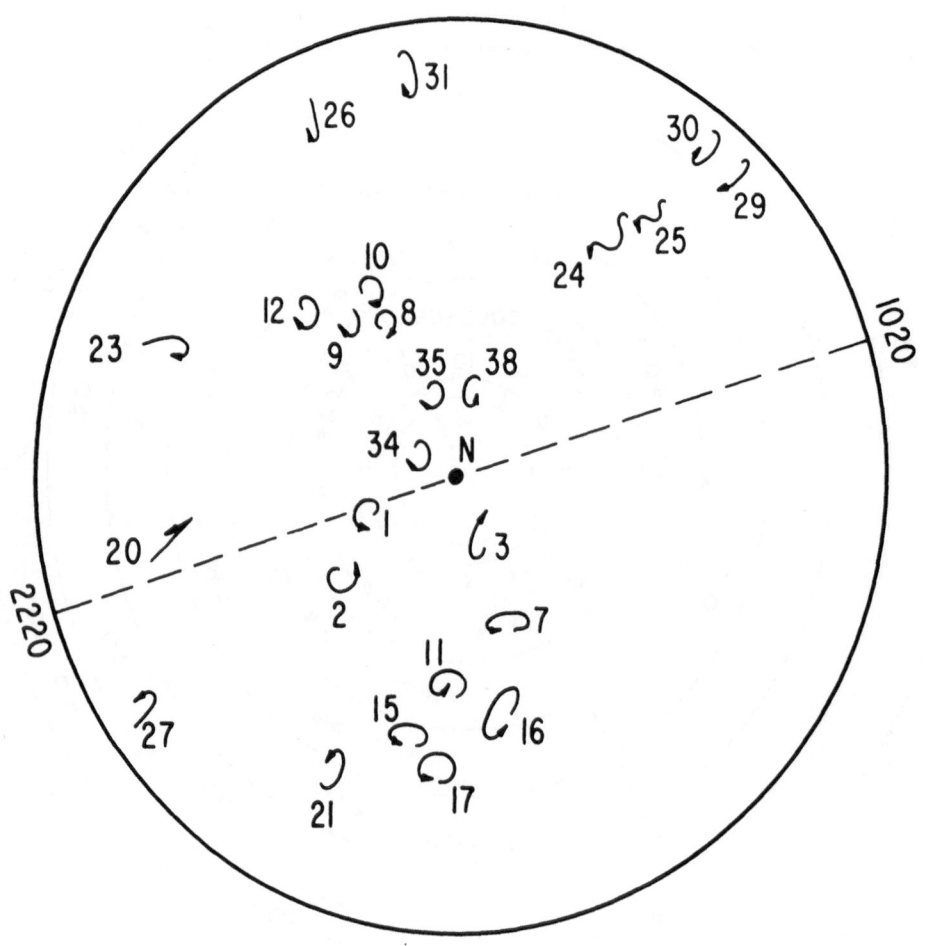

Fig. 6. Polarization of the horizontal perturbation vector
for the ssc of 0315 U.T., October 22, 1958. The
polarization diagrams are drawn for 20 northern stations at their projected positions on the equatorial
plane. The meridian plane separating the two zones
of opposite polarization is indicated by broken lines.

When the magnetosphere boundary is struck by the solar plasma, a hydromagnetic shock wave is created. The shock wave then sweeps across the magnetosphere essentially in the direction parallel to the sun-earth line. The shock is strongest in the equatorial region where the magnetic field is perpendicular to the direction of the solar plasma flow.

As the hydromagnetic shock wave advances in the magnetosphere it "blows" the lines of magnetic force toward the dark side of the earth.

The lines of force are blown more easily at large distances from the earth than near it. Thus, if we take the lines of force near the sunrise meridian, their equatorial portion is bent toward the dark side of the earth more than the portions of the lines of force farther away from the equatorial plane. Therefore, the projection of each line of force on to the equatorial plane will be a curve whose tip (i.e. the furthest end from the earth) is bent toward the night hemisphere. This initial blow causes the lines of force there to rotate in the clockwise sense when viewed from above the north pole.

This oscillation of the lines of force will be transmitted along the lines of force to the earth in high latitudes both in the northern and southern hemispheres. The direction of the rotation of the magnetic perturbation vector in the northern hemisphere will be counterclockwise when viewed downward at ground level. In the southern hemisphere the sense of rotation is opposite when likewise viewed downward.

In the evening meridians the sense of rotation of the lines of force will be opposite to that described above for the morning meridians.

Extending these arguments for the representative meridians near 06 and 18 hours local time to other meridians the polarization distribution shown in Fig. 5 can be qualitatively understood.

Matsushita (1962, 1963) criticized our view on the ground that only in a limited number of cases the polarization rules described above are perfectly satisfied. However, as is true with many geophysical phenomena, the problem dealt with here is an extremely complex phenomenon, and a complete agreement between the observed results and those derived from an idealized model is not expected.

It is noted in this connection that giant pulsations observed in the auroral zones follow a similar polarization pattern; see Fig. 7. These giant pulsations can be understood as due to oscillations of the lines of magnetic force whose mode of oscillation is similar to that of ssc's. The polarization rule for giant pulsations has been recently confirmed by Nagata et al (1963) with the data from Syowa Base and Reykjavik.

Kato and Saito (1958), Kato (1959), and Benioff (1960) have noted damped pt-type pulsations accompanying ssc's. According to Kato and Saito, these pulsations are observed during the daytime and are very weak or not noticeable in the dark hemisphere; the periods of these damped oscillations are about 20 seconds.

Kato and Saito attributed the pt-type pulsations accompanying ssc's to a poloidal hydromagnetic oscillation of the magnetosphere boundary caused by the impact of the solar plasma flow.

Berthold et al (1960) reported their observations of pulsations of period of about 30 seconds, coincident with ssc's, and they also attributed the pulsations to the resonant oscillation of the magnetosphere.

There is another type of oscillation associated with ssc's. These oscillations are of longer periods than the type discussed above, and are observed more frequently

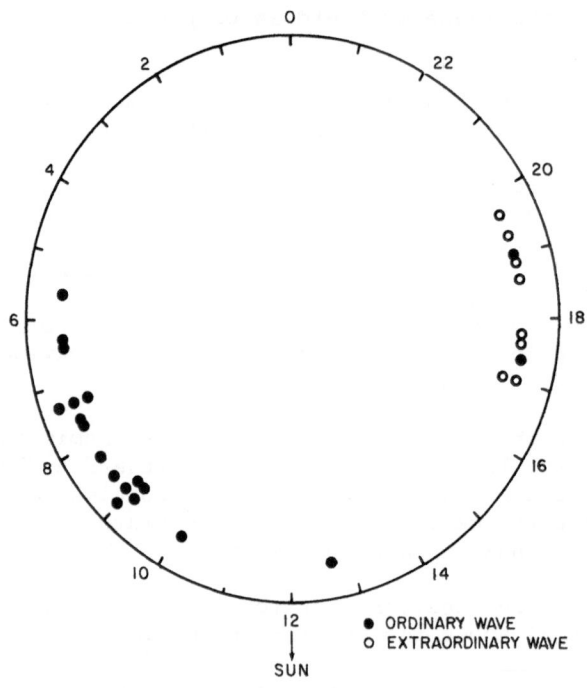

Fig. 7. Polarization rules for giant pulsations
observed at College. Black circles represent counterclockwise rotation and open
circles clockwise rotation.

in high latitudes than in low latitudes. In fact, in the auroral zones most of ssc's
are followed by an oscillation. It may be said that in high latitudes the oscillation is an essential part of ssc's.

This oscillatory nature of ssc's in high latitudes is expected from our model.
The lines of magnetic force extending to the outer regions of the magnetosphere
receive a strong blow from the shock wave, and continue to oscillate until the
oscillation is damped out.

This second type of oscillation has been investigated by Watanabe (1956), Wilson
and Sugiura (1961), Sano (1962), and Wilson (1963).

Wilson showed that the power spectra of the ssc oscillations have a harmonic
structure. If the ssc oscillations are due to resonant oscillations of the lines of
force, the oscillations must be simultaneous at magnetically conjugate points. This
conjugate nature of the ssc oscillations has been demonstrated by Wilson and Sugiura
and Wilson.

In this connection, an important remark is made on the magnetic variations observed at a pair of conjugate points when the line of force connecting the pair
oscillates. If the field configuration and the ion density are assumed to be symmetric with respect to the magnetic equator, the following conclusion can be drawn

58

regarding the symmetry of the magnetic variations at the conjugate points.

If the field line displacement is symmetric with respect to the equator as in oscillations of odd harmonics, the variations in H at the conjugate points are parallel and those in D antiparallel; whereas if the field line displacement is antisymmetric with respect to the equator as in oscillations of even harmonics, the variations at the conjugate points are antiparallel in H and parallel in D.

Therefore, if the displacement of the field line is symmetric with respect to the equator (odd harmonics), the polarization diagrams, as viewed downward at the conjugate points, are mirror images of each other with respect to the magnetic meridian plane; whereas if the displacement of the field line is antisymmetric (even harmonics), the polarization diagrams are mirror images of each other with respect to the magnetic equator.

Although the model presented here is much idealized, the present author believes that the model can explain the essential features of ssc's. To account for the more detailed morphology of ssc's, improvements and refinements of the model are desired.

Nishida (1963) investigated the screening effect of the ionosphere and regarded the initial reversed impulse as an ionospheric modulation.

Vestine and Kern (1962) discussed a mechanism in which distortions of the outer magnetosphere due to the compression by the solar stream can produce the initial reversed impulse of an ssc. In their theory, the distortions of the magnetosphere set up a polarization electric field, which is transported to the polar ionosphere along the lines of magnetic force and drives Hall current. The average current system for the initial part of ssc's, which was earlier derived by Nagata and Abe (1955) and Jacobs and Obayashi (1956), can be accounted for by the Hall current generated in this way.

However, the rotation of the magnetic perturbation vector in high latitudes makes it doubtful if the representations of the ssc variations by an average current system or by a few such systems are meaningful, or are possible at all. This objection, however, does not imply the rejection of the mechanism proposed by Vestine and Kern described above.

There is a peculiar feature in ssc's that has not been adequately explained. This is the enhancement of ssc's at the magnetic equator during the day, the effect being most notable near noon (Ferraro and Unthank, 1951; Sugiura, 1953; Forbush and Vestine, 1955; and others).

In a narrow belt over the magnetic equator the effective conductivity in the east-west direction is greatly enhanced during the day, and the potential electric field of the solar daily magnetic variation, Sq, drives an electrojet along the magnetic equator; thus, Sq is greatly augmented at the magnetic equator; likewise the lunar magnetic variation, L, is augmented there. (Hirono, 1952; Baker and Martyn, 1952, 1953; Ratcliffe and Weeks, 1960; for the augmentation of Sq and L, see Chapman and Bartels, 1940, Chap. VII and VIII; and Chapman, 1948, 1951; Forbush and Casaverde, 1961; Rastogi, 1962).

There is little doubt that this enhanced conductivity at the magnetic equator is also the cause of the augmentation of ssc's there during the day. However, it is not certain whether the electric field that drives the electric current along the magnetic equator at the time of an ssc is the electric field contained in the hydromagnetic wave or it is a field originating in the polar regions.

On the basis that an ssc is the result of the impact of a solar stream front on the geomagnetic field, Francis, Green and Dessler (1959) and Dessler, Francis and Parker (1960) studied the transit time of the hydromagnetic wave from various positions on the magnetosphere boundary to the earth. The latter authors showed that because of different transit times from different originating points on the boundary to a given point on the earth, the rise time of an ssc is several minutes even if the initial impact is abrupt. Their argument has been criticized on the ground that the ray path theory used in their theory is not strictly applicable because the wavelength involved is long compared with the dimension under consideration.

INITIAL PHASE

The sudden commencement of a magnetic storm is in itself an outstanding feature of the storm variation and can be studied separately from the later developments of the storm.

For the study of the magnetic storm variations following the sudden commencement it is convenient to analyze the storm variations into two parts: Dst and DS. One of the notable features of the storm variations is the dependence of their characteristics on geomagnetic latitude. Thus, at any station P in latitude θ, the magnetic storm variation D is a function of the storm time, T, reckoned from the storm commencement, and of the longitude, λ, of P measured eastward from the midnight meridian (i.e. local time of P). The average of D round the parallel of latitude θ, at time T, is denoted by Dst (T). It is a function of T and is called the storm time variation at latitude θ. The difference D - Dst (T) is denoted by DS (T,λ). It is a function of both T and λ, and is called the disturbance longitudinal inequality. These variations are defined for the three magnetic components. (Sugiura and Chapman, 1960)

When the average Dst is determined from many storms, it is an increase in the horizontal component H above its pre-storm level during the first few hours. This positive change in H is followed by a large decrease. The average Dst for declination is small throughout the storm, and that for the vertical component Z is smaller than in H and reversed in sign when Z is measured positively downward.

The phase in which Dst in H is positive is called the initial phase. The statistical study of Dst (and DS) has been made extensively; see e.g., Chapman and Bartels (1940), Vestine et al (1947), Sugiura and Chapman (1960).

The storm field in the initial phase is approximately a uniform field parallel to the earth's magnetic axis. The initial phase has been interpreted by Chapman and Ferraro as the compression of the earth's field by the solar stream; for a summary of their theory (Chapman and Bartels, Chap. XXV, 1940; Ferraro, 1960a).

As has been shown statistically by Chapman (1952a) and Sugiura and Chapman, DS begins to develop immediately after the sudden commencement and reaches its maximum earlier than Dst. DS is mainly of polar origin. Thus, even during the initial phase the polar disturbance may reach a high activity, and may attain its peak well before the Dst main phase.

In this regard, the _initial_ phase refers only to the early (positive) stage of the development of Dst and does not refer to any particular phase of polar part of the storm variations.

60

MAIN PHASE

The main phase of a magnetic storm is characterized by a large decrease in H in low and middle latitudes; as in the initial phase, it refers to the Dst part of the storm variations.

It is now generally thought that the main-phase decrease of the geomagnetic field is due to a ring current formed in the magnetosphere.

According to Chapman (1952b), the first suggestion of such a current was made by Störmer in 1911, though it was not a complete ring, but a stream of electrons deflected eastward round the earth on its afternoon side. However, the repulsive electrostatic forces between the electrons will not hold them together. Chapman further refers to the study made in 1924 by Adolf Schmidt. Schmidt suggested a neutral ring current that exists permanently and is enhanced at the times of magnetic storms.

Chapman and Ferraro (1933, 1941) studied a mechanism of the formation of a ring current, its equilibrium, stability and decay; for a summary, see e.g., Chapman (1952b). In their theory the centrifugal force on the charges in the ring current is balanced by the Lorentz force on the charges due to their motion in the earth's magnetic field and the electrostatic force due to the polarization within the ring. Thus, the charged particles in the ring are in circular orbits about the earth on and near the equatorial plane.

Störmer (1955) made extensive calculations of the orbits of a charged particle in a dipole field. His work shows that charged particles can drift longitudinally while spiraling about the lines of magnetic force.

Alfvén (1950) developed a perturbation method by which the trajectories of charged particles in a magnetic field can be calculated without integrating the equations of motion. Since the particles gyrate about the lines of magnetic force, the movement of the center of the gyration, called the guiding center, describes the drift motion of the particle. Alfvén gave a method of calculating the drift velocity of the guiding center. For this reason his method is called the guiding center approximation.

Singer (1957) proposed that a solar stream perturbs the geomagnetic field at large distances and that part of the solar particles are captured by the earth's field. These particles then drift longitudinally in the magnetic field in the manner shown by Störmer and Alfvén. Namely, positive ions drift westward and electrons eastward, producing a net westward current. Singer identified this westward current as the ring current that produces the main-phase decrease.

Dessler and Parker (1959) presented a hydromagnetic theory of magnetic storms. According to these authors, during the initial phase, instability causes small plasma clouds to become imbedded in the earth's magnetic field, and the particles diffuse into the magnetic field to about 3 to 5 earth radii and form a belt of trapped particles consisting chiefly of protons and electrons. The trapped protons exert stresses mainly due to centrifugal force, which cause the decrease of the magnetic field during the main phase. Instead of expressing the effect of the ring current in terms of its magnetic field, these authors express it by the stresses.

Dessler, Hansen and Parker (1961) proposed that the hydromagnetic shock waves generated in the magnetosphere by the impact of solar plasma heat the ambient protons in the regions 3 to 5 earth radii to thermal velocities comparable to the Alfvén velocity and that these protons form the ring current.

Kern (1962a) gave a model for a mechanism of accelerating particles by hydromagnetic shock waves to produce energetic particles that can form a ring current.

The magnetic field of model ring currents has been calculated by Akasofu and Chapman (1961), Akasofu, Cain and Chapman (1961), and Apel, Singer and Wentworth (1962).

The magnetic field measurements by the satellite Mechta (Dolginov and Pushkov, 1959) and the satellite Explorer 6 (Smith et al, 1960) indicated deviations of the measured magnetic field from the dipole field, but the exact position of the ring current has as yet not been determined definitively by any satellite or space probe. However, the magnetic field measurement by Vanguard 3 definitely indicates that the current responsible for the main phase must be above a few thousand kilometers from the earth's surface (Cain et al, 1962).

Piddington (1960) suggested a theory for the main phase in which the lines of force of the geomagnetic "tail" become connected with the passing solar stream and are pulled out in the direction away from the sun due to the momentum of the ions in the stream, resulting in a decrease of the geomagnetic field.

RECOVERY PHASE

After reaching a minimum, Dst in H decays rather rapidly at first and then more slowly toward the normal level. This recovery phase takes from several days to sometimes two weeks.

Statistical studies of the recovery phase have been made by Chapman (1918); see also Chapman and Bartels, Chap. IX, (1940), Sugiura and Chapman (1960) and others.

Recently Akasofu, Chapman and Venkatesan (1963) showed that the main-phase ring current of great storms is composed of two parts with different decay rates, and proposed that during a great storm two ring current belts are formed at different geocentric distances.

The studies of the recovery phase made in the past are mainly based on the average Dst variations determined from many storms. This was due to the circumstance that there were not enough observatories in low latitudes to determine Dst accurately for individual storms.

The IGY afforded an opportunity to obtain Dst not only for the storms but as a function of universal time on a continuous basis. Such data, if published regularly, should be valuable to workers of geomagnetism, cosmic rays, radiation belts, and other fields of geophysics. Several years ago Professor Chapman proposed to the International Association of Geomagnetism and Aeronomy to publish the Dst data on a regular basis like Kp indices. The present author pursued this proposal and determined the hourly values of equatorial Dst for the IGY (Sugiura, 1963) and for 1961. At the Thirteenth General Assembly of the International Union of Geodesy and Geophysics held at Berkeley, California, August 19-31, 1963, the International Association of Geomagnetism and Aeronomy passed a resolution to recommend the publication of the hourly values and a graphical representation of the equatorial Dst variation on a regular basis.

The hourly Dst data for IGY (1957-1958) and 1961 provided material for the determination and comparison of the recovery rate of Dst for an epoch of high solar activity (1957-1958) and for an epoch of low solar activity (1961). The details of this study are discussed in a separate paper, and only a summary of the results is given below.

The present method of determining the recovery rate of Dst differs from those used by other authors in the following respect. In most cases Dst recovers in several steps with temporary increases or with a brief pause of recovery between successive steps. These temporary breaks in the otherwise smooth recovery are interpreted as being due to a replenishment of the ring-current particles either by a fresh penetration of solar particles or by acceleration of ambient particles. In the present method only smoothly recovering portions are used to determine the rate of recovery.

It is then found that the recovery rate is remarkably similar in almost all the storms studied, although the temporary enhancements of Dst mentioned above give quite different appearances to different storms.

As has already been mentioned, the recovery phase consists of two phases with different recovery rates. The early phase, in which this rate is large, is here called Phase I and the later phase, in which the rate is less, is called Phase II.

Fig. 8 illustrates an example of the change in the recovery rate from Phase I to Phase II.

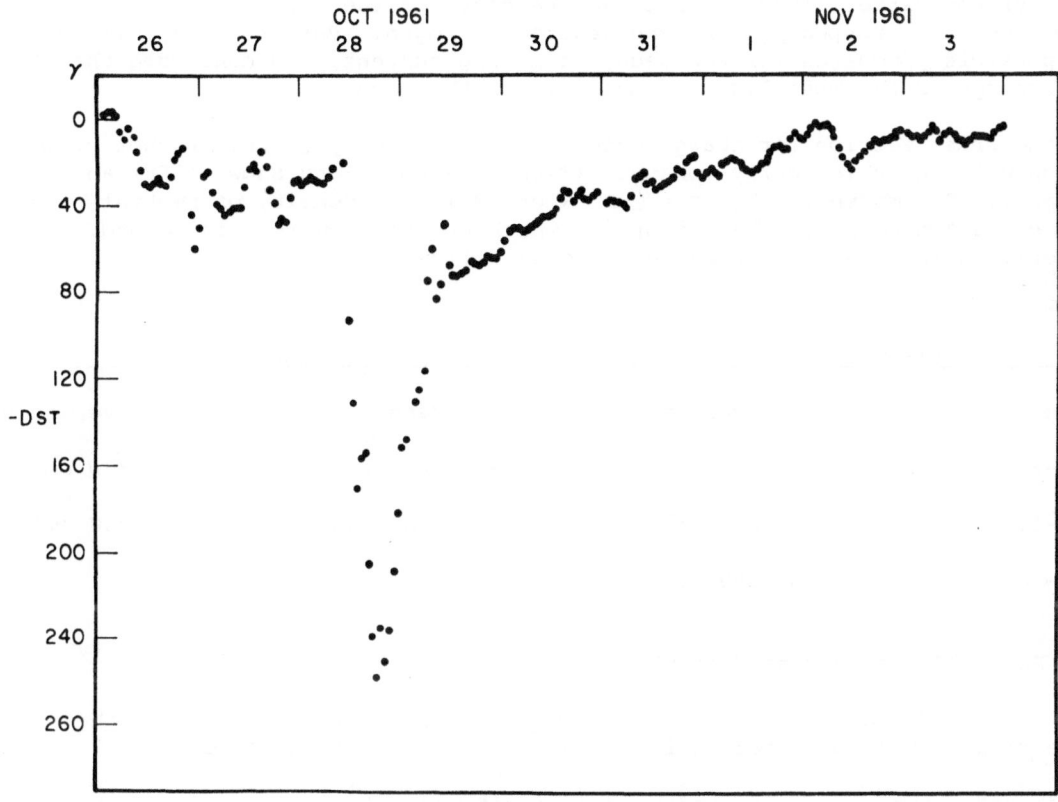

Fig. 8. Illustrating the sudden change in the recovery rate;
the rate decreases suddenly after the middle of Oct. 29.

In both phases the decay of Dst is approximately exponential. Let the time variation of Dst be expressed in the form $Ae^{-\alpha t}$, where A is constant and t is time. The factor α, which varies with the range of Dst, was found to have changed appreciably from 1957-1958 to 1961.

Since the storms that occurred in the IGY were on the whole more intense than those which occurred in 1961, the determination of α was made for a wider range of Dst for 1957-1958 than for 1961. Values of α for Phase I are tabulated below.

1957-1958		1961	
Dst range	α	Dst range	α
-400 to -200γ	2.1×10^{-5} sec^{-1}	-200 to -100γ	3.2×10^{-5} sec^{-1}
-200 to -100	1.9×10^{-5} sec^{-1}	-100 to - 50	2.5×10^{-5} sec^{-1}
-100 to - 60	1.5×10^{-5} sec^{-1}		

The value of α for Phase II was about 0.6×10^{-5} sec^{-1} for both 1957-1958 and 1961.

The ratio of α for 1961 to α for 1957-1958 is 1.7 for the range -200 to -60γ.

Dessler and Parker (1959) attributed the decay of the ring current to charge exchange between the trapped protons and neutral hydrogen. Parker (1962) discussed various possible processes for the decay of a ring current, and concluded that the charge exchange is the most likely mechanism for the decay.

If the cross section for charge exchange is denoted by Q, the product of Q and the proton velocity v is nearly constant (Fite, Brackmann and Snow, 1958) and Qv is approximately 10^{-7} cm^3/sec. Then $\alpha = QvN$, where N is the density of neutral hydrogen. Thus, we can estimate N for the regions in which the ring current is formed. Values of N determined in this way for Phase I are as follows.

1957-1958		1961	
Dst range	Average N	Dst range	Average N
-400 to -200γ	210 cm^{-3}	-200 to -100γ	320 cm^{-3}
-200 to -100	190 cm^{-3}	-100 to - 50	250 cm^{-3}
-100 to - 60	150 cm^{-3}		

For Phase II the average N is 60 cm^{-3}.

The average N above means the average density of neutral hydrogen in the region which the ring current occupied while Dst was in the specified range.

These estimates are crude, and need more refined calculations of the proton density variation and of the magnetic field variation. But the above results indicate that if the charge exchange is the main mechanism for removing the protons in the ring current, the density of ambient neutral hydrogen must be approximately 60 to 300 in the regions where the ring current is formed. The above results for Phase I show that

64

the hydrogen density increased from the epoch of peak sunspot activity to the epoch of much lower sunspot activity.

In the above estimates, the induction effect of the earth is not taken into account. When corrections are made for this effect, the above numbers will change to some extent, but the trend of increase from 1957-1958 to 1961 will not be altered since the induction effect will be the same for both epochs.

When the storm is intense (i.e. when Dst is large in magnitude), the inner boundary of the ring current must be much closer to the earth, probably about 3 earth radii or nearer, than when the storm is weak. Thus, the larger the magnitude of Dst the greater the value of α and hence, the greater the average N.

It is interesting to note that the average N for Phase II did not change from 1957-1958 to 1961. This seems to mean that the hydrogen density in the outer region of the ring current did not change with solar cycle. This can be interpreted in the following way.

In an epoch of low solar activity the hydrogen density does increase, but because of the lower temperature the scale height becomes less and hence, the density decreases with height more steeply than in high solar activity. These two trends compensate each other, and the hydrogen density in the outer region of the magnetosphere may not vary appreciably with solar cycle. This explanation was first suggested by Johnson (1961).

POLAR DISTURBANCES

In the auroral zones the storm variations are very large, sometimes reaching 1000γ or 1500γ or even more. Over the polar caps the storm variations are of smaller amplitude than in the auroral zones, but are much larger than below the auroral zones.

In the preceding sections, the storm field D was decomposed into two parts Dst and DS, recognizing that Dst contains both the field variations due to the solar plasma pressure and those due to the ring current, and that DS is mainly of polar origin.

Akasofu and Chapman (1961a, 1961b) more explicitly divided D into three parts, DCF, DR, and DP, representing the above three parts, respectively. The symbol D stands for disturbance; CF for corpuscular flux; R for ring current; P for polar.

Since DS appears to be mainly due to electric currents flowing in the ionosphere, it is convenient to represent DS by an equivalent current system. Such a current system has been drawn by many authors: see e.g., Chapman and Bartels, Chap. IX, Fig. 25, 26 (1940); Vestine et al (1947); Fukushima (1953) and Vestine (1960a).

The main features of the DS current system are: (1) An intense westward electrojet flows along the auroral zone in the early morning side. (2) An electrojet of less intensity than the westward electrojet flows eastward along the auroral zone in the afternoon side. (3) A nearly uniform return current flows across the polar cap approximately toward the longitude of nine hours local time. (4) Weak return currents flow in latitudes below the auroral zone. These features refer to both the northern and southern hemispheres.

The intensity of the auroral electrojets does not grow and decay smoothly like Dst. Intense electrojets are often generated for a short duration of time but repeatedly during a storm. In spite of these irregular variations in intensity the general pattern of the DS current system appears to be roughly maintained. At times the westward electrojet is so strong that the eastward current is nearly or completely masked.

As was already mentioned in the discussion of the Initial Phase, DS develops more rapidly than Dst. After DS reaches a maximum, it recovers more rapidly than Dst. These features were shown for average storms of three different intensities by Sugiura and Chapman (1960). These trends are also evident in the comparison of Dst and Ap for individual cases (Sugiura, 1963).

Fukushima (1953) and Vestine (1954) and others supposed that the auroral electrojets (and polar cap currents) are generated by a dynamo mechanism as in Sq and that the greatly enhanced conductivity along the auroral zone causes the current to be concentrated there. However, as was pointed out by Akasofu (1960), the electromotive force by the dynamo mechanism seems to be inadequate for intense electrojets. The oscillatory characteristics are also difficult to explain by the dynamo mechanism (Chapman, 1962).

Martyn (1953) proposed that since the conductivity along the lines of magnetic force is large, the radial polarization field in the equatorial ring current is transported along the lines of force to the ionosphere over the auroral zones, there creating a meridional electric field. This electric field drives a Hall current in the east-west direction along the auroral zones. However, Martyn based his theory on the ring current proposed by Chapman and Ferraro, and since the time when Martyn presented his theory the picture of the ring current has been greatly changed.

Akasofu (1960) considers an equatorward electric field due to a difference in position of the precipitation zones for electrons and protons; the electric field then drives a Hall current along the auroral zone.

Akasofu emphasizes that the polar DS develops abruptly and lasts for only a few hours, and that such short-lived disturbances occur repeatedly during a storm. He suggests that "several polar disturbances occur quite independently in a course of a storm, rather than that the same DS persists with some change in its intensity during a storm."

However, though the intensity of the polar disturbance in a storm may vary violently and may have large irregular time-variations on a global basis, the present author believes that the concept of DS is essentially correct. The DS variations in individual storms are certainly more irregular than the statistically determined DS, but even when there are large fluctuations in DS there appears to be a "background DS" on a global scope throughout a magnetic storm. As has already been pointed out, the westward electrojet may dominate over the eastward electrojet; this aspect has been studied by Knapp (1961).

Vestine (1960b) applied the concept of the integral invariant to the magnetosphere distorted by solar plasma. The lines of force near the midnight meridian are stretched outward more than near noon. The particles drifting in the geomagnetic field therefore must move equatorward near the midnight meridian, thereby lowering their mirror points. Vestine associated this effect with the aurora and the electrojets. Chamberlain, Kern and Vestine (1960) extended Vestine's idea and suggested that the above effect is greater for electrons than for protons and that the protons may precipitate in higher latitudes than the electrons, thus creating a meridional electric field in the ionosphere.

Kern (1961, 1962b) proposed that the distortion of the magnetosphere by the solar stream creates longitudinal magnetic-field gradient and causes the protons and electrons to separate in meridional planes while drifting longitudinally. Initial irregularities in the plasma number density in the magnetosphere and the drift separation of charges provide a mechanism for a continuous supply of polarized particle fluxes to the auroral latitudes.

Axford and Hines (1961) discussed a convection system in the magnetosphere that can explain the geomagnetic DS variations, the aurora and other associated phenomena. The details of their theory are discussed by Axford in the chapter for the magnetosphere.

Piddington (1960) considered an electromagnetic friction between the solar wind and the lines of force of the geomagnetic field near 06 and 18 hours local time, and proposed that these lines of force are twisted and produce space charges in the lower ionosphere driving a Hall current; during the main phase the twists are maintained by the deformation of the magnetosphere.

Cole (1961a) discussed a localized dynamo mechanism for the auroral electrojets and the auroral movements. He re-examined Piddington's theory and combined it with the dynamo mechanism (Cole, 1961b).

Professor Alfvén has criticized the theories of magnetic storms that ascribe DS to ionospheric currents. He argues that there has been no definitive proof that the currents in the auroral zones actually close their circuits entirely within the ionosphere. This point must be admitted; there are not enough stations to establish the ionospheric origin of DS. With the magnetic data from the Second Polar Year, Vestine and Chapman (1938) compared the current system of the type Chapman had proposed and the one proposed by Birkeland (similar to Alfvén's current system) with the observational results, and concluded that the former agrees better with the observational data.

In the electric field theory of magnetic storms which Professor Alfvén proposed (Alfvén, Chap. VI, 1950), the neutral solar stream is electrically polarized due to its motion in the solar magnetic field. The motions of the electrons and ions are studied in the combined system of this electric field and the geomagnetic field. It is then shown that the drifting particles form a sort of a ring current in the equatorial plane. Because of the different patterns of drift motions for the ions and electrons, a positive space charge is built up on the day side and a negative space charge on the night side. These space charges then neutralize each other by a discharge along the lines of magnetic force. The current, therefore, flows from the equatorial plane to the polar ionosphere along the lines of force, then it flows along the auroral zones and over the polar caps and returns to the equatorial plane following the lines of force on the opposite side of the earth.

Alfvén's theory has been criticized by Cowling (1942), who raised an objection that the particle trajectories proposed by Alfvén are not possible because of the diamagnetic repulsion.

When magnetic field measurements by a satellite in a polar orbit are made, the question of whether the DS current system is entirely within the ionosphere or, as Birkeland and Alfvén have supposed, there are current intake from, and outflow to, higher altitudes will be settled.

ACKNOWLEDGEMENT

While this paper was prepared, the author held a NASA-National Academy of Sciences-National Research Council Senior Post-Doctoral Resident Research Associateship on leave of absence from Geophysical Institute, University of Alaska.

REFERENCES

Akasofu, S.-I.: "Large-scale auroral motions and polar magnetic disturbances. I. A polar disturbance at about 1100 hours on 23 September 1957," J. Atmosph. Terr. Phys. 19, 10 (1960)

Akasofu, S.-I., J. C. Cain, and S. Chapman: "The magnetic field of a model radiation belt, numerically computed " J. Geophys. Res. 66, 4013 (1961)

Akasofu, S.-I., and S. Chapman: "The sudden commencement of geomagnetic storms," Variana 250, (1960)

"The ring current, geomagnetic disturbance, and the Van Allen belts," J. Geophys. Res. 66, 1321 (1961a)

A neutral line discharge theory of the aurora polaris, Phil. Trans. Roy. Soc. London, A, 253, 359 (1961b)

Akasofu, S.-I., S. Chapman, and D. Venkatesan: "The main phase of great magnetic storms," J. Geophys. Res. 68, 3345 (1963)

Alfvén, H.: "Cosmical Electrodynamics," Oxford Univ. Press (1950)

Apel, J. R., S. F. Singer, and R. C. Wentworth: "Effects of trapped particles on the geomagnetic field," Advances in Geophysics 9, 131 (1962)

Axford, W. I., and C. O. Hines: "A unifying theory of high-latitude geophysical phenomena and geomagnetic storms," Can. J. Phys. 39, 1433 (1961)

Axford, W. I.: "The interaction between the solar wind and the earth's magnetosphere," J. Geophys. Res. 67, 3791 (1962)

Baker, W. G., and D. F. Martyn: "Conductivity of the ionosphere," Nature 170, 1090 (1952)

"Electric currents in the ionosphere. I. The Conductivity," Phil. Trans. A, 246, 281 (1953)

Beard, D. B.: "The interaction of the terrestrial magnetic field with the solar corpuscular radiation," J. Geophys. Res. 65, 3559 (1960)

"The interaction of the terrestrial magnetic field with the solar corpuscular radiation. 2. Second-order approximation," J. Geophys. Res. 67, 477 (1962)

Beard, D. B., and E. B. Jenkins: "The magnetic effects of magnetosphere surface currents," J. Geophys. Res. 67, 3361 (1962)

Benioff, H.: "Observations of geomagnetic fluctuations in the period range 0.3 to 120 seconds," J. Geophys. Res. 65, 1413 (1960)

Berthold, W. K., A. K. Harris, and H. J. Hope: "Correlated micropulsations at magnetic sudden commencements," J. Geophys. Res. 65, 613 (1960)

Biermann, L.: "Kometenschweife und solare Korpuskularstrahlung," Z. Astrophys. 29, 274 (1951)

Cahill, L. J., and P. G. Amazeen: "The boundary of the geomagnetic field," J. Geophys. Res. 68, 1835 (1963)

Cain, J. C., I. R. Shapiro, J. D. Stolarik, and J. P. Heppner: "Vanguard 3 magnetic field observations," J. Geophys. Res. 67, 5055 (1962)

Chamberlain, J. W., J. Kern, and E. H. Vestine: "Some consequences of local acceleration of auroral primaries," J. Geophys. Res. 65, 2535 (1960)

Chapman, S.: "The abnormal daily variation of horizontal force at Huancayo and in Uganda," Terr. Mag. 53, 247 (1948)

"The equatorial electrojet as detected from the abnormal electric current distribution above Huancayo, Peru, and elsewhere," Arch. Met. Geophys. u. Bioklimat. 4, 368 (1951)

"The morphology of geomagnetic storms. An extension of the analysis of DS, the disturbance local-time inequality," Annali di Geofisica 5, 481 (1952a)

"Theories of the aurora polaris," Annali di Geofisica 8, 205 (1952b)

"Idealized problems of plasma dynamics relating to geomagnetic storms," Rev. Modern Phys. 32, 919 (1960)

"Earth storms: Retrospect and prospect," J. Phys. Soc. Japan 17, Suppl. A-I, 6 (1962)

Chapman, S., and J. Bartels: "Geomagnetism," 1, 2, Oxford Univ. Press (1962)

Chapman, S., and V.C.A. Ferraro: "A new theory of magnetic storms," Terr. Mag. 36, 77, 171 (1931)

"A new theory of magnetic storms, Part II, the main phase," Terr. Mag. 38, 79 (1933)

"The geomagnetic ring current. I. Its radial stability," Terr. Mag. 46, 1 (1941)

Cole, K. D.: "A dynamo theory of the aurora and magnetic disturbance," Aust. J. Phys. 13, 484 (1961a)

"On solar wind generation of polar geomagnetic disturbance," Geophys. J. 6, 103 (1961b)

Coleman, P. J., Jr., C. P. Sonett, D. L. Judge, and E. J. Smith: "Some preliminary results of the Pioneer V magnetometer experiment," J. Geophys. Res. 65, 1856 (1960)

Cowling, T. G.: "On Alfvén's theory of magnetic storms and of the aurora," Terr. Mag. <u>47</u>, 209 (1942)

Dessler, A. J., W. E. Francis, and E. N. Parker: "Geomagnetic storm sudden-commencement rise times," J. Geophys. Res. <u>65</u>, 2715 (1960)

Dessler, A. J., W. B. Hanson, and E. N. Parker: "Formation of the geomagnetic storm main-phase ring current," J. Geophys. Res. <u>66</u>, 3631 (1961)

Dessler, A. J., and E. N. Parker: "Hydromagnetic theory of geomagnetic storms," J. Geophys. Res. <u>64</u>, 2239 (1959)

Dolginov, S. S., and N. W. Pushkov: Dokl. Akad. Nauk SSSR <u>129</u>, 77 (1959)

Dungey, J. W.: "The steady state of the Chapman-Ferraro problem in two dimensions," J. Geophys. Res. <u>66</u>, 1043 (1961)

Ferraro, V.C.A.: "On the theory of the first phase of a geomagnetic storm: A new illustrative calculation based on an idealized (plane not cylindrical) model field distribution," J. Geophys. Res. <u>57</u>, 15 (1952)

"Theory of sudden commencements and of the first phase of a magnetic storm," Revs. Modern Phys. <u>32</u>, 934 (1960a)

"An approximate method of estimating the size and shape of the stationary hollow carved out in a neutral ionized stream of corpuscles impinging on the geomagnetic field," J. Geophys. Res. <u>65</u>, 3951 (1960b)

Ferraro, V.C.A., and H. W. Unthank: Geofisica pura e appl. <u>20</u>, 27 (1951)

Fite, W. L., R. T. Brackmann, and W. R. Snow: "Charge exchange in proton-hydrogen-atom collisions," Phys. Rev. <u>112</u>, 1161 (1958)

Forbush, S. E., and M. Casaverde: "Equatorial electrojet in Peru," Carnegie Inst. Wash. Publ. <u>620</u>, Washington, D. C. (1961)

Forbush, S. E., and E. H. Vestine: "Daytime enhancement of size of sudden commencements and initial phase of magnetic storms at Huancayo," J. Geophys. Res. <u>60</u>, 299 (1955)

Francis, W. E., M. I. Green, and A. J. Dessler: "Hydromagnetic propagation of sudden commencements of magnetic storms," J. Geophys. Res. <u>64</u>, 1643 (1959)

Fukushima, N.: "Polar magnetic storms and geomagnetic bays," J. of Faculty of Sci., Univ. of Tokyo, Sect. II, Geology, Mineralogy, Geography, Geophysics <u>8</u>, Part V, 293 (1953)

Gold, T.: "Gas Dynamics of Cosmic Clouds," North-Holland Publishing Co., and Interscience Publishers, Inc., New York (1955)

Heppner, J. P., N. F. Ness, C. S. Scearce, and T. L. Skillman: "Explorer 10 magnetic field measurements," J. Geophys. Res. <u>68</u>, 1 (1963)

Hirono, M.: "A theory of diurnal magnetic variations in equatorial regions and conductivity of the ionosphere E region," J. Geomag. Geoelect. <u>4</u>, 7 (1952)

Hurley, J.: "Interaction of a streaming plasma with the magnetic field of a two-dimensional dipole," Phys. Fluids <u>4</u>, 854 (1961)

Jacobs, J. A., and T. Obayashi: "The average electric current system for the sudden commencements of magnetic storms," Geofisica pura e appl. <u>34</u>, 21 (1956)

Johnson, F. S.: "Structure of the upper atmosphere," Satellite Environment Handbook, Stanford Univ. Press (1961)

Kato, Y.: "Investigation of the geomagnetic rapid pulsation," Sci. Rep. Tôhoku Univ., Ser. 5, Geophys. <u>11</u>, 1 (1959)

Kato, Y., and T. Saito: "Investigation on the magnetic disturbance by the induction magnetograph. Part VII, On the damped type rapid pulsation accompanying ssc," Sci. Rep. Tôhoku Univ., Ser. 5. Geophys. <u>9</u>, 99 (1958)

Kern, J. W.: "Solar-stream distortion of the geomagnetic field and polar electrojets" J. Geophys. Res. <u>66</u>, 1290 (1961)

"A note on the generation of the main-phase ring current of a geomagnetic storm," J. Geophys. Res. <u>67</u>, 3737 (1962a)

"A charge separation mechanism for the production of polar auroras electro-jets," J. Geophys. Res. <u>67</u>, 2649 (1962b)

Knapp, D. G.: "Some features of magnetic storms in high latitudes," J. Geophys. Res. <u>66</u>, 2053 (1961)

Martyn, D. F.: Proc. Roy. Soc., A. <u>218</u>, 1 (1953); Phil. Trans., A. <u>246</u>, 304 (1953)

Matsushita, S.: "Studies on sudden commencements of geomagnetic storms using IGY data from United States Stations," J. Geophys. Res. <u>65</u>, 1423 (1960)

"On geomagnetic sudden commencements, sudden impulses, and storm durations," J. Geophys. Res. <u>67</u>, 3753 (1962)

"Reply," J. Geophys. Res. <u>68</u>, 3320 (1963)

Mead, G. D.: "Numerical solutions to the Chapman-Ferraro problem," (Abstract), Trans. Am. Geophys. Union <u>43</u>, 459 (1962)

"The geomagnetic field near the magnetopause," (Abstract), Trans. Am. Geophys. Union <u>44</u>, 82 (1963)

Midgeley, J. E., and L. Davis, Jr.: "Computation of the bounding surface of a dipole field in a plasma by a moment technique," J. Geophys. Res. <u>67</u>, 499 (1962)

"Calculation by a moment technique of the perturbation of the geomagnetic field by the solar wind," J. Geophys. Res., in press (1963)

Nagata, T., and S. Abe: "Notes on the distribution of SC* in high latitudes," Rept. Ionosphere Research, Japan <u>9</u>, 39 (1955)

Nagata, T., S. Kokubun, and T. Iijima: "Geomagnetically conjugate relationships of giant pulsations at Syowa Base, Antarctica, and Reykjavik, Iceland," J. Geophys. Res. <u>68</u>, 4621 (1963)

Neugebauer, M., and C. W. Snyder: "The mission of Mariner 2: Preliminary observations," Science <u>138</u>, 1095 (1962)

Newton, H. W.: "'Sudden commencements' in the Greenwich magnetic records (1879-1944) and related sunspot data," Monthly Notices Roy. Astron. Soc. Geophys. Suppl. <u>5</u>, 159 (1948)

Nishida, A.: "Ionospheric screening effect and storm sudden commencement,"
        (Abstract), Trans. Am. Geophys. Union <u>44</u> (1963); also private communi-
        cation.

Parker, E. N.: "Interaction of the solar wind with the geomagnetic field," Phys.
        Fluids <u>1</u>, 171 (1958a)

        "Dynamics of the interplanetary gas and magnetic fields," Astrophys. J.
        <u>128</u>, 664 (1958b)

        "The hydrodynamic treatment of the expanding solar corona," Astrophys. J.
        <u>132</u>, 175 (1960a)

        "The hydrodynamic theory of solar corpuscular radiation and stellar winds,"
        Astrophys. J. <u>132</u>, 821 (1960b)

        "Sudden expansion of the corona following a large solar flare and the
        attendant magnetic field and cosmic-ray effects," Astrophys. J. <u>133</u>, 1014
        (1961)

        "Dynamics of Geomagnetic Storms," Space Sci. Rev. <u>1</u>, 62 (1962)

Piddington, J. H.: "Geomagnetic storm theory," J. Geophys. Res. <u>65</u>, 93 (1960)

        "A theory of polar geomagnetic storms," Geophys. J. <u>3</u>, 314 (1960)

Rastogi, R. G.: "Longitudinal variation in the equatorial electrojet," J. Atmosph.
        Terr. Phys. <u>24</u>, 1031 (1962)

Ratcliffe, J. A., and K. Weeks: "The ionosphere," Physics of the Upper Atmosphere,
        Ed. by J. A. Ratcliffe, Academic Press, New York, Chap. 9, 392 (1960)

Sano, Y.: "Morphological studies on sudden commencements of magnetic storms using
        the rapid-run magnetograms during the IGY," J. Geomag. Geoelect. <u>14</u>, 1
        (1962)

Singer, S. F.: "A new model of magnetic storms and aurorae," Trans. Am. Geophys.
        Union <u>38</u>, 175 (1957)

Slutz, R. J.: "The shape of the geomagnetic field boundary under uniform external
        pressure," J. Geophys. Res. <u>67</u>, 505 (1962)

Smith, E. J., P. J. Coleman, D. L. Judge, and C. P. Sonett: "Characteristics of the
        extraterrestrial current system: Explorer 6 and Pioneer 5," J. Geophys.
        Res. <u>65</u>, 1858 (1960)

Sonett, C. P., E. J. Smith, and A. R. Sims: Space research, I. Proc. Intern.
        Space Sci. Symp., 1st Nice, ed. by H. Kallmann-Bijl, North-Holland Pub.
        Co., Amsterdam, 921 (1960)

Spreiter, J. R., and B. R. Briggs: "Theoretical determination of the form of the
        boundary of the solar corpuscular stream produced by interaction with the
        magnetic field of the earth," J. Geophys. Res. <u>67</u>, 37 (1962a)

        "On the choice of condition to apply at the boundary of the geomagnetic
        field in the steady-state Chapman-Ferraro problem," J. Geophys. Res. <u>67</u>,
        2983 (1962b)

Störmer, C.: The Polar Aurora, Oxford Univ. Press (1955)

Sugiura, M.: "The solar diurnal variation in the amplitude of sudden commencements of magnetic storms at the geomagnetic equator," J. Geophys. Res. **58**, 558 (1953)

"Hourly values of equatorial Dst for the IGY," to be published in the Annals of the International Geophysical Year, Pergamon Press, (1963)

Sugiura, M., and S. Chapman: "The average morphology of geomagnetic storms with sudden commencement," Abh. Akad. Wiss. Göttingen, Math.-Physik. Kl., Sond. **4**, 53 (1960)

Vestine, E. H.: "Winds in the upper atmosphere deduced from the dynamo theory of geomagnetic disturbance," J. Geophys. Res. **59**, 93 (1954)

"The upper atmosphere and geomagnetism," Physics of the Upper Atmosphere, Chap. 10, edited by J. A. Ratcliffe, Academic Press, New York (1960a)

"Polar aurora, geomagnetic and ionospheric disturbances," J. Geophys. Res. **65**, 360 (1960b)

Vestine, E. H., I. Lange, L. Laporte, and W. E. Scott: "The geomagnetic field, its description and analysis," Carnegie Inst. Wash. Publ. **580**, Washington, D. C. (1947)

Vestine, E. H., and S. Chapman: "The electric current system of geomagnetic disturbance," Terr. Mag. **43**, 351 (1938)

Vestine, E. H., and J. W. Kern: "Cause of the preliminary reverse impulse of storms," J. Geophys. Res. **67**, 2181 (1962)

Watanabe, T.: "Studies on p.s.c. after the Ashour-Price's model for the ionospheric shielding effect," Sci. Rep. Tôhoku Univ., Ser. 5, Geophys. **8**, 9 (1956)

Wilson, C. R.: "Hydromagnetic interpretation of sudden commencements of geomagnetic storms," Ph.D. Thesis, Univ. of Alaska, (May 1963)

Wilson, C. R., and M. Sugiura: "Hydromagnetic interpretation of sudden commencements of magnetic storms," J. Geophys. Res. **66**, 4097 (1961)

"Discussion of our earlier paper 'Hydromagnetic interpretation of sudden commencements of magnetic storms,'" J. Geophys. Res. **68**, 3314 (1963)

Zhigulev, V. N., and E. A. Romishevskii: "Concerning the interaction of currents flowing in a conducting medium with the earth's magnetic field," Dokl. Akad. Nauk SSSR, **127**, 1001 (1959); English translation, Soviet Physics, Doklady **4**, 859 (1960)

APPENDIX I

MICROPULSATIONS ASSOCIATED WITH POLAR DISTURBANCES

In the Session on Magnetic Storms, Dr. W. H. Campbell of National Bureau of Standards, Boulder, Colorado, discussed his work on micropulsations. In particular, he discussed geomagnetic effects associated with electron precipitation in the auroral zone as evidenced by the balloon measurement of bremsstrahlung X-rays.

Campbell showed examples of simultaneous occurrences of micropulsations and X-ray events, and proposed that the micropulsations and fluctuations in magnetic bays are caused by changes in conductivity rather than by changes in the electric field that drives the electrojet.

He showed remarkable examples of simultaneous oscillations in the auroral emission at 3914A and micropulsations and in the cosmic radio noise absorption and micropulsations.

Campbell discussed the study, made by him and J. M. Young of National Bureau of Standards, of infrasonic phenomena observed in Alaska.

References relevant to Dr. Campbell's discussion are listed below.

Campbell, W. H., and H. Leinbach: "Ionospheric absorption at times of auroral and magnetic pulsations," J. Geophys. Res. 66, 25 (1961)

Campbell, W. H., and M. H. Rees: "A study of auroral coruscations," J. Geophys. Res. 66, 41 (1961)

Campbell, W. H.: "Magnetic micropulsations and electron bremsstrahlung," J. Geophys. Res. 66, 3599 (1961)

"Some auroral zone disturbances at times of magnetic micropulsation storms," J. Phys. Soc. Japan 17, Suppl. A-1, 112 (1962)

Campbell, W. H., and S. Matsushita: "Auroral-zone geomagnetic micropulsations with periods of 5 to 30 seconds," J. Geophys. Res. 67, 555 (1962)

Campbell, W. H., and J. M. Young: "Auroral zone ionospheric disturbances and associated infrasonic phenomena," Abstract, Trans. Am. Geophys. Union 44, 42 (1963)

Campbell, W. H.: "A study of geomagnetic effects associated with balloon evidence of auroral zone electron precipitation," (preprint)

APPENDIX II

OTHER DISCUSSIONS

Most of the discussions made in this Session have been incorporated in the main text. Additional discussions are given below.

<u>Dr. J. Heirtzler</u> of Lamont Geological Observatory, commented that the power spectra for sudden commencements are very unstable and that the spectral peaks depend strongly on exactly which section of the record is taken.

<u>Professor J. A. Jacobs</u> of University of British Columbia, called attention to the following studies made by him and his colleagues.

"Jacobs and Nishida (1962) analyzed magnetograms obtained during the IGY looking for world-wide changes of the geomagnetic field (other than ssc's and si's). World-wide changes are observed quite frequently; at least 20% of every one-hour period and at least 90% of all days contained at least one during a three month period near sunspot maximum. They represent changes in the 'solar weather' not so obvious as ssc's and si's. The form of the change varies depending both on local time and on latitude. The distribution of the magnitude and the mode of propagation have been studied. They show a pronounced similarity to those of ssc and si and indicate that there is a permanent interaction between the solar wind and the magnetosphere (Nishida, A., and J. A. Jacobs, World-wide changes in the geomagnetic field, J. Geophys. Res. <u>67</u>, 525 (1962))."

"Jacobs and Sinno (1960) observed long period, continuous pulsations of about 6 minutes period which they called LPc's. They are usually observed in polar regions and are frequently associated with Pc's but have larger amplitudes. They are also nearly synchronous over a very wide area from the arctic to the antarctic and, on occasions, in equatorial regions as well. The polarization is usually nearly linear at each station; when non-linear, it appears as a counter-clockwise flat-shaped ellipse. It is suggested that LPc's might be the expression of the leakage of the fundamental mode of standing waves in the magnetosphere between about 1000 km and the outer boundary. (Jacobs, J. A., and K. Sinno: World-wide characteristics of geomagnetic micropulsations, Geophys. J. <u>3</u>, 333 (1960))."

THE AURORA

A. G. McNish

National Bureau of Standards
Washington, D.C.

The content of this paper does not differ much from what I would have presented if I had been discussing this subject prior to the recent Geophysical Year. The results of the Geophysical Year have not altered any of our basic opinions about auroral phenomena. They have tended to confirm many of the things which had been thought for a long time.

A particular new concept bearing on the aurora concerns the magnetosphere and the idea that a continual particle flux proceeds from the sun toward the earth and other planets and envelopes them. This is in contrast to the concept that prevailed a few years ago that the space between the sun and the earth was a great vacuum, that the sun's outer atmosphere stopped at some distance and the earth's atmosphere as well. This is one concept that altered some of our thinking. Another concept with respect to aurora involves a better understanding of corpuscular streams through the work of Willard Bennett who repeated some of the early experiments on a much more elaborate basis. As you may remember, some years ago Lindemann pointed out that a stream of corpuscles coming from the sun, as hypothesized by Birkeland at the beginning of this century, cannot hold together during the flight from the sun to the earth because of mutual repulsion. Our concept of this has also been changed, because the stream can refocus itself if it is going through a region where there is ionization. Particles of opposite charge would be drawn into that stream without partaking of its motion so the stream would be electrically neutral. Particles of the same sign in motion in that region would be drawn together by magnetic forces. Thus, what was worked out mathematically for the trajectory of a single particle is also the trajectory of the stream.

I am not going into the mathematical theory. We know today that observations have confirmed what was hypothesized many years ago. One example is the existence of certain regions, which we have now identified as the Van Allen Belts, although the earlier theories regarding them were rather vague. It was not clear how charged particles could get into these regions, but since this time we know we learned about neutrons (remember neutrons were only discovered about thirty years ago). A neutron has a half-life of about thirteen minutes, and being uncharged can get into these belts and decay, its products peopling the Van Allen belts. Thus, such a belt of radiation can be produced by neutrons coming in and disintegrating there leaving electrons and protons resulting from the disintegration to follow their spiral paths from one mirror point to another, back and forth in the earth's upper atmosphere.

Another new development which has helped us understand these phenomena is the recognition that the sun not only produces corpuscular radiation, a radiation which would explain the aurora, but also produces x-rays. I remember the days when we rediscovered solar flares and their magnetic effect. In the early thirties there was almost complete agreement among people working in this field that there was no one-to-one correspondence between solar phenomena and terrestrial phenomena. A number of observations in the mid-thirties left no doubt that a bright chromospheric eruption of a certain intensity would produce ionization in the atmosphere and that the radiation which produced this ionization reached the earth with the speed of light. We were hard pressed to try to account for this because what we believed was this ionizing radiation was identified with the Lyman continuum and it was difficult to explain the magnitudes involved. Now we know the Lyman continuum was not the cause of this ionization which we attribute to x-rays which have been observed during solar flares. Except for these few changes in our thinking the picture remains very much the same as it was years ago.

There are many aspects of the aurora. We can talk about the composition of the air, as viewed by spectrometry of auroral lines. This is interesting in showing the composition of the upper atmosphere, but I do not believe it is particularly pertinent to the aspects of the aurora which may concern us here. It is obviously of importance to those who are concerned with the composition of the upper atmosphere. Of course, we recognize the great advance which was made in the understanding of aurora and of the whole outer atmosphere when the auroral green line was found to be due to atomic oxygen. Preceding that time there has been much speculation, such as the "Stickstoffstaubhülle" theory of auroral light which attributed it to frozen nitrogen related to the familiar nitrogen afterglow. These are the types of things that one encounters in a spectral study of the auroral light.

There has also been a great deal of work done on night-sky light, or airglow, quite apart from the aurora; spectrally the light is very similar in nature to that of the aurora. Personally, I cannot believe that the two phenomena are of the same origin even though the light coming from them may be much the same.

Then there were other people in the early days of aurora research who spent all their time classifying aurora according to their shapes, draperies, curtains, arcs, coronas, etc. It seems to me that such investigations had little real meaning and added little to our understanding.

The most interesting aspect of the aurora, to me, and I am going to concentrate on this, is the relation between auroral phenomena and other geophysical phenomena, in the fields of aeronomy and geomagnetism. This seems to me to be the most mean-ingful thing from the standpoint of this conference. First, let us consider the geographical distribution of the aurora. This may be graphically represented by isochasms which show the estimated percentage of days in various localities on which aurora may be seen on cloudless nights. The first chart of this sort was prepared for the Northern Hemisphere by Fritz in 1872. A similar distribution has since been obtained for the Southern Hemisphere. These distributions have been confirmed but not appreciably altered by additional data including those from the International Geophysical Year. It seems inconceivable that any new data would alter this picture. The centers of these systems of isochasms are near the geomagnetic poles (Fig. 1), so this distribution of lines of auroral frequency is much more closely related to the geomagnetic axis of the earth than to its rotational axis. Their oval shapes reflect the effects on auroral events of some of the higher harmonics of the earth's field, that is, those involving terms of inverse powers of r higher than three. One of these ovals in the northern system which denotes a region where aurora may be ex-pected to be seen on the average about once in ten years reaches almost to Panama. There is a great difficulty in making a map like this, one must do a lot of sketch-ing in because there are not enough clear nights to afford an adequate statistical sample of auroral occurrence. But we can construct a chart which is very similar from geomagnetic observations, and this points to the very close connection between the two phenomena.

When we construct a chart (Fig. 2) with lines connecting regions of equal inten-sity of magnetic disturbance as given by the range in daily magnetic variation measured in gammas, it closely resembles the auroral chart. Such a chart has been constructed from old data, not taken during the International Geophysical Year, but I think that by the addition of more data it would be essentially unchanged. The magnetic data which we have is abundant and it gives us the same pattern as the

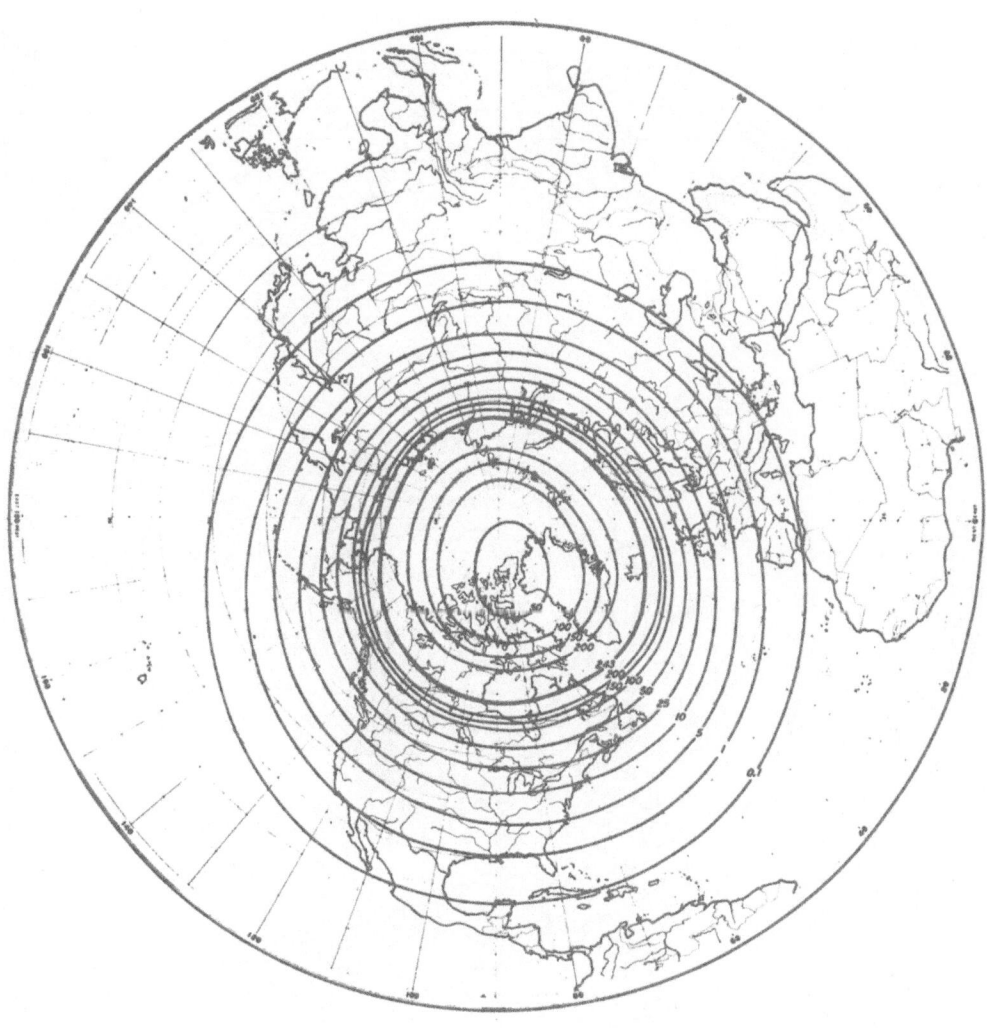

Fig. 1. Estimated annual frequency of days of visible auroral display, Northern Hemisphere (based on Fritz's data, 1700-1872, and later data, 1872-1942).

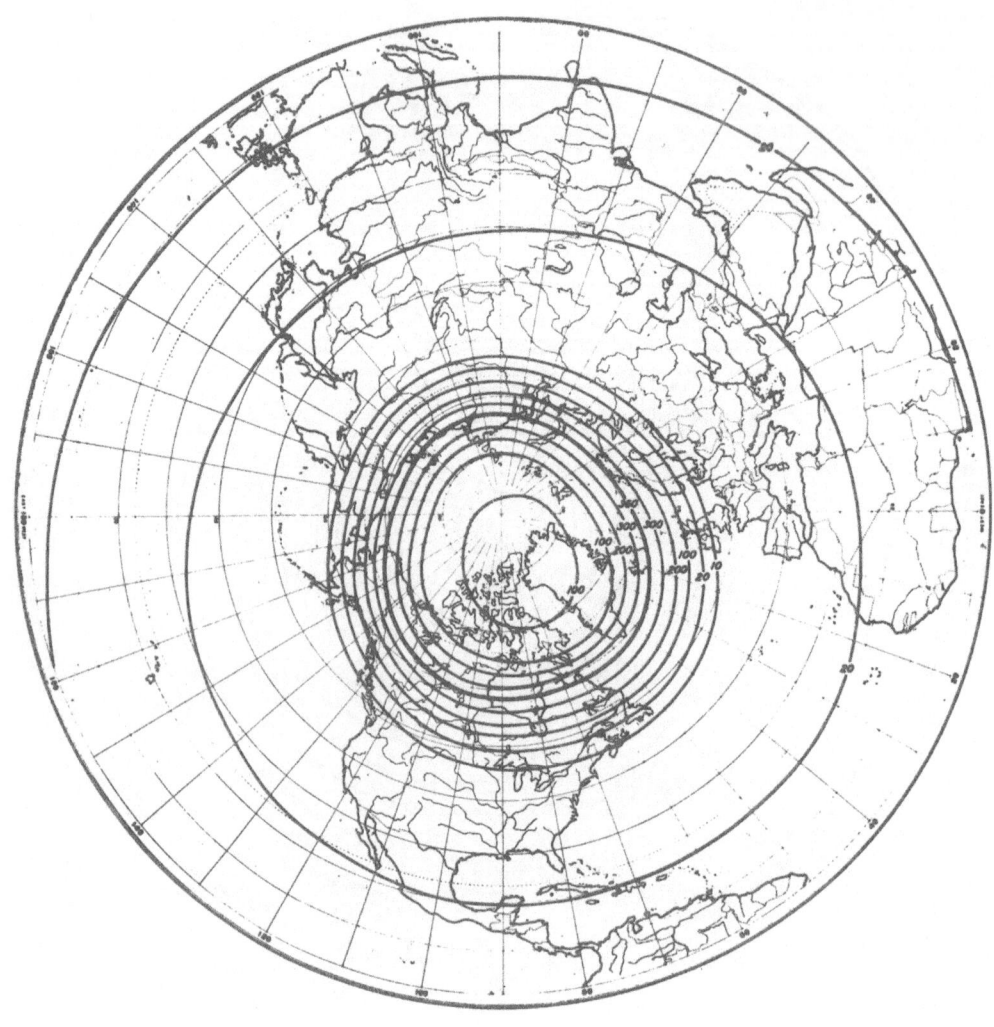

Fig. 2.   Lines of average range of total-intensity disturbance
          in gammas for 60 selected international disturbed days,
          Second International Polar Year, 1932-1933, Northern
          Hemisphere.

auroral data. If the light does not go out on the magnetograph or some other equipment failure occurs, we have a record for 24 hours a day, 365 days a year, and year after year, from many places on the earth. Furthermore, the magnetic records supply us with data even when sunlight or moonlight would interfere with visible observations of aurora.

If we consider the effects of local time on the distribution of aurora other interesting parallelisms with geomagnetic phenomena can be found, as may be seen on Fig. 3. The distribution of isochasms of auroral frequency drawn for midnight Greenwich time reveals a maximum of auroral activity shortly before midnight local

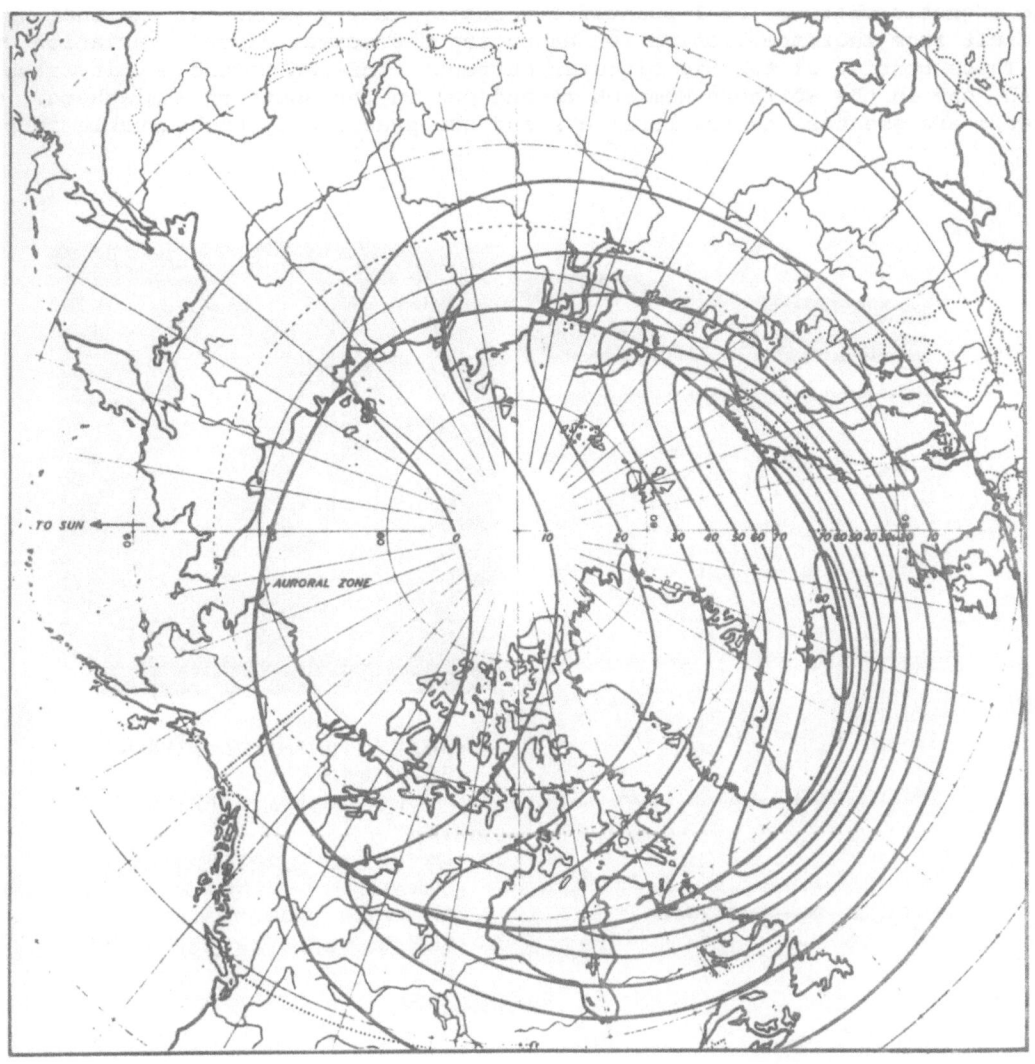

Fig. 3. Estimated annual average lines of equal hourly percentage-frequency visually observed aurora, ray-forms, clear, dark, nights, Northern Hemisphere, basis stations various years, 1881-1942 (for 0^h GMT).

time, somewhere about 23 hours. This is of particular interest in relation to geo-magnetic disturbances since there are certain types of disturbances, such as would be caused by a strong westward current flow paralleling the auroral zone which tends to occur at this time.

What are these types of disturbances which we may associate with the aurora? There are typical forms of disturbances which have been recorded. One of these is a daytime disturbance which is not associated with the aurora. It was first identified by Birkeland while studying records obtained during the first solar year, in 1882, the parent of our geophysical year. He called it a cyclo-median storm.

The distinctive feature about disturbances of this type is that they are con-fined to the daylit portions of the earth. We now know that these disturbances occur simultaneously with the appearance of bright chromospheric eruptions on the sun (see Fig. 4). The pattern of the currents flowing in the ionosphere which would cause the observed magnetic changes, as I pointed out nearly thirty years ago, is that which would result from increased ionization enhancing the normal diurnal variation cur-rents. These consist of two big circular currents confined to the daylit areas of the earth, one in the Northern Hemisphere and one in the Southern Hemisphere. Thus, the effects are greatest in low latitudes and are practically unobservable in polar regions.

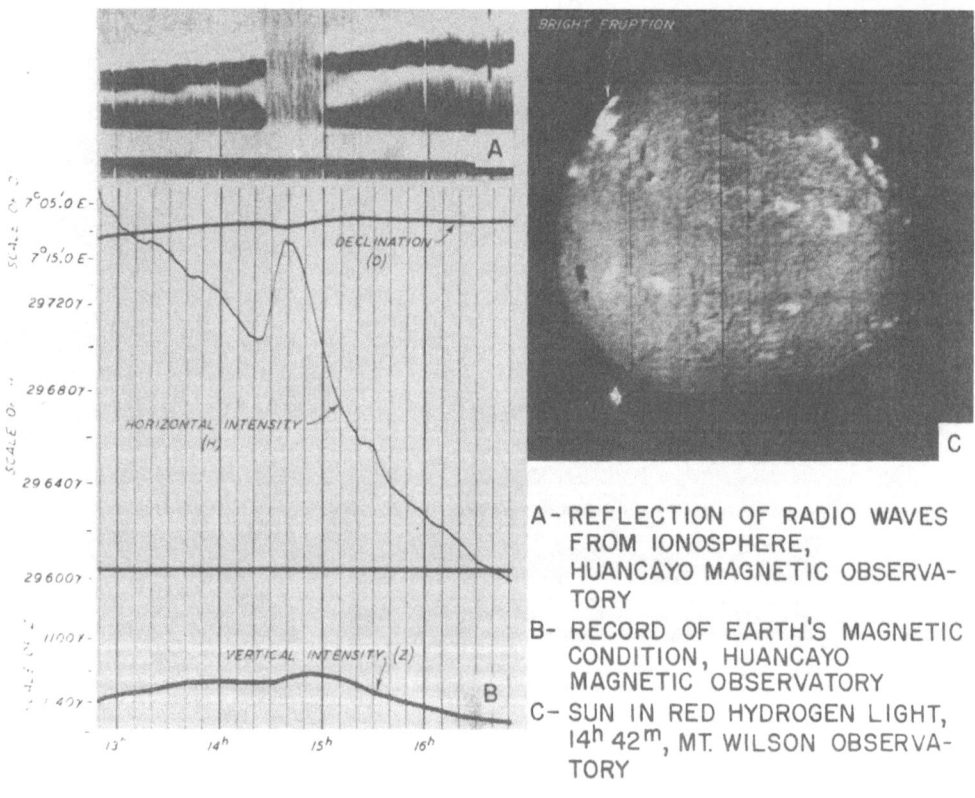

A- REFLECTION OF RADIO WAVES
   FROM IONOSPHERE,
   HUANCAYO MAGNETIC OBSERVA-
   TORY
B- RECORD OF EARTH'S MAGNETIC
   CONDITION, HUANCAYO
   MAGNETIC OBSERVATORY
C- SUN IN RED HYDROGEN LIGHT,
   14^h 42^m, MT. WILSON OBSERVA-
   TORY

Fig. 4. Changes in earth's magnetism and cessation of radio reflections from iono-sphere accompanying bright eruption in solar chromosphere, Aug. 28, 1937, (75° west meridian times). (a) Reflection of radio waves from ionosphere, Huancayo Magnetic Observatory; (b) Record of earth's magnetic condition, Huancayo Magnetic Observatory; (c) Sun in red hydrogen light 14^h 42^m Mt. Wilson Observatory.

The ionizing radiation responsible for these effects must be electromagnetic, not particle radiation, because of the prompt onset after beginning of the chromospheric eruption and because it is undeviated by the geomagnetic field. An increase in absorption of radio waves accompanies these disturbances and again is confined to the daylit areas, in the lower ionosphere, which ordinarily is ionized by ultraviolet radiation. At first these effects were thought to be due to solar ultraviolet radiation, but the presence of soft x-rays have since been detected during chromospheric eruptions and it now seems more reasonable to attribute the effects to these. One important fact is that aurora do not ordinarily accompany these events, although chance coincidences might be observed sometimes.

On the other hand, there are types of disturbances which are definitely associated with aurora. These are the bay-type disturbances (Fig. 5) which can be profitably studied in their simplest form. The magnetic field changes are such as would be due to strong electric currents flowing along the auroral zone, or parallel with it, just shortly before the local midnight meridian, the return current being by a circuitous path. The strength of these effects is a maximum in high latitudes and diminishes toward the equator, increasing again in high southern latitudes. Thus, when a strong current flow occurs along the auroral zone in one hemisphere a corresponding current flow occurs along the auroral zone in the other, the effects in the two hemispheres being approximately mirror images of each other. This corresponds to a fact about the aurora. We have long known that if a very intense aurora appears in the Northern Hemisphere a corresponding effect is also observed in the Southern Hemisphere.

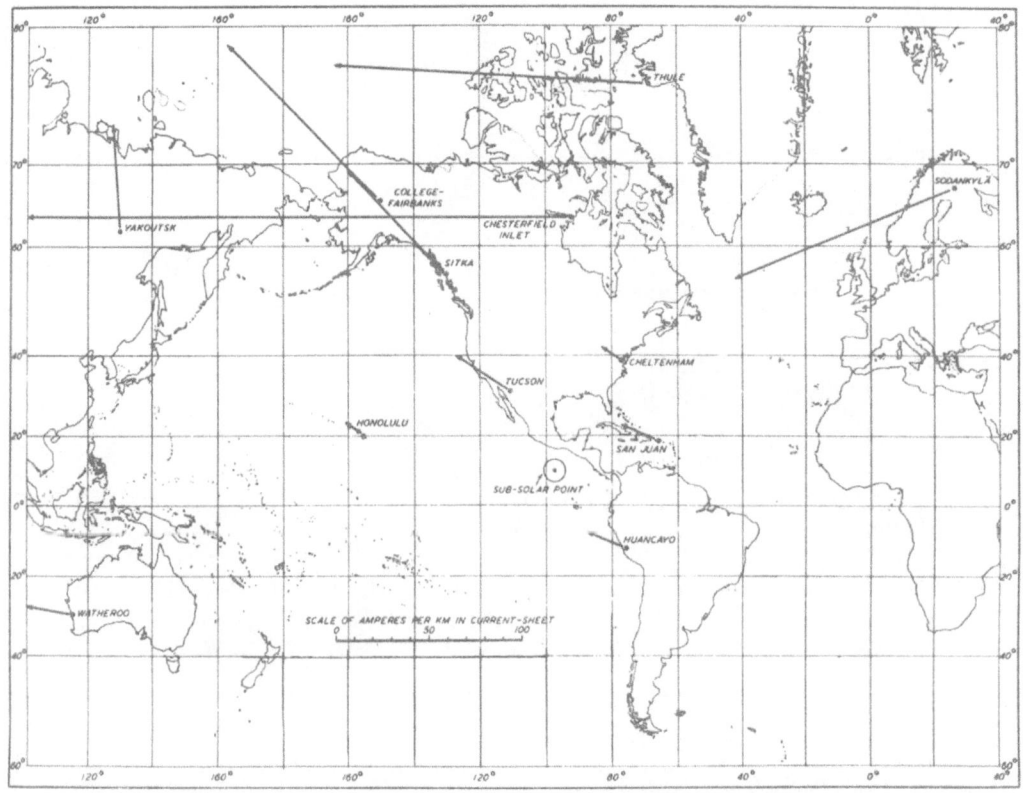

Fig. 5. Overhead currents necessary to produce magnetic changes at time of "bay"-type disturbance at 19^h 35^m GMT, April 16, 1933. This type frequently occurs during magnetic storms.

Now let us look at the whole geomagnetic field during the time of magnetic disturbances. There is a very familiar representation (Fig. 6) due to Chapman of the

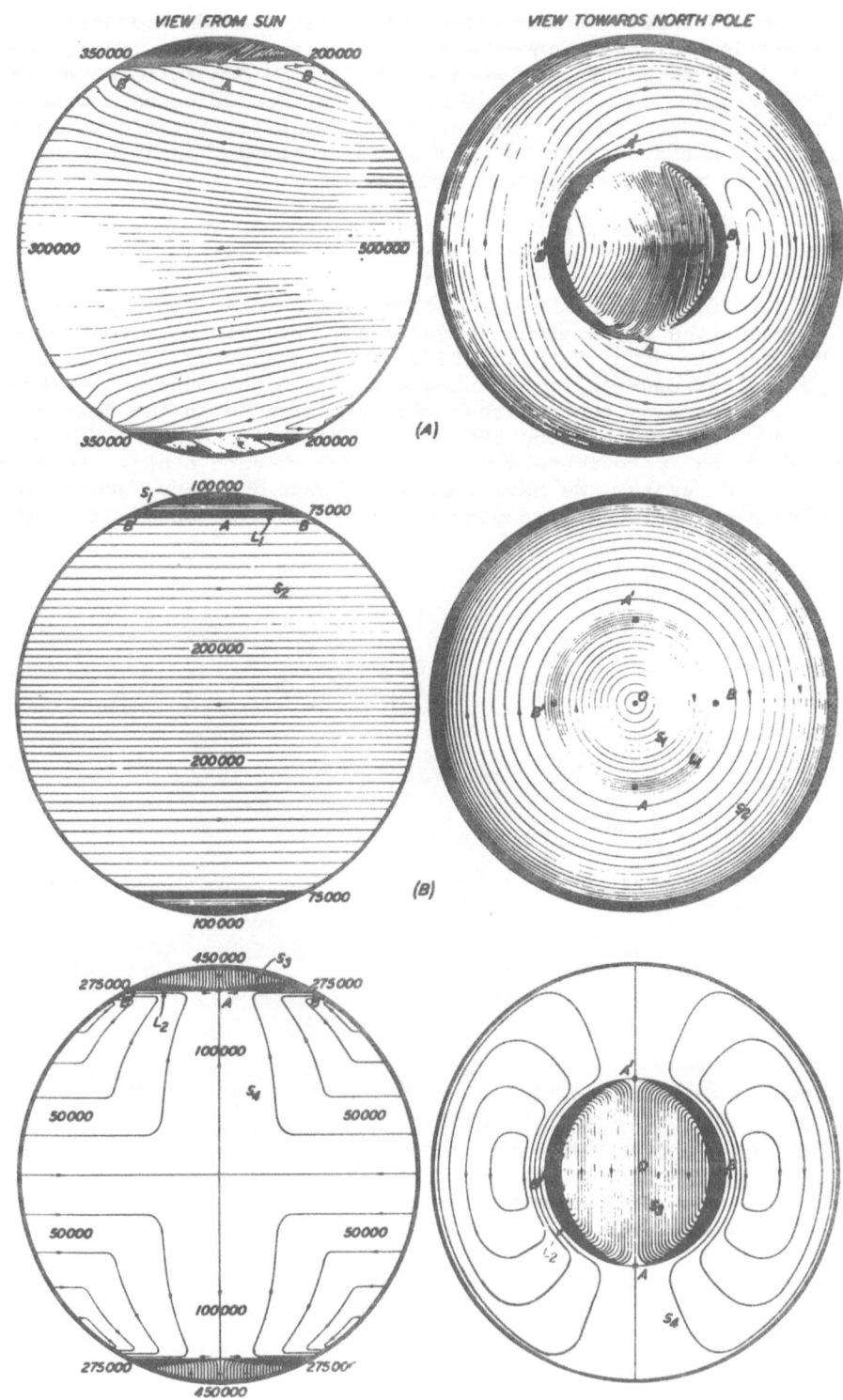

Fig. 6. (a) Electric current-system of geomagnetic disturbance; (b) and (c) respectively, partial current-systems D_{st} and S_D comprising (a).

electric currents in the ionosphere required to produce an average magnetic storm. He presented a composite picture of the entire current system which he divided into symmetrical and unsymmetrical parts. One of these parts is a zonal system; that is, a system that can be accurately represented by zonal harmonics. The other sub-system involves no zonal harmonics and thus varies with local times. Its aspects are recognized empirically as the diurnal variation on disturbed days. This representation indicates a strong westward current along the auroral zones, northern and southern, between midnight and noon and an eastward current between noon and midnight. Because of the zonal flow the westward currents are enhanced. This region of strong westward flow we identify with the region of maximum auroral activity. In the Chapman representation it appears somewhat later in local time than that pre-midnight time which we have identified as the time for most frequent appearance of the aurora, but this discrepancy may be the result of averaging.

These are general aspects of the aurora which we see almost every day in high latitudes. What is really interesting is what occurs during great magnetic storms. I should like to discuss the events which were observed during the great magnetic storm on March 24, 1940, Easter Sunday. I remember the occasion well because all intercontinental radio communication was out and I found myself on a news broadcast in a national radio hook-up in competition with Jack Benny's program trying to explain why the sponsor could not deliver an on-the-spot broadcast from Europe. We have a record of the earth current flowing in a telegraph line between New York City and Binghamton, New York (Fig. 7) which shows that at one time the voltage induced in the

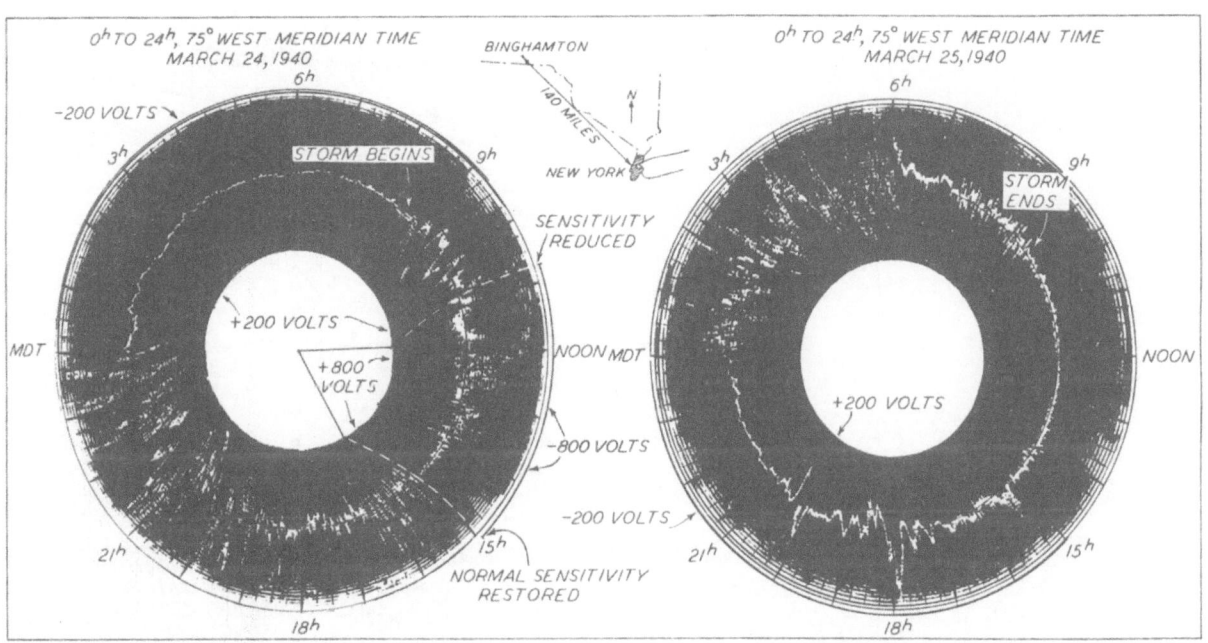

Fig. 7. Earth-potential records on Western Union Telegraph line, New York, N. Y., to Binghamton, N. Y.

line by the geomagnetic field changes was 800 volts. The effects occurred not only in the telegraph lines but also in the power lines. Circuit breakers went out, and power failures were caused by magnetic storms. Because of this power lines with this circuitry are no longer used so that one has to go back some time to find data of this kind.

If we assume the field changes are due to a current flowing overhead and measure magnetic field changes at certain observatories along a line, then we can, by means of these measurements, estimate the strength of the current, determine its location and calculate the induced voltages in the ground. Such calculations are consistent with the currents observed in telegraph and power lines, as shown in Fig. 8. At some places the calculated voltages ran as high as 8 volts per kilometer. Protective devices of AT T installed to safeguard instruments against ground current induced by such magnetic storms were operated in many parts of North America during this particular disturbance.

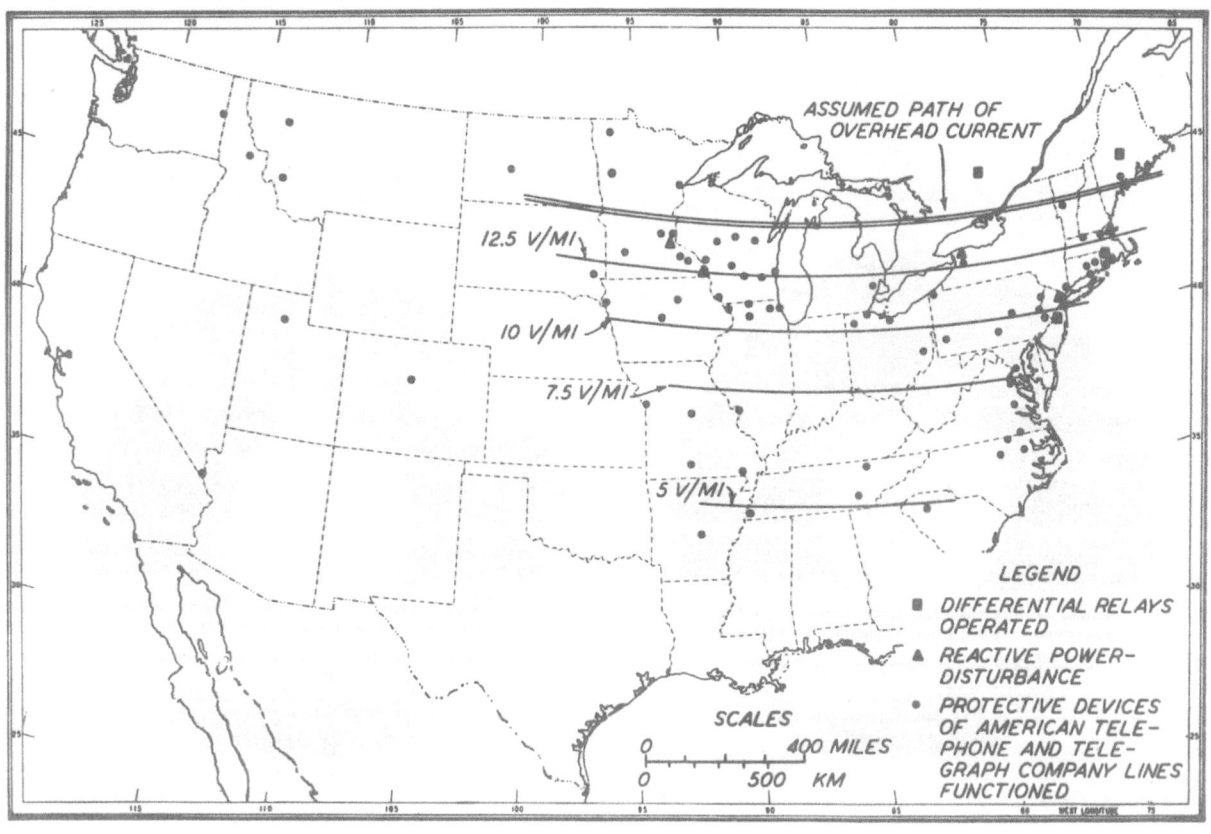

Fig. 8. Perturbing force during bay at 21^h 45^m GMT, January 15, 1933.

An interesting thing about this current is its location. The auroral zone where such currents ordinarily flow is far to the north of the place where this particular current was located which is essentially a narrow path running from east to west essentially coinciding with the mirror points for the inner Van Allen Belt. Here we are seeing some connection of particular interest which attaches a special meaning to these inner radiation belts. Such currents as these are observed only during great magnetic storms.

The location of such currents with respect to geographic coordinates can be obtained easily. If we imagine the current to be of limited extent and transverse to the direction of flow then it will be encircled by magnetic field lines. North of the area over which the current is flowing these field lines will have a downward direction and south of the area, upward, for a westward current flow. Directly under the current these field lines will be horizontal. If several magnetic observatories are located north and south of the line of flow its geographic position can be located with considerable certainty.

A determination of the height at which the current flows is a more difficult task, but one of particular interest in connection with auroral phenomena. Strictly speaking, one cannot infer any unique height for these currents from observations conducted on the earth's surface, but reasonable limitations on the height can be set. Most geomagnetic fluctuations are too complicated to admit of attack by the method I shall now describe, but some results can be obtained by selecting a simple case for mathematical treatment.

For this purpose I selected a case of disturbance attributable to a westward current flowing along the auroral zone (Fig. 9). The form of this disturbance and its geographical location were such that several operating magnetic observatories were located close to an approximately north-south line on the earth's surface perpendicular to the current flow. A rigorous mathematical analysis on the spherical surface of the observed field changes during the maximum of this disturbance is not easily accomplished because of the high order harmonics involved, but if a few simplifying assumptions are made an approximate mathematical treatment may be carried out.

These assumptions are that the earth is flat and that the current is of infinite extent in the direction of its flow. Such assumptions are allowable if the height of the current above the earth's surface is small compared with the other dimensions involved, if the cosine of the angular geographic distance of the observatories from the current path does not differ much from unity, and if the return flow of the current is sufficiently far from the observatories that its effect is negligible. These assumptions are sufficiently approximated that the following treatment should give meaningful results to within about 5 or 10 per cent.

The magnetic potential at the earth's surface due to the current may be represented in the x-z plane, perpendicular to the earth's surface and the auroral zone, by a harmonic series $W = A_{e,n} e^{-2\pi n j x} e^{nz} + A_{i,n} e^{-2\pi j n x} e^{-nz}$, the terms with A_e referring to those in the potential arising from the overhead current and the terms with A_i to those arising from currents induced in the earth by the overhead current. Values for the coefficients may be calculated from the observed horizontal and vertical components (Fig. 10) of the magnetic disturbance given by $\partial W/\partial x$ and $\partial W/\partial z$, respectively. Contributions arising from the currents below the earth's surface and above the earth's surface are separable (Fig. 11) because of a difference in sign of the two contributions appearing in $\partial W/\partial z$.

One device employed in this analysis is to assure that the harmonic coefficients derived to represent the data are well behaved. This is to assume that the values of $\partial W/\partial x$ and $\partial W/\partial z$ are zero for considerable distances beyond the places where their observed values are approximately zero to the north and south of the axis of the disturbance. This is the mathematical equivalent of our physical assumption that the

87

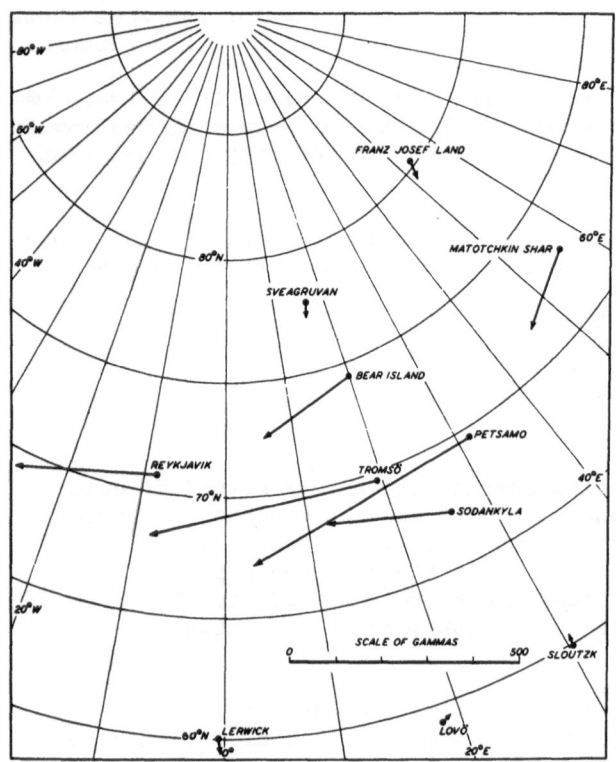

Fig. 9
Current arrows for disturbance at
21^h 45^m GMT, January 15, 1933.

Fig. 10.
Observed effects of magnetic storm
of March 24, 1940, and calculated
distribution of induced electromotive forces in eastern United
States.

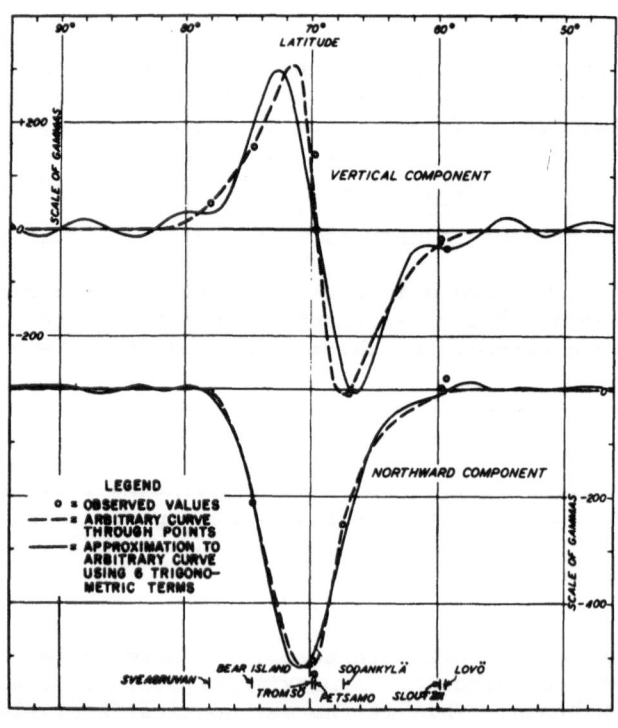

88

return flow is sufficiently far from the axis of the disturbance as to have little effect on the field changes at the observatories.

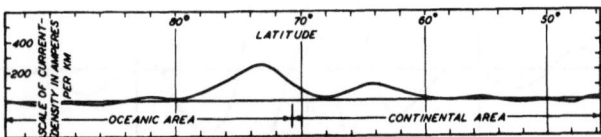

Fig. 11.
Current distribution at earth's surface giving rise to internal portion of disturbance at 21^h 45^m GMT, January 15, 1933.

The potential which fits the observed data may be calculated for any plane above the earth's surface (Fig. 12) provided all currents contributing to the potential in that half-space flow above that plane. Also the currents flowing in any plane which would give rise to the potential may be calculated.

As the calculation is carried out for greater and greater heights the higher order terms become greater with respect to the first term because of the term e^{nz} (see Fig. 13). For a height of 200 km several higher order terms in the current function exceed the first term implying that if the currents giving rise to the magnetic field changes flow at that height they must have a banded structure, that is east-west and west-east currents must be flowing side by side. Since such a complicated structure seems unlikely from physical considerations one is inclined to believe that the currents are flowing at a lesser height. The behavior of the extrapolation indicates a height of about 100 km for the current flow, agreeing closely with the most common height for auroras determined from visual triangulation.

Thus, we see that the aurora is the visual manifestation of the electrical processes which perhaps are better understood through the magnetic effects involved. Truly, the similarity between auroral displays and the visual phenomena observed in the cathode-ray tubes led to the first comprehensive theory of the nature of the processes involved proposed nearly three-fourths of a century ago. Progress in understanding them has advanced steadily with few brilliant new ideas or discoveries. But some views which were firmly held for many years by experts in the field have recently been disturbed.

There have been many reports of people having "heard" the aurora. These have been dismissed by scientists as old trapper's tales, induced probably by the long and lonely arctic nights. Recent evidence has conclusively shown that pressure waves in the lower atmosphere are frequently associated with the geomagnetic and auroral activity. These pressure waves, first discovered by Cook, Chrzanowski, Greene, Lemmon and Young at the National Bureau of Standards, lie in the infrasonic region of the spectrum. (That audible signals accompany them is still open to question.)

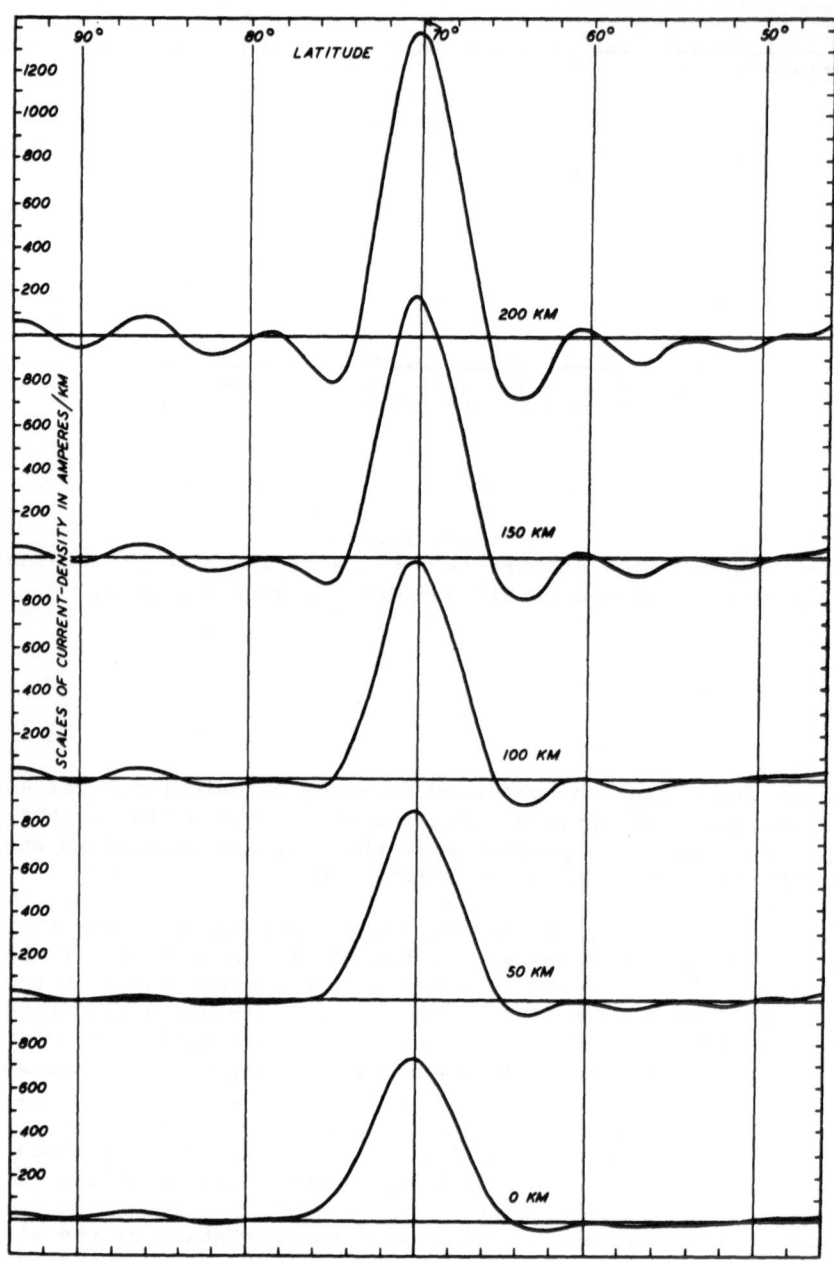

Fig. 12. Equivalent westward current sheet at various assumed
heights causing bay at 21^h 45^m GMT, January 15, 1933.

    These waves were discovered incidental to investigation of other infrasonic
waves in the atmosphere. By observations at three stations both the direction and
speed of propagation of these infrasonic waves could be ascertained. It was found
that the speed of propagation of the wave fronts along the earth's surface was
greater than that of sound, suggesting that the propagation vector was not parallel
with the earth's surface (like sea waves breaking on the shore); that their source
was to the north for middle latitude stations and shifted from east to west; and

90

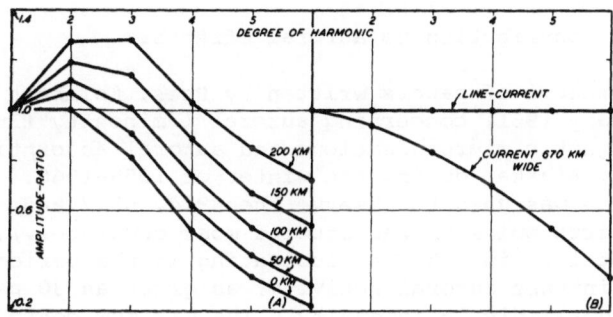

Fig. 13.
Amplitude ratio of higher harmonics to first harmonic:
(a) equivalent current distribution at various heights
to produce observed effects; (b) hypothetical distributions.

that occurrence of these waves was closely correlated with the occurrence of geomagnetic disturbances.

In these infrasonic waves a clear indication of a connection between phenomena
observed in the troposphere and those attributed to the ionosphere. Other claimed
relationships of the aurora and other ionospheric phenomena to tropospheric phenomena
are subject to dispute except for the well substantiated claim that aurora are most
frequently seen when the sky is not overcast.

91

W. H. CAMPBELL - First contribution to auroral effects.

This is a brief resumé of papers written by Rees, Leinbach and myself (J. Geophys. Res. 66, January 1961) concerning auroral luminosity fluctuations and their relationships to geomagnetic micropulsations and auroral absorption. The study was carried out at College, Alaska, during the winter of 1959-1960. A photometer with a 70° to 90° field of view was used to observe the 3914, 5577 and 6300 A emissions from the aurora (the first emission was studied most completely). On the average, a few per cent of the total luminosity are fluctuating in the period range of 25 to 0.5 seconds. At times of intense auroral activity, as great as 30 per cent of the emission may be varying rapidly in intensity. The maximum contribution to the emission fluctuation rises shortly before midnight and continues to rise until dawn concludes the observation. A study of the periods shows a predominance of the slow variations early in the evening and of the shortest period variations before dawn. There is a close relationship of the luminosity pulsations with geomagnetic micropulsations of the same oscillatory period range (individual oscillation correspondence is often observed) and with the ionospheric absorption as measured by the riometer. The generally assumed process for ionization and emission

$$\left[e + N_2 \rightarrow 2e + N_2^+ \rightarrow 2e + N_2^+ + h\nu\,(3914\ \text{etc.}) \right]$$

and dissociative-recombination type removal of N_2^+ shows the resultant increased electron density to be proportional to the square root of the luminosity. With the absorption roughly proportional to the electron density, the observed relationship between luminosity and riometer absorption is consistent with the expected processes. One must require the geomagnetic micropulsations also to be proportional to the electron density to understand the observed relationship of luminosity and field fluctuation data. Energy considerations led to the attitude that the "storm type" class of geomagnetic micropulsations result from small perturbations of the auroral electrojet. These perturbations may be brought about by small transient conductivity variations which occur when bombarding electrons in irregular bundles are stopped in the atmosphere, ionizing the region and exciting the observed nitrogen emissions.

Using a two-station triangulation, particular investigation of a luminosity profile was found from which an electron density profile could be associated and thus the absorption to cosmic noise sources determined. It was found that at least 50 per cent of the observed 30 Mc/R absorption occurred in the region between 90 and 110 kilometers.

W. H. CAMPBELL - Second Contribution to Auroral Effects.

A review of a paper "Auroral Zone Observations of Infrasonic Pressure Waves Related to Ionospheric Disturbances and Geomagnetic Activity" by Campbell and Young (J. Geophys. Res. 68, November 1963) is presented. The study concerned an experiment at Fort Yukon, Alaska, in August 1962, measuring the 10 to 110 second period infrasonics pressure waves and associated auroral type disturbances. The experiment confirmed the supposition that many infrasonic pressure fluctuations have their source in the auroral zone and are directly associated with the ionospheric absorption and magnetic effects attending the aurora. The nightly arrival direction was consistent with the geomagnetic midnight progression of maximum in activity from east to west. Propagation from such non-local sources at relative low acoustic velocities and low attenuation could explain the observed general appearance of pressure variations with respect to the local activity: preceding it in the early evening, together with it near midnight and extending later into the morning hours. The inability to find identical pulses on the micropulsation and infrasonic records was also anticipated because of the above effects. The pressure fluctuations were thought to arise from the joule heating of the varying electrojet current system. A study was also made of a unique event, not of the auroral type, which gave evidence that a large, primarily tropospheric, pressure front produced a magnetic field fluctuation and radio wave absorption disturbance in the ionosphere. In this case it was assumed that the transportation of ionization by the pressure disturbance could account for the related effects.

THE IONOSPHERE: PROPAGATION THEORY

J. J. Gibbons
The Pennsylvania State University
Ionosphere Research Laboratory
University Park, Pennsylvania

The propagation of low frequency electromagnetic waves in the atmosphere of the earth is a function of the degree of ionization of the constituent gases. The ionization is produced (Friedman, 1962; Ratcliffe, 1960) by the short-wave electromagnetic spectrum of the sun, X-rays and ultra-violet; and to some extent by particles, cosmic rays, meteors, and ion clouds from the sun. In general, a given ionizing agent will produce a layer with maximum ionization rate at some level, for the ionizing flux finds very few atoms or molecules to ionize at very great heights in the atmosphere. Then as the radiation penetrates more deeply, more ions $sec^{-1}cm^{-3}$ are produced, but in this process the ionizing radiation is absorbed so that at lower heights the rate of ionization again falls to lower values, even though more ionizable molecules are present. Thus, any specific ionization process tends to produce a layer with a maximum ionization rate at some level.

The many different ionization processes produce thus a superposition of many such layers, smoothing out these maxima so that sometimes only one or two broad maxima are in evidence. The radio sounding records, one should be warned, exaggerate the variations of electron density. The low group velocity due to high dispersion near reflection makes a slight change of slope in the electron density profile look like a definite maximum.

The lowest of the ionospheric regions, D, has been explained by Nicolet and Aikin (1960) as due to cosmic rays below 70 km, to Lyman α, (1216 Å) the resonance line of atomic hydrogen, which gets through a window in the upper atmosphere down to 75 km, and to X-rays < 10Å at the bottom of E, 85 to 100 km. The ions formed are principally O_2^+ and NO^+. Densities of 100 to 1000 electrons cm^{-3} normally occur in this layer.

Above this, say 100 to 140 km, lies the E region, attaining densities of about 10^5 electrons cm^{-3}. The ionizing agents here are X-rays 10 to 100 Å, Lyman β (1026 Å), C_{III} (977 Å), and the Lyman continuum 910 to 800 Å. This region, as explained above, may hardly show as anything more than a slight slope change on a real height electron density profile.

Finally in the F region, $\sim 10^6$ electrons/cm^{-0} maxima, we have a wide range of the ultra-violet spectrum of the sun contributing to ion production. The daytime splitting of this layer into two is caused by ion diffusion and variation of loss rates with height. The principal ion becomes O^+ as one goes to the top of F.

The opposing processes leading to electron disappearances, or at least of removing them from the effects of radio frequencies in the medium may be direct recombination, attachment to a heavy neutral to form a negative ion, dissociative recombination e.g. $e^- + NO^+ \rightarrow 0$, or bodily removal by mass motions or diffusion. A conservation equation for the electron would then read

$$\frac{\partial n}{\partial t} = Q - L + \nabla \cdot (n \vec{v})$$

where n is the number of electrons/cm^{-3}, Q the number per second per cm^{-3} produced by processes described above, L the effective loss of electrons per second per cm^{-3} due to recombination, attachment, dissociate recombination; and $\nabla \cdot (n \ \bar{v})$ where $\bar{v}$ is the mass velocity of electrons is the loss due to mass motion. Q will have the form of a sum of ionization cross sections times radiation intensities. For recombination L will have the form αn^2, for attachment, βn. The velocity v includes that of diffusion due to concentration gradients as well as that due to mass motion of the medium.

In any event, the propagation of an electromagnetic signal through this ionized medium depends on the electron density n, the applied magnetic field $\bar{H}$ (earth's field), and some damping factor such as a virtual collision frequency ν, such that the damping force on an electron of mass in moving with velocity $\bar{r}$ is $- m\nu\bar{r}$. Then neglecting the effects of the magnetic field of the signal itself we may write Newton's law on the electron in the field E of the signal

$$m \ \ddot{\bar{r}} = - e \ \bar{E} - m \ \nu \ \dot{\bar{r}} - \frac{e}{c} \ (\dot{\bar{r}} \times \bar{H})$$

Dealing with one frequency component ω and using the time factor $e^{j\omega t}$ we obtain

$$- \omega^2 \ m \ \bar{r} = - e \ \bar{E} - j\omega m\nu\bar{r} - \frac{j\omega e}{c} \ (\bar{r} \times \bar{H})$$

Writing the polarization as $- n \ e \ \bar{r}$, we can solve for $\bar{E}$ in terms of $\bar{P}$

$$\bar{E} = \left[\frac{- m\omega^2 + j \omega m\nu}{n \ e^2} \right] \ \bar{P} + \frac{j\omega e}{ne^2 c} \ [\bar{P} \times \bar{H}]$$

$$= a \ P + b \ [\bar{P} \times \bar{H}]$$

or solve for $\bar{P}$ in terms of $\bar{E}$

$$\bar{P} = \frac{a \ \bar{E} + b \ \left[\frac{b}{a} \ (\bar{H} \cdot \bar{E}) \ \bar{H} - \bar{E} \times \bar{H} \right]}{a^2 + b^2 \ H^2}$$

Then with $\bar{D} = \bar{E} + 4 \ \pi \ \bar{P}$ and the equation

$$\nabla \times \nabla \times \bar{E} = \omega^2 / c^2 \ \bar{D}$$

from Maxwell's equations we obtain the vector propagation equation for $\bar{E}$

$$\nabla \times \nabla \times \bar{E} - \frac{\omega^2}{c^2} E = \frac{\omega^2}{c^2 \ X^2 \left[(-1 + j \frac{\nu}{\omega})^2 - (\frac{\omega_H}{\omega})^2 \right]} \left\{ (-1 + j \frac{\nu}{\omega}) \ \bar{E} \right.$$

$$\left. - \frac{(\frac{\omega_H}{\omega})^2}{(-1 + j \frac{\nu}{\omega})} \ (\bar{h} \cdot \bar{E})\bar{h} - j \ (\frac{\omega_H}{\omega}) \ \bar{E} \times \bar{h} \right\}$$

where $\frac{\omega^2}{X^2} = \omega_o^2 = \frac{4 \pi n e^2}{m}$ $\omega_H = \frac{e|H|}{m_c}$

and $\bar{h}$ is a unit vector in the direction of the earth's magnetic field. Note that $\nabla \cdot E$ is not zero even for constant n, as long as there is a magnetic field present, so that in general, there will be a component of E in the direction of propagation.

In the form of the propagation equation above there is no mention of index of refraction or of modes although, of course, the dielectric tensor is involved in the coefficients. But for useful interpretation, one generally breaks down this equation into components or modes. One sees that in general the vector operations have the effect of coupling the component equations, i.e., a given scalar component equation contains more than one component of the E field. There are any number of ways of resolving the propagation equation into components, and therefore, any number of "indices of refraction." The commonest modes are perhaps those arising from the equations as resolved by Försterling (1942) and others, which have the unique property that the component equations become uncoupled in a uniform medium (const. n, ν, $\overline{H}$). In the actual nonuniform ionosphere, however, even if idealized to a horizontally stratified medium, there are regions where the coupling terms have large coefficients and make the solution difficult to obtain. Particularly is this the case for the Försterling modes near "critical coupling," that is when the $\omega = \omega_0$ level is near the level of critical collision frequency

$\nu = \left| \dfrac{\omega_H \sin^2 \theta}{2 \cos \theta} \right|$ where θ is the angle between earth's field and direction of propagation.

The two Försterling (1942) modes have different polarizations, generally elliptical. The "indices of refraction" for these two modes are sketched in Fig. 1 as functions of

$$\omega_0 = \frac{4\pi n e^2}{m}$$

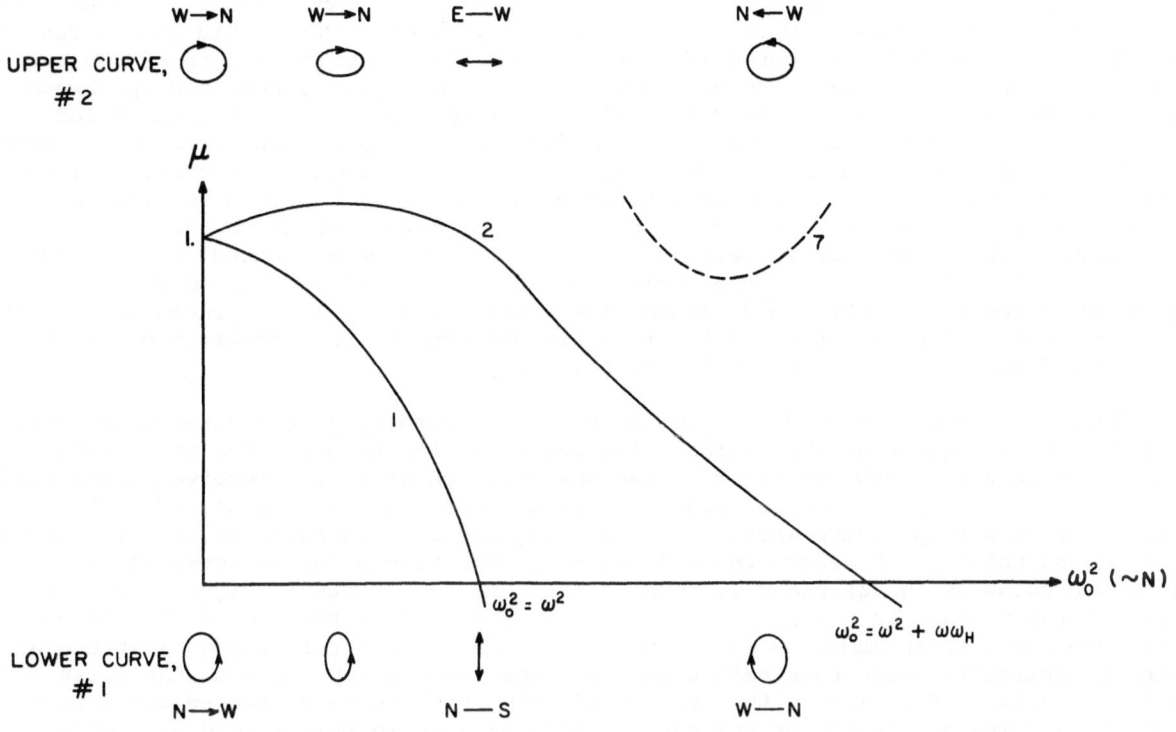

Fig. 1. μ^2 vs. N below gyro (no collisions).

The polarizations of the two modes, sometimes called "ordinary and extraordinary," after crystal optics, are indicated on the diagram for vertically incident or reflected plane waves. The absolute direction of rotation of the electric vector for each mode is the same whether the wave is going up or down.

Instead of using the "normal" Försterling components, one may diagonalize the dielectric tensor. In the resulting modes, critical coupling no longer occurs and the equations may easily be solved in the Försterling coupling region, but this method, used by Davids (1953), will have the disadvantage that the coupling instead of being highly localized as in the Försterling modes now occurs over the whole propagation region.

The effect of coupling terms, no matter what modes are used, is to generate one mode from the passage of the other. That is, positive and negative waves of, say, an extraordinary mode are generated by an ordinary mode going in either single direction. This may give the level of high coupling the appearance of a reflection region, although, of course, the "reflected" wave will not be of the same mode as the incident wave. This appears fairly clearly in the method of Gibbons and Bellas (1952) of selecting new modes as linear combinations of the original Försterling modes, selected to minimize the coupling in a given region. When this is done the new indices (and, of course, there must be new indices every time a new linear combination is chosen) definitely show a variation in one of them which would give a reflection, while the other index goes through the region at more or less constant value (Fig. 2), giving no reflection.

Using the Försterling components, let us examine the process of reflection of an electromagnetic wave for, say $\omega < \omega_H$. An originally linearly polarized wave impinges on the lower ionosphere, where we imagine it to be separated into the ordinary and extraordinary components. As these go through a relatively narrow coupling region, around 85 to 90 km, both the E (extraordinary) and O (ordinary) modes send back signals, the E sending back an O and the O sending back an E. This would give a resultant linear except that a slight difference of absorption and phase path of the two modes gives a thin ellipse. The original upgoing E and O waves continue up through the coupling region to where the O is reflected at $\omega_O^2 = \omega^2 + \omega \omega_H$, but the E for reasonably low collision is transmitted through (this is the "whistler mode"). When the O wave comes down again after reflection, it again traverses the coupling layer, adding an E mode to itself. The wave received at the ground is then an ellipse, the vector sum of the O and E wave. One sees that this process explains the common observation at these low frequencies: the polarization of the reflected wave is independent of the polarization of the incident wave. It should be noticed that an expected reflection of the E mode at the lower level $\omega = \omega_O$ is prevented because the high collision frequency at these lower levels smooths out the variation of index which would exist there at low collision frequency.

There are several methods of probing the ionosphere by ground based transmitters. In Fig. 3 we see group height against time record at 150 kc/sec. The transmitter sends up pulses of a few hundred microseconds duration which are received after reflection from the ionosphere and recorded by continuously running film as intensity variations in a scope trace whose sweep is triggered on each pulse in time to record part of the pulse as it leaves the transmitter. The time delay in receiving the reflected pulse at the ground will then be recorded as a virtual height on the photograph of the trace. In the figure the lower bright band is the end of the ground pulse and above it at about 90 km is the first echo. Where reflection is strong an echo at apparently twice this height is seen, the "second hop," due to the first reflected pulse reflecting on the ground and going back to be reflected again in the ionosphere. The complexity of the echo is evident even in this record of a quiet night. One can see the effects of sunrise in the lowering of the echo between six and seven o'clock. Fig. 4 shows in the lower part a record of a night disturbed by a magnetic storm.

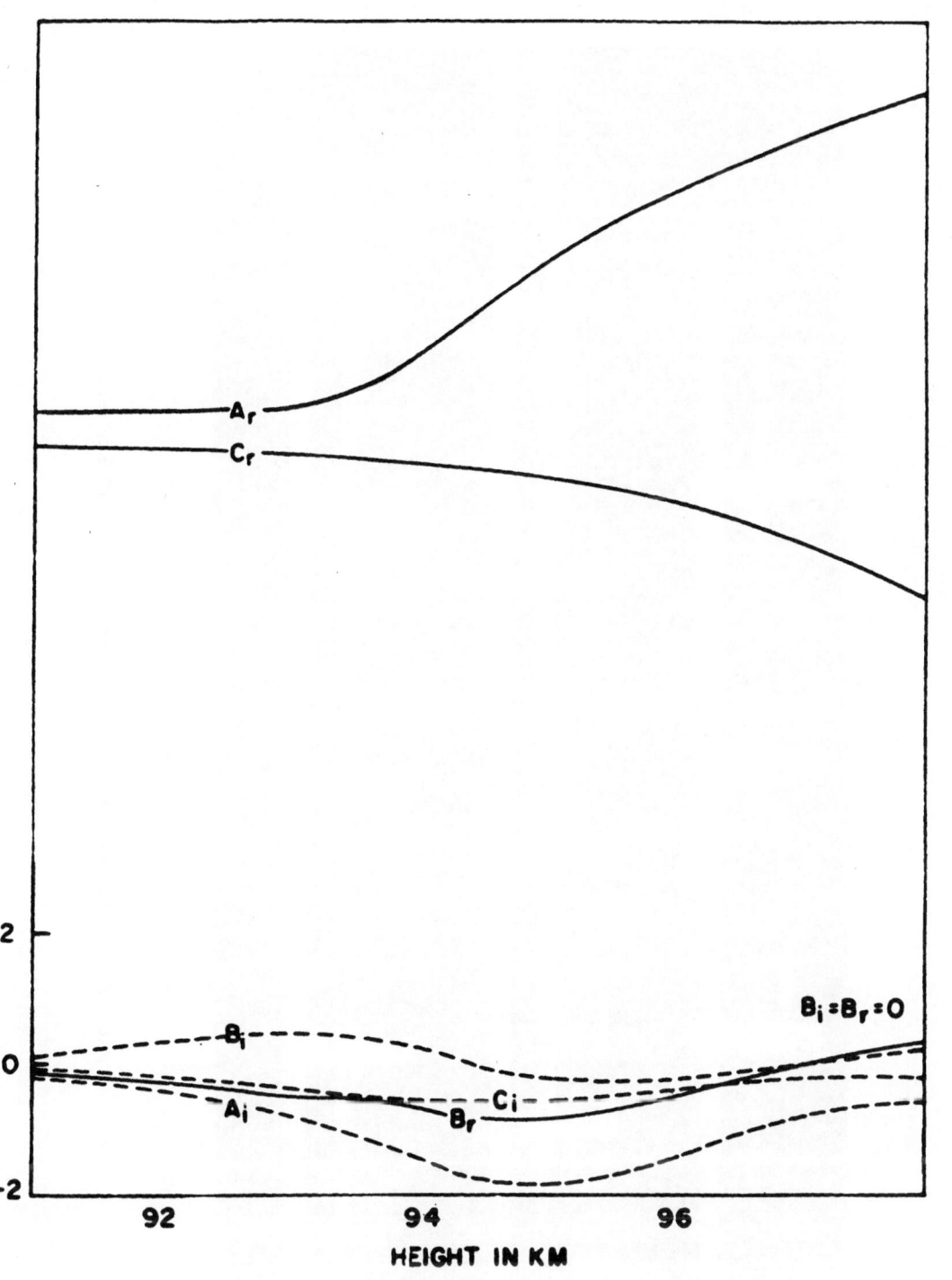

Fig. 2.   Indices $A_r + i A_i$, $C_r + i C_i$, and Minimized Coupling Factor $B_r + i B_i$.

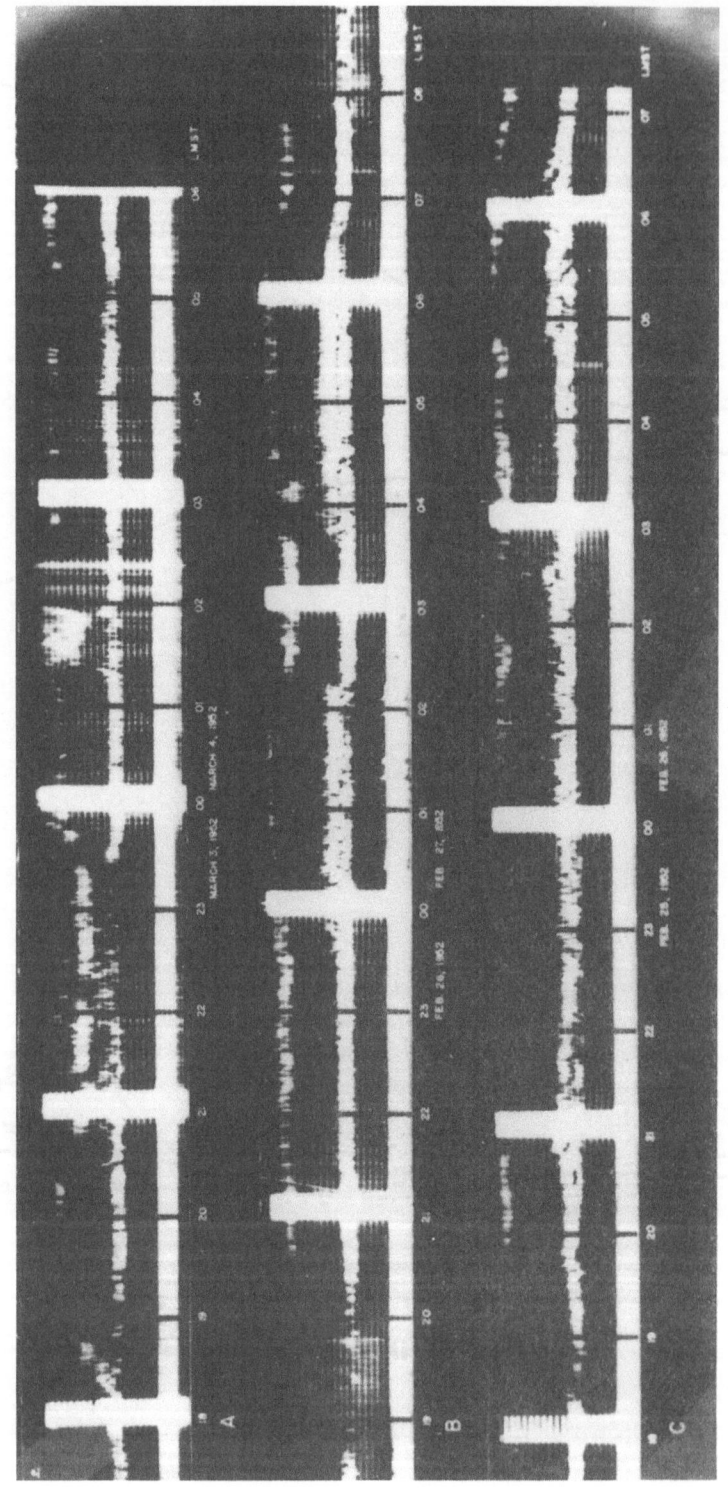

Fig. 3. Group height vs. time record at 150 kc/sec.

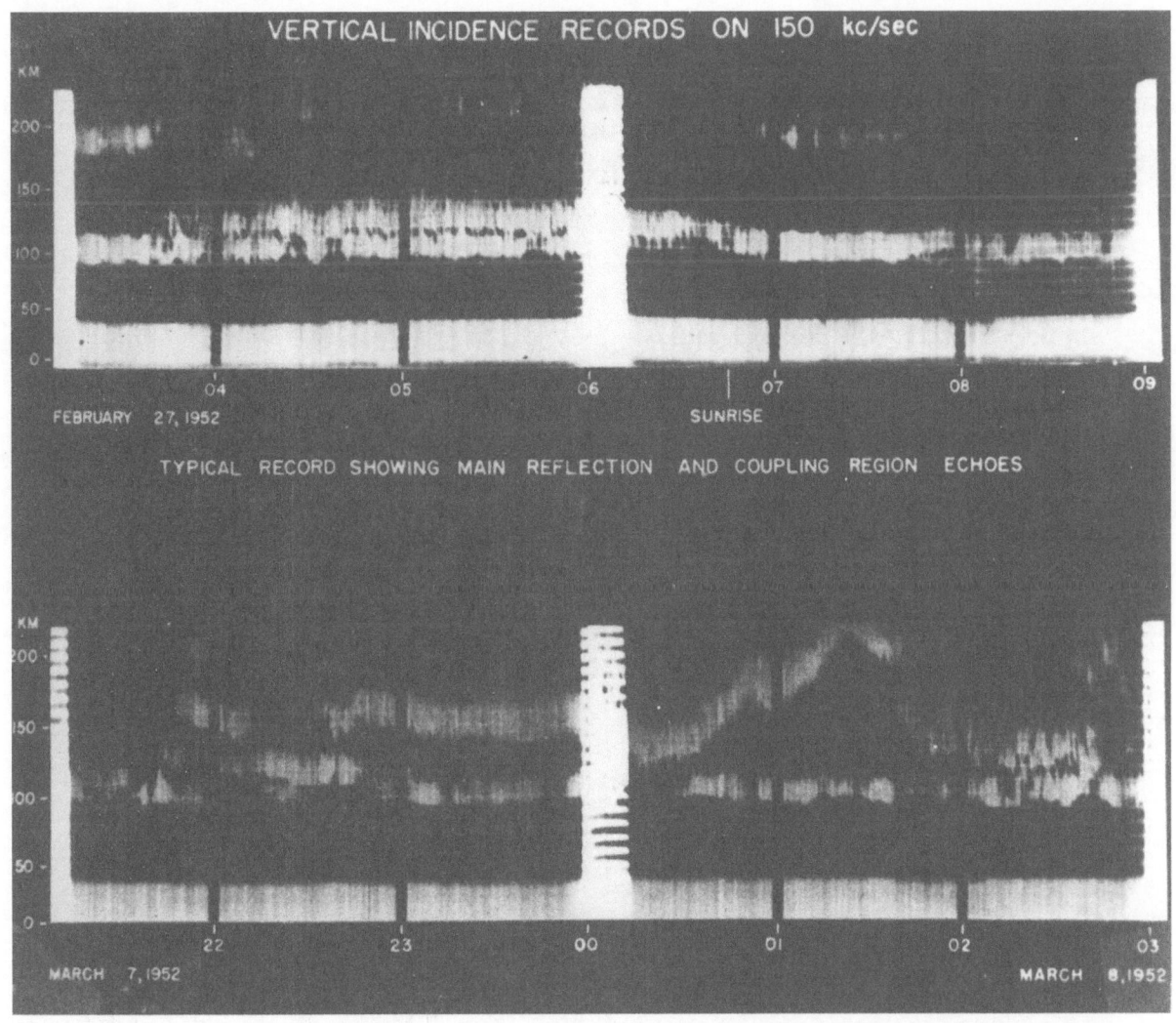

Fig. 4. Typical record obtained during a magnetic storm.

In Fig. 5, one has a record of the phase height variations. The fringes are produced by interference between the down coming group and a continuously running oscillator locked to the phase of the transmitter as each pulse begins to be radiated. The motion of the fringes then corresponds to change of phase height. As one of the lines moves down by the distance between fringes, the phase height is lowered by one wave length. One should notice the independent motion of the phase waves as compared to that of the group. This is due to the two entirely different indices, group index and phase index.

Measurements at the Pennsylvania State University, Ionosphere Research Laboratory, over a period of two years on 150 kc have yielded a rather curious result, also noticed by the workers at the Cavendish Radio Laboratory: - The phase height on records on which phase can be followed without interruption for 24 hours or more does not cycle back to its original value, say every midnight. The phase height, in fact, drops about an average of 0.5 km a day. This result could not occur in an ordinary isotropic medium. It is fairly certain that this effect is due to the presence in the

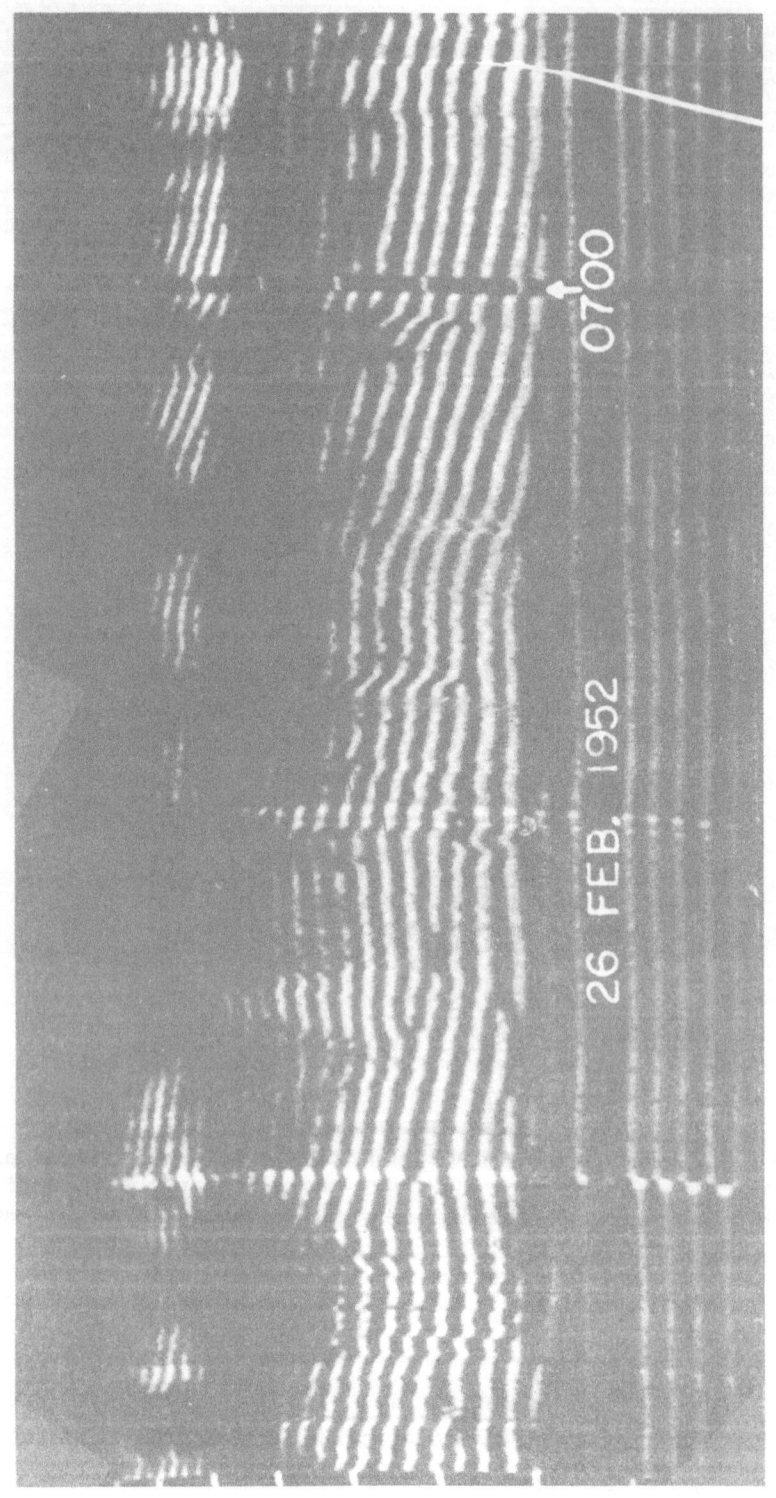

Fig. 5.
Sample of phase height variation vs. time recording,
showing split echo and second hop.

coupling factor of the derivative of the electron density with height. In general, the ionospheric profile starts in the morning and may come back to the same profile at night, but this is done by going through a set of profiles from morning to noon different from those it goes through from noon to night.

Note that because of indices different from unity, neither the phase nor group records give "true height" variations. The difference between group and phase indices from the expression

$$\mu_g = \mu + f \frac{\partial \mu}{\partial f}$$

is a function of the dispersion, which can be very high near reflection, where the phase index drops to low values. As an example, in Fig. 6 is presented a graph of the variation of group index as a function of electron density N and collision frequency ν at 500 kc/sec (Gibbons and Rao, 1957). The value N = 12,000 is very near the reflection level for this mode.

Since the group velocity of the pulse is slowed down in this way as it approaches and recedes from the reflection level, records such as that in Fig. 3 give heights of reflection, and therefore, corresponding electron densities, which are always too high, and this error may be of order 50 to 100 per cent for those frequencies just approaching the penetration frequency of the layer. The true height of the reflection can be obtained by essentially the inversion of an integral. For the apparent, or virtual, height for a given frequency, f would be given by

$$h'(f) = \int^{h_R(f)} \mu g(f, n, \nu) \, dh$$

where μg is the group index and $h_R(f)$ is the real height of reflection. The problem is like the solution of a very large number of simultaneous linear equations. It can be solved analytically for $\nu = 0$ (Appleton, 1930), and with the earth's field H = 0, but for H = 0 it must be solved by machine methods, and/or various laboratory short-cut procedures (Budden, 1954; Schmerling, 1957).

The methods described above for sounding the ionosphere are all incapable of giving information on the ionosphere above the level of maximum - n. Radio sounders placed on satellites, observations on Faraday rotation of satellite signals can be used for this purpose, and there is one ground based experiment which in principle can be (and is) used, but the theory is not yet completely clear. This is the method of incoherent scatter by the ionospheric electrons of signals sent from the ground (Bowles, 1958). This scattered radiation, unlike the coherent scatter giving rise to reflection, will be returned from region where there are free electrons.

The usual theory involved in interpretation of these ionospheric propagation phenomena is a first order one, quite accurate for ordinary electromagnetic wave amplitudes. When second order effects are examined, one must consider the heating of the electrons by a wave passing through, and the consequent change of index. With relaxation times of the order of 10^{-3} sec, the heating and cooling can follow the audio modulation of a signal, and the medium can then impress this modulation on any other wave passing through it (Luxembourg Effect), and indeed can affect its own transmission.

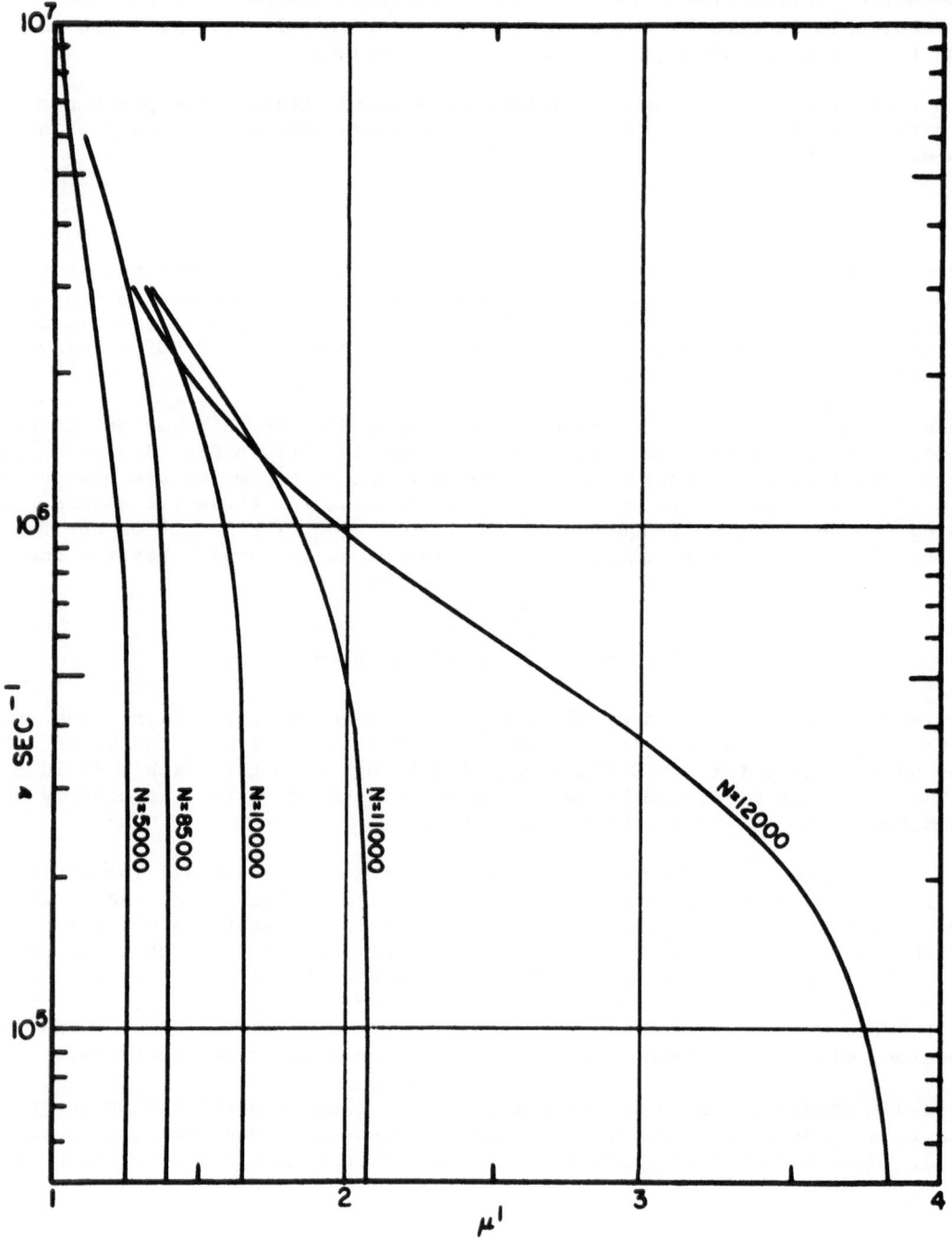

Fig. 6.
Group indices at 500 Kc/s as a function of ν for various values of N.

104

REFERENCES

Appleton, E. V.: Proc. Roy. Soc. 42, 321 (1930)

Bowles, K. L.: Phys. Rev. Letters, 1 454 (1958)

Budden, K. G.: Cambridge Conf. on Ionospheric (Phys. Soc.) 332 (1954)

Davids, N.: J. Geophys. Res. 58, 311 (1953)

Försterling, K.: Hochfrequency vs. Elek Akus 59, 10 (1942)

Friedmann, H.: Proc. of Int. Conf. on Ionosphere, London (Inst. of Physics and the
        Physical Society) 3 (July 1962)

Gibbons, J., and F. Bellas: Pennsylvania State Univ. Ionosphere Research Report No.
        42 (1952)

Gibbons, J., and R. Rao: J.A.T.P. 11, 151 (1957)

Nicolet, M., and A. J. Aikin: J. Geophys. Res. 65, 1469 (1960)

Ratcliffe, J. A.: "Physics of the Upper Atmosphere," New York Academic Press (1960)

Schmerling, E. R.: Pennsylvania State Univ. Ionosphere Research Report No. 94 (1957)

EXPERIMENTAL RESULTS AND TENTATIVE EXPLANATIONS
FOR GEOMAGNETIC EFFECTS CAUSED
BY HIGH ALTITUDE NUCLEAR DETONATIONS

E. Selzer
Institut de Physique du Globe
de Paris

INTRODUCTION

High altitude nuclear detonation experiments have been performed a certain number of times since 1958 and have produced a great variety of geophysical effects. From the study of all these effects, we will restrict our present investigation to the discussion of the magnetic perturbations caused by such detonations. We feel that this limitation is justified, not only for the sake of coherency, but also for the two subsequent reasons:

1) The magnetic effects associated with experiments of this kind have been less extensively studied than other ones.

2) As far as these magnetic effects have been investigated, no definite understanding has been gained yet on the kind of mechanism involved in their generation.

We can even remark that, at the beginning (Argus Experiment, 1958), the experimental material that could be assembled seemed to be quite contradictory: the complexity of the effects prevented their full exact registration at each station of observation, thus making the comparison of the different records very difficult to perform according to a common scale of interpretation. In this respect, the situation has lately improved and even if our theoretical understanding of the phenomena involved is still very poor and hazardous, we feel now that a common assembly of coherent experimental results can be adopted as a base of new efforts toward a comprehensive approach to the problem.

One can ask what degree of importance the study of such apparently small magnetic perturbations - like those we are going to examine here - can have for any scientific point of view. The answer is that the study of any geophysical effect, however small it is, may be very important when it appears to be a world-wide effect. Moreover, the similarity between these artificial magnetic phenomena and the natural magnetic perturbations like the S.S.C. (Storm Sudden Commencements), S.I. (Sudden Impulses), and S.F.E. (Solar Flare Effects), can be of paramount importance when making progress in the determination of the natural condition governing the space around the earth, especially when dealing with interactions between solar streams and the earth's outer atmosphere.

Experiments with high altitude nuclear detonations represent a first step in "Experimental Terrestrial Magnetism" (cf. with Experimental Seismology). Unfortunately, it is difficult to tell whether more experiments of this kind will be performed in the future. However, it is not our task here to examine the political, diplomatic and even social problems involved with the continuation of such experiments. As pure scientists, we are confronted with the fact that some of these experiments were performed. It is our responsibility to get all the knowledge which <u>may be extracted</u> from their effects.

* Contributions and remarks made during the Institut and after have been included in the written paper. See Acknowledgements.

We might, of course, some day in the future - and always as pure scientists - have to make a decision on the desirability of performing more experiments of that kind. We can imagine, for example, that all international political problems have been solved satisfactorily. Then we will be asked (or we will ask ourselves) if we want to prepare a program for more high altitude nuclear detonation experiments. If the answer is "Yes," then under what conditions, with what powers, of what burst altitudes and, last but not least, in what proportion we will be willing to perturb the natural conditions of our environment in order to acquire some new knowledge about it. This last question can also, as the preceding ones, be examined as a pure problem of physics. The experiment need not be carried too far in changing the previous non-perturbed conditions which are to be studied. Even so, we will sometimes have a rather difficult problem to solve. Besides, we might get into competition in this line with physicists belonging to another discipline, and our criteria for experimenting under "non-perturbed conditions" generally will not be exactly the same for all of us.

All this is quite well known already, but the reason to recall attention to it here is that we are going to examine the results given by a very limited number of experiments and we might wish to dispose of more experimental points than the few we presently have, so we ascertain our findings.

INFORMATION MATERIAL

Our first intention was to bring out here all experimental evidences referring to our subject with, if possible, a critical examination of their degree of validity and importance. But we had to recognize soon that such a project was much too ambitious for our means. For example, even the apparently simple task of listing all publicly known atmospheric and high atmospheric nuclear explosions required extensive, and often deluded, searches through an innumerable amount of newspapers and press releases. The reproduction of magnetograms and of any kind of graphic registrations of the effects (magnetic and associate ones) of the detonations was quite feasible some time ago when very few such effects had been detected, but not any more now.

In order to palliate the consequences of this situation, we have prepared, as complete as we could, the references which are to be found at the end of this report.

Table 1 is specifically concerned with the few experiments which have been of paramount importance for our studies, and covers a very limited choice of remarkable registrations. In addition, in the Appendix two tables of nuclear detonations relative to the 1961 and 1962 series are reproduced, without any modification, as a facsimile of a French publication.

Copies of Original Records.

The necessary choice we had to make resulted in the following procedure:

1) We eliminated all "normal" - slowly run - magnetograms, in spite of the very valuable informations they contained (which we have used for drawing the vector distribution map shown on Fig. 1), simply because of the technical difficulties met in trying to bring out distinctly on the copies the (generally) very small perturbations visible on the original documents.

TABLE 1
THE MOST IMPORTANT AND CHARACTERISTIC HIGH-ALTITUDE NUCLEAR DETONATIONS*

Project Name	Date	Time (U.T.)	Coordinates	Place	Altitude (Approximate)	Yield
TEAK	1 Aug 1958	1050	17°N, 169°W	Johnston Is.	60 km	1 MT
ORANGE	12 Aug 1958	1030	17°N, 169°W	Johnston Is.	30 km	1 MT
ARGUS I	27 Aug 1958	0230	38°S, 12°W	So. Atlantic	480 km	1-2 KT
ARGUS II	30 Aug 1958	0320	50°S, 8°W,	So. Atlantic	480 km	1-2 KT
ARGUS III	6 Sep 1958	2210	50°S, 10°W	So. Atlantic	480 km	1-2 KT
STARFISH	9 Jul 1962	0900	17°N, 169°W	Johnston Is.	400 km	1.4 MT
RUSSIAN TESTS	22 Oct 1962	0341	?	Central Asia	?	100 KT?
	28 Oct 1962	0441	?	Central Asia	?	?

Note: We have not included in this Table a few other experiments whose high-altitude specific character, as defined by their magnetic effects, was questionable for a certain length of time, but has been definitely rejected in the scope of our present study. These experiments were the ones of 5 Aug 1962, 26 Oct 1962, and 1 Nov 1962 (two: one of the Soviet Series and one U.S.). The reader may consult the Tables I and II given in the Appendix.

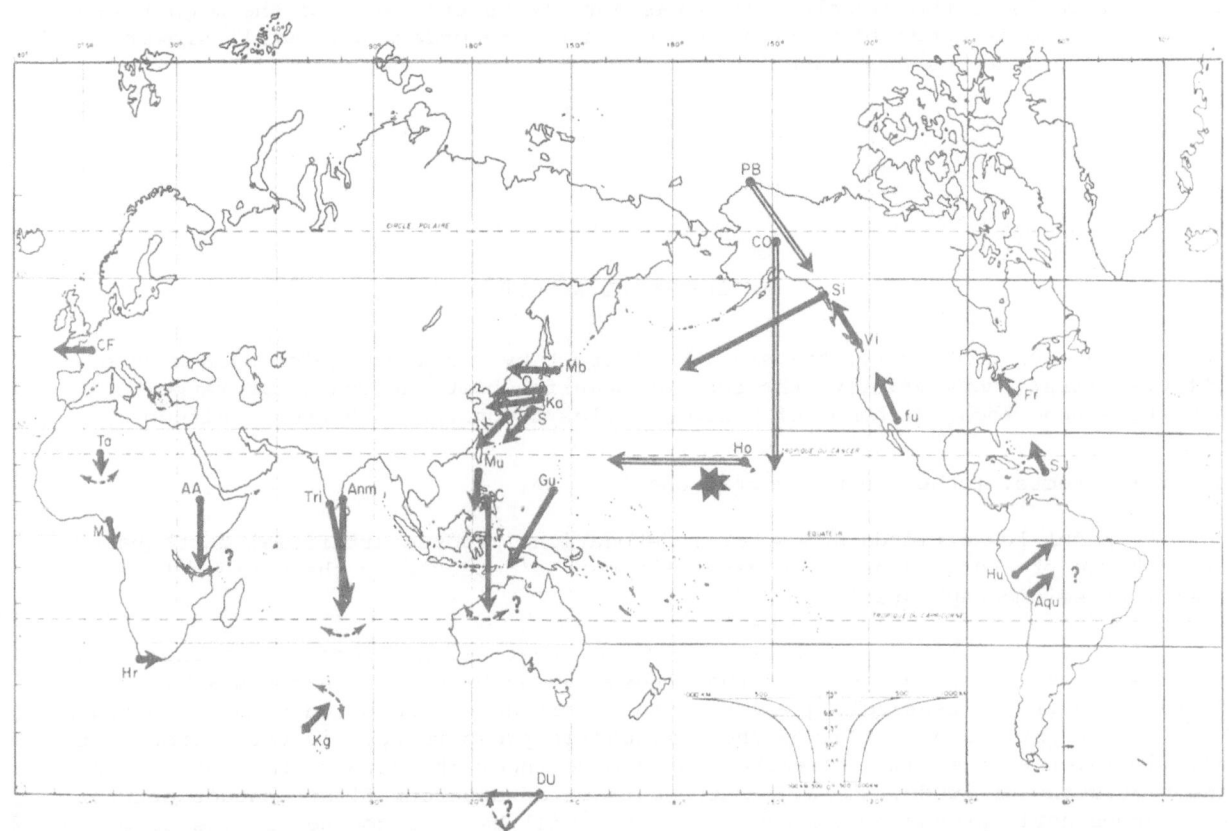

Fig. 1. Tentative representation of the world distribution of the maximum magnetic horizontal perturbations for STARFISH, deduced from "normal magnetograms."

2) Concerning rapid run magnetograms (or tellurograms), we have retained a collection of French records obtained at the "twin stations" Chambon-la-Forêt and Garchy (situated about 100 and 200 km south of Paris, respectively), and at Kerguelen (South Indian Ocean). These two groups of stations have the advantage to have been equipped with very similar equipment and to provide an idea, at first sight, of the world, local, or regional character of any magnetic or telluric perturbation. All of the examples which are given this way are related to the 1962 experiments (viz. July 9, October 22 and October 28). They are followed by two very remarkable records obtained July 9, 1962, at Byrd Station in the Antarctic by the Pacific Naval Laboratory of Canada. These records, recently released, have been forwarded to us thanks, to the courtesy of J. E. Lokken. Thanks to international cooperation of this kind, we hope that one day a World Atlas of all recorded effects - a very valuable and important enterprise - will be produced. It is important to have in mind when one is comparing various registrations of a particular event the diversity of the equipment which was used. This would be a part of the heavy duty in constituting a World Atlas.

The French records reproduced here are either magnetograms obtained by a) an integrated, induction method, or b) tellurograms. In a), the frequency response of the system is linear ($\pm$10%) from 2 to 100 sec. period for the "barre-fluxmètre" equipment type A, and from 1/3 to 3 sec. period for the "barre-fluxmètre" type B. The corresponding sensitivities are 0.05 gamma/mm for type A and (roughly) 0.002 gamma/mm for type B. In b), the sensitivity has been adjusted in order to reach the same order of value as given above for type B, when transposed from telluric to magnetic registration around the central period of one second.

It is evident when studying records obtained by other devices, that many of them were affected by time constants which were much too large to permit the detection of the first parts of the signals. This was not always the fault of the magnetometers themselves and it appeared that in many cases the recorders were still slower and smoothed the graphics unnecessarily.

EXPERIMENTAL RESULTS

For the sake of continuity, the study will start by considering for each kind of effects: first, very briefly, the case of powerful bombs detonated at rather low altitudes, and then, of bombs of various yields detonated at higher altitudes.

General Effects. (other than magnetic ones)

Powerful low altitude atmospheric detonations produce artificial earthquakes and atmospheric pressure waves which generally can be detected by the world network of classical seismograph stations.

Without entering into details, let us say that seismic records obtained in such circumstances show larger periods than those caused by natural earthquakes. These differences are increasingly greater as the altitude of the detonation is increased. This can be explained by the way the atmospheric pressure wave is transformed into seismic waves at the time it reaches the ground under the site of the detonation: the higher the altitude, the bigger the surface of apparent first contact between this atmospheric pressure wave and the ground will be. Of course, at the same time, the energy transmitted to the ground decreases and for detonations of a given yield, no seismic record can be obtained when the altitude is higher than a certain value - depending on this given yield.

It might have been expected that some other effects - like the magnetic ones that we are going to study later on in detail - would be able to add some supplementary indications to the ones deducted from seismic records.

It is a pity, of course, for this kind of consideration that we have no experiment in which were registered at the same time seismic and magnetic effects. It looks like: When the altitude is low enough so that the seismic effect can be detected, it is then too low for producing any magnetic perturbation, and inversely, when the altitude is high enough so that a magnetic perturbation can be detected, then it is too high for producing any seismic effect. Of course, the total number of experiments which we can take into consideration in this discussion is so restricted that the conclusion we have given above is very crude. Further, there is a big range of altitudes for which absolutely no information is available.

The ionospheric effects have been discussed at great length in numerous papers dealing with all kinds of nuclear detonations. Let us give a brief summary of deductions that can be brought out in this respect. In the case of very powerful atmospheric detonations, (a great number of MT) there will be a power-disturbing-effect of the ionosphere each time the summit of the ionized cloud reaches the lower boundaries of the ionospheric layers. The disturbances which are generated in this case can travel at great distances and even reach all parts of the world but through various kinds of propagation processes involving speeds of the order of magnitude comparable with the speed of sound. No very rapid transient ionospheric perturbation has ever been put into evidence very clearly on such occasions. On the other hand, smaller detonations at greater heights have produced ionospheric shock waves that were able to propagate with much greater speed. This kind of perturbation being closely connected with the magnetic ones will be examined later.

We will mention only very briefly the production of the artificial aurora which have been abundantly reported and described at the occasion of various atmospheric detonations. Many different kinds of luminous phenomena can be produced by the sudden release of a large amount of energy in the high atmosphere. The discussion of such phenomena is of a scientific interest only if we are able to define, with some precision, the exact conditions of the emission of light in these artificial aurora. We cannot do this here. We will simply retain from the informations which have been published concerning these effects that their appearances can be associated with many kinds of high atmospheric detonations. We may add the important remark that every time an artificial magnetic perturbation has been observed on the "normal" magnetograms of an observatory, and that an aurora caused by the same detonation could be observed even at another place, these two phenomena were nearly coincident in their time of appearance and of disappearance.

The situation is quite different with radiation belts effects; we mean by that, the creation of artificial radiation belts around the earth which have been associated with a few of the higher altitude detonations. We will quote in this respect various papers devoted to this subject, published by Van Allen (1959 and 1962), Van Allen et al (1959) and a special issue of the Journal of Geophysical Research (1959). We will examine later the close association of the creation of such radiation belts with the production of world-wide magnetic effects.

Geomagnetic Perturbations Observed on Normal Magnetograms.

Powerful bombs detonated at low altitudes like the one belonging to the TEAK and ORANGE tests (Johnston Island in 1958) have produced visible - and sometimes important - perturbations on the normal magnetograms of the observatories which were geographically conveniently situated. In the case of the two tests mentioned above, as well as in some British tests of the same kind conducted in the surroundings of Christmas Island in the Pacific, these perturbations have been very pronounced at places situated not too far (a few thousand kilometers) from the site of the explosion.

But they could not be found on the magnetograms of observatories which were at distances greater than 3000 or 4000 kilometers from this site. That means that for these powerful explosions detonated at altitudes lower than the under boundary of the ionosphere (about 70 kilometers), the magnetic perturbations which were created suffered a strong attenuation with increasing distances. The only ambiguous case in this kind of deduction comes from Japanese reports concerning the ARGUS experiment (Maeda, 1959).

If we consider the effects of the bombs which were detonated at much greater altitudes and which were generally as far as we know of much smaller yields (e.g. the three detonations belonging to the ARGUS series), these effects were too small and too rapid to be registered on normal magnetograms at any place. If now we consider the few bombs which were detonated more recently (STARFISH experiment, July 9, 1962, above Johnston Island in the Pacific, and also some of the Russian Tests of the Autumn Series 1962, particularly the ones of October 22 and 28), at altitudes comparable to the one of the ARGUS series, but with yields of higher values (100 KT to a few MT), we find that their magnetic effects have the following two properties:

1) due to the high altitudes these effects could be observed all over the world.

2) due to the rather high yields, part of them could be detected even on normal magnetograms.

It is interesting in these last cases to collect records obtained at well distributed observatories around the world and to use them in order to draw a map of the distribution of the magnetic perturbation vectors and, as a further step, of the electrical current in the upper atmosphere which could, more or less conventionally, be determined from these vectors. We have tried to do this in the case of the STARFISH experiment, thanks to the cooperation of many of our colleagues of various countries. We are well aware that the collection we have made is still very incomplete, but taking into account the slow geographical modification of the perturbation all along the world, we think that we have collected enough informations to draw a map giving a true indication of the world distribution law of the phenomena. This map is shown as Fig. 1.

We can add the following comments:

1) in every observatory the perturbation shows about the same shape. A kind of small bay disturbance of approximately five to ten minutes duration, this depending on the local general intensity of the signal and the sensitivities of the various apparatus which were used.

2) the maximum deflection from the unperturbed traces occurs more or less simultaneously for all the magnetic components, which is an indication that the perturbation is linear polarized, at least roughly.

3) this simultaneity can be extended, very roughly, to many stations of observations situated far apart on the earth's surface, with the exception of the stations situated at only a few thousands of kilometers from the detonation.

4) at comparable distances the perturbations are much more intense to the west than to the east of the detonation.

The remarks 1) to 3) explain, without having to go into more detailed explanations, how it was possible to draw the vectors representing the most intense phase of the perturbations.

Magnetic Perturbations, As Observed on Rapid-Run and Ultra-Rapid-Run Magnetograms.

Perturbations of this kind have taken place when the nuclear detonations have been at a sufficient high altitude. There are few such detonations: The three ARGUS experiments (August 27 and 30 and September 6, 1958), the STARFISH experiment of the DOMINIC series (July 9, 1962), and most probably two of the Russian Tests (October 22 and 28, 1962). From what we said earlier, we know that among these experiments the three ARGUS ones did not make any perturbations on normal magnetograms. In the STAR-FISH experiment and in the two similar Russian Tests such magnetic perturbations were present. This difference did not seem to have a direct influence on the type of the very small and very rapid magnetic perturbations which were found to be present in all these cases. An interesting remark is that when both kinds of perturbations occurred, on the normal magnetograms and on the rapid run, they were recognized as a first approximation to be in synchronism. Of course, in the case of the STARFISH experiment, we were able to recognize that the perturbations on the normal magnetograms (beginning of the bay-like shape disturbance) started about forty seconds later than the one detected by the rapid-run devices. This can be interpreted as a proof that the mechanisms involved in the two effects, although related to one another, are dependent on different processes. We will bear this in mind during our later examination of the interpretation of all results.

Let us now examine with some more details the structure of the signal which has been recorded on rapid-run magnetogram in all these high altitude events.

The ARGUS Experiment.

A lot has already been written about this structure and about its first instant of appearance. According to Berthold et al (1960), which were first to detect effects outside of conjugate area, on large horizontal loops installed on the American Continent (Arizona and New Jersey) by the U. S. Army Signal Research and Development Laboratory (as reported by Newman, 1959), the signal did not arrive exactly at the same time at the different places. Moreover, two distinct phases of arrival were reported by them. From this, they inferred that the signal had propagated at two distinct velocities, respectively of the order of 3,000 and 600 (more exactly: "between 430 and 760") km/sec.

From the records obtained at the French network of stations put into operation for the International Geophysical Year, 1957-1958, it was inferred that the signal which was recorded at all these stations had traveled in all directions at speeds probably much higher than 1,000 km/sec (Selzer, 1959; Eschenbrenner et al, 1960).

A little later Troitskaya (1960) published the results of the Soviet network of telluric stations which showed that at all these stations, distributed over a great longitudinal and latitudinal range, the front of the signal had been received everywhere within the same second. This was an indication that some parts, at least, of the signal could travel at a speed not so very different from that of light. This result which could be found comparable - with a better precision - to that of the French stations, was first considered by some authors as being in contradiction with the one given by Berthold et al. But we do not believe so and, in fact, similar results from later experiments (STARFISH experiment as reported by Roquet et al, 1962a), while confirming Troitskaya's finding for ARGUS, showed that due to the complexity of the signal, and the numerous possible mechanism for its propagation, the conclusion first reached by Berthold et al could not be dismissed altogether and could be applied to some specific of the signal.

The STARFISH Experiment.

We can discuss with much more precision the same facts. Most of the stations and observatories which were equipped adequately had been openly informed in advance of the occurrence of the event and therefore they were able to register the signals which had been expected with accurate timing and high sensitivities. As an example, let us recall here the results obtained in France, in the Kerguelen Islands, and on the Antarctic Continent (Roquet et al, 1962a; 1962b). In these two notes were given the exact times at which the first impulse due to the detonation was recorded at Chambon-la-Forêt (for examples of records, see Fig. 2, 2[1], 3 and 4) and at Port-aux-Français (Kerguelen Islands, South Indian Ocean), remarking that these two times (respectively, 0900:08.8 $\pm$ 0.2 and 0900:08.7 $\pm$ 0.5 UT) differed by less than a few tenths of a second. Later on, these authors, Roquet et al (1963) found that with the same approximation these instants were in fact coincident with the zero time of the detonation (0900:09.0 UT). From this they have come to the conclusion that the first impulse had traveled with a velocity far greater than the fastest that can be calculated for hydromagnetic waves, and that it had been propagated either by some electromagnetic means or by some other very fast process, such as extremely fast neutrons as suggested by Crain and Tamarkin (1961) and Field (1963).

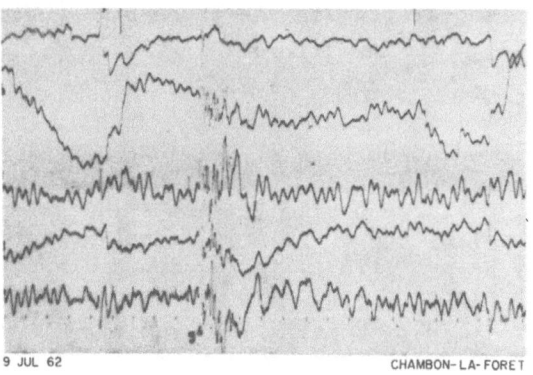

9 JUL 62   CHAMBON-LA-FORET

Fig. 2. STARFISH, as recorded at Chambon-la-Forêt on the regular "barre-fluxmètre" equipment, type A. From up to down the traces are: Z (vertical magnetic component), T_H (West-East telluric component), H (Horizontal meridian magnetic component), T_D (South-North telluric component), D (declination component). -- Recording Petiau-Selzer.

The speeds involved would be at least greater than one twentieth the speed of light. Apart from these findings, many other stations, including the French station Dumont d'Urville in the Antarctic, instead of detecting the signal at practically the zero time of the detonation, recorded its first effect at approximately 0900:11 UT, that is, about two seconds later. (In this respect, it is of special interest to note that earth current fluctuation was observed at Scott Base in the Antarctic, with a delay of 2 $\pm$ 0.4 sec. (Unwin, R. S., private communication).

These findings have been confirmed more recently still, and thanks to the release of some new data and records, these results can be regarded as definitely confirmed. This means that we can accept now as a proven fact that two distinct impulses, the first one at about zero time, the second at two seconds after zero time, occurred in most parts of the world. (Similar results related to other nuclear detonations recently have been brought to our knowledge by Santirocco, R. A., private communication.) One first conclusion of these findings is that the controversy which had been

114

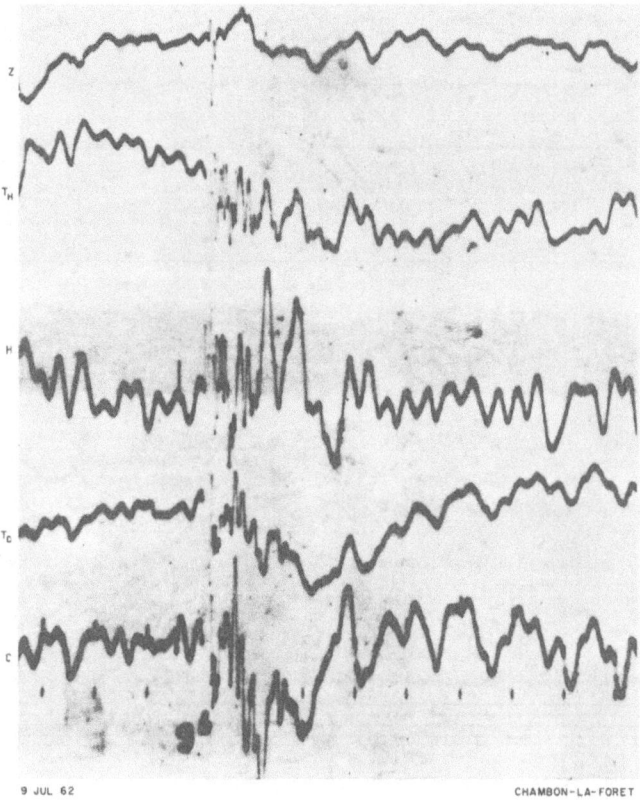

9 JUL 62   CHAMBON-LA-FORET

Fig. 2^1. Same as Fig. 2, enlarged.

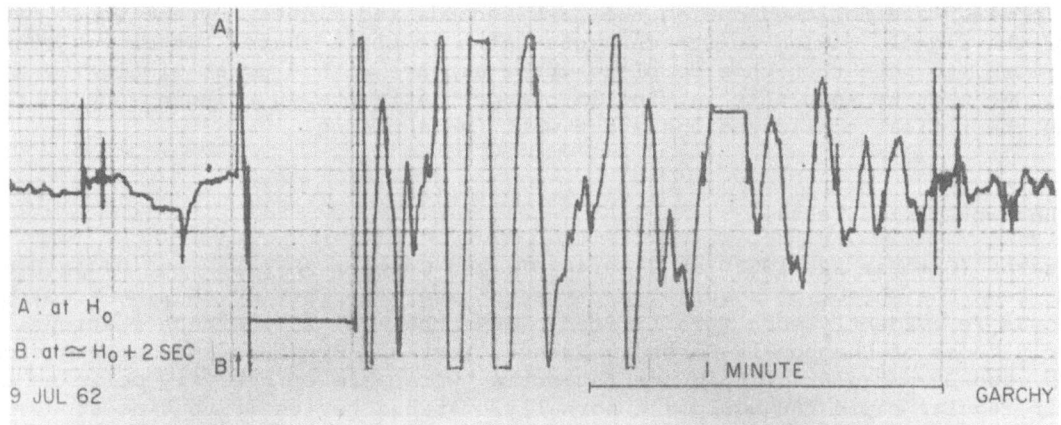

Fig. 3. STARFISH, as recorded at Garchy with a large horizontal loop and
         equipment type B (Z component). --Recording Gilbert-Roquet.

open about ARGUS can be considered closed, i.e. some components of the signal can
reach most parts of the world with practically the speed of light, while some other
components appear at the same places two seconds later, uniformly. Then, of course,

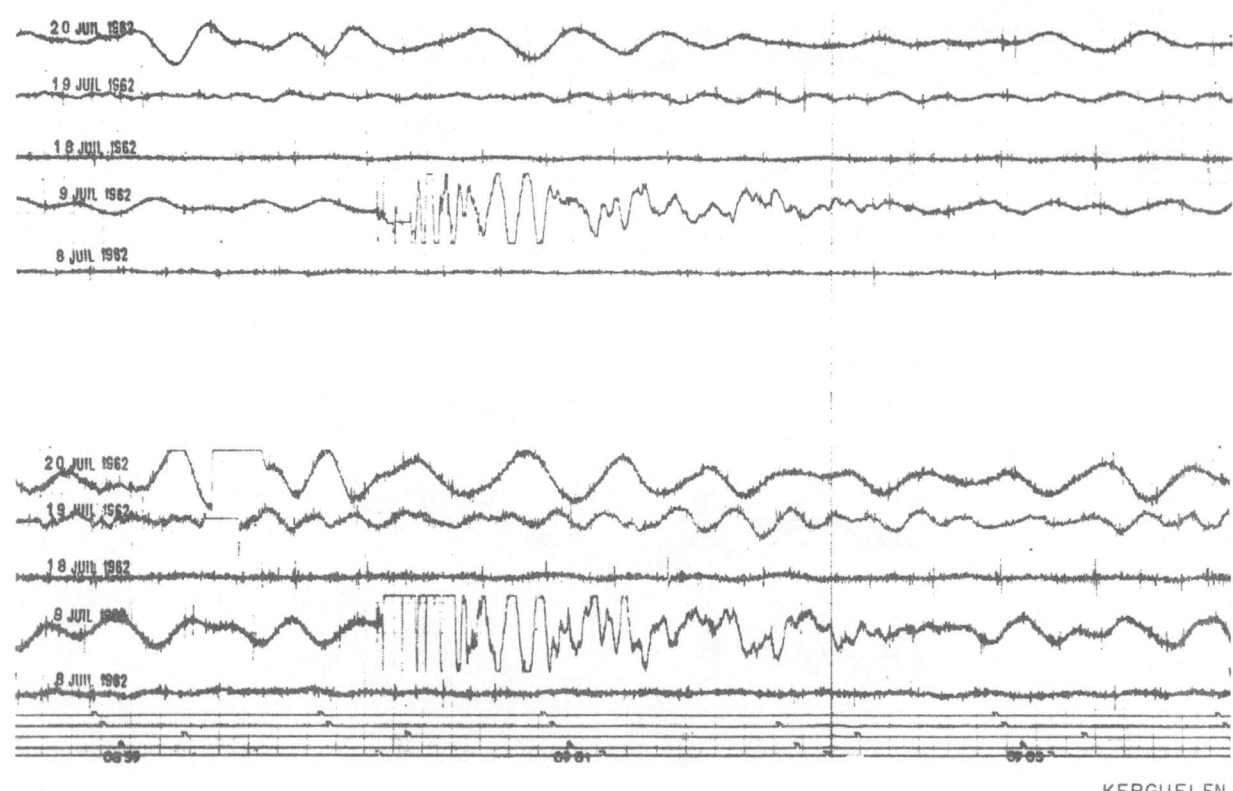

KERGUELEN

Fig. 4. STARFISH, as recorded at Kerguelen on the telluric components:
Top - South-North; Bottom - West-East.
--Recording Maillard-Plessard-Schlich.

these impulses are followed during the next seconds and minutes by the whole development of the signal, giving it its characteristic complete shape. It is remarkable,
furthermore, to note that this complete shape appears with a great similarity at most
places. We want to keep this in mind in our next attempts in bringing out tentatively some theoretical considerations about what is happening.

        The two Russian Tests.

These tests (October 22, 1962, at 0340:46 UT, and October 28, 1962, at 0441:18 UT)
had not been the object of special warnings to observatories and thus, our collection
of records is evidently much more limited. Nevertheless, some French stations, particularly those of Chambon-la-Forêt in France, Port-aux-Francais in the Kerguelen
Islands, and Dumont d'Urville in the Antarctic, were able to get very definite records
on their regular rapid-run equipment normally operated for their routine study of
natural micropulsations. (See, for example, Fig. 5, 6, 7, 8, 9 and 10.) These findings were supplemented with other records made in Upsala, Sweden, and in Germany by
Bath and Ehmert, respectively, as reported by Egeland et al (1963). From these
records we can infer that the magnetic perturbations caused by these two detonations
are, in a broad sense, quite similar to the one produced by STARFISH. On some of the
French records for October 22 it is even possible to distinguish relatively clearly
the arrival of two successive impulses as in STARFISH: One difference being that the
time delay between them seems to be a little longer (five or six seconds instead of
two). From all this, we can infer that these two Russian tests have been performed as

116

STARFISH at very high altitudes. So far, however, there has been no information released permitting us to check this assumption. It may even be that one, or both, of
these Russian tests took place at altitudes far greater than that of STARFISH, and
that might be the cause of the changing of time differences between the two impulses.

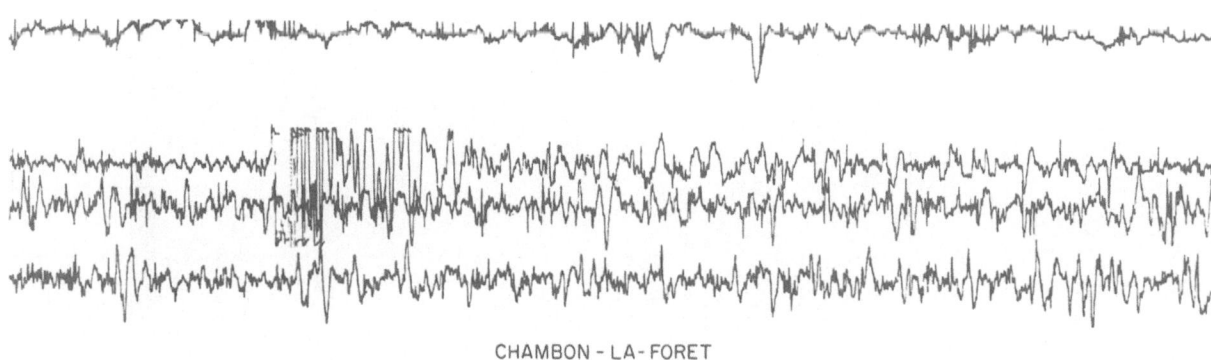

CHAMBON-LA-FORET

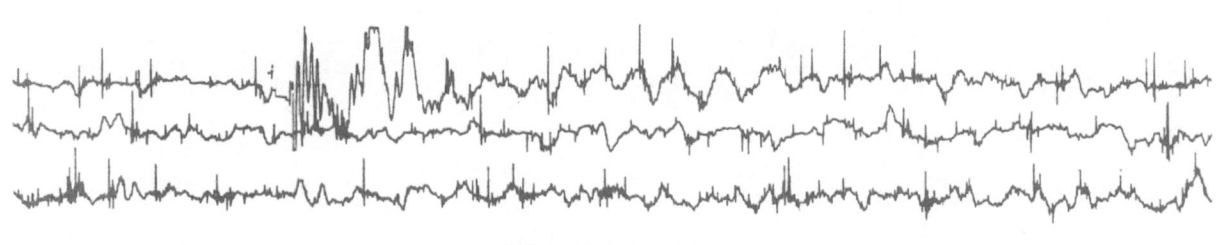

CHAMBON-LA-FORET

Fig. 5.   October 22, 1962, event, as recorded at Chambon-la-Forêt
on the D (equipment type B) and T_D traces.
--Recording Petiau-Selzer.

So far in the review of all the preceding events, we have focused our attention
on the initial manifestations of the signals. We now want to give some description
of the constitution of the body of the signal. One way of doing it is, of course, to
put the records under the eyes of the reader. This unfortunately is, for practical
reasons, not feasible because the selection of the graphs has to be limited. So, let
us try to explain this signal appearance by a description bringing out its most
important features. This can be simplified by the fact mentioned before that these
features appear to be quite the same at all places of recording. Also, our investigation being still in an elementary state, we want first to direct our attention
toward the common part of these appearances, leaving the examination of their particularities for further studies. In this respect, we can say that for all these
high altitude detonation experiments (ARGUS, STARFISH, RUSSIAN TESTS), the most
characteristic structure of the body of the signal is constituted by irregular
pulsations. These pulsations are at the beginning part of the impulses, their pseudo-
periods being at this time much shorter than one second, making their full registration very unclear or even unreadable on most records. We call attention, in this
connection, to the comments by Lokken et al (included at the end of our report) in

117

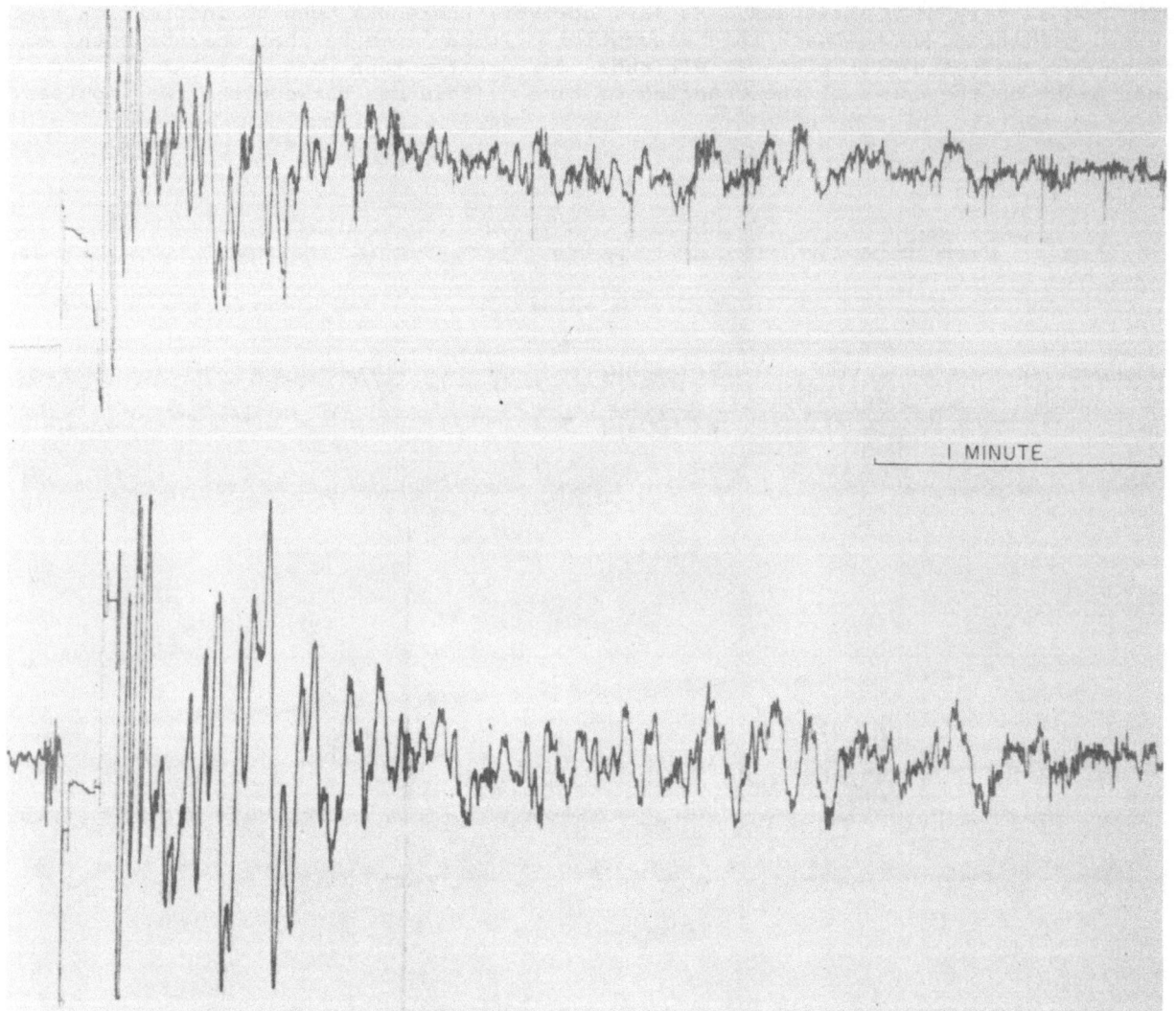

I MINUTE

GARCHY   22 OCT 62

Fig. 6.
October 22, 1962, event, as recorded at Garchy on two components: Top - telluric
South-North; Bottom - magnetic Z (large horizontal loop and equipment type B). It
is interesting to underline that in Garchy, due to geological particularities, the
magnetic Z component is very similar to the magnetic D component and so makes a
conjugate pair with the telluric South-North component T_D. Another particularity
of this record is that the top trace was off-scale when the signal came unexpectedly
and put it back on the right track.   --Recording Gilbert-Roquet-Selzer.

118

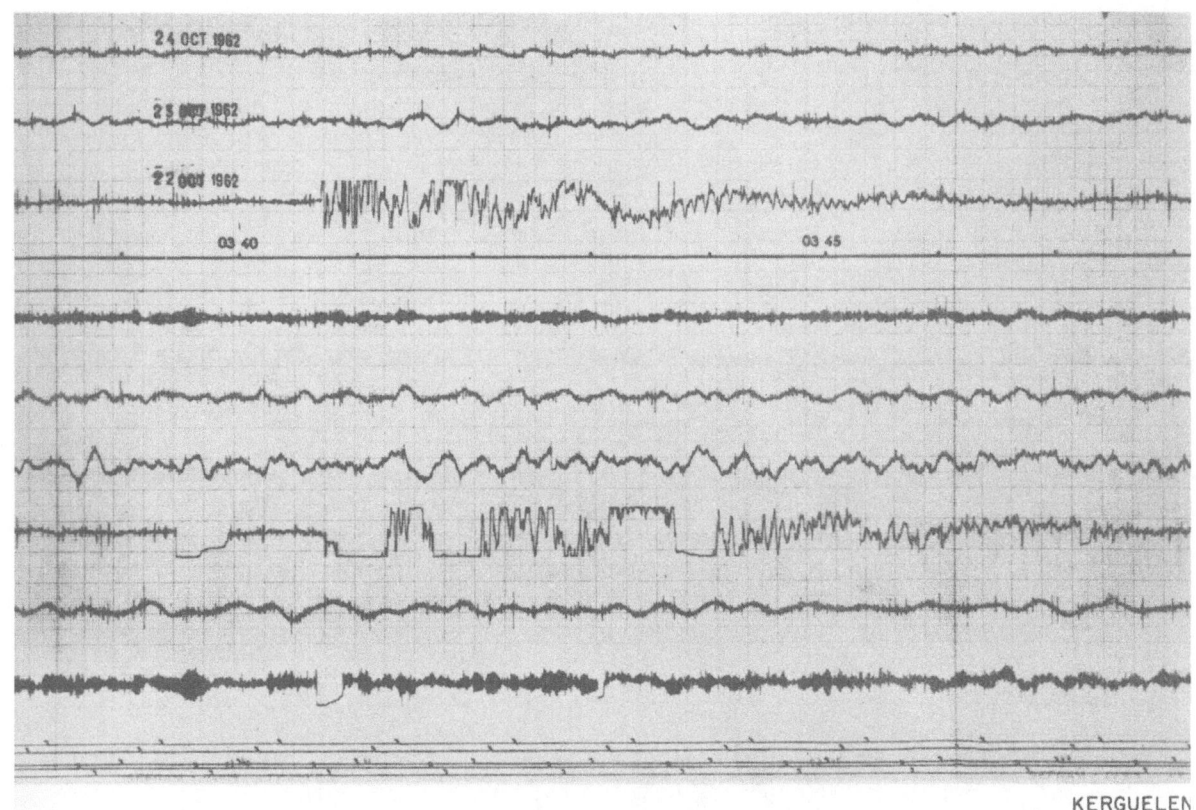

Fig. 7. October 22, 1962, event, as recorded at Kerguelen
on the two telluric traces (same as Fig. 4).
--Recording Maillard-Plessard-Schlich.

which pulsations having a period 1/8 of a second are reported as having been observed
at the beginning of the signal in the STARFISH case. After the first seconds the
values of the periods of these pulsations increase and may reach values as high as
one minute or more toward the end of the signal as they vanish. This process of the
period increasing along the progress of the signal can be observed systematically in
all the experiments. One difference, however, is that in each case some dominant
range of periods reveals itself with increased amplitude in some parts of the signal,
and that these dominant periods don't seem to have been the same in all cases. For
example, in the ARGUS case, the dominant period was around one to two seconds, while
in the STARFISH case, it was extended from about two to eight seconds. This change
can give rise to more speculation as to some variations in the factors relative to
these two experiments. We remember that the STARFISH detonation was about a thousand
times more powerful than each of those of ARGUS and the volume of the outer atmosphere which was disturbed by the fireball was certainly much larger in the first case.
That would correspond to an augmentation of all time-constants involved in the dynamics of the STARFISH event, as reported by Bomke and Harris (private communication,
1963).

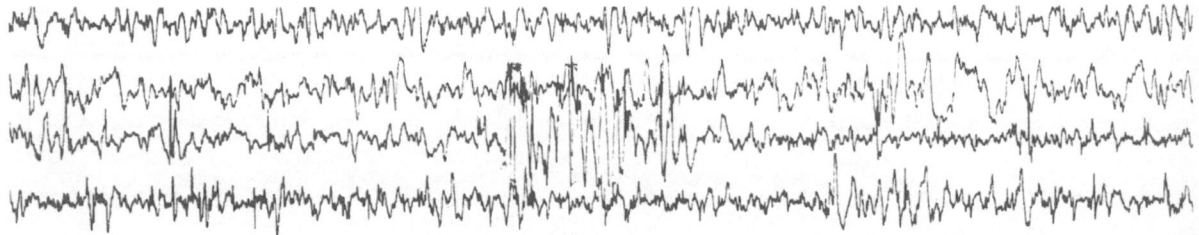

CHAMBON-LA-FORET

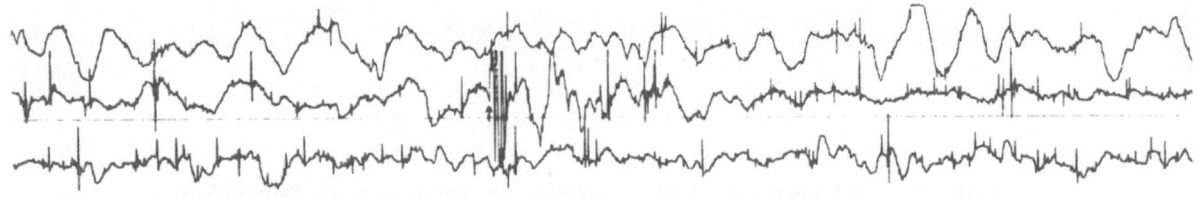

28 OCT 62   CHAMBON-LA-FORET

Fig. 8. October 28, 1962, event, as recorded at Chambon-la-
Forêt on the D and T_D traces (same as for October 22,
Fig. 5).   --Recording Petiau-Selzer.

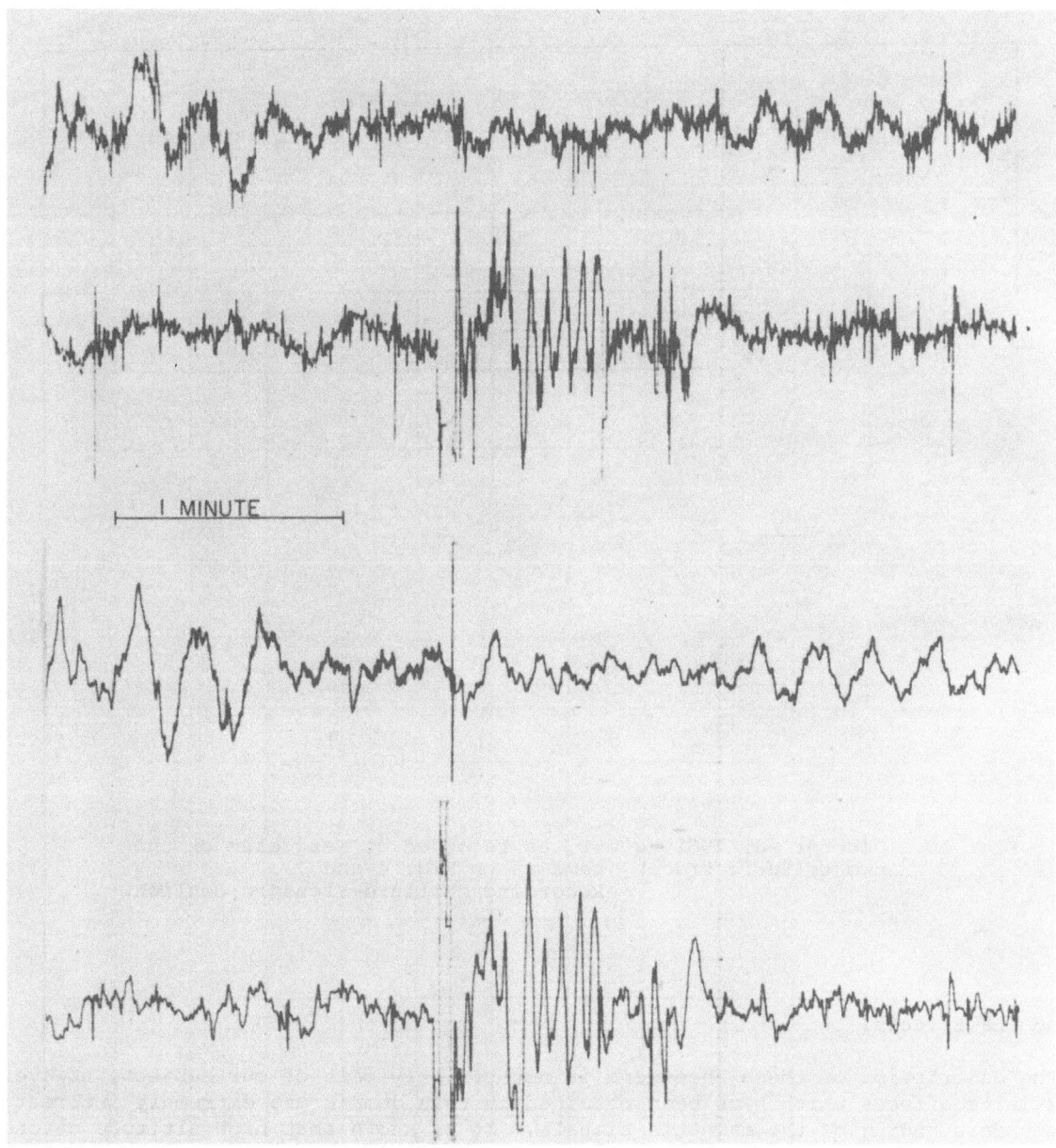

GARCHY   28 OCT 62

Fig. 9. October 28, 1962, event, as recorded at Garchy on
        the two telluric traces: Top - South-North; Bottom -
        West-East.   --Recording Gilbert-Selzer.

121

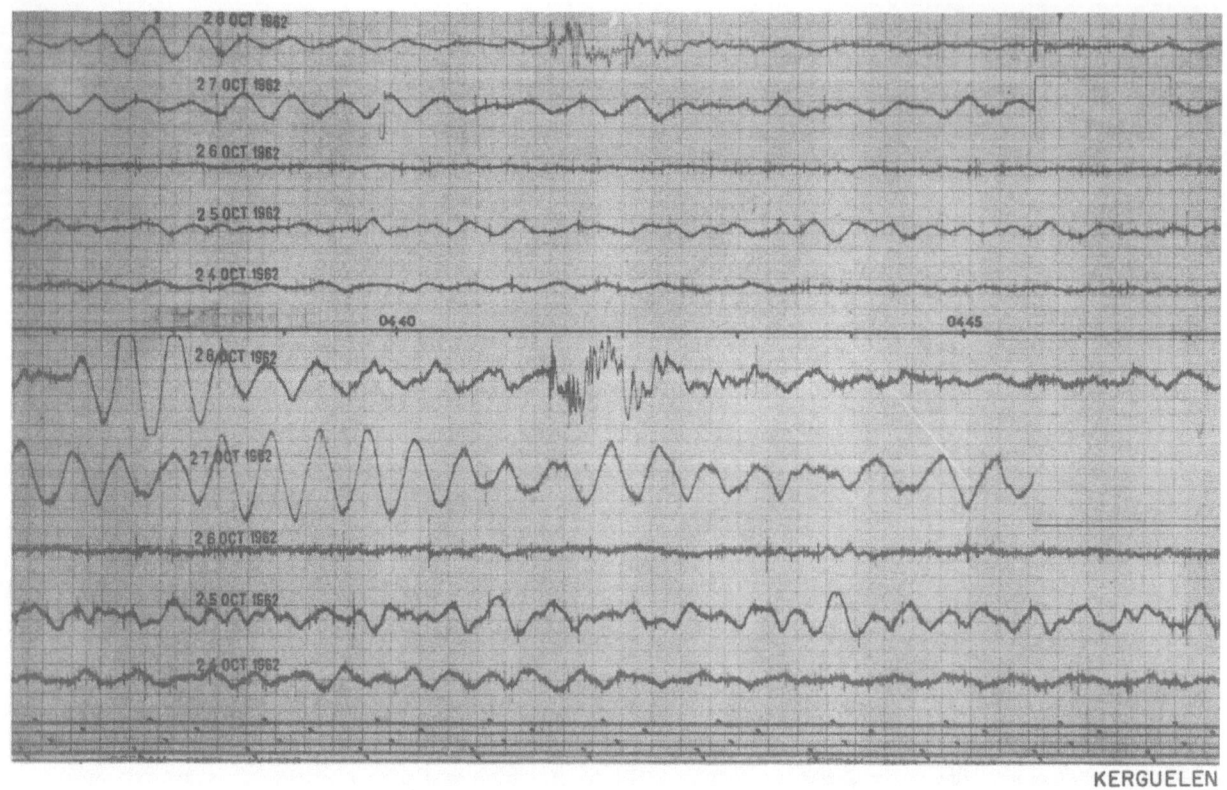

Fig. 10.   October 28, 1962, event, as recorded at Kerguelen on the
two telluric traces (same as on Fig. 4 and 7).
--Recording Maillard-Plessard-Schlich.

VLF and ELF Effects.

The description of these phenomena is not strictly part of our subject, however, the definite effects which have been obtained in this domain are extremely interesting for an understanding of the magnetic signals. It is known that high altitude detonations may cause pronounced disturbances of radio-wave propagation. These disturbances have been especially well recorded in the STARFISH case on wave lengths of the order of 15 to 16 kc/s used in regular world telecommunications system. Communications along paths crossing the region of the detonation were, as expected, the most disturbed. The perturbations began in quasi-synchronism with the zero-time of the detonation, but this with the reserve of a lack of accuracy in the definition of the time amounting to one or more seconds. It is important to note that even paths that were far from crossing the region of the detonations were also affected and precisely at the same time.

Lokken has kindly supplied some ELF records (2-30 cps) of the STARFISH detonation, taken at Byrd Station in Antarctica (See Fig. 11a and b).

122

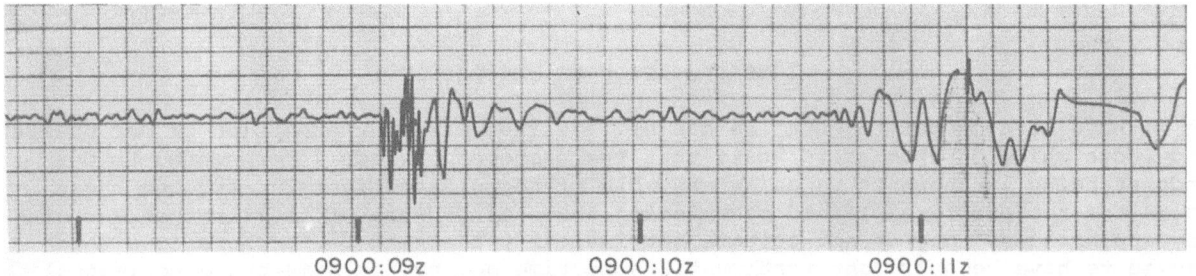

0900:09z   0900:10z   0900:11z

ELF 2-30 CPS BAND, VERTICAL COMPONENT, RECORDED AT BYRD STATION 9 JUL 62

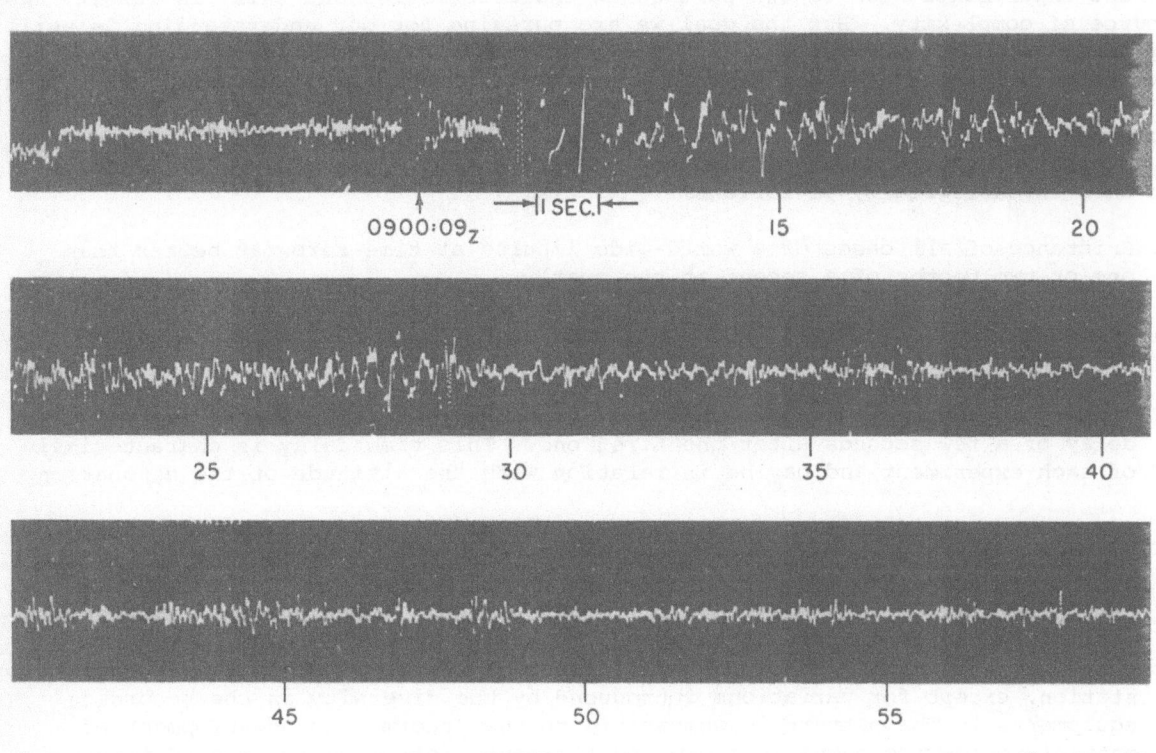

0900:09z |→|I SEC.|←| 15   20

25   30   35   40

45   50   55

BYRD STATION   9 JUL 62

Fig. 11.
STARFISH, as recorded at Byrd Station by the Pacific Naval
Laboratory of Canada (communicated by J. E. Lokken).
a) vertical component in ELF 2-30 band;
b) similar registration on 35 mm film.

INTERPRETATION OF THE RESULTS

As we have already pointed out, we can be satisfied with the summation of our present knowledge which enables us to state that the phenomenon which is presented to our understanding is known to us as an ensemble of coordinated facts. Although these facts are not easily explained by existing theories because of their rather astonishing appearances, they do not contain intrinsically the kind of incoherencies which seemed to have been brought forth until some time ago by the numerous experimental reports which had been given, particularly about "ARGUS." We do not pretend that the full results of these kinds of experiments could be reduced to perfect coherency. Each of these experiments had its own particular characteristics and this, in itself, was a source of complexity. But the goal we are pursuing for our understanding is still very crude, i.e. a comprehension of a first order approximation to the fundamental facts which allows us at least at this first stage of our examination of the problem to discard much of the complexity. This situation enables us to concentrate our attention on a few experimental points which are to be explained fundamentally. Although other ways of presenting them could be reasonably adopted, we are going to list them chronologically as follows:

1st. Existence of all cases of a world-wide impulse at time-zero (at better than one or two tenths of a second at the most).

2nd. Detection of small oscillations immediately following this impulse, or even belonging intrinsically to it (cf. comments by Lokken, J. E., loc. cit.).

3rd. Synchronous world-wide appearance of a second more intense impulse with a time delay of a few seconds after the first one. This time delay is characteristic of each experiment and may be in relation with the altitude of the detonation.

4th. Synchronous world-wide reception of a series of irregular, fast-oscillations, lasting from half a minute to about ten minutes, this duration depending not only on the sensitivities of the receptors, but also on some of the factors of the detonation (altitude, yield, location, etc.). For each definite experiment, this series of pulsations seems, to a first approximation, to follow everywhere the same kind of pattern. (By everywhere, we mean at every recording station, except for variations introduced by the diversity in the recording equipments.) The pattern is comparable to the "storm time" development of a natural world-wide magnetic storm and starting, following the second impulse with the shortest periods (one or a few seconds) and finishing with the longest ones (a few tens of seconds) while the signal is at the same time fading away. These limits seem to be dependent on some factors characteristic of each experiment, for example, the periods were, on the whole, longer for STARFISH than for ARGUS.

5th. A small bay, looking like a magnetic "crochet" ("S.F.E.") is recorded, equally on a world-wide scale, if the detonation is powerful enough (STARFISH, some of the Russian Tests, but not ARGUS). Like the signal registered on fast-run records, this small perturbation visible on normal magnetograms seems to follow, at least to a first approximation, the same development everywhere, and independently of day or night local conditions. Its beginning is delayed compared with the beginning of the rapid variations (by a few tens of seconds) but its total duration seems to be about the same, although there can be not much precision in such kind of statement.

Let us examine now, one by one, the successive effects:

The first impulse is not, in itself, hard to understand. Fundamentally, it must be electromagnetic. Difficulties come when we want to know the exact paths followed by this first part of the signals and the precise mechanism responsible for its propagation. We can then imagine the propagation to be:

a) above the ionosphere, along the discontinuity ionosphere-magnetosphere,

b) inside the ionosphere (but this would meet serious difficulties),

c) under the ionosphere, along the discontinuity ionosphere - low atmosphere, or better, along the resonant cavity earth-ionosphere, and

d) along some conductive layers in the earth's crust.

Of all these possibilities, c) seems to be the most probable one, especially* if we take into account the comments of Lokken (loc. cit.).

So, as far as this first impulse is concerned, the difficulty is not to find a plausible explanation for the phenomenon, but to choose the exact solution. An important remark at this point is that even the powerful nuclear bombs (tenths of megatons), which produce under all conditions an electromagnetic initial pulse (such effects are always present anyway in all kinds of detonations, nuclear or not, cf. Rocard, Y. and Steinberg, J.L.; and Proces-Verbaux, Genève, 1958), do not permit the capture of the slow-tail - a few cycles per second - of this e.m. pulse, all over the world, if these bombs are detonated lower than a specific critical height. How can we explain such a decisive effect due to the altitude of the detonation? We shall make the very general remark that some process has to start above the ionosphere (or at least at an altitude greater than the one corresponding to the layer of maximum density) so that a mechanism of propagation under the ionosphere (if we adopt case c) might be effectively set up. This necessary preliminary process might be the creation of a huge volume of "emptiness" above the ionosphere (the magnetic lines of force being suddenly pushed away by the detonation by a kind of hydromagnetic effect, thanks to the sudden intense release of mechanical and thermal energy together with an enormous flux of ionized particles).

In the case of a detonation of about the same yield - or even of a greater yield - occurring under the ionosphere, such a volume would be much smaller. In this connection and considering only the experiments performed above the ionosphere, we might try to find for a bomb of a given yield - the relations between the altitudes of detonation, the dimensions of the diamagnetic volume created by the detonation and the properties of the electromagnetic signals so generated, principally in what concerns their distribution around the globe.

We are still very far from being able to present, even in a crude way, a mechanism which might explain the general excitation of the earth-ionosphere cavity as observed by Lokken (loc. cit.). This effect is connected with the excitation of the same cavity by lightning flashes and is an important and complex subject which still needs better understanding.

In a short paper recently published (Roquet et al, 1963) we proposed, as an alternative, the kind of process suggested by Crain and Tamarkin (1961) and Field (1963). But, as pointed out to us by Lokken ("Comments" by Lokken, loc. cit.), there is some difficulty in admitting sufficiently rapid action for such a process which could better apply to the second impulse than to the first one. Because of the scarcity of available data concerning detailed registrations of the first impulse, it is difficult to decide about this alternative. We can remark that this impulse occurs more or less in synchronism with other rapid transient perturbations of the ionosphere and that all these effects seem to stem from a single mechanism.

* Remark made during the revision of the manuscript.

In our present state of knowledge, it seems easier to connect these ionospheric perturbations with the arrival of fast electrons spiraling down the earth's magnetic lines of forces as described by Crain and Tamarkin, than with the excitation of the earth-ionosphere resonance cavity by electromagnetic waves. But this second kind of action has the advantage over the first one in that its effect cannot be anything but world-wide as is required by the observations. In contrast to this, the Crain and Tamarkin effect appears to be limited in its direct geographic extension to a definite "illuminated" portion of the earth's surface (see Fig. 12). The exact determination

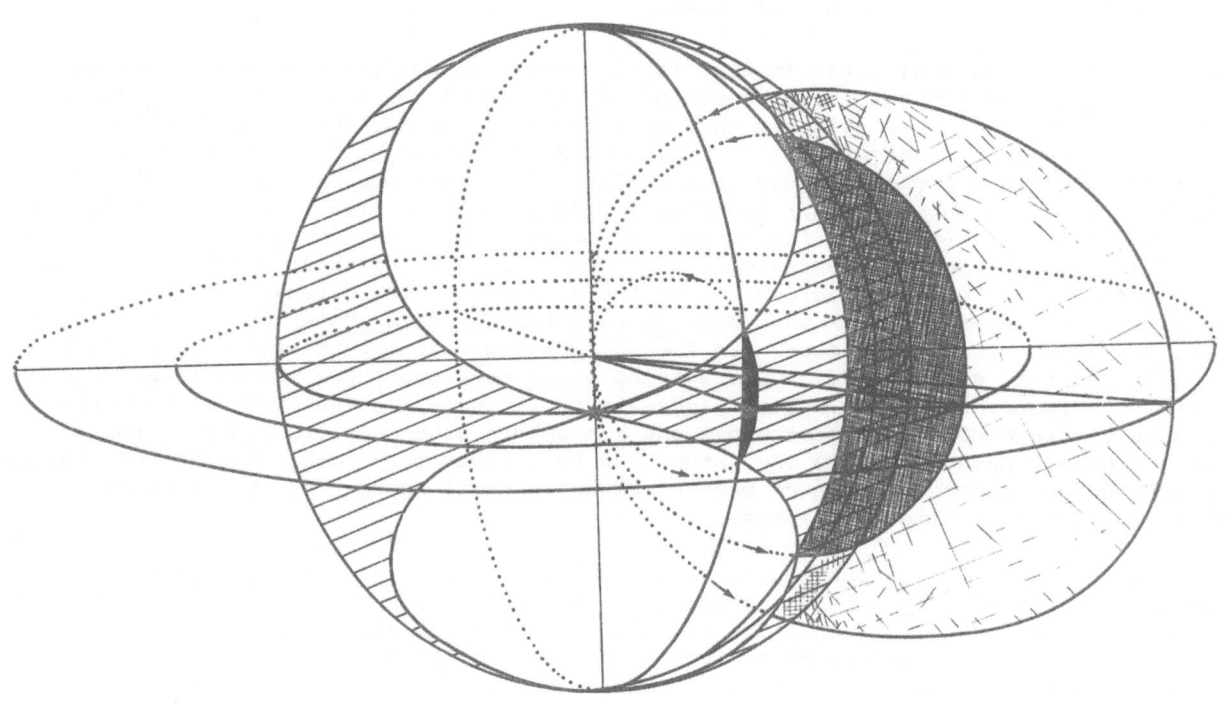

Fig. 12. Geometrical determination of an initial "illuminated zone" deduced from the neutron-decay model proposed by Crain and Tamarkin. The graph illustrates the case of an equatorial detonation.

of the "illuminated" and of the "shadowed" zone is a point of pure geometry - at least for a given configuration of the magnetic field lines. Its general solution can be expressed quite exclusively in function of the (geomagnetic) latitude of the detonation (supposing some reasonable limitations for the detonation's altitudes).

Fig. 1 and 2 give an idea of the results for a few characteristic cases. A complete detailed determination of this geometrical question will be published.

Going further into our investigations, we need to find a plausible explanation for the world appearance of a second impulse. We will first consider again our former statement (Roquet et al, 1963) that two very general possible explanations could be taken as a first base of thinking. (We were then referring to STARFISH, but except for some numerical values, they can apply to the other cases.) Firstly, that this second impulse had originated at (around) instant zero plus two seconds at the zero point itself, and had traveled with great speed (an important fraction, at least, of the speed of light) all over the world, and secondly, that the first impulse, or a synchronously similar perturbation having reached, in practically no time, all various local ionospheric regions of the world, had determined in each of these places local, but simultaneous, effects. Then the delay of two seconds would have been the time necessary for the process, involved in each of these local excitations, to be performed. This alternative seemed to be able to cover all eventualities. But the experimental findings revealed by Lokken ("Comments" loc. cit.) still bring something new: the possibility that the delay of two seconds (if we still, as an example, follow the case of STARFISH) did not take place at the starting point nor at the arriving point, but in the earth-ionosphere cavity considered as a whole; in other words, by a process using the resonance properties of the intermediate medium. If we want to examine the possibilities offered by such a process, we have to explain the two fundamental modes of transfer of energy which are involved: the first one, already mentioned from the detonation into the cavity, and the second one from the cavity into the different ionospheres. We cannot at the present time have any clear idea on the conditions governing these two modes of transfer, or even of their degree of reality, but it will be very important to push their study.

Another alternative is to have a similar role to be played by some other cavity situated above the ionosphere instead of underneath it. We can think, for example, of different space domains situated between the ionosphere and various other surfaces of discontinuity to be found inside the magnetosphere. The excitation of such outer ionospheric cavities might be then produced by the same neutron-decay mechanism as suggested by Crain and Tamarkin and mentioned earlier.

In spite of the difficulty of deducing all the observed facts by making use of the Crain and Tamarkin process, this process is able to furnish a coherent explanation for many of the principal characteristics of the signal as it is observed in many places. Thus, if we consider all the stations situated inside the illuminated area, we are able to understand:

1st. The quite simultaneous strong reinforcement of the ionization in all the corresponding local ionospheres and all the related phenomena, including the beginning of the most intense part of the magnetic perturbation.

2nd. The successive appearance of first fast, then slower, oscillations which are found in this perturbation everywhere. The augmentation of the (pseudo) periods of these (irregular) oscillations corresponds most likely to the increase of the dimensions of the magnetospheric domain from which the excitation by the neutron decay products (protons and electrons, these last ones being alone to play a direct role in this process) has taken place. At the end, when these neutrons are too far from the earth's surface, or have gone outside the limits of the magnetosphere, these oscillations, which by then have attained a period of one minute, or even more, are fading away.

3rd. Concerning the shadow zone, the same process can still be taken as the one intervening in a primary action. But we need to figure out how it can be followed and complemented by some other one. We believe that such a complementary process could be assumed by some kind of longitudinal hydromagnetic coupling between the outside limit in space of the illuminated magnetic shell and the shells situated underneath and which are in direct contact with the ionosphere above these shadow zones.

If this is really the total process, that would explain why, even inside the shadow zone, the succession of oscillations of increasing periods is about the same as outside: the signal having reached, in practically no time, the frontier in space

between the two zones has just at this time adopted another mode of locomotion a little slower than the first one but which makes possible its arrival on the local
ionospheric layers, and on the ground, everywhere.

As long as this new mode of transmission by hydromagnetic longitudinal waves
does not introduce any dispersion in these waves (the speed of such waves being independent of the frequencies), it does not bring any important changes in the succession of oscillations which compose the signal which is only delayed by a few seconds
compared with its appearance in the illuminated zone.

The most important change could be introduced by the differences in the quick-
ness of the coupling effect which must exist inside and outside the illuminated zone
considered in space. Inside this zone these coupling effects must act very rapidly
producing a mixing of oscillations of various frequencies which originated in different parts of this illuminated space. Quite to the contrary, in order to be effective
outside this zone, these couplings need various time delays and parts of the signals
can reflect the spreading in time of the corresponding frequencies.

Without changing radically the model discussed in the previous section, we can
think of another mechanism, as an alternative to the neutron-decay process, to
explain the first part of the propagation. For this mechanism, we need to rely on an
agent capable of spreading at great speeds along straight lines, but insensitive to
the anisotropic effect of the earth's magnetic field. Furthermore, this agent,
without being sensitive to the magnetic lines of forces, should be able to perturb
them when it crosses their paths. Besides neutrons, a very steep shock front, traveling in the same approximate mode as the longitudinal hydromagnetic wave, could be
this agent. It would be very little disturbed by the ambient geomagnetic field
because of the disparity between its own energy and the energy density of this field.
Along its path, the lines of forces would be perturbed and would effectuate forced
oscillations, first in an hydromagnetic toroidal transverse mode, then in a more
complex pseudo-poloidal motion. As a result, the excitation process would have spread
all over the "illuminated zone." It may be interesting to note that such a mechanism
would be quite comparable to the one which is generally adopted to explain the first
phase of a magnetic storm (storm sudden commencement, or S.S.C.) except for a reversal of the initial propagation for the phenomenon. Such a process would justify
the term "artificial storm-sudden-commencement" which has been attributed by many
commentators to the nuclear detonation geomagnetic effect.

We come now to the subject of how to explain the bay-like or crochet-like perturbation which is to be noted on normal magnetograms, practically everywhere on the
earth's surface for STARFISH, in a more restricted area for the Russian Tests, and
for TEAK and ORANGE, and which did not appear at all for ARGUS, this last fact raising
no new problem when we think of the relatively very small yield of the ARGUS detonations compared with the other ones. In fact, in contrary to the pulsed part of the
signal, this bay-effect is mainly associated with the power of the detonation.

But many different mechanisms are still, of course, offering plausible explanations for this "power-effect," and they might not be the same in all cases. For
example, in TEAK or ORANGE we could associate the strong reinforcement of the ionization with strong radial "winds" caused by the direct thermal and mechanical expansion
of the detonation, as suggested by Matsushita (1959a). Or we might prefer the purer
"solar-flare-effect" theory supported by McNish (1959) (cf. also the "Comments").

In the case of STARFISH, that is of a detonation at a much higher altitude, the
theory exposed by Pisharoty (1962) seems to furnish a plausible answer for most of
the observed facts. It is based on the well-known property of the protons and electrons produced in great quantity at the instant and place of the detonation. They
are trapped in oscillatory orbits between north and south mirror points and drift
around the earth's magnetic axis, the protons drifting westward and the electrons
eastward. But, according to Pisharoty (loc. cit.) the electrons would have been

practically all absorbed around the lower mirroring places above South America and the protons would be the real cause of the overhead electrical currents passing over India on their way from the Pacific area to Africa and Europe. The delay of many tens of seconds which has been found between the instant of first manifestation of the pulsed part of the signal (that is, around the zero-time of the detonation) and the one corresponding to the beginning of the "bay," would represent then the drift-time of the protons to come from the geomagnetic longitude of the detonation to the one of the observing station. We will not insist any further on the analysis of this bay-like effect, which seems to be less characteristic than the rapid pulsated part of the signal, at least when dealing with the special case of a real high-altitude nuclear event.

ACKNOWLEDGEMENTS

Very valuable remarks and contributions by W. H. Campbell, A. Egeland, J. E. Lokken, A. G. McNish, and many other members of the Faculty, which have been communicated to us, either orally during the Faculty session, or later on by written Comments, are thankfully acknowledged.

They have been taken into account when we have given this report its definitive written presentation, as it had become a logical impossibility to ignore them any longer.

We have quoted individually these new sources of inspiration each time we have been able to do so materially.

Furthermore, these Comments will be given here following our own report, three in their original form and the fourth one as an abstract.

A Note referring to an important paper by Bomke et al, which was released at the occasion of the I.U.G.G. General Assembly in Berkeley, following our NATO Faculty Meeting in Bad Homburg, has been added as a further step toward a better comprehension of our subject.

DISCUSSION AND COMMENTS

<u>W. H. Campbell</u>: The Geophysical Institute of Peru which joins with us of the National Bureau of Standards in the operation of a geomagnetic micropulsation unit near Lima, has published in JGR, not long ago, the record of the signal received during the STARFISH high-altitude nuclear explosion. The record is about identical to that described by you for the middle latitudes. If your mechanism is real, isn't it difficult to explain such data in the "shadow zone" at the equator?

E. Selzer: It was not my intention to present the neutron decay mechanism as solving all difficulties, at least in its simpler form. In fact, having been already convinced that the signal had appeared everywhere on the earth's surface with about the same aspect (and Dr. Campbell's Comment brings another proof of this observation), I have been looking for a plausible explanation of this surprising fact. If the neutron decay mechanism has specially retained my attention, that is because it explains remarkably well the quite identical appearances of the signals outside of the shadow zone. What happens inside this zone certainly needs much further investigation.

A. Egeland: The only magnetic effect at Kiruna associated with nuclear weapon tests July to December 1962 which may be attributed with any degree of confidence to a nuclear explosion is the magnetic perturbation at approximately 0900.20 Z on July 9, 1962. The time of United States high altitude test at Johnston Island on July 9 has been reported (e.g. by Durney et al; Nature, Vol. 195, p. 1245, September 29, 1962) as 0900.15 Z, which is only five seconds before the observed magnetic effect at Kiruna. As the timing accuracy of the quick-run magnetograph is of the order of seconds, a quite accurate estimate of the time of travel of the disturbance is not possible.

On the original recording the perturbation in question is seen to be a sharp commencement of very rapid pulsations (amplitude of about 10γ), with a small decrease of the horizontal north-south component, lasting for a few minutes. There was, however, no apparent effect on the vertical component. The time scale of the pulsations is outside the effective limits of the equipment response.

Significant effects on very-low-frequency and low-frequency propagation after the high altitude tests have been observed at Kiruna Geophysical Observatory, cf. Nature, Vol. 198, p. 1076, June 15, 1963.

J. E. Lokken: The records (Fig. 11) made at Byrd Station, Antarctica, on the Pacific Naval Laboratory ELF equipment in the nominal pass band 0.2 to 30 cps show an exceptionally high amplitude burst of approximately 8 cps activity, commencing 30 milliseconds after T_O and lasting for about one second. The amplitude of this burst is one, or perhaps two, orders of magnitude larger than we have ever seen at Byrd Station from thunderstorm activity. Dr. Katsufrakis of Stanford University has recordings here of the amplitude of several VLF stations received at Stanford University. These recordings are from stations in North America and thus the path does not directly cross the geomagnetic longitude of Johnston Island. The change in signal strength of the VLF transmissions commences at approximately T_O + 50 milliseconds and reaches its maximum value in about two seconds - the time of commencement of worldwide "micropulsations." The ELF equipment at Byrd, responding to higher frequencies than the micropulsation equipment, recorded the onset of "micropulsation" activity at 0900: 10.5Z, while the micropulsation equipment showed a commencement at about 0900: 11.1z.

In view of the extraordinarily large 8 cps burst (the fundamental cavity resonance mode) and the abrupt change in signal strength of the VLF transmissions - requiring near velocity of light propagation to the ionosphere between the transmitters and the receiving station in North America - we should not rule out 'a priori' the possibility that the lower ionosphere, say D region, has been severely disturbed by the cavity resonance or higher frequency electromagnetic radiation. This disturbance would then change the VLF reflection height and generate micropulsations of about one second period which are observed about two seconds after burst time everywhere.

The disturbance in the ionosphere due to electromagnetic radiation would at most explain the high frequency initial phase of the geomagnetic disturbance which is observed at all stations, including those in the shadow zone of your theory. An explanation such as yours would adequately explain the longer period "geomagnetic" activity occurring after the first few seconds.

The high amplitude cavity oscillation is undoubtedly due to the large "fire-ball" produced by the prompt gamma radiation. The mean free path is sufficiently great at 100 km or higher to be limited only by the denser material at low altitudes within line of sight of the burst. At lower altitudes the fire-ball is too small to produce significant energy in the cavity resonant mode, therefore, only high altitude tests - above a critical height as you pointed out - could produce sufficient energy in the cavity to cause an ionospheric disturbance.

The suggestion that electromagnetic energy might disturb the D-region sufficiently to affect the reflection height and generate micropulsations of about one second periods may be extreme, but some calculations should be carried out before eliminating the possibility entirely. So far, we have not done any calculations.

A. G. McNish: As the reader may have noticed, the author of the preceding report has not examined with much detail the effects of the experiments like TEAK and ORANGE (1 and 12 August 1958, respectively, above Johnston Island in the Pacific) as well as those of earlier British Tests over Christmas Island (in the Pacific also), which have been performed at "intermediate altitudes": 60 kilometers for TEAK, 30 kilometers for ORANGE and around 10 kilometers for the British Tests.

These experiments have certainly not been lacking of interest for our study, this specifically because of the intermediate conditions they were presenting, but they have been quite extensively studied already, notably by A. G. McNish (1959) and many other authors like Mason and Vitousek (1959) for the British Tests, and Maeda (1959), Maeda and Ondoh (1960), Matsushita (1959), Lawrie, Gerard and Gill (1959)and Obayashi (1963) for TEAK and ORANGE). Without going into details already published (cf. McNish 1959), Mr. McNish has made some new remarks which give strong support to the thesis according to which each time a small "bay-like" or "crochet-like" disturbance appears on the (normal) magnetograms as an effect of a nuclear detonation this perturbation can be explained as being a consequence of the strong reinforcement of ionization produced by the detonation. This means that this effect can be considered as an "artificial solar-flare-effect" originated in the portion of the ionosphere more or less directly "illuminated" by the detonation. As a result the direction of this effect will be governed by the day - or night - time conditions existing in the regions surrounding the detonation. For example, in the case of STARFISH, it was nighttime over Johnston Island when the experiment took place (around 23 h. local time) and this would explain why the small bay-like perturbation visible on normal magnetograms appeared in its geometrical characteristics as a "reverse solar-flare effect."

Of course (this last remark by Selzer), if this explanation does not meet any special difficulty in the understanding of the phenomenon as it appears inside the nighttime zone (and this was the only case to be considered for TEAK and ORANGE, McNish, 1959), it takes a little more complex thinking to follow it for the sunlit part of the earth. This is the case, because we do not know exactly what kind of coupling exists in this respect between the night and day zones. At this point of our investigations, we can try to draw another comparison between a natural solar-flare-effect and this kind of artificial one: for the natural phenomenon there is practically no coupling between the sunlit part of the ionosphere and the obscure part (one knows that no solar-flare effect has ever been observed with any degree of certainty way inside the dark side of the earth's surface); the reason for a coupling in the case of the artificial flare produced by a nuclear explosion might reside in the most intense bruskness of the illumination imposed this way to a portion of the ionosphere, which can determine the formation of a shock front in the transmission and the propagation of the ionization process. (Other kinds of shock front entirely different in their mechanism from the one just suggested can also be caused by a nuclear detonation as it has been proposed by various authors.)

In any case, this model of an artificial reverse solar-flare-effect presented by Mr. McNish in order to explain the small bay-like disturbance, is markedly different from the one suggested by Pisharoty (1962) already reported, but it may be that these two modes, which do not exclude each other, intervene both.

Note concerning the release of a report by H. A. Bomke, I. A. Balton, H. H. Grote, and A. K. Harris, from the U. S. Army Electronic Research and Development Laboratory, Fort Monmouth, New Jersey, on: "Near and Distant Observations of the 1962 Johnston Island High-Altitude Nuclear Tests".

This report discloses various precise experimental findings which confirm fundamental assumptions we had adopted already as a basic start for our discussion.

For the 9 July shot (STARFISH), they have noted on records obtained at Hawaii and at Samoa (situated respectively at about 1500 km east and 3400 km south from the explosion's site), "an extremely sudden onset at zero time and also a large sudden change occurring at about two seconds post-shot."

At other stations situated along the East Coast of the United States, the onset at zero time could not be found on the records, but the large sudden change at two seconds post-shot was still present. (At 1.9 seconds post-shot, more precisely.) So there is no doubt that the two-second time delay was independent of the distance separating the observing station from the detonation's site and this confirms our own finding that it was world-like.

In what concerns the sudden onset at zero time and its absence at certain stations, we can recall our former statement that this discrepancy might be caused either by an anisotropic propagation of this part of the signal, or more simply by some inadaptation of the registering equipment to detect the very rapid fluctuations which are the main constituents of this sudden onset. We do not believe - and that is our only small divergence with Bomke et al - that it was caused by a more rapid attenuation of this part of the signal with increasing distances, compared with the behavior of the second onset. (This, because of our own registrations in France and in Kerguelen of this first onset.)

In any case, the general interpretation given by Bomke et al of all the observed facts takes for granted the world existence of two impulses, one at the zero time and another one about two seconds later. They suggest that these two impulses have been generated both by some kind of electromagnetic process initiated around the explosion site and that they have been traveling as electromagnetic waves (and-about-the-speed-of light) in the earth-ionosphere resonant cavity.

Here are the exact mechanisms supposed to be responsible for these impulses: the first one would have been created by the sudden "illumination," by prompt gamma rays, of the local ionosphere situated just under the detonation; the second one would have been preceded by a kind of longitudinal hydromagnetic wave issued from the detonation and directed downward from the explosion into the same local low ionosphere, producing oscillating electric charges at the lower boundary there. These charges would then act as secondary source for an electromagnetic disturbance propagating in the low non-conductive atmosphere, all over the world, at the speed of light. The two seconds would represent, then, the duration of the crossing of the ionosphere by the longitudinal hydromagnetic process.

132

One can remark that these quite simple mechanisms can fit very well with our own (more general) conceptions as exposed in Roquet et al (1963).

Another very interesting finding of Bomke et al is that the "instantaneous portion of the signal" (in other words, the first impulse contains a broad spectrum from near zero frequency to many kilocycles, and that during the first second after zero time, weak oscillations of the order of the fundamental Schumann's resonant frequency (8 to 10 c/sec) are seen at some stations. This confirms Lokken's findings (cf. Comments by J. L. Lokken, loc. cit.).

REFERENCES

Allcock, G. M., C. K. Branigan, J. C. Mountjoy and R. A. Helliwell: Whistler and other very low frequency phenomena associated with high-altitude nuclear explosion on July 9, 1962, J. Geophys. Res. 68, 12 (1963)

Allen, L., J. L. Beavers, W. A. Whitaker, J. A. Welch and R. B. Walton: Project Jason measurement of trapped electrons from a nuclear device by sounding rockets, J. Geophys. Res. 64, 893 (1959)

Armstrong, R. J. and A. E. B. Wharton: Effects of the high-altitude thermonuclear explosion of July 9, 1962, 09.00 UT, observed at Jamaica, J. Geophys. Res. 68, 1779 (1963)

Ashburn, E. V., J. P. Lee, and R. N. Francis: Observations of the changes in the earth's magnetic field induced by the high-altitude nuclear explosion of July 9, 1962, J. Geophys. Res., 67, 4933 (1962)

Baker, R. C. and W. M. Strome: Magnetic disturbance from a high-altitude nuclear explosion, J. Geophys. Res. 67, 4927 (1962)

Berthold, W. K., A. K. Harris and H. J. Hope: World-wide effects of hydromagnetic waves due to ARGUS, J. Geophys. Res. 65, 2233 (1960)

Bomke, H. A., W. J. Ramm, S. Goldblatt and V. Klemas: Global hydromagnetic wave ducts in the exosphere, Nature, London 185, 299 (1960).

Breiner, S.: Effects of nuclear detonation on the geomagnetic field at Palo Alto, California, J. Geophys. Res. 68, 335 (1963)

Caner, B. and K. Whitham: A geomagnetic observation of a high-altitude nuclear detonation, Canadian J. of Phys., i.e. 40, 12 (1962)

Casaverde, M., A. Giesecke and R. Cohen: Effects of the nuclear explosion over Johnston Island observed in Peru on July 9, 1962, J. Geophys. Ref. 68 2603 (1963)

Christofilos, N. C.: The ARGUS Experiment, J. Geophys. Res. 64, 869 (1959)

Congress of the United States: Joint Commission on Atomic Energy, Technical aspects of detection and inspection controls of a nuclear weapons test ban, Summary and analysis of hearings of April 19-22, 1960 (1960)

Crain, C. M. and H. G. Booker: The effects of nuclear bursts in space on the propagation of high-frequency radio waves between separated earth terminals, J. Geophys. Res. 68, 2159 (1963)

Crain, C. M. and P. Tamarkin: A note on the cause of sudden ionization anomaly in regions remote from high-altitude nuclear bursts, J. Geophys. Res. 66, 35 (1961)

Crook, G. M., E. W. Greenstadt, and G. T. Inouye: Distant electromagnetic observations of the high altitude nuclear detonation of July 9, 1962, J. Geophys. Res. 68, 1781 (1963)

Cullington, A. L.: A man-made or artificial aurora, Nature, Lond. 182, 1365 (1958)

Cummack, C. H. and G. A. M. King: Disturbance in the ionospheric F-region following
        the Johnston Island nuclear explosion, Nature, Lond. 184, BA 32-33 (1959)

Davidson, D.: Nuclear burst effects on long-distance high-frequency circuits,
        J. Geophys. Res. __, 331 (1963)

Delloue, J.: L'éclair magnétique du test nucléaire du 13 Février 1960 à Reggane -
        Comptes Rend. Acad. Sciences, Paris, 250, 2536 (1960)

Dessler, A. J.: Large amplitude hydromagnetic waves above the ionosphere, J.
        Geophys. Res. 63, 507 (1958)

Dieminger, W. and H. Kohl: Effects of nuclear explosions on the ionosphere, Nature,
        Lond. 193, 963 (1962)

Dixon, J. M.: Attenuation of medium frequency sky-wave signals in Australia follow-
        ing the Mid-Pacific high-altitude nuclear explosions in August 1958,
        J. Geophys. Res. 67, 123 (1962)

Eckart, C.: Hydromagnetics of oceans and atmospheres, Pergamom Press, New York
        (1960)

Egeland, A., R. Lindquist, A. Pedersen and W. Riedler: Effects of nuclear explosions
        on very-low-frequency and low-frequency propagation, Nature 198, 1076
        (1963)

Elliot, H. and J. J. Quenby: The Samoan artificial aurora, Nature, Lond. 183,
        810 (1959)

Eschenbrenner, S., L. Ferrieux, R. Godivier, R. Lachaux, H. Larzilliere, A. Lebeau,
        R. Schlich, and E. Selzer: Analyse expérimentale des effets magnétiques et
        telluriques de l'expérience "ARGUS" enregistrés par les stations françaises.
        Annales de Geophys. 16, 264 (1960)

Ferrieu, J. F., and Y. Rocard: Mesure du courant électrique total fourni par une
        explosion nucléaire, Comptes Rend. Acad. Sciences, Paris 253, 2931 (1961)

Fowler, P. H., and C. J. Waddington: An artificial aurora, Nature, Lond. 182, 1728,
        (1958)

Fraser, B. J.: Geomagnetic micropulsations from the high-altitude nuclear explosion
        above Johnston Island, J. Geophys. Res. 67, 4926 (1962)

Gendrin, R. and R. Stefant: Effet de l'explosion thermonucléaire a tres haute alti-
        tude du 9 Juillet 1962 sur la résonance de la cavité terre-ionosphère,
        a) Résultats expérimentaux. Comptes Rend. Acad. Sciences, 255, 2272 (1962)
        b) Interprétation. Comptes Rend. Acad. Sciences, 255, 2493 (1962)

Gregory, J. B.: New Zealand observations of the high-altitude explosion of 9th July
        at Johnston Island, Nature, 196, 508 (1962)

Heacock, R. R. and V. P. Hessler: Telluric current micropulsations bursts,
        J. Geophys. Res. 68, 953 (1961)

Herman, J. R.: AVCO Corporation, Mass. Tech. Report, RAD-TR-62-2 (1961)

Hines, C. O.: Internal atmospheric gravity waves at ionosphere heights, Can. J. Phys. 38, 1441 (1960)

Hoerlin, H.: Los Alamos Sci. Lab., Univ. of California, LAMS-2536 (1959)

Hulqvist, B. et al, Sci. Report Kiruna Geophys. Obs. KGO-611 (1961)

Karzas, W. J. and R. Latters: Electromagnetic radiations from nuclear explosion in space, Phys. Rev. 126, 1919 (1962)

Kellogg, P. J., E. P. Ney, and J. R. Winckler: Geophysical effects associated with high-altitude explosions, Nature, Lond. 183, 358 (1959)

Kimpara, A.: Perturbations ionosphériques à début brusque causées par une explosion atomique, C. R. Acad. Sci. Paris 248, 2117 (1959)

Launay, L. and M. Beccaria: Perturbations magnétiques et telluriques enregistrées à Tamanrasset à la suite des explosions nucléaires à haute-altitude dites "Expérience "ARGUS" - Ann. Géophys. 16, 289 (1960)

Lawrie, J. S., V. B. Gerard and P. G. Gill: Magnetic effects resulting from two high-altitude nuclear explosions, Nature, Lond. 184, BA 34 (1959)

Lichtman, S. W. and E. J. Andersen: Ionospheric effects of nuclear detonations in the atmosphere, Communication presented at the International Conference on the Ionosphere, London (July 2-6, 1962)

Lippman, B. A.: Bomb excited "whistlers," Proc. Inst. Radio Engrs. 48, 1778 (1960)

McNish, A. G.: Geomagnetic effects of high-altitude nuclear explosions, J. Geophys. Res. 64, 2253 (1959)

Maeda, H.: Geomagnetic disturbances due to nuclear explosions, J. Geophys. Res. 64, 863 (1959)

Maeda, H. and T. Ondoh: Evidence of quasi-perpendicular propagation of hydromagnetic waves caused by nuclear explosions over Johnston Island, Nature, Lond. 188, 1018 (1960)

Malville, J. M.: Artificial auroras resulting from the 1958 Johnston Island nuclear explosions, J. Geophys. Res. 64, 2267 (1959)

Martyn: Cellular atmospheric waves in the ionosphere and troposphere, Proc. Roy. Soc. Series A, 201 216 (1950)

Mason, R. G. and M. J. Vitousek: Some geomagnetic phenomena associated with nuclear explosions, Nature, Lond. 184, BA 34 (1959)

Matsushita, S.: On artificial geomagnetic and ionospheric storms associated with high-altitude explosion, J. Geophys. Res. 64, 1149 (1959a)

Geomagnetic and ionospheric phenomena associated with nuclear explosions, Nature, Lond. 184, BA 33 (1959b)

On the simultaneity of geomagnetic sudden commencements and sudden impulses, 43rd Annual Meeting American Geophys. Union, Washington, D. C. (April 25-28, 1962)

Morlet, B.: Phénomènes Géophysiques provoqués par l'explosion nucléaire en haute altitude du 9 Juillet 1962, Publica. GRI/NT/2 - Issy-les-Moulineaux (1962)

Newman, P.: Optical, electromagnetic and satellite observations of high-altitude nuclear detonations, Part 1, J. Geophys. Res. 64, 923 (1959)

Nishida, A. and J. A. Jacobs: World wide changes in the geomagnetic field, J. Geophys. Res. 67, No. 2, 525 (1962)

Obayashi, T.: Upper atmospheric disturbances due to high altitude nuclear explosions, Planetary and Space Science, 10, 47 (1963)

Obayashi, T., S. C. Coroniti, and E. T. Pierce: Geophysical effects of high altitude nuclear explosions, Nature, Lond. 183, 1476 (1959)

Odencrantz, F. K.: Electromagnetic effects from high altitude nuclear explosions, J. Geophys. Res. 68, 2057 (1963)

Peterson, A. M.: Optical, electromagnetic and satellite observations of high altitude nuclear detonations, Part II, J. Geophys. Res. 64, 933 (1959)

Pisharoty, P. R.: Geomagnetic disturbances associated with the nuclear explosion of July 9, Nature, Lond. 196, 822 (1962)

Pomeroy, P. and J. Oliver: Seismic waves from high altitude nuclear detonations, J. Geophys. Res. 65, 3445 (1960)

Porter, R. W. (chairman), Symposium on Scientific Effects of Artificially Introduced Radiations at high-altitude, J. Geophys. Res. 64, 865 (1959)

Proces-Verbaux de la Conférence des Experts, Geneve, (1958)

Rawer, K, and K. Suchy: Whistlers excited by sound waves, Proc. Inst. Radio Engrs., 49, 968 (1961)

Riedler, W., A. Egeland, R. Lindqvist, and A. Pedersen: Effects of nuclear explosions on very low frequency propagations, Nature, Lond. 198, 1070 (1963)

Rocard, Y., and J. L. Steinberg: Bruit électromagnétique des nuages orageux, Comptes Rend., Acad. Sciences, Paris, 228, 1960 (1949)

Roquet, J., R. Schlich and E. Selzer: Perturbations transitoires mondiales du champ magnétique terrestre observées en France lors de l'explosion nucléaire spatiale du 9 juillet 1962, C. R. Acad. Sciences, Paris 255, 549 (1962a)

Réception quasi-simultanée, C. R. Acad. Sciences, Paris, 255, 1226 (1962b)

Evidence of two distinct synchronous world impetus for the magnetic effects of the nuclear high altitude detonation of July 9, 1962, J. Geophys. Res. 68, No. 12 (June 15, 1963)

Rose, G., J. Oksman, and E. Kataja: Round the world sound waves produced by the nuclear explosion on October 30, 1961, and their effect on the ionosphere at Sodankyla, Nature, Lond. 192, 1173 (1961)

Rothwell, P., and C. E. McIlwain: Magnetic storms and the Van Allen Radiation Belts-Observations from Satellite 1958[e] (Explorer IV), J. Geophys. Res. 65, 799 (1960)

Rothwell, P., J. H. Wager, and J. Sayers: Effects of the Johnston Island high altitude nuclear explosion on the ionization density in the topside ionosphere, J. Geophys. Res. 68, 947 (1963)

Samson, C. A.: Effects of high altitude nuclear explosion on radio noise, J. Res. Nat. Bur. of Stand., Radio Propagation, 64D, 37 (1960)

Sandford, B. P.: Arctic Inst. of North America, Res. Paper No. 18 (1961)

Selzer, E.: Enregistrements simultanes en France, à l'équateur et dans l'Antarctique, des effets magnétiques engendrés par l'"Expérience Argus", Comptes Rend. Acad. Sciences, Paris 294, No. 13 (1959)

Steiger, W. R., and S. Matsushita: Photographs of the high altitude nuclear explosion "Teak", J. Geophys. Res. 65, 545 (1960)

Störmer, C.: The Polar Aurora, Clarendon Press, Oxford (1955)

Thomas, L., and R. E. Taylor: Sporadic-E phenomena associated with the high altitude nuclear explosions over Johnston Island, J. Atmos. Terr. Phys. 21, 205 (1961)

Troitskaya, V. A.: Izv. Acad. Nauk, Ser. Geophys. No. 9 (1960)

Pulsation of the earth's electromagnetic field with periods of 1 to 15 seconds and their connection with phenomena in the high atmosphere, J. Geophys. Res. 66, 5 (1961)

Unterberger, R. R., and P. E. Byerly: Magnetic effects of a high altitude nuclear explosion, J. Geophys. Res. 67, 4929 (1962)

Utlaut, W. F.: U. S. Nat. Bur. of Stand. Report No. 6050 (1959)

Uyeda, H., H. Maeda, A. Kimpara, T. Obayashi, S. Ishikawa, and Y. Kawabata: Geophysical effects associated with the high altitude nuclear explosion, J. Geomag. Geoele. 11, 39 (1959)

Van Allen, J. A., C. E. McIlwain, and G. H. Ludwig: Satellite observations of electrons artificially injected into the geomagnetic field, J. Geophys. Res. 64, 877 (1959)

Van Wijk, A. M.: Magnetic effects of high altitude nuclear detonations, J. Geophys. Res. 66, 647 (1960)

Magnetic effects of high altitude bomb, J. Geophys. Res. 67, 5352 (1962)

Willard, H. R., and J. F. Kenney: Ionospheric effects of high altitude nuclear tests, J. Geophys. Res. 68, 2053 (1963)

Wilson, C. R., and M. Sugiura: Hydromagnetic waves generated by the July 9, 1962, nuclear weapons test observed at College, J. Geophys. Res. 68, 3149 (1963)

Zmuda, A. J., B. W. Shaw, and C. R. Haave: Very low frequency disturbances caused by the nuclear detonation of October 26, 1962, J. Geophys. Res. 68, 4105 (1963)

Very low frequency disturbances and the high altitude nuclear explosion of July 9, 1962, J. Geophys. Res. 68, 745 (1963)

TABLEAU I*
EXPLOSIONS NUCLÉAIRES SOVIÉTIQUES
Années 1961 et 1962

Date 1961	Heure T.U. h. mn. s.	Puissance	Altitude	Lieu**
24.06	−	−	Explosion camouflée	−
29.06	−	35 à 40 kT	Souterraine	T.
12.07	−	Plus. MT.	H.	N.Z.
01.09	−	75 kT	Atmosphère	S.P.
04.09	−	25 kT	Atmosphère	S.P.
05.09	−	35 kT	Atmosphère	S.P.
06.09	−	25 kT	Hte Atmosphère	K.I.
06.09	−	10 kT	Hte Atmosphère	S.P.
10.09	09.00.10	4,7 MT	Atmosphère	N.Z.
10.09	−	10 kT	Atmosphère	N.Z.
10.09	−	20 kT	Atmosphère	N.Z.
12.09	10.08.20	2 MT	Atmosphère	N.Z.
13.09	−	15 kT	Atmosphère	N.Z.
13.09	−	75 kT	Atmosphère	S.P.
14.09	09.56.30	2 MT	Très Hte Altitude	N.Z.
16.09	09.08.15	1,6 MT	Atmosphère	N.Z.
17.09	−	35 kT	Atmosphère	S.P.
18.09	−	2,3 MT	Atmosphère	N.Z.
19.09	−	15 kT	−	S.P.
20.09	−	1,5 MT	Atmosphère	N.Z.
21.09	−	10 kT	−	S.P.
22.09	−	750 kT	Hte Altitude	N.Z.
02.10	−	580 kT	Atmosphère	N.Z.
04.10	−	10 kT	Atmosphère	S.P.
04.10	07.31.00	2.9 MT	Hte Altitude	N.Z.
06.10	07.00.10	4,6 MT	Atmosphère	N.Z.
06.10	−	200 kT	−	K.I.
08.10	−	25 kT	Atmosphère	N.Z.
11.10	−	10 kT	Souterraine	S.P.
12.10	−	35 kT	Atmosphère	S.P.
17.10	−	10 kT	Atmosphère	S.P.
19.10	−	10 kT	Atmosphère	S.P.

+ Please see Note between "TABLEAU" I and "TABLEAU" II.

TABLEAU I (Continued)

Date 1961	Heure T.U. h. mn. s.	Puissance	Altitude	Lieu**
20.10	08.07.10	2,7 MT	Atmosphère	N.Z.
21.10	–	10 kT	65 km	K.I.
23.10	08.31.26	25 MT	Atmosphère	N.Z.
23.10	–	20 kT	Sous-marine	N.Z.
25.10	–	850 kT	Atmosphère	N.Z.
27.10	–	10 kT	65 km	K.I.
27.10	–	15 kT	Atmosphère	N.Z.
30.10	08.33.30	58 MT	4 km	N.Z.
31.10	–	5,1 MT	Atmosphère	N.Z.
31.10	–	1,5 MT	Atmosphère	N.Z.
01.11	–	20 kT	Atmosphère	S.P.
02.11	–	60 kT	Atmosphère	N.Z.
02.11	–	300 kT	Atmosphère	N.Z.
03.11	–	10 kT	Atmosphère	S.P.
04.11	–	Faible	Atmosphère	N.Z.
04.11	–	3,2 MT	Atmosphère	N.Z.
02.02	–	–	Souterraine	S.
05.08	09.08.46	30 à 40 MT	Atmosphère	N.Z.
10.08	–	3 MT	Atmosphère	N.Z.
20.08	09.02.14	12 MT	Atmosphère	N.Z.
22.08	09.00.06	10 MT	Atmosphère	N.Z.
25.08	–	11 MT	Atmosphère	N.Z.
25.08	–	–	–	S.P.
27.08	09.00.51	15 MT	Atmosphère	N.Z.
02.09	–	20 kT à 1 MT	–	N.Z.
08.09	10.18.06	1 MT	–	N.Z.
15.09	08.02.13	15 MT	–	N.Z.
16.09	10.59.11	23 MT	–	N.Z.
18.09	08.29.10	9 MT	–	N.Z.
19.09	11.01.01	20 MT	–	N.Z.
21.09	08.01.13	8-10 MT	–	N.Z.
25.09	13.02.40	30 MT	–	N.Z.
27.09	–	30 MT	–	N.Z.
07.10	–	10 kT à 1 MT	–	N.Z.
14.10	–	–	–	S.P.
22.10	03.40.--	100 kT	–	A.C.

TABLEAU I (Continued)

Date 1961	Heure T.U. h. mn. s.	Puissance	Altitude	Lieu**
22.10	-	Plus. MT	Atmosphère	N.Z.
27.10	-	-	Atmosphère	N.Z.
30.10	-	-	-	N.Z.
01.11	-	1 MT	Atmosphère	N.Z.
01.11	-	-	Hte Altitude	A.C.
03.11	-	4 MT	-	N.Z.
03.11	-	-	-	N.Z.
04.11	-	-	-	S.P.

**Legend

```
    T.......TIENSHAN

  N.Z.......NEW-ZEMBLE

  S.P.......SEMIPALATINSK

  K.I.......KAPOUSTIN  IAR

  S.........SIBERIA

  A.C.......CENTRAL ASIA
```

The times are ones communicated by the International Central Bureau of Seismology except for the 22.10.62 event given by this I.P.G. - Paris.

+ "TABLEAU" I and "TABLEAU" II are FACSIMILE of PUBLICATION GRI/NT/2 of the "GROUPE DE RECHERCHES IONOSPHERIQUES--CNET, CNRS, IPG. (Issy-les-Moulineaux, 13 Décembre 1962).

TABLEAU II*
EXPLOSIONS NUCLÉAIRES AMÉRICAINES EN 1962
(Liste provisoire)

Date	Heure T.U. h. mn.	Puissance	Altitude	Lieu
11.05	–	–	–	–
27.05	–	–	–	–
08.06	17.00	20 kT à 1 MT	Atmosphérique	Christmas
09.06	15.30	"	"	"
10.06	16.00	1 à 5 MT	"	"
12.06	15.30	20 kT à 1 MT	"	"
15.06	16.00	"	"	"
17.06	16.00	100 kT	"	"
19.06	15.00	Moyenne	"	"
22.06	16.00	"	"	"
27.06	15.30	Environ 1 MT	"	"
30.06	15.30	"	"	"
06.07	17.00	100 kT	Souterraine (-300m)	Nevada
07.07	–	1 kT	Sol + 2 m	"
09.07	09.00	1,4 MT	400 km	Johnston
10.07	16.30	Moyenne	Atmosphérique	Christmas
11.07	11.30	1 à 5 MT	"	"
20.10	08.30	20 kT	32 à 48 km	Johnston
01.11	12.10	1 MT	10 km	"
04.11	07.40	"	32 à 48 km	"

+ Please see Note between "TABLEAU" I and "TABLEAU" II.

OBSERVATIONS OF ENHANCED RADIATION
IN THE kc/s-BAND (5 AND 8 kc/s) DURING AURORAL
AND GEOMAGNETIC DISTURBANCES

L. Harang and R. Larsen

INTRODUCTION

A number of observers have reported the existence of a naturally occurring electromagnetic radiation in the audio-frequency range. It is now apparent that at least in regions near the auroral zones we may separate this electromagnetic radiation into two components: 1) a certain level of "hiss" which may vary from day to day but appears to be a phenomenon of fairly constant character varying with the time of the year and the place of observation, and 2) sudden bursts of "enhanced radiation" which may occur during short time intervals from several minutes up to hours.

In the following, we will report on this "enhanced radiation" based on observations made in the vicinity of the Auroral Observatory, Tromso (70°N), lying close to the auroral zone.

The observations of Ellis (1957, 1959a, 1959b, 1959c, 1960a, 1960b and 1961) in Australia convincingly showed that periods of enhanced radiation of 4.5 kc/s were connected with geomagnetic disturbances and in some cases also with variation in the intensity of the red line 6300 Å from the night sky and aurorae.

Concerning observations inside the auroral zone, Martin, Helliwell and Marks (1960) have reported on the association between aurorae and VLF-hiss observed at the Byrd Station, Antarctica.

Morozumi (1962) working at the South Pole in the frequency range 150 c/s to 30 kc/s demonstrated in a number of cases the simultaneous appearances of enhanced radiation and aurorae.

Helms (1963) working at the Byrd Station inside the southern auroral zone has given a detailed analysis of the connection between hiss in the frequency range 1 to 20 kc/s and the appearance of aurora.

Stockflet Jorgensen and E. Ungstrup (1960) observing at Godhavn, Greenland (lying inside the auroral zone) also have stated examples of increase in hiss on 8 kc/s during appearance of aurorae.

The observations in Tromso indicate a similar connection. The close connection between the start of enhanced radiation and the appearance of quiet auroral arcs low on the northern horizon is especially apparent. We also state in a number of cases that brilliant aurorae in the zenith and on the southern sky late in the night are not or only slightly accompanied by appearance of enhanced radiation. In the analysis of the Tromso recordings, the appearance of enhanced radiation will be compared with aurorae and geomagnetic disturbances and also simultaneous riometer recordings.

The work was started in the autumn of 1961. It soon turned out that the appearance of bursts of enhanced radiation was especially strong during the winter months. We have, therefore, restricted our discussion of the recordings to two periods: December to February 1961-1962 and December to February 1962-1963. During these periods, the appearance of bursts was almost a daily occurrence. In summer, especially during the two months of the midnight sun, the hiss level was dominated by continuous noise due to atmospherics, and only a few cases of clear appearance of geomagnetically determined enhanced radiations could be identified.

INSTRUMENTAL EQUIPMENT

Narrow-band amplifiers centered at the frequencies 4.8 and 8 kc/s with bandwidths from 1 to 1.7 kc/s were used. The detector had a double time constant with a slow build-up and a fast decay time in order to discriminate between impulse signals. Single coil frames of 100 m^2 were connected to a preamplifier placed close to the aerial and from this shielded cables went to the main amplifier*. Two sites for observations were used: 1) the Auroral Observatory, Tromso, where the local noise level was fairly high, and 2) at Lavangsdal, about 30 km from the observatory, where the local noise level was very low. Two frames were in use on each frequency with their planes in E-W and N-S directions and the frames were connected alternatively to the amplifier over periods of five minutes. One thus gets a picture of a possible preferred direction in azimuth of the incoming radiation. An auxiliary aerial was used for calibration, and assuming the radiation to be of noise character within the bandwidth of the amplifier, the energy can be given in watts/m^2 c (assuming horizontal propagation with the E-plane vertical).

RECORDINGS

In the following examples of the connection between enhanced radiation and recordings at the Auroral Observatory, Tromso of 1) geomagnetic perturbations, 2) riometer records on 28 Mc/s and 3) appearance of aurorae recorded on an all-sky camera will be discussed. The discussion will be restricted to two periods of observation at midwinter time, December to February 1961-1962 and 1962-1963, during which the sun is below the horizon (Tromso: $\varphi = 70°N$) and the diurnal variation of the normal ionospheric layers is especially simple.

It soon turned out that connection between burst of enhanced radiation and geomagnetic disturbances depended highly on the type of the geomagnetic disturbance, and in the following, examples of this effect will be discussed.

VLF Bursts and Small Geomagnetic Disturbances.

In Fig. 1 is shown a record of VLF bursts on 8 kc/s during a small geomagnetic storm. The H-record shows a small positive deflection during the first phase of the disturbance, which at 21.30 MET reverses to a negative value. This represents a normal development of a disturbance which on the record is indicated by the broken

---

* We wish to express our sincere thanks to Professor Ellis for valuable discussions concerning construction of the amplifier.

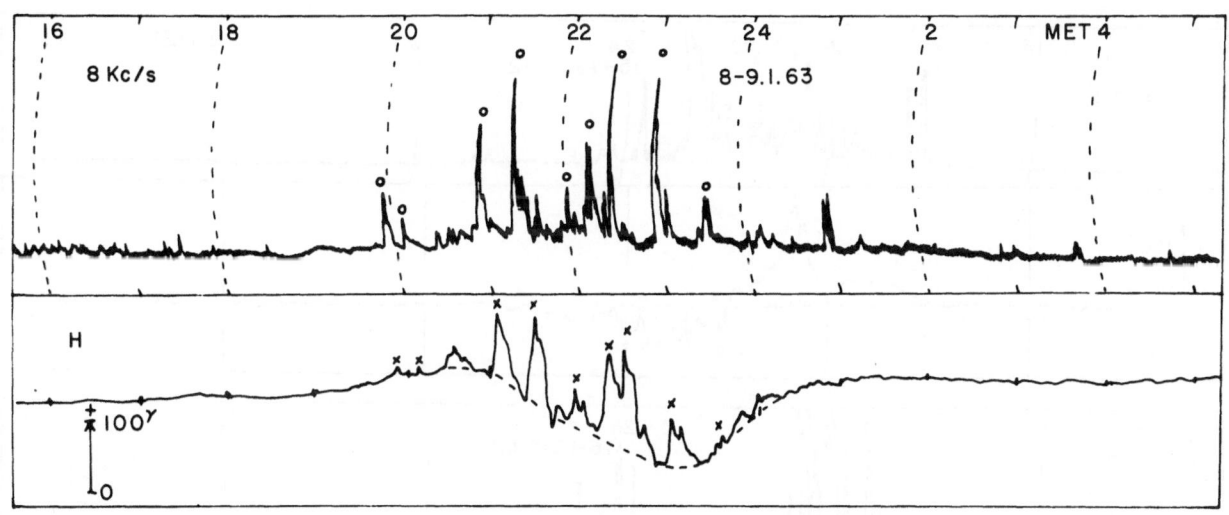

Fig. 1.   VLF bursts on 8 kc/s during a <u>small</u>
          geomagnetic disturbance (only the H-
          component is indicated).

line. On this main development there are, however, a series of secondary positive
maxima, and it is apparent that there is a close connection in time between each of
these peaks and bursts of VLF radiation. This close connection often appears during
<u>small</u> disturbances.

VLF Radiation With Increasing Magnitude of Geomagnetic Disturbance.

     In Fig. 2 are shown records from three days representing increasing magnitudes of
disturbances. It is apparent that in the beginning and during the positive phase of
the geomagnetic disturbance the VLF bursts are often strong. With increasing strength
of the geomagnetic disturbance the VLF bursts often decrease in intensity and often
disappear during the strong negative phase of the storm.

VLF Radiation During Great Geomagnetic Disturbances.

     In the following, a series of examples will be demonstrated during which a strong
<u>negative</u> deflection in <u>H</u> starts as a sudden commencement (SC). The records are shown
in Fig. 3. The shape of the <u>H</u>-record indicates the following character of the geomagnetic disturbance: During the positive phase in <u>H</u> there will flow a current (of
electrons) toward <u>W</u> along the auroral zone. At the moment indicated as SC a negative

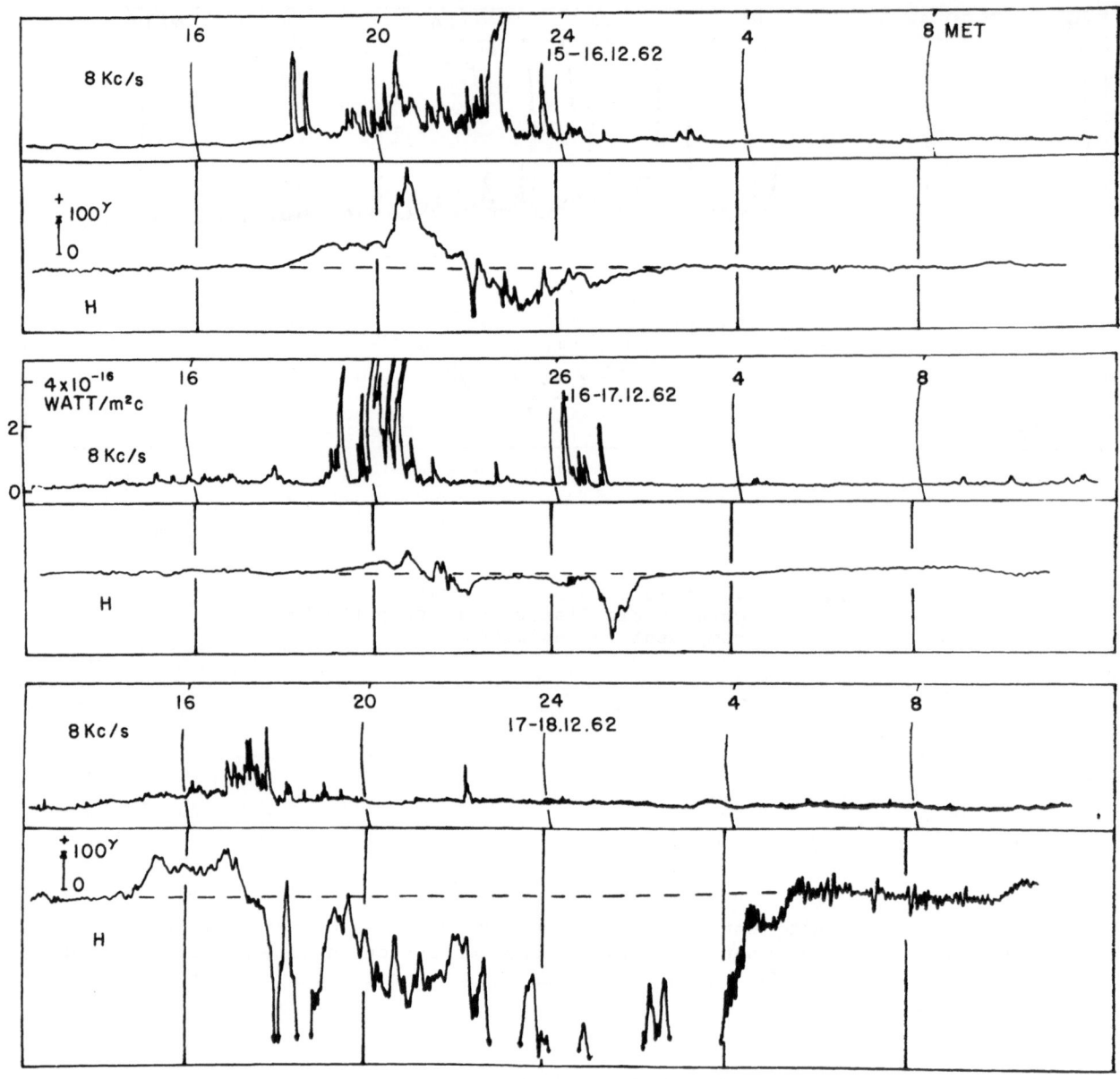

Fig. 2. VLF radiation on three days with increasing
magnitude of geomagnetic disturbance.

disturbance appears. The current reverses, and at the same moment the auroral zone is
displaced laterally toward S. This reversal of current and displacement toward S is
accompanied by VLF radiation. We must assume that the change in character of the current system is due to creation of electric fields in the outer space, which apparently
is accompanied by bursts of VLF radiation.

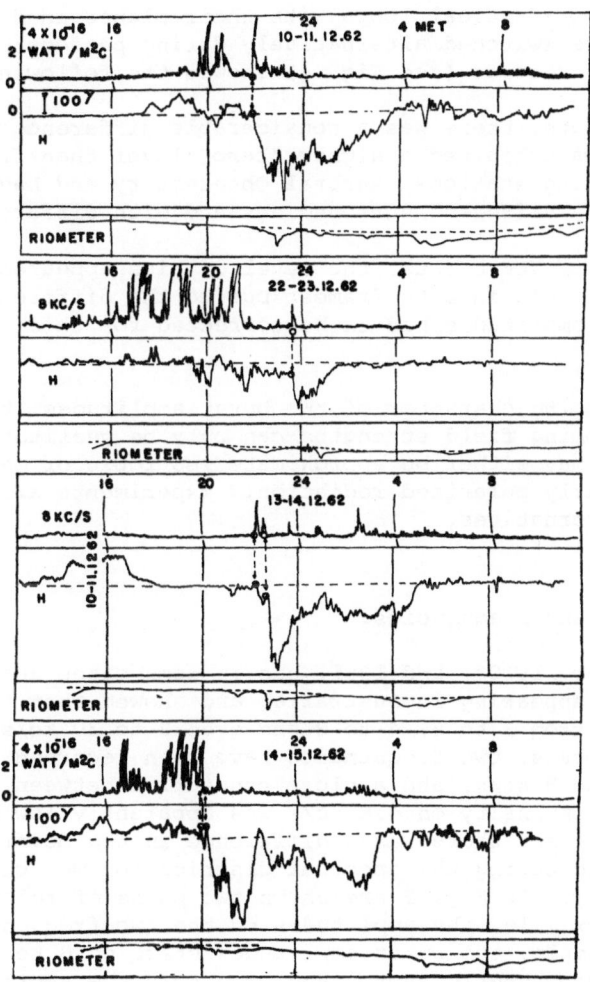

Fig. 3.   VLF radiation during great geomagnetic disturbances. At the
moment when the positive deflection in H turns negative (as a
SC), VLF bursts appear.

The Direction of Arrival (In Azimuth) of the VLF Radiation.

The appearance of VLF bursts during geomagnetic disturbances indicates that the
appearance of VLF radiation is connected with the auroral zone. Ellis (1959a, 1959c,
1960a, and 1961) has made directional observations in Australia of VLF bursts on
5 kc/s using a goniometer arrangement. He concludes that in a number of cases the
bursts came from apparent southern sources, and in some cases he considers discrete
noise sources to appear.

147

In order to look for directional effects of the bursts recorded at Tromso, recordings were made using two identical loops with their planes in N-S and E-W directions, the inputs of which were switched alternatively during periods of five minutes to the amplifier channel. The records (see Fig. 4) showed the following effects:

1) during quiet hours, there was a considerable difference in the level of hiss, the E-W reception always exhibited a higher "zero" level than from N-S. This effect appeared at both recording stations, Auroral Observatory and Lavangsdal, lying 30 km apart, and in completely different surroundings and with different local noise levels.

2) during periods of VLF bursts, the level of hiss appeared with amplitudes of the same order of magnitudes in both frames. Due to the different levels of background "zero" hiss, the burst amplitudes had to be corrected for this effect, as indicated in Fig. 4.

Due to the rapidly changing character of the burst amplitudes, this alternative recording of directional incoming field strengths can only be qualitative, but it indicates the incoming radiation may either be approximate isotropic or may be coming in vertically as a circularly polarized radiation. Experiments are being planned for investigating these alternatives.

VLF Radiation on Different Frequencies.

Ellis (1959a, 1959c, 1960c, and 1961) has in some cases analyzed the frequency spectrum of the bursts appearing in Australia, and showed that they covered various ranges of frequencies, from 3 to 5, 4 to 8 and even 2 to 30 kc/s for great bursts. Mostly for control purposes, two frequencies have been recorded simultaneously over several periods, 4.8 and 8 kc/s, and a close connection between the amplitude variations was stated. The intensity on 4.8 kc/s was apparently lower than on the higher frequency, and in some cases there was a difference in the development of amplitude variation with time, indicating the spectral distribution was changing during the period of burst activity. In Fig. 5 are shown two pairs of recordings on 4.8 and 8 kc/s. In the first example, the amplitudes on the two frequencies go closely parallel. In the second example, the burst amplitudes start on 8 kc/s during the slight positive phase of the geomagnetic storm, and on 5 kc/s the amplitudes start one to two hours later.

VLF Radiation and Riometer Recordings.

Simultaneous riometer recordings of the cosmic radiation on 28 Mc/s were available at the Auroral Observatory, and in some cases also from Spitzbergen (Longyear-byen). The riometer records are the most sensible indication of intrusion of charged particles in the region overhead, and a riometer deflection indicates the presence of strong absorption in the lower part of the underline{E}-layer due to the intrusion of these particles. From previous studies of records from a net of riometers, it is concluded that the areas of absorption appearing during a riometer record deflection are of the dimensions of 200 to 300 km in the region of the auroral zone.

It would be of special interest to investigate if the sudden dips in the riometer records, indicating sudden intrusion of charged particles overhead, were accompanied by VLF bursts. Fig. 6 shows two examples of isolated riometer dips and VLF bursts. The two records illustrate an effect often stated. An isolated sudden riometer dip may not be accompanied by a simultaneously appearing VLF burst, but the bursts may (as in this case) appear before the riometer dip. During a bay-like dip, we may observe a simultaneous VLF radiation. On Fig. 3 the riometer records are drawn up during the days of great geomagnetic disturbances. During the positive phase of the geomagnetic disturbance with only slight riometer dips, the VLF bursts are strong, whereas during

148

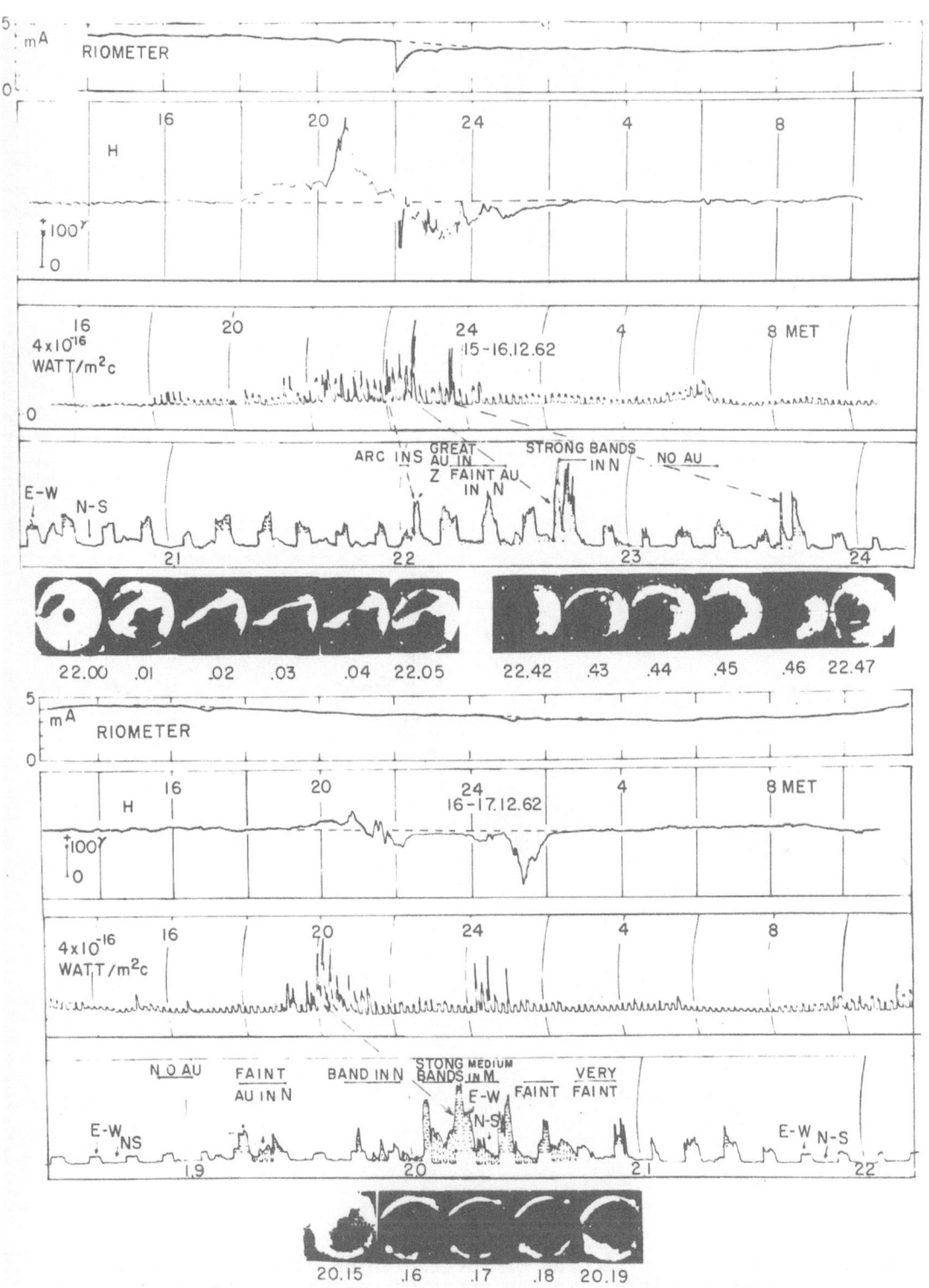

Fig. 4.

Records of directions of arrival of VLF bursts. Two frames with planes lying in N-S and E-W are switched alternatively to the amplifier. The records indicate two effects: 1) during quiet periods the "zero" hiss level is different for E-W and N-S reception. The E-W level is higher; 2) during bursts, amplitudes of about the same order of magnitude appear in both frames. (These amplitudes have to be corrected for different "zero" levels of background hiss in N-S and E-W.)

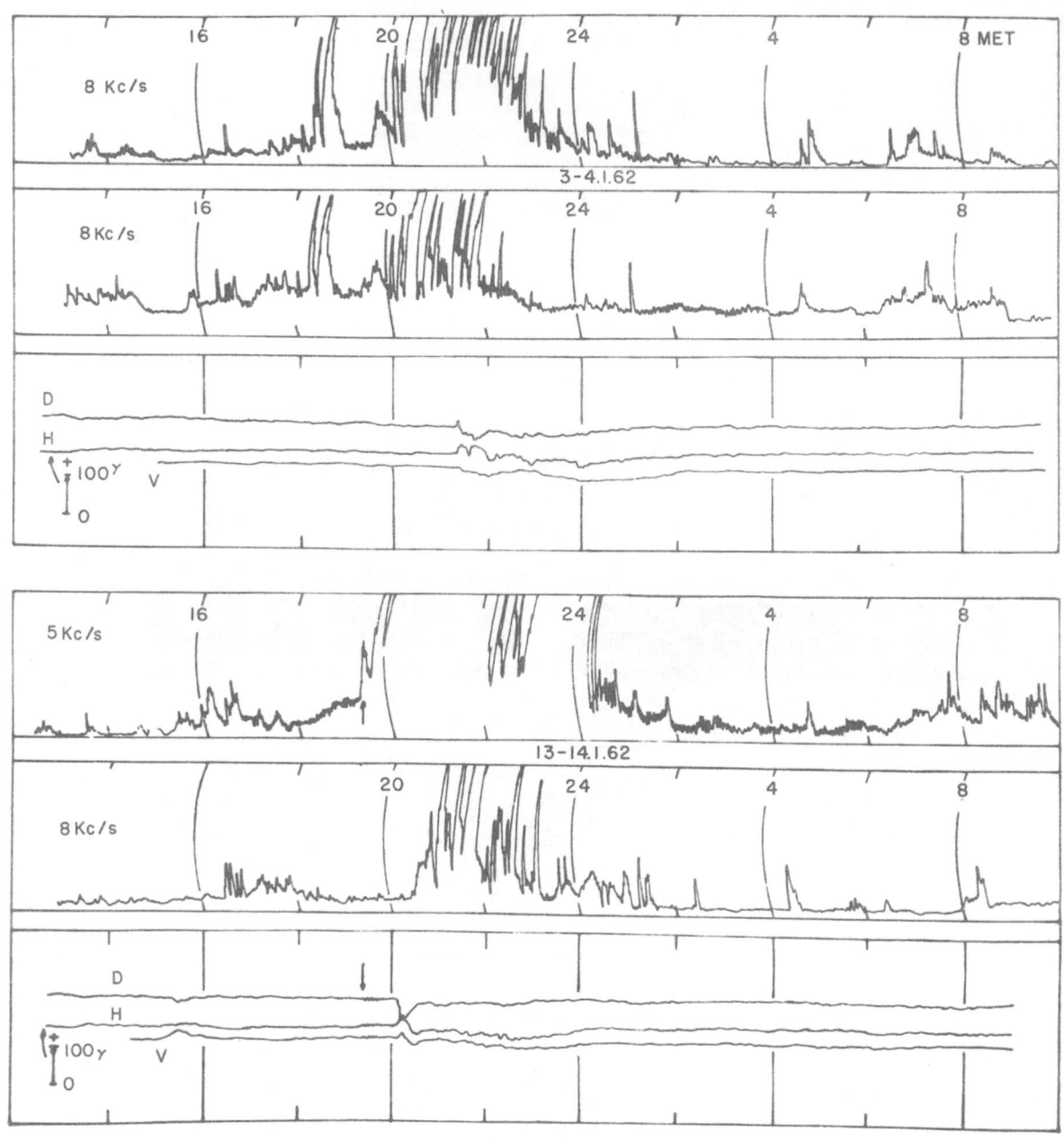

Fig. 5. Simultaneous records of VLF radiation on two frequencies,
8 and 4.8 kc/s. In the upper curves there is a parallel
variation in amplitudes on both frequencies. In the lower
curves the VLF radiation on 4.8 kc/s starts 1 to 2 hours
after the 8 kc/s radiation.

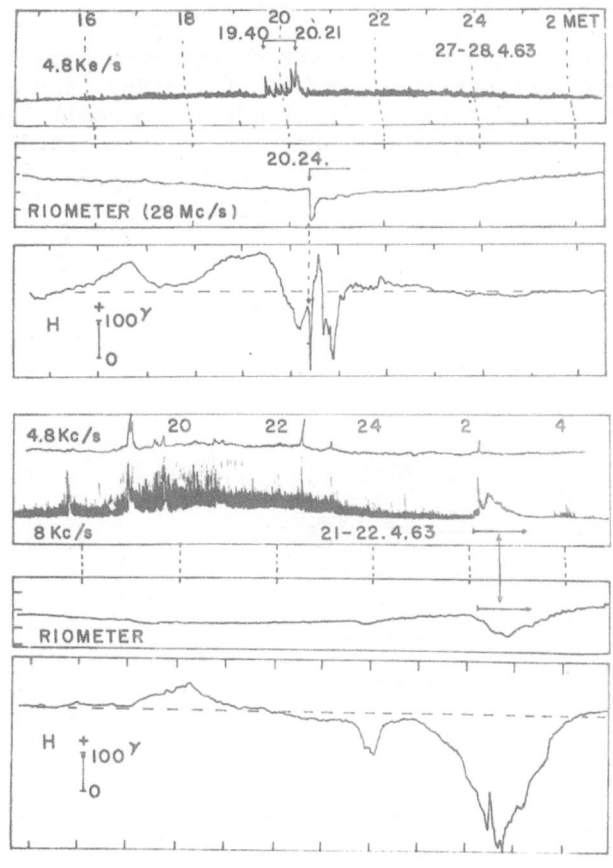

Fig. 6.   VLF bursts during riometer deflections. On the upper record,
a series of VLF bursts appear several minutes before the sudden dip in the riometer record. On the lower record the bay-
shaped riometer dip and VLF burst appear simultaneously.

the intense negative phase of the geomagnetic storm with great riometer dips, no VLF
bursts appear*.

VLF Radiation and Aurorae.

As stated by previous observers, the strongest VLF bursts often appear during
periods of auroral bands and arcs appearing on the low northern horizon. This effect
has also been confirmed in a number of cases. On Fig. 2 the observations from simultaneous all-sky photos have been indicated. It is apparent that the big and luminous
displays appearing overhead have only slight effects, whereas arcs and bands appearing
in the low northern horizon are accompanied by periods of strong bursts. It is also
apparent that in a number of cases the burst activity follows the general development
of auroral luminosity.

---

*In a discussion of this paper at the Symposium, Dr. Gallet drew attention to the fact
that these records of VLF bursts were made on a single, fixed frequency and may be of
limited value for this comparison of simultaneous appearance of bursts and riometer
deflection. Using sweep-frequency display of the bursts, Gallet showed examples of
bursts in which there was a close coincidence between the bursts on a low VLF frequency. It is apparent that for a closer study of the coincidences in time a continuous spectrum analysis of the bursts is necessary.

The Diurnal Variation of the VLF Bursts in Wintertime.

On Fig. 7 the hours of the day when VLF bursts have been recorded have been indicated for the two winter periods 1961-1962 and 1962-1963, and also the mean geomagnetic disturbance deflection in H for the same periods. The complicated appearance of VLF burst compared with the geomagnetic disturbances, as demonstrated in the preceding sections, make such a presentation of statistics of limited value for the analysis of the phenomenon.

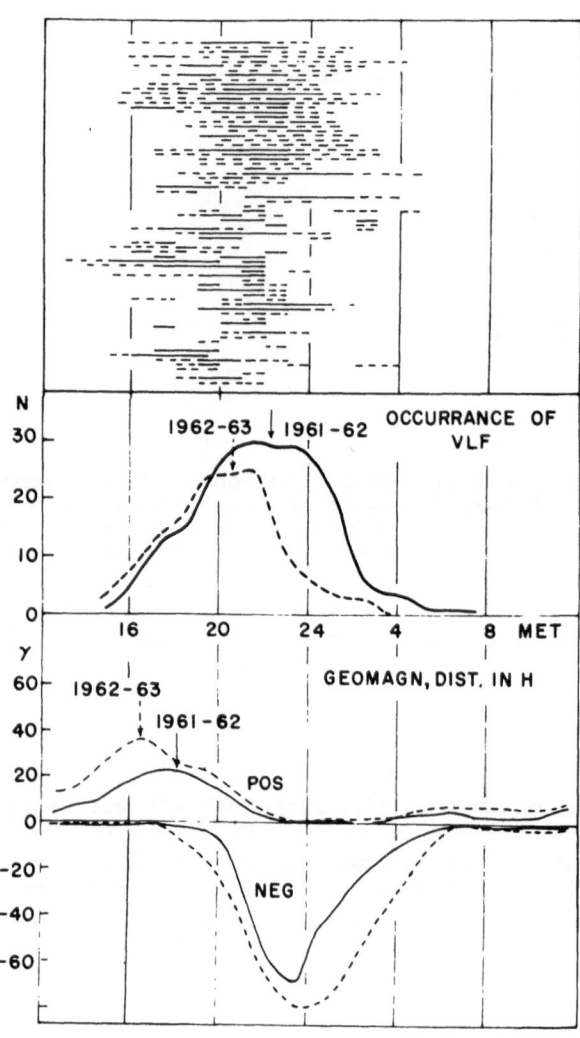

Fig. 7. The diurnal variation of VLF-enhanced radiation on 8 kc/s during the two winter periods 1961-1962 and 1962-1963 at Tromso. The mean geomagnetic disturbance (in H) as posigive or negative deflections are indicated below.

The Diurnal Variation of Continuous Hiss in Summertime.

As previously stated, the appearance of VLF bursts is most frequent during the winter months. During the summer months, especially in June and July - the midnight sun period - the level of continuous hiss increases strongly, and it is difficult to isolate separate bursts which may be assumed to be due to cosmic activities from the level of noise. Fig. 8 shows the diurnal variation of the continuous hiss during periods in May and June 1963. This "background" hiss is obviously a summer phenomenon and a diurnal variation of this type and magnitude is not recorded in wintertime.

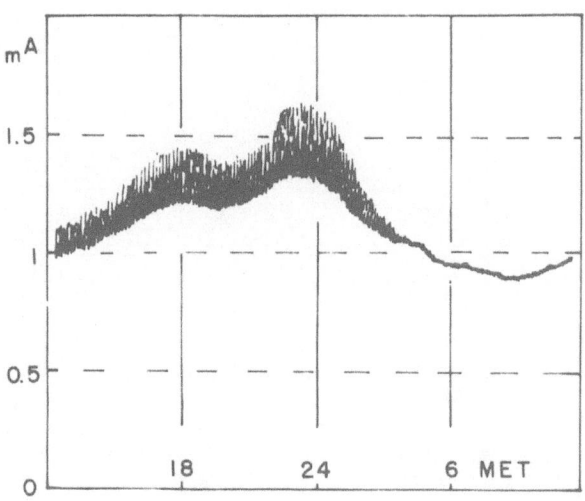

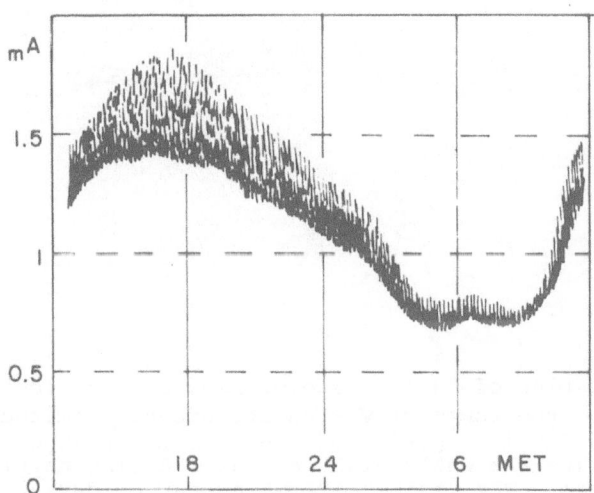

Fig. 8. Top: Continuous "background" hiss on 5 kc/s, mean value 1-7 June 1963, Tromso. Bottom: Continuous "background" hiss on 8 kc/s, mean value 26-29 June 1963, Tromso.

153

Noise Storms.

The diurnal variation of the appearance of VLF bursts demonstrated in Fig. 6 indicates that the hours of greatest intensity appear before midnight and during the hours 19-22 MET, and no burst activity is recorded during the day. On some few days, however, a continuous burst activity has been recorded, covering the whole night and even during some of the hours of the day. In all cases, these "noise storms" have appeared during strong geomagnetic disturbances. Fig. 9 shows three examples of such "storm" days. The character of these "storms" has been different. In some cases, the first two on Fig. 9, there appeared burst amplitudes lasting for several minutes, which decreased and increased. In the last example, it was a continuous increase in a "hiss" level with very rapidly changing amplitudes.

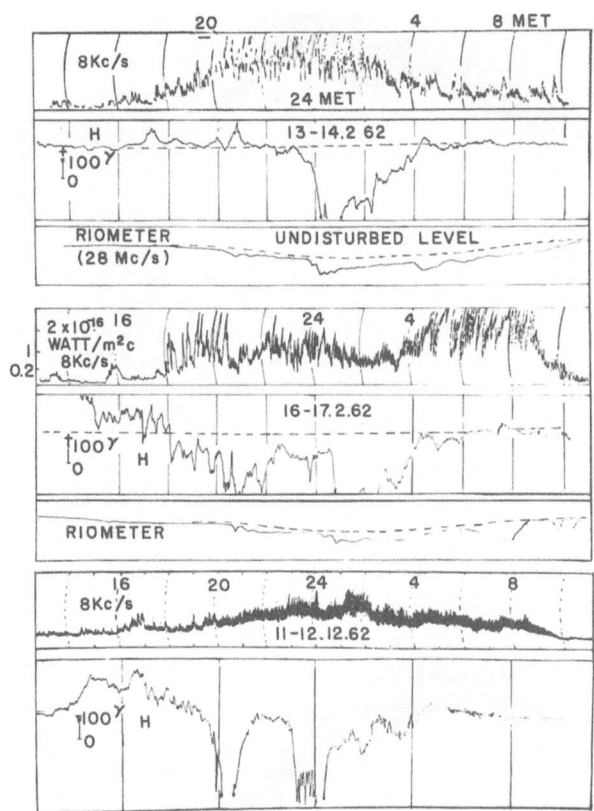

Fig. 9. Examples of "noise" storms recorded at Tromso.
The appearance of VLF bursts are not limited to
the "normal" period of 18-24 MET, but appear
during the whole night and also during daytime
hours. The "noise" storms appear on days with
great geomagnetic disturbance.

CONCLUSION CONCERNING THE ANALYSIS OF OBSERVATIONS

The appearance of enhanced VLF radiation on 8 kc/s (and also on 5 kc/s) has been stated during auroral displays and geomagnetic storms. The flux of the bursts is of the order of 10^{-16} watt/m^{2}c.

The appearance of the VLF bursts during various types of geomagnetic disturbances has been discussed. The VLF bursts have their greatest amplitudes during the positive phase of the geomagnetic storm and often during only slight disturbances. During great, and negative, disturbances, when the auroral zone is displaced laterally toward south and strong absorption overhead appears indicated by the considerable riometer deflections, the VLF-burst activity decreases.

There is often a close parallel in appearances of quiet auroral arcs and bands appearing on the low northern horizon and strong VLF bursts, whereas strong auroral displays overhead may not be accompanied by intense VLF bursts.

Simultaneous records on 5 and 8 kc/s show a closely parallel amplitude variation, but the relative magnitude of the amplitudes may change during the period of burst activity, indicating a change in the spectral composition of the bursts.

Reception on E-W and N-S orientated frames shows that the mean "zero" level of hiss during undisturbed conditions is different. It is stronger from E-W than from N-S direction. Upon this directional "zero" level, the VLF bursts arrive with amplitudes of the same order of magnitude in both frames.

The appearance of VLF radiation is normally confined to the night hours. In some cases, during strong geomagnetic disturbances, examples of "noise" storms lasting almost one whole day have been recorded.

DISCUSSION

Sugiura: You showed that 8 kc/s radiation stops at the beginning of large bays commencing suddenly. Could this be interpreted as absorption?

Harang: I assume that this may be due to two effects: to absorption, and to a displacement of the area of entry of the VLF emission, the last is perhaps of greatest importance. The sudden, great negative value of $\underline{H}$ indicates that the whole current system producing the geomagnetic disturbance is displaced toward south.

Sugiura: If so, what are the noise storms that occur during bays?

Harang: I can give no simple explanation of these, not frequently occurring days. The variation during the day indicates that the ionosphere must act as a reflector; there is an indication of a sunrise effect.

Sugiura: You showed close relation between emission and the aurora. Usually when there is a large bay, we usually see auroral activity. Isn't there some inconsistency here?

Harang: No. During a great negative bay, the auroral zone has been displaced toward
           the south and we are observing inside the auroral zone.

Sugiura: Have you estimated the electron density needed for absorption of 8 kc/s
           radiation?

Harang: No.

REFERENCES

Ellis, G. R.:   J. Atmos. Terr. Phys. 10, 302 (1957); Nature 184, 1307 (1959a);
                 Nature 184, 1391 (1959b); Pl. Sp. Sci. 1, 253 (1959c); J. Geophys.
                 Res. 65, 839 (1960a); J. Geophys. Res. 65, 1705 (1960b);
                 J. Geophys. Res. 66, 19 (1961).

Helms, W. J.:   Thesis, Department of Electrical Engineering, Univ. of Washington
                 (USA) (1963).

Jorgensen, T. S., and E. Ungstrup: Report No. 11, Ionosphere Laboratory, Royal
                 Technical University of Denmark, Copenhagen (1960).

Martin, L. A., R. A. Helliwell, and K. R. Marks: Nature 187, 751 (1960).

Moruzumi, H. M.: Thesis, Dept. of Physics and Astronomy, Iowa State University
                 (USA) (1962).

ELF (500-1000 CPS) EMISSIONS AT HIGH LATITUDE

Alv Egeland
Kiruna Geophysical Observatory
Kiruna C., Sweden

ABSTRACT

Naturally, occurring electromagnetic radiation in the audio- and subaudio-frequency range have been studied in recent years at Kiruna Geophysical Observatory (geomagnetic latitude 65.3°N). Many emission bands, which are quite distinct from the normal background signals and are obtained as strong enhancements in portions of the spectrum, have been recorded. The main results observed for the emissions between 500 and 1000 cps are:

1) The frequency of the maximum signal strength of the emission band is almost constant in time; e.g. at 650 cps ± 150 cps.

2) The amplitude of the emission at 650 cps is often ten times higher than the normal background noise level at this frequency range.

3) The average half-intensity bandwidths of these 650 cps emissions are ± 200 cps.

4) A pronounced morning maximum (between 04 and 10 MET) is found for the different seasons.

5) The diurnal variation, as well as the correlation with micropulsations, indicate that the behavior of the 650 cps enhancements are different from all other audio-frequencies investigated at the Kiruna Observatory.

6) It seems likely that the 650 cps radiation is an auroral zone phenomenon.

7) A marked Doppler broadening band shift is found for many of the emission events.

It is considered probable that these emissions consist of electromagnetic radiation generated outside the E-layer by arrival of protons, which is propagated in the extraordinary mode along the lines of force of the earth's magnetic field.

INTRODUCTION

Investigations of naturally-occurring electromagnetic radiation in the audio-frequency range (10 to 10,000 cps) have been carried out at Kiruna Geophysical Observatory (geographic latitude 67.8°N; geomagnetic latitude 65.3°N) since 1958. The first measurements were carried out at Poikkijärvi, about 7 kms east of the Kiruna

Observatory. The results of these recordings are described and discussed by Gustafsson, Egeland and Aarons (1960). In November 1961, new measurements in the same frequency range, but with a somewhat modified, extremely low frequency spectrum-analyzer, were started near Paksuniemi, about 25 kms east of the geophysical observatory at Kiruna, at a place which is almost free from man-made contamination. The nearest 50 cps power lines are more than 7 kms from the observation site. Spectrograms of the electromagnetic energy in the frequency region between 10 cps and 10 kc/s have been monitored continuously from November 14, 1961, to January 29, 1962; March 10 to April 15; June 5 to July 31; September 15 to December 20, 1962; and March 15 to May 15, 1963. Furthermore, the measurements have also been extended during the two last recording periods to include recordings of the background noise in the range 2 to 40 cps.

The ELF data obtained from the measurement at Paksuniemi are discussed in the two reports of Egeland (1962) and Egeland, Olsen and Gustafsson (1963).

The main purpose of this long-time investigation has been to study naturally-occurring broad band noise emissions in the audio-frequency range. Secondly, the diurnal variation has been investigated as a function of time and frequency. In this report only the broad-band noise emissions between 500 and 1000 cps will be discussed.

INSTRUMENTATION

The main part of the data is obtained with the slow frequency spectrum-analyzer. This equipment operates in the frequency range 10 to 10,000 cps, divided into two bands, 10 to 1000 cps, and 500 to 10,000 cps. The most important characteristics of the system are listed in Table 1. [For further details see Egeland (1962).]

TABLE 1

	Frequency Range	Gain
Maximum gain with 1μV input at 20°C	10 to 1,000 cps	120 db
	500 to 10,000 cps	123 db
Time of 1 sweep		$\approx$44 min
Bandwidth	10 to 1,000 cps	2 cps
Bandwidth	500 to 10,000 cps	40 cps

All data presented here have been obtained with a 10-turn horizontal loop with a diameter of 32 m.

BROAD-BAND NOISE EMISSIONS BETWEEN 500 AND 1000 CPS

That the noise emissions in the audio-frequency range may exhibit one, or sometimes two peaks in the spectrum, has been observed earlier in the auroral zone by Egeland (1959, 1962), Aarons, Gustafsson and Egeland (1960), Gustafsson et al (1962) and Egeland, Olsen and Gustafsson (1963). A strong and very marked band emission centered around 650 cps have often been recorded at Paksuniemi. It has been suggested that the peak between 600 and 700 cps, which has been rather stable in frequency, might be due to gyro-radiation from protons of moderate to low velocity. Here it should only be mentioned that the gyro-resonance frequency of protons is about 680 cps at an altitude of 300 km, for the average magnetic field lines through the Kiruna area. Some original examples of such broad band noise recordings are shown in Fig. 1. For comparison, two normal examples of the quiet background noise can be seen in Fig. 2(a) and 2(b).

Occurrence and Duration.

During a 20-day period in March and April 1962, at Kiruna, 22 per cent of all frequency sweeps contained 650 cps radiation with an amplitude well above the quiet level. During a three-hour interval, centered on 0900 L.T., the contribution was as high as 46 per cent for this spring period. For the spring period in 1963 the corresponding figure is nine per cent. For the summer period in 1962, nine per cent of all sweeps showed a marked resonance effect between 600 and 700 cps, while for the autumn period in 1962 about eight per cent of all sweeps contain a marked peak at the same frequency range. In the winter period 1961-1962, only two per cent of all sweeps show a resonance peak about 650 c/s (see Fig. 3). As the sensitivity of the equipment has been almost exactly constant during all ELF measurements at Paksuniemi from 1961 to 1963, the maximum frequency of occurrence during the spring of 1962 must be due to the fact that these emissions were more likely to occur during this spring period.

Most of the events of 650 cps radiation lasted for several sweeps and the longest one remained almost steady at constant intensity for 12 sweeps, which means almost nine hours.

Diurnal Variation.

The diurnal variation of the 650 cps emission for the spring, summer and autumn, 1962, as well as for the spring of 1963 is shown in Fig. 3. This figure clearly demonstrates that the 650 cps-radiation has a diurnal variation which is different from the background noise of the other audio-frequencies (Egeland, Olsen and Gustafsson, 1963). The maximum number of occurrences always appear between 0600 and 1100 local time. The number of 650 cps peaks are probably still more pronounced between 09 and 11 MET than Fig. 3 indicates, due to the fact that calibration and other checks of the equipment are made at that time. Furthermore, there is little seasonal variation of the time of maximum activity.

Diurnal variation of the average intensity of 27.6 Mc/s cosmic noise absorption for the equinoxes at Kiruna is shown in Fig. 4. [Months in which strong PCA's occurred are excluded in Fig. 4.] The marked maximum in the D-region absorption coincide very closely in time with the maximum number of 650 cps emission peaks. So one arrives at the somewhat surprising conclusion that the strong broad-band radiations between 500 and 1000 cps are most probably observed during the time of maximum absorption.

As all existing emission theories assume that ELF and VLF emissions are generated by streams of charged particles somewhere well above D-region heights, these waves

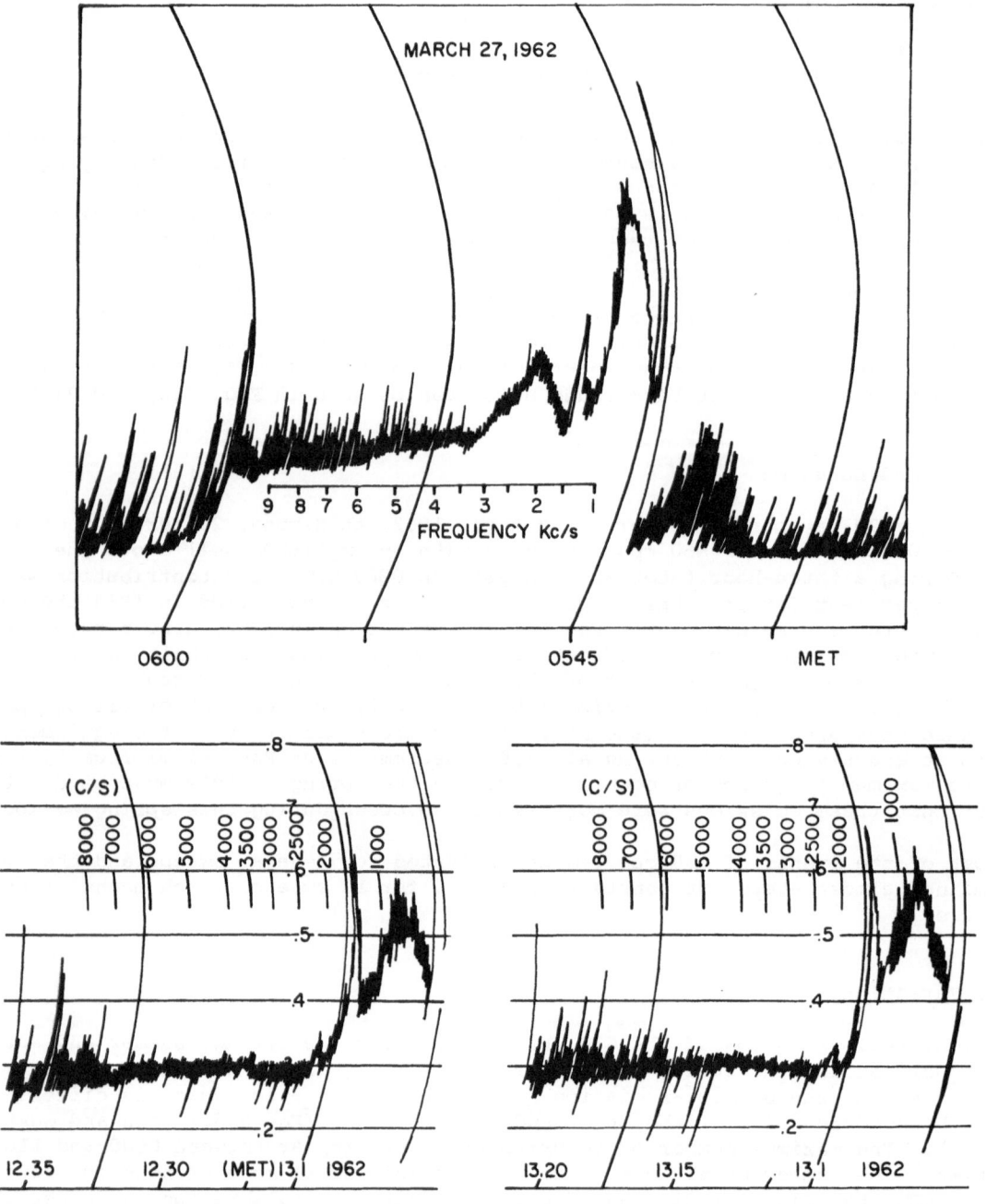

Fig. 1. Example of sweeps showing 750-radiation and hiss (Paksuniemi).

should therefore also be subject to the marked daytime absorption in their passage through the D-layer. It is difficult to accept this idea in view of the diurnal and seasonal behavior of broad-band noise emissions.

The maximum value of the intensity of the 650 cps bands for a few events exceeded 4.0µV, which means that this radiation is sometimes more than ten times stronger than the normal background, as measured at 700 cps.

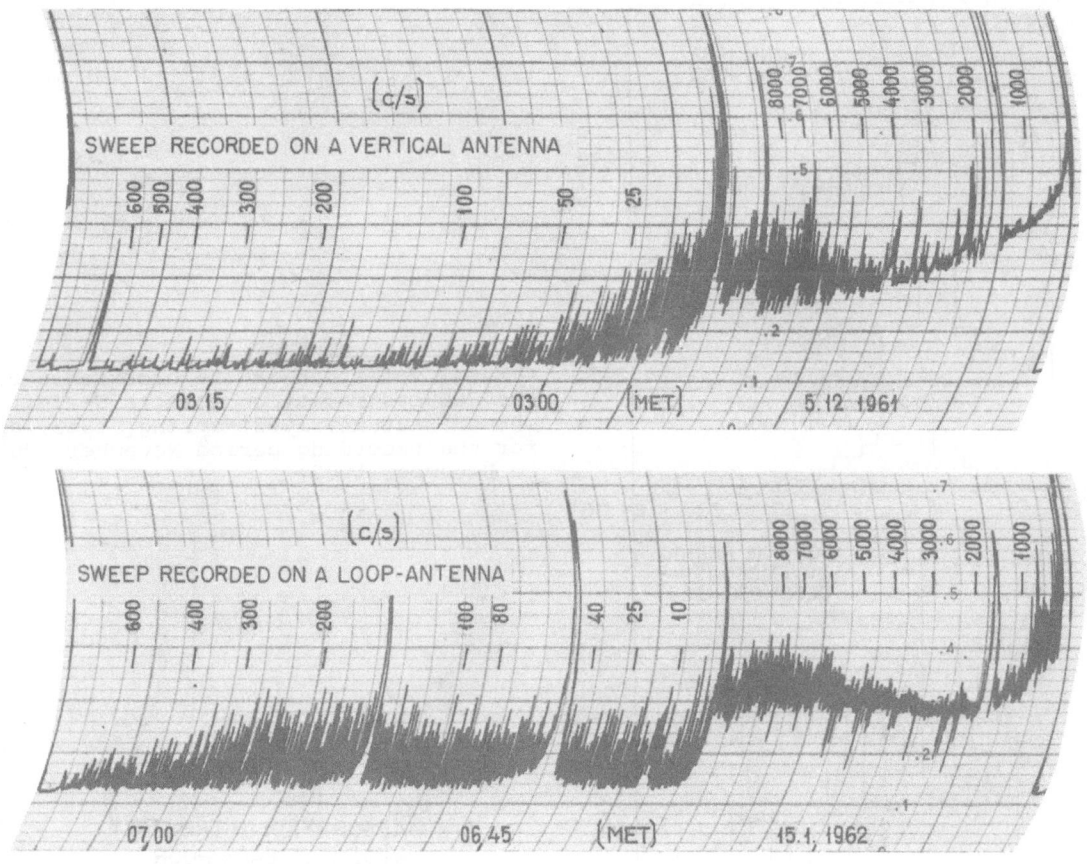

Fig. 2.   Two normal frequency sweeps recorded at Paksuniemi.

Frequency Stability and Bandwidths of the 650 cps Emissions.

Usually it has been found that the frequency of the maximum amplitude of the emission band is centered at 650 cps and that it is fairly constant, i.e varies less than about ± 150 cps.

The average half-intensity bandwidths of all 650 cps radiation in summer and autumn, 1962, and spring, 1963, are shown in Fig. 5. Normally, the average half-intensity width is ± 200 cps, but the bands seem to be somewhat broader for strong emissions events.

Fine Structure.

A few measurements have been made of this broad-band radiation with the sweep disconnected, and with the amplifier operating on a fixed frequency. An example of these types of recordings is shown in Fig. 6. It can be seen on the record that an almost sinusoidal type of oscillation occurred between 0940 and 0950, with a frequency of 25 ± 3 seconds per period. The maximum peak-to-peak amplitude of the oscillation was 1.8µV which corresponds to a field strength of the order of 1µV/m. These oscillations occurred during a magnetically quiet period, and on the magnetic record there could be seen during the same time-interval, a variation of the X-component with approximately the same period (25 seconds per period) and with an amplitude of about

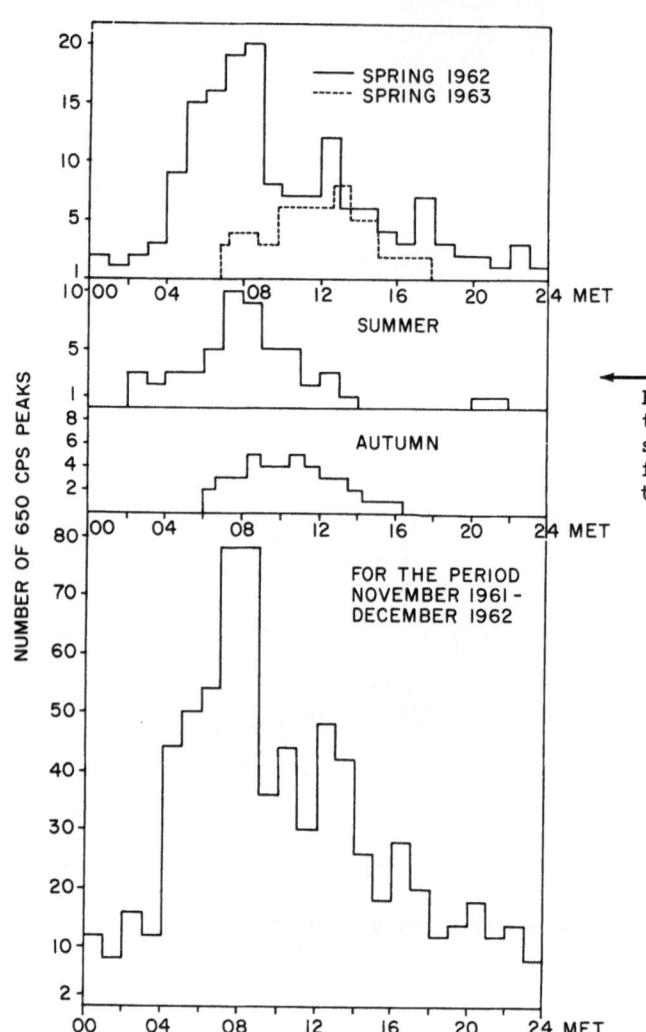

Fig. 3.
Dirunal variation in the occurrence of
the 700 cps radiation for the spring,
summer, autumn, 1962, spring, 1963, and
for the recording period November 1961
to December 1962.

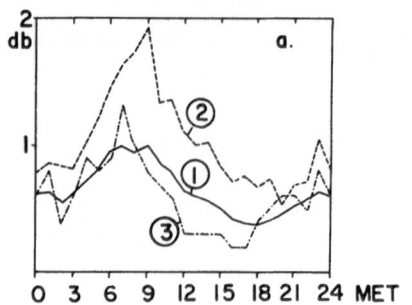

ABSORPTION IN THE FIRST
MINUTE OF EACH HOUR

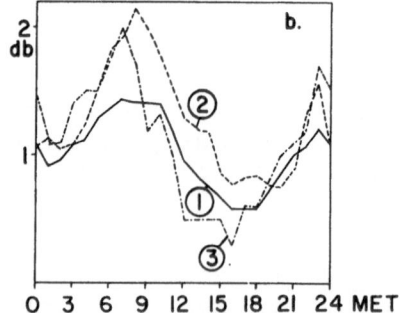

PEAK ABSORPTION FOR EACH
HOUR

Fig. 4.
Diurnal variation of average intensity
of 27.6 Mc/s cosmic noise absorption
for the equinoxes at Kiruna.   Curve 1
represents the average for April,
September and October 1959; March,
October 1960; March, April, September
and October 1961.   Curve 2 is for
September 1961 alone and curve 3 for
September 1959.

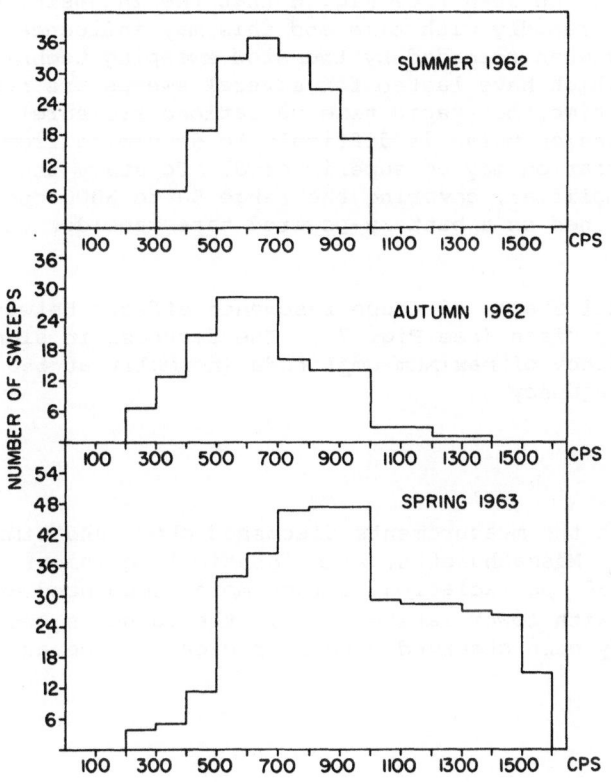

Fig. 5. The average bandwidths of the 700 cps radiation, plotted on a linear frequency scale for the summer and autumn period in 1962 and spring 1963.

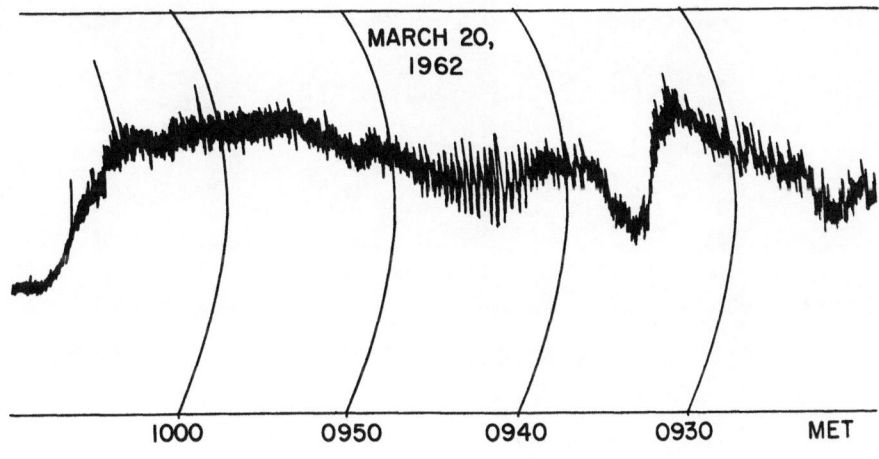

Fig. 6. A fixed frequency record of the 700 cps radiation. The bandwidth is 40 cps. [Paksuniemi]

two gammas. It can also be seen from Fig. 6 that the intensity of the 650 cps radiation may change rather rapidly with time and this may influence the shape of the emission band markedly when recorded by the slow sweeping technique. Furthermore, the radiation events which have lasted for several sweeps are not believed to be constant for such a long time, but rapid time variations are surely superimposed. The exact shape of the emission pulse is difficult to determine from our recordings because rapid time variation may be superimposed. To study the shape of these emissions, a broad-band amplifier, covering the range 50 to 3000 cps, has been built. The spectra have been obtained on a battery-powered tape recorder and analyzed with a sonagraph.

Marked asymmetrical shapes of these resonance effects between 500 and 1000 cps have been recorded very often (see Fig. 7). The decrease in signal strength is much slower above the frequency of maximum amplitude (normally at 650 cps) than the decrease below this frequency.

Latitude Dependence.

By comparison with the measurements discussed above and similar recordings at Sagamore Hill, Bedford, Massachusetts, with identical equipment, it seems reasonable to conclude that the 650 cps radiation is much more pronounced at the auroral zone latitudes as compared with lower latitudes. To the author's knowledge, the 650 cps resonance peak has only been observed within or close to the auroral zone.

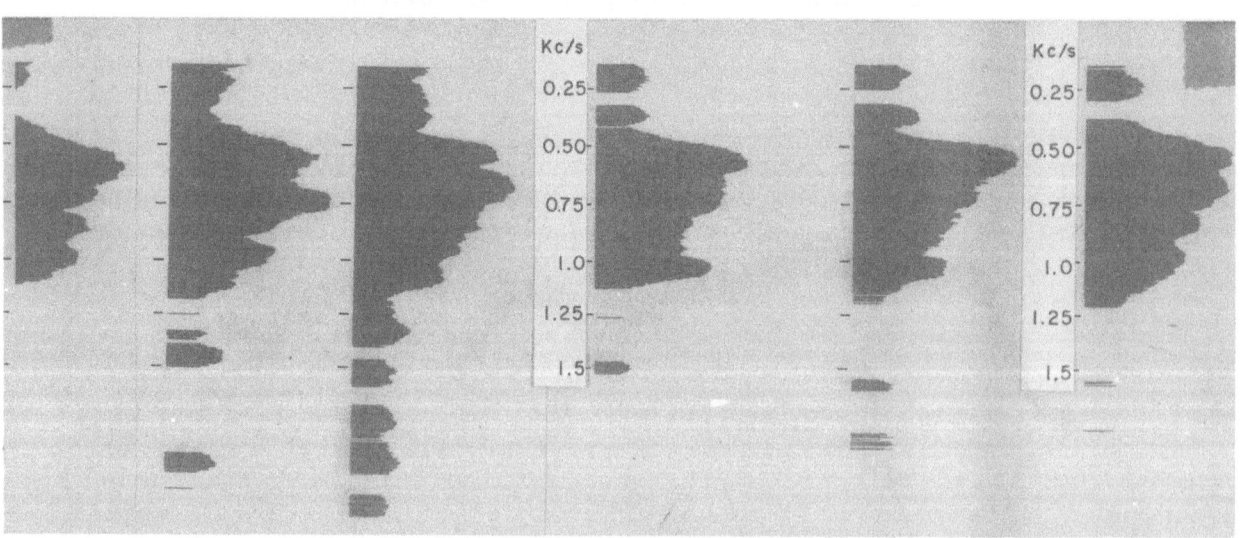

Fig. 7. Power spectra of the background noise between 25 to 1500 cps.

164

Correlation with Geomagnetic Disturbances.

It was found at Kiruna by Aarons, Gustafsson and Egeland (1960) that the 650 cps emission was correlated with micropulsations in the geomagnetic field. The sensitivity of the ELF equipment has been increased since 1960 and the instrument has been placed at a site where the local noise level is lower (Paksuniemi), so it is now possible to detect 650 cps emissions of much smaller amplitude. During the last recording periods it has not always been possible to find any deflection on the magnetograms (which means that they are less than about one gamma) at times when there have been weak 650 cps emissions. The reason for this lack of correlation may be either that the sensitivity of the magneto-variograph is too low, about one gamma, or that a strong disturbance was occurring at the same time, making the reading of very small amplitudes impossible.

ACKNOWLEDGEMENT

I am deeply indebted to Dr. J. Aarons and Mr. W. Barron at the Radio Astronomy Branch, Air Force Cambridge Research Laboratory, Bedford, Massachusetts, for stimulating discussions on the subject. My thanks are also due to Mr. S. Olsen for operation of the measuring equipment. For processing and rendering of the data, the author is greatly indebted to Mrs. G. Hultqvist for drawing the figures, Mr. A. and Mr. V. Bjornstrom for reduction of the observational material and Mrs. C. Lindgren for secretarial work. This work has been supported by the U. S. Air Research Development Command (Contract AF 61(052)-678).

DISCUSSION

Sugiura: Concerning the normal variation of 700 c/s radiation, is the sharp drop after the six to ten maximum due to absorption?

Egeland: No. This is a real drop. The phenomenon is essentially a quiet-time phenomenon.

REFERENCES

Aarons, J., G. Gustafsson, and A. Egeland: "Correlation of audio-frequency electromagnetic radiation with auroral zone micropulsations," Nature 186, 148 (1960)

Egeland, A.: "An investigation of the microstructure of the perturbations in the geomagnetic field in the auroral zone," Inst. Theoretical Astrophysics,
Blindern, Oslo, Report No. 6 (1959)

"Noise emissions in the audio-frequency range," Final Report, Task 3,
Contract No. AF 61(514)-1314 (1962)

Egeland, A., S. Olsen, and G. Gustafsson: "Audio- and subaudio-frequency electromagnetic radiation at high latitude," Final Report, Tasks 1 and 2, Contract
No. AF 61(052)-600 (1963)

Gustafsson, G., A. Egeland, and J. Aarons: "Audio-frequency electromagnetic radiation in the auroral zone," J. Geophys. Res. 65, 2749 (1960)

Gustafsson, G., A. Egeland, W. Barron, and J. Aarons: "Band emissions at gyro frequencies of ionospheric ions and hiss frequencies," AGARDOgraph (Munchen
meeting) Germany (1962)

WHISTLER MODE AND THEORY OF VLF EMISSION

R. M. Gallet
National Bureau of Standards
Boulder, Colorado

INTRODUCTION

The object of the present lectures is to give an account of the theory of the natural VLF emissions. One of the main characteristics of the VLF emissions is that they are narrow band. The frequency at a given instant is generally very well defined, therefore, requiring a very specific mechanism for their emissions. The emissions lie principally in the range from 2 to 10 Kc/sec. This is remarkably low, corresponding to a ratio $f/f_H = 2.7 \times 10^{-3}$ near the earth's surface (typical values: $f = 4$ Kc/sec, $H = 0.4$ to 0.5 gauss, $f_H \simeq 1.5$ Mc/sec). Since the two natural frequencies for a plasma are the plasma frequency f_p, which in the Earth's exosphere and ionosphere is everywhere larger than 100 Kc/sec, and the electron gyrofrequency f_H, both considerably larger than the observed frequencies almost everywhere in the magnetosphere, the VLF emissions are certainly not due to resonance phenomena by local excitation of these frequencies in the ambient plasma surrounding the earth. Far away from the Earth the resonance at the gyrofrequency could give right values for the frequencies of VLF emissions, but the absorption for electromagnetic waves propagating at the gyrofrequency in a plasma is so large that there is no propagation.

Similarly, the emissions are also probably not related to resonance of the ions in the magnetosphere. Protons are the lightest ions and their gyrofrequency is already 1840 times smaller than for electrons. This means at most frequencies of the order of 700 cps near the Earth. For oxygen ions in the ionosphere, the gyrofrequency is 16 times smaller.

These considerations, and also empirical information such as the general correlation between the occurrence of VLF emissions and geomagnetic activity, led to the idea that all VLF emissions result from the interaction between high energy electrons and electromagnetic waves, propagating in the whistler mode in the ambient plasma of the magnetosphere. This interaction results from the fact that the electromagnetic waves in the whistler mode have a slow phase velocity (typical refractive index between 10 and 100), and that high energy electrons exist in the magnetosphere with velocities comparable or equal to that of the wave. The kind of resulting interaction is similar to that existing in the general class of traveling wave tubes, and takes place continuously over a relatively long distance measured in terms of wave lengths of the electromagnetic wave (here the wave length within the plasma, not in vacuum). The combination of the magnetic field and of the ambient plasma, at rest in the magnetosphere, plays the role of the slow-wave circuit or wave guide for the electromagnetic wave. Also, it will become clear that all the interactions which can take place are essentially collective interactions between a beam of particles and the E.M. wave, and do not result only from emission by independent particles moving through the plasma (Cerenkov emission, and Doppler-shifted gyrofrequency of fast-moving electrons). To build a sufficient intensity from emission by independent particles would require fluxes of particles many orders of magnitude too intense to be compatible with the known fluxes.

PRELIMINARY CONSIDERATIONS

In order to understand the mechanisms by which the natural VLF emissions are generated, several types of information are necessary. First, the mode of propagation of the electromagnetic wave should be specified. Second, a source of energy must exist, and third, a coupling mechanism is required by which a certain fraction of the energy from the source is transformed into electromagnetic energy. A successful and complete theory should predict quantitatively the frequency, or frequencies, emitted and their time variation at the receiving point. If possible, it should also account quantitatively for the intensity of the observed VLF emissions, although this problem is both much more complicated and less important. A simple analogy permits to show why: it is much more easy to calculate precisely the frequency produced in a resonance circuit knowing the circuit parameters than to calculate the H.F. power which can be produced. In a less trivial case, the same remarks could be made for the excitation of plasma oscillations in an ionized gas, knowing the electron density.

Several years ago the first theory of the origin of VLF emissions was produced by R. A. Helliwell and the present author (Gallet and Helliwell, 1959; Gallet, 1959). In many ways, this theory was elementary and did not attempt to obtain a deep understanding of the coupling mechanism involved. But the basic elements were found and the theory was quite successful in predicting correctly the frequencies and their time variation. It is still believed that the theory found at that time (during 1957) is essentially correct, and that the basic mechanism, called by the authors the "traveling wave-tube mechanism" of the VLF emissions, does effectively apply and account for some classes of VLF emissions. However, the progress of the observations as well as numerical applications of the theory have shown that there are many other classes of natural VLF emissions whose spectra are not accounted for by the first elementary theory. Probably other processes exist, and several have been proposed by different authors in recent years. There is at present considerable activity in this field and the literature is in a state of flux. When the different physical processes which can produce VLF emissions will have been clearly defined, a future task will be to find the circumstances under which one particular process is at work and not the others, since the shape of the spectrum of a given class of VLF emissions is very specific. A first phase in this task will consist of a systematic comparison between the observed shapes and the theoretical shapes predicted by the theory of a given process. One of the aims of the theory is to furnish quantitative information about the dynamics of particles in the magnetosphere (energy, bunching, rate of occurrence, lines of force and their L-values on which this activity takes place) using the spectra of VLF emissions as a very sensitive tracer. Ultimately, these studies should contribute to the understanding of the mechanisms of acceleration of the high energy electrons.

In these lectures the intent is not to make a systematic survey of all the processes which have been put forward. Rather, an account will first be given of the elementary theory because many of the elements of this theory will also of necessity be common to all other mechanisms. In recent years new developments have been made by the present author, somewhat along the original lines, but with a much more detailed analysis of the physical process involved. These developments furnish new specific formulas for the frequencies emitted and predict new effects, such as emissions of electromagnetic waves in a direction counter to the direction of the beam of particles which generate the VLF emissions, and the splitting of the frequency emitted in two or more separate frequencies. Also, this new theory contains in a very natural way several of the results proposed by other authors in recent years, such as the frequencies predicted from the Doppler-shifted cyclotron radiation from electrons discussed by Dowden (1962; 1963). A detailed account of this new theory will not be given here, but the main lines and all the new results for predicted frequencies will be presented.

In all that follows, it is assumed that:

a) the electromagnetic wave propagates in the whistler mode, and follows lines of force in the magnetosphere.

b) the source of energy for the VLF emission is a stream, or a discrete (but quite extended) bunch of high energy electrons following a line of force (probably from few kiloelectron volts to few tens of kev).

These assumptions rest on experimental evidence which will not be discussed here. Also the mechanisms by which the streams or the discrete bunches of high energy electrons are produced somewhere in the outskirts of the magnetosphere will not be discussed, but this is essentially the same process which produces the Van Allen belt and the observed precipitation of large fluxes of high energy electrons upon the upper atmosphere and probably also the aurora. What is really remarkable is the sharpness and the rapidity of the process producing discrete bunches, sometimes at a high rate and during several hours. Another important question which is not discussed here is the mechanism by which a bunch of particular characteristics (energy and pitch angles of the electrons) does not disperse rapidly in its motion along the line of force. In many remarkable cases the bunch seems to be mirroring a large number of times along a particular line of force, periodically re-emitting the same shape of VLF emission. In such cases the pitch angle of the electrons cannot have a large dispersion, and probably is progressively modified by the electromagnetic field of the wave accompanying the motion, in such a way as to maintain the synchronism between the bunch of particles and the wave.

The theory will first consider the frequencies which can be emitted at a given position in the magnetosphere, along with a discussion of the mechanisms of emission. Second, by following the motion of the beam of particles along a particular line of force, it will be shown how the shape of the spectra of VLF emissions (frequency versus time at the observing point at the earth's surface) can be computed.

From the viewpoint of the energies involved, it is interesting to intercompare the relative amounts of energy in the principal phenomena taking place in the magnetosphere. For the present purpose, only orders of magnitude need to be considered. We estimate the energy density (in ergs/cm^3) for the following phenomena.

-<u>Magnetic energy</u> density of the geomagnetic field at places where the intensity is 10^{-2} and 10^{-3} gauss (10^3 to $10^2 \gamma$)

$$\frac{H^2}{8\pi} = 4 \times 10^{-6} \text{ and } 4 \times 10^{-8} \text{ erg/cm}^{-3}$$

-Kinetic energy of the <u>ambient plasma</u>, with $N_e \simeq 10^3$ e/cm^3, T = 1500°K

$$3kTN_e = 6.2 \times 10^{-10} \text{ erg/cm}^{-3}$$

$$\left(\beta = 3kTN_e \Big/ \left(\frac{H^2}{8\pi}\right) = 1.6 \times 10^{-4} \text{ to } 1.6 \times 10^{-2} \right)$$

-Energy density of "typical" peak values of observed fluxes of <u>high energy electrons</u>:

     a) in the heart of the outer radiation zone, from Explorer XII data of 5 September 1961 (O'Brien, Van Allen et al, 1962).

        -omnidirectional intensity of electrons with E > 40 kev: 10^8 e(cm^{2}sec)$^{-1}$

        -one deduces an energy flux >6.4 ergs (cm^{2}sec)$^{-1}$, an energy density $\sim 5 \times 10^{-10}$ erg/cm^{-3} and a particle density $\simeq 10^{-1}$ e cm^{-3}

b) above 1000 km in auroral zone, from INJUN I data of 25 September 1961 (O'Brien, 1962) for illustrating exceptional cases.

-a very large energy flux >400 erg (cm^2 sec steradian)$^{-1}$ for electrons E > 1 kev was observed. The corresponding energy density is <2x10^{-8} erg/cm^{-3} (but at a position where the energy density of the geomagnetic field is 10^{-2} erg/cm^{-3}), and the particle density is < 100 e cm^{-3}

-Energy density of the electromagnetic field of a typical whistler or VLF emissions.

The highest measured values of the energy flux from whistlers and strongest VLF emissions at ground are estimated of the order of 10^{-15} watts/m^2/cps = 10^{-12} erg (sec cm^2 cps)$^{-1}$. If no losses (by absorption and by partial reflection did occur at the ionosphere boundary (D region) the flux will be invariant = Energy density in the magnetosphere times U_{group}. With the typical values of $U_{group} = \frac{c}{n'}$, and n' = 10 to 30, and for a signal occupying a bandwidth 10^2 to 10^3 cps, the energy density is of the order of 10^{-19} to 10^{-18} erg/cm^{-3}. One should account for the transmission losses at the ionosphere boundary, perhaps 20 decibels, but the above value was a high value to start with.

It is clear from these comparisons that the geomagnetic field is the dominating factor in the magnetosphere, accounting for its stability. It is equally clear that the energy of the electromagnetic waves, either propagating (whistlers) or emitted (VLF emissions) in the whistler mode, represents a very small fraction, at best of the order of 10^{-6}, of the kinetic energy of the fluxes of high energy electrons. One sees that the E.M. waves can hardly produce very slight perturbations of the motion of the charged particles, such as changing the phase or modifying the pitch angle without change of energy, and could not be responsible for accelerating particles to their high energy in less than many days of continuous action. On the other hand, the radiation of VLF emissions by high energy electron streams represents only a very minor perturbation and loss of energy.

The real physical problem of the VLF emissions is in the radiation mechanism, specific enough to produce well defined spectra; it is not an energetic problem. The VLF emissions are very sensitive messages, yet to be still deciphered, on the dynamics of the magnetosphere.

WHISTLER MODE FOR ELECTROMAGNETIC WAVES
PROPAGATING IN THE EXOSPHERE

At a given place in the magnetosphere the propagation of electromagnetic waves depends on the following parameters:

Plasma Parameters

ω_P such that $\omega_P^2 = \dfrac{4\pi Ne^2}{m}$ plasma frequency

numerically, $N_e = 1.24 \times 10^{-8} f_P^2$ (N_e in electrons per cm^3)

ω_H $\omega_H = \dfrac{eH}{mc}$ electron gyrofrequency

numerically, $f_H = 2.800 \times 10^6 H$ (H in gauss)

$b = \dfrac{\omega_P}{\omega_H}$ is a parameter (without dimension) characteristic of the plasma, and independent of the waves.

ν collision number is assumed to be zero.

Wave Parameters

ω wave frequency.

θ direction of the wave normal relative to the direction of the magnetic field H.

Everywhere in the exosphere and the ionosphere $b^2 \gg 1$ (typically b is between 3 and 10). For present needs, it will be assumed that the wave propagates strictly along the magnetic line of force; therefore, $\theta = o$. Under these conditions, electromagnetic waves of only one polarization can propagate in the frequency range

$$o < \omega < \omega_H$$

and propagation is not possible (for any polarization) for the large frequency range between ω_H and frequencies near ω_P. The dispersion law for the plane wave approximation is:

$$n^2 = 1 + \frac{\omega_P^2}{\omega(\omega_H - \omega)}$$

$$\text{or } n^2 = 1 + \frac{b^2}{\sigma(1-\sigma)} \qquad \text{or even} \qquad n^2 \simeq \frac{b^2}{\sigma(1-\sigma)} \quad \text{when } b^2 \gg 1$$

171

n   is the refractive index $n = \dfrac{c}{V_{phase}}$

$\sigma = \dfrac{\omega}{\omega_H}$ is the <u>reduced</u> frequency, relative to the gyrofrequency and will be used many times in what follows.

More details about the dispersion law and particularly about the anisotropy effects when θ is not zero could be found in another publication (Gallet, 1963).

In fact, the propagation along magnetic lines of forces, both for the whistlers and for the VLF emissions, is most of the time due to a "ducting" effect. A <u>nonuniformity</u> of the plasma density across a thin magnetic shell acts as a true dielectric wave guide within which the electromagnetic wave is ducted and follows the curvature of the plasma shell. From observations of the discrete components of whistlers, it is seen that in general there exist several, and sometimes many, discrete shells in the magnetosphere which somehow have a structure similar to that of an onion. The mechanism which produces this sort of fibrous structure is not known; it is possible that it is related to the production of high energy particles on well defined, discrete and thin magnetic shells, as evidenced in shapes of auroras and patterns of precipitation upon the upper atmosphere.

The ducting has an important consequence. A ducted mode requires that the envelope of the wave normal (phase-trajectory) meanders in a snake-like fashion around the axis of the duct. As a result of the finite transverse extension of the ducted wave, it can also be considered in general as resulting from the superposition of a distribution of plane waves, the wave normal of each making a certain angle with the axis of the duct. (This is a classical decomposition in the theory of wave guides, but it is considerably more difficult to perform for a <u>nonuniform</u> medium, with no sharp or discontinuous boundary, which in addition exhibits a strong anisotropy as a function of θ.) Under these conditions for which θ, while small, is not strictly zero, the resulting wave will always have a small <u>longitudinal</u> component of the electric field vector, directed along the magnetic line of force, besides its transverse electric field, which is generally more important. It is this small longitudinal component of the electric field which provides the coupling for interacting with the longitudinal velocity of electrons from a beam moving along the magnetic line of force. This will be discussed further in these lectures.

A detailed theory should discuss the structure of the wave propagating under such conditions, and give the relations between the various components of the magnetic and electric vectors of the wave. The wave propagating in the whistler mode is circularly polarized and its transverse components of electric and magnetic fields rotate in the same direction as the electrons (but at a frequency ω different from ω_H). Suppose that the wave propagates in the same direction as the magnetic field, taken as the direction for the positive z. An observer moving with the wave and looking in the same direction as the direction of propagation will see the vectors rotating in a <u>clockwise</u>, or <u>negative</u> rotation. The transverse components E_x, E_y of the electric field can be written:

$$E_x = E_o \cos(kz - \omega t)$$

$$E_y = -E_o \sin(kz - \omega t)$$

in which $k = \dfrac{2\pi}{\lambda} = n\dfrac{\omega}{c}$ is the (angular) wave-number,

or using the unit vectors e_1, e_2, e_3 along the axes x, y, z, the equation of the plane, uniform wave can be written

$$E(z,t) = E_o(e_1 - ie_2)e^{i(\omega t - kz)}$$

(In more modern notation, such a circularly polarized wave is said to have a <u>positive helicity</u>.)

Similarly, consider the whistler wave propagating in the direction opposite to that of the magnetic field: toward the negative z. The electric vector of the wave is still rotating <u>with</u> the electrons. Only k changes sign and the equation of the wave is now:

$$E_x = E_o \cos(kz + \omega t)$$

$$E_y = E_o \sin(kz + \omega t)$$

$$\text{or} \quad E(z,t) = E_o(e_1 - ie_2)e^{i(\omega t + kz)}$$

This time the observer moving with the wave and looking forward sees the electric and magnetic fields rotating counterclockwise or in the positive rotation. The wave is in this case circularly polarized in the reverse direction = negative helicity.

The ratio of the magnetic field to the transverse electric field of the wave $\frac{H}{E}$ is n times this ratio for a plane wave in vacuum, n being the refractive index: $H = \mp inE$ (upper sign for propagation in the positive z direction).

For a ducted wave of <u>finite</u> extent in the x and y direction, the amplitude of the electric and magnetic components is necessarily <u>nonuniform</u> in an x, y plane, and becomes a function of x and y.

From div D = o a non-zero longitudinal component E_z necessarily appears. If the variation of the amplitude, away from the duct axis, is sufficiently small, and maintaining the circular polarization, it can be shown that in first approximation

$$E(x,y,z,t) \simeq \left[E_o(x,y)\ (e_1 \pm ie_2) + \frac{i}{k}\left(\frac{\partial E_o}{\partial x} \pm i\frac{\partial E_o}{\partial y}\right) e_3 \right] e^{i(kz-\omega t)}$$

This formula permits to evaluate the longitudinal component of the electric field and its variation in the x, y plane. There is no longitudinal component of the wave magnetic field $H_z = o$.

The speed of energy propagation is given by the group velocity

$$U_{group} = \frac{c}{n'}$$

in which n', the group-refractive index, is:

$$n' = n + \omega\frac{dn}{d\omega} = n + \sigma\frac{dn}{d\sigma}$$

With the present approximation:

$$n' = \frac{n}{2}\left[\frac{1}{1-\sigma}\right]$$

At a particular frequency the time of travel of the signal through the medium is:

$$dt = \frac{ds}{U} = \frac{1}{c} n' ds$$

Therefore:

$$t_\omega = \frac{1}{c}\int n'_\omega(s)ds$$

the integral being taken along the line of force from the place of origin of that particular frequency to the earth's surface.

ELEMENTARY THEORY OF THE VLF EMISSIONS

This theory was based on the simplest concept of a self-excited traveling wave tube, used as a wave generator. Without analyzing in details the mechanism of interaction, one just accepts the idea that the longitudinal component of the electric field of the wave provides the coupling with the motion of the particles along the direction of the magnetic field. The source of energy for the growing wave is, of course, the kinetic energy of the electrons in the beam, which ultimately is slightly reduced. Interchange of energy takes place progressively and is the strongest when

phase velocity of the E.M. wave equals the velocity of particles.

$$V = u_o$$

Emission will take place only for those frequencies for which this condition is satisfied.

Since $\frac{V}{c} = \frac{1}{n}$, using the expression for the refractive index n, one obtains:

$$\left(\frac{u_o}{c}\right)^2 = \frac{\sigma(1-\sigma)}{b^2}$$

$$\text{or} \quad b\frac{u_o}{c} = \left[\sigma(1-\sigma)\right]^{\frac{1}{2}}$$

This quadratic equation in σ furnishes the two emitted frequencies

$$\sigma = \frac{1}{2}\left[1 \pm \sqrt{1 - \left(2b\frac{u_o}{c}\right)^2}\right]$$

or in absolute value:

$$\omega = \frac{\omega_H}{2}\left[1 \pm \sqrt{1 - 2\left(\frac{\omega_P}{\omega_H}\frac{u_o}{c}\right)^2}\right]$$

Discussion of the Properties of the Emitted Frequencies.

The emitted frequencies can be represented on the accompanying diagram:

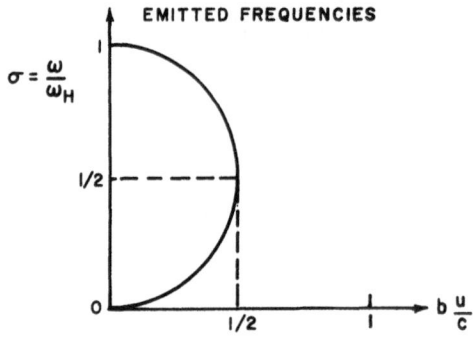

Fig. 1.

174

a) **No emission** when $2b\dfrac{u_0}{c} > 1$.

Velocity of the beam is _too_ _fast_ for this particular plasma. (It will be seen later that this property will not be maintained at least in a strict sense in a more developed theory.)

b) Remark the role of $\dfrac{\omega_H}{2}$; the two emitted frequencies are placed symmetrically relative to $\sigma = \tfrac{1}{2}$, which is the (double) frequency emitted when $b\dfrac{u}{c} = \tfrac{1}{2}$.

c) Remark that often in the exosphere and ionosphere conditions $\left(2b\dfrac{u}{c}\right)^2 \ll 1$, case of a slow beam.

Then a good approximation is:
$$\omega_1 = \left(\dfrac{u}{c}\right)^2 \dfrac{\omega_P^2}{\omega_H}$$

$$\omega_2 = \omega_H \left[1 - \left(\dfrac{u}{c}\,b\right)^2\right] \qquad \text{a value near } \omega_H$$

The lowest frequency ω_1 is of particular interest in the exosphere, since the whistler studies have shown that a good empirical model for the exosphere is such that $\dfrac{Ne}{H}$ is approximately constant along a line of force, or $\dfrac{f_P^2}{f_H} = F$ in which F is a constant having the dimension of a frequency (best value at present $F \simeq 1$ to 0.8×10^6 cps). That such a model represents fairly well the exosphere was obtained for the first time by the present author (Gallet, 1959); this has been well confirmed by several subsequent studies (Smith, 1961, Carpenter, 1962, 1963). Consequently, if the velocity of the particles stays approximately constant along a line of force, the emitted frequency ω_1 will also stay approximately constant. In this way, one can understand the class of VLF emissions called the "quasi-horizontal tones." If the observed $f_1 \simeq 4$kc/sec, $\left(\dfrac{u}{c}\right)^2 \simeq 4$ to 5×10^{-3}, or $u = 2 \times 10^9$ cm sec^{-1}, corresponding to an energy of 1.1 kev.

It is seen that the expression without dimension $b\dfrac{u}{c}$, characterizing the combination of properties of the ambient plasma and of the beam, is the basic parameter for the emission of VLF emissions. This parameter maintains its importance in a more elaborate theory.

A CLOSER APPROACH TO THE MECHANISM OF THE EMISSION

Until now, attention has been paid only to one property of the beam, namely, its velocity u_O. Also only a purely kinematical concept has been used: the equality of two velocities, one characterizing the beam, the other characterizing the E.M. wave. Many important dynamic and energetic questions have been disregarded. Consider, for example, the effect of the longitudinal component E_z of the electric field of the wave upon an electron beam initially of uniform density moving at a speed identical to the phase velocity of the wave. In the frame of reference moving with the beam, the electrons are at rest, but they "see" a weak periodic electric field along the z axis. Electrons one-half wavelength apart see electric vectors of opposite directions, and as a consequence are accelerated in opposite directions. It is obvious that they will tend to form bunches with a periodicity scaled to the wavelength of the E.M. wave. Very soon the effect of the wave upon the electron beam is to produce a periodic density modulation at the same time that it has also produced a periodic velocity modulation. The periodic space-charge resulting from the density modulation in turn produces a periodic longitudinal electric field (electrostatic origin).

In truth, the resulting longitudinal electric field pertains as much to the E.M. wave as to the periodic electrostatic space charge which has developed on the beam: it does not "know" to which it belongs, and it is common to both. Since the relations between the different components of the electric and magnetic fields of the E.M. wave are imposed by the ambient plasma in which it propagates and have to be respected, the other components adjust to the new value of the longitudinal component of the electric field. The energy for so doing is extracted from the original kinetic energy of the electrons of the beam (remember the order of magnitude of the different energies involved). It can be shown that a small velocity modulation results in large density modulation, and therefore in a large space charge effect: this permits to understand the growth in the amplitude of the E.M. wave.

While the above qualitative description of the mechanism of coupling between a beam and an E.M. wave is basically correct, the detailed theory may become extremely involved, particularly the question of the growth of the E.M. wave. In addition, a similar, but separate, coupling resulting from the transverse rotating component of the electric field upon the transverse velocity component of the electron may take place. Thus from a dynamical viewpoint, interaction can take place in many different ways, each one being satisfied at one or several discrete frequencies. In what follows, such a dynamical analysis will not be given, but only the necessary conditions for coupling will be examined in the same spirit as in the preceding section (these conditions may not be sufficient: the growth rate should be examined quantitatively).

An electron beam, moving along a magnetic field, is capable of supporting several kinds of "beam waves" in a way similar to waves on a moving string. These beam waves are independent of any electromagnetic wave superposed to the beam, and result from the general fact that the beam possesses several "proper frequencies." A beam of infinite transverse dimensions has two basic proper frequencies which should be measured in the moving frame of reference of velocity u_O, at which the beam is at rest. They are:

- the "Beam Plasma Frequency" F_P^{Beam} ($\Omega = 2\pi F_P^{Beam}$) solely resulting from the finite electron density of the beam N_e^{Beam}. This exists even in the absence of an external magnetic field.

- the gyrofrequency f_H ($\omega_H = 2\pi f_H$) in presence of an external magnetic field, and independent of the beam density.

A beam of _finite_ transverse dimensions, such as a cylindrical beam, would possess additional proper frequencies corresponding to deformation of the boundary transverse compression and expansion, etc.

Numerical Values of the Beam Proper Frequencies for the Conditions in the Magnetosphere.

The observed largest fluxes of electrons that bombard the earth during auroral and geomagnetic disturbances (see for example O'Brien and Laughlin, 1962) correspond to a space density of high energy electrons between about 10^{-2} to 1 particle per cm^3 (exceptionally somewhat higher). This should be compared to the density of the ambient plasma, of the order of 10^3 particle cm^{-3} in the outer exosphere. However, such low beam densities correspond to appreciable beam plasma frequencies, as shown in the following table:

N_e Beam (e/cm^3)		F_P Beam
1.24×10^{-2}	$\longrightarrow$	1 Kc/sec
0.1	$\longrightarrow$	3 Kc/sec
1.24	$\longrightarrow$	10 Kc/sec

Even very modest beam electron densities correspond to beam plasma frequencies comparable to the frequencies of the observed VLF emissions. Obviously, one should expect the beam plasma frequency to play some important role for the emitted frequency.

A New Viewpoint for the Production of VLF Emissions.

The new viewpoint which emerges from these studies is that the emissions result from a coupling between two kinds of waves: beam waves and electromagnetic waves. The beam waves propagate and may be growing independently of any electromagnetic waves; their phase velocities will be comparable to and will be measured in terms of the beam velocity u_0; their properties depend on the kinematics and dynamics of the particles composing the beam. The properties of the electromagnetic waves propagating in the whistler mode depend on the ambient plasma. The interchange of energy between the two waves takes place via either the longitudinal or the transverse component of the electric field of the E.M. wave, and the energy necessary for the production of both waves comes ultimately from the kinetic energy of the undisturbed beam. The necessary condition for coupling is the equality of the phase velocities of these two separate waves. Other conditions on the polarization, when the coupling takes place by the transverse electric field component, should be respected. These conditions are not sufficient for insuring that the E.M. wave will grow. This question requires a deeper study of the dynamics of the beam waves.

BEAM WAVE PROPERTIES

In this section the properties of beam waves necessary for the present treatment are discussed independently of any electromagnetic wave.

General Dispersion Law for Beam Waves and Its Properties.

Relative to the observer at rest, beam waves of any frequency ω can propagate on the beam. To each "proper frequency" of the beam Ω corresponds at a given ω a <u>pair</u> of beam waves of different phase velocities v_{phase}

$$v_{phase} = \frac{u_o}{1 \pm \frac{\Omega}{\omega}}$$

This is obtained purely from the Doppler effect and results only from kinematical reasons. The frequency ω, seen by the laboratory observer on a wave passing him with the phase velocity v, is simply the Doppler-shifted proper frequency Ω, which another observer moving with the beam at the velocity u_o will observe for oscillations stationary relative to him. A demonstration using only the Doppler effect is given in Appendix A.

An equivalent form of the dispersion law is, using the wave number $k = \frac{2\pi}{\lambda}$, since $v_{phase} = \frac{\omega}{k}$:

$$\omega = ku_o \mp \Omega$$

which also furnishes immediately the group velocity

$$u_{group} = \frac{d\omega}{dk} \longrightarrow u_{group} = u_o$$

This is a remarkable result. The dispersion law of beam waves is such that for any arbitrary frequency and whatever the phase velocity may be, the wave energy is always transported at the beam velocity, illustrating the fact that this energy consists of stored oscillatory motion at the proper frequency of the beam.

The properties of the dispersion law are represented on two different diagrams in Fig. 2.

The upper sign in both formulas for the dispersion law corresponds to the so-called "slow wave" (S branch), because the phase velocity is always inferior to the beam velocity u_o. The lower sign furnishes the other branch consisting of two parts:

- for $\omega > \Omega$:   F branch, for "fast wave" because the phase velocity is always larger than the beam velocity u_o.

- for $o < \omega < \Omega$: B branch, for which the phase velocity is <u>negative</u>, the phase velocity (relative to the laboratory observer) is of a direction <u>opposite</u> to the direction of the beam velocity (and consequently to its group velocity), and can take any absolute value between zero and infinity.

178

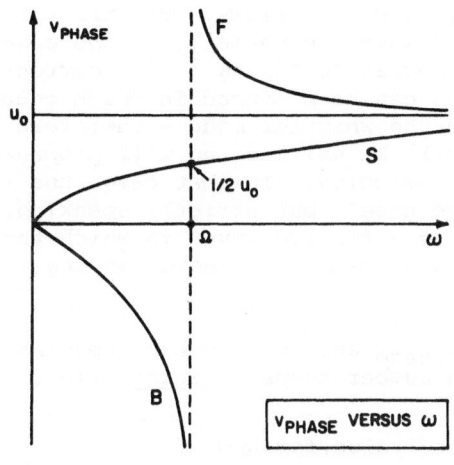

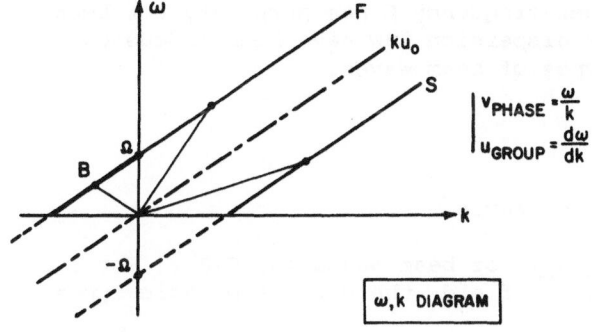

Fig. 2.
Properties of the dispersion law of beam waves. In the (ω,k) diagram the positive direction for k is taken in the direction of the beam velocity.

A wave for which the group velocity is of a direction opposed to that of the phase velocity is called a backward wave. While a beam is capable of supporting backward beam waves, the study of the sign of $\frac{d\omega}{dk}$ for plane electromagnetic waves propagating in a plasma (Appleton-Hartree equation) shows that in contrast there are no frequencies for which the E.M. wave is a backward wave. It is very easy to verify the absence of a backward E.M. wave for the whistler mode, the case of interest in the present study, using the formulas of the third section. Of course, backward electromagnetic waves are produced and used in certain types of microwave tubes, but the required dispersion law, such that $\frac{d\omega}{dk}$ is negative, is obtained by means of special waveguide structures (helices, periodic cavity structure, etc.).

It is recommended to use the nomenclature for beam waves:

Sign in formula for v_{phase}

+   S, for "slow" (slower than u_O), and always a "forward" wave.

−   F, meaning both "forward" and "faster" than u_O; only for $\omega > \Omega$.

      B, "backward," a part of the "fast" branch, but the use of this last term would be somewhat improper here since in absolute value the phase velocity could be much "slower" than u_O. Frequency range $o < \omega < \Omega$.

179

Such a nomenclature will be used systematically in what follows. The present writer has found the use of the terms "slow" and "fast" waves, especially in the case of backward waves, not always very explicit and somewhat confusing in the current literature. It will be seen below that VLF emissions can be produced in which coupling takes place between an E.M. wave propagating in the whistler mode - therefore always forward - and a B-beam wave; the VLF emission will be emitted and will propagate in a direction opposite to the direction of the beam velocity. In that case, the result is similar to that of a "backward traveling-wave tube"; but strictly speaking, the radio wave propagating in the exosphere is always a forward wave, in which the transport of <u>radio</u> energy is counter to the direction of beam-wave energy moving with the speed and direction of the beam.

From the dispersion law remark that both v_{phase} and ω go to zero together in such a way that for the limit of $\omega = o$ the wave number keeps a <u>finite</u> value:

$$\lambda_{limit} = \frac{2\pi u_o}{\Omega} \qquad , \text{ a finite value for } \omega = o.$$

Until this point, the nature of the proper frequency Ω has purposely not been specified, and only properties related to the dispersion law have been discussed. It is now necessary to consider the different types of beam waves.

Types of Beam Waves: Space Charge and Cyclotron Waves.

To each proper frequency Ω is attached <u>a pair</u> of beam waves (S; F-B). For the infinite electron beam, in presence of a magnetic field, there are two basic types of beam waves.

-<u>Space charge waves</u>, for which $\Omega = \Omega_P$, plasma frequency of the beam, related to the electron density of the beam.

-<u>Cyclotron waves</u>, for which $\Omega = \omega_H$, gyrofrequency, related to the external magnetic field.

The space charge waves are <u>longitudinal</u> waves, similar to the propagation of pressure waves in a gas.

The cyclotron waves are <u>transverse</u> waves, similar to the propagation of torsional waves in a solid.

The space charge waves consist of periodic displacements of the electrons relative to their equilibrium position in the beam in absence of wave. The structure of the wave consists of crests and troughs of electron density, accompanied by corresponding periodic velocity fluctuations of the electrons. The maxima and minima of electron density create a periodic electrostatic potential, and therefore, a periodic longitudinal electric field in the direction of the beam velocity (which in the present problem is also in the direction of the external magnetic field). The energy of each wave has two contributions: the potential energy of the periodic space charge fluctuations (this plays the same role as the pressure fluctuations for a sound wave in an ordinary gas), and the kinetic energy of the harmonic motion of the electrons.

The structure of the cyclotron waves is less easily visualized. It consists of the <u>organization in phase and magnitude</u> of the <u>transverse</u> component of the electron velocity $u_\perp$; the longitudinal component stays unchanged and equal to the beam velocity u_O. In a given z plane the transverse velocity at a given instant is the same for all the electrons of the beam, both in magnitude and in direction (the beam is supposed of infinite extension in the z plane).

The components of the transverse velocity $u_\perp$ along the x and y axes are given by the following equations which describe the cyclotron waves.

-When u_O is positive:

F and B waves (forward and backward parts of the "fast" branch)

$$u_{(x,y)} = u_\perp (e_1 - ie_2) \; e^{i(\omega t - kz)} \qquad \text{with} \qquad k = \frac{\omega - \omega_H}{u_O}$$

S wave

$$u_{(x,y)} = u_\perp (e_1 + ie_2) \; e^{i(\omega t - kz)} \qquad \text{with} \qquad k = \frac{\omega + \omega_H}{u_O}$$

-When u_O is negative:

F and B waves

$$u_{(x,y)} = u_\perp (e_1 - ie_2) \; e^{i(\omega t - kz)} \qquad \text{with} \qquad k = \frac{\omega - \omega_H}{u_O}$$

S wave

$$u_{(x,y)} = u_\perp (e_1 + ie_2) \; e^{i(\omega t + kz)} \qquad \text{with} \qquad k = \frac{\omega + \omega_H}{u_O}$$

At a given frequency ω the two cyclotron beam waves are circularly polarized in opposite directions. An observer in the laboratory frame sees the vector transverse velocity in that z plane rotating at the wave frequency ω, either in one direction of rotation or the opposite depending on which cyclotron wave S or F-B is excited, but as a function of time he is always observing new electrons. The beam electrons themselves, which cross the plane at the velocity u_O, "see" only the proper frequency ω_H (see Appendix A), and rotate only in their natural clockwise direction. All electrons describe parallel spiral paths with a pitch angle α, such that $\tan \alpha = \frac{u_\perp}{u_O}$; all spiral paths intersect a given z plane at the same phase angle. At any point in space there is no density fluctuation of any kind resulting from cyclotron waves, and therefore, no space charge effect and no electric field. The energy of wave is entirely in form of kinetic energy of the transverse motion $\frac{1}{2} N_e^{beam} mu_\perp^2$ per cm^3, and the modulation of the wave energy results only from the change in $u_\perp$. Of course, the change in energy will result from the work of transverse electric fields, which are present only at the time during which $u_\perp$ changes. This remark permits to understand that the coupling and interchange of energy between the cyclotron beam wave and the whistler mode E.M. wave will take place mainly via the transverse component of fields, electric and magnetic, while by contrast, the coupling with space charge waves takes place exclusively via the longitudinal component of the electric field. For coupling with the cyclotron wave, the longitudinal electric field, if any is present, also participates, in a subtle way, in the interaction. By slightly modifying the longitudinal electron velocity, it modifies the phase of its transverse motion.

Perhaps the easiest way to visualize the wave structure is to isolate by thought the electrons initially on a line parallel to the propagation axis z when the transverse velocity is zero. The wave will consist of an infinity of such lines deforming exactly in the same way and in the same phase. When the wave is present the line is now shaped as an helix, either right-handed or left-handed.

1. __F wave at frequencies__ $\boxed{\omega > \omega_H}$

The phase velocity v_{phase} is greater than the beam velocity u_o.

SHAPE OF A SINGLE LINE IN THE F CYCLOTRON WAVE ($\omega > \omega_H$)

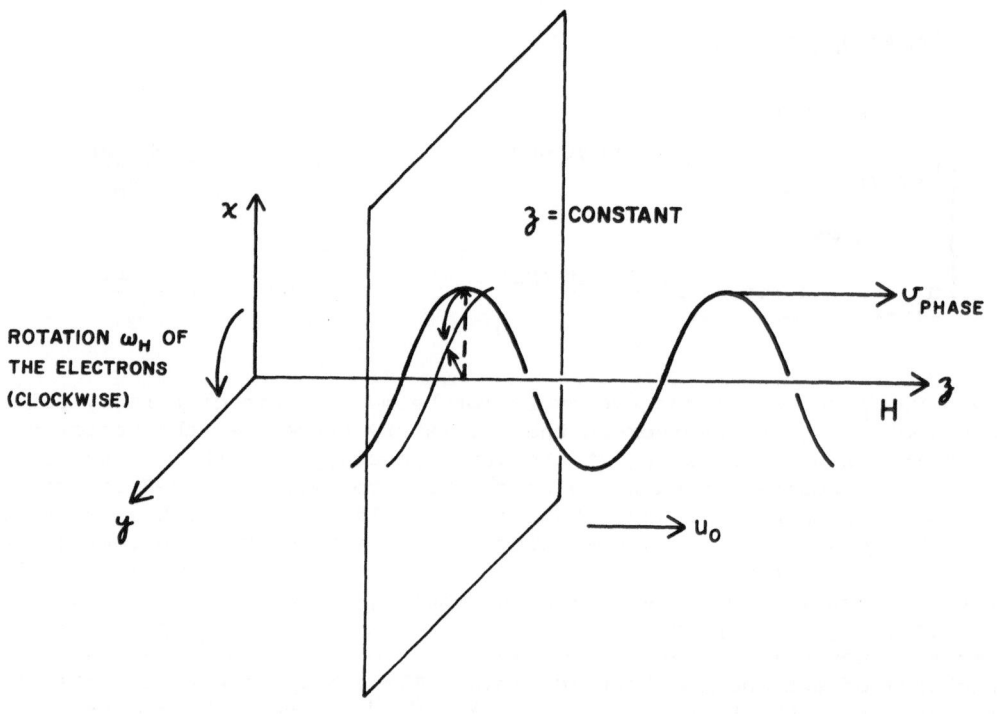

Fig. 3.

The line is shaped as a __left-handed helix__. The pattern moves forward, without rotating, with the velocity v_{phase}. This left-handed helix striking a constant z plane has its point of intersection rotating in the __clockwise__ direction, for an observer looking in the same direction than H and u_o.

2.   S-wave $\boxed{\text{any } \omega}$ $\underline{\text{and}}$ B-wave $\boxed{\omega < \omega_H}$

In both cases the phase velocity is smaller (algebraically) than the beam velocity. The line is now shaped as a <u>right-handed helix</u>.

SHAPE OF A SINGLE LINE IN THE S AND B CYCLOTRON WAVES

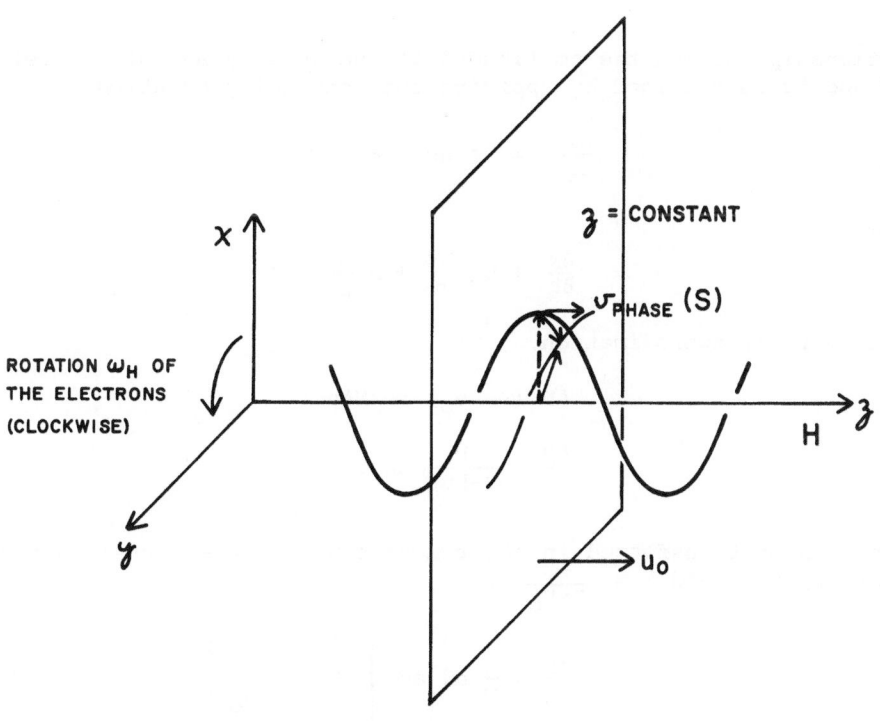

Fig. 4.

For the S-wave the helical pattern moves forward with the velocity v_{phase}. The rotation of the intersection with a plane z = constant is <u>counterclockwise</u> (looking in the same direction as H).

For the B-wave v_{phase} is negative. The helical pattern moves backward, and now the rotation of the intersection with a constant z plane is <u>clockwise</u>. This polarization is, of course, in agreement with the same property given above for the F-wave, since the B-wave is a branch of the complete "fast" beam wave. The important point here (of which use will be made later in this text) is that for the frequency range o < ω < ω_H (to which the E.M. wave propagating in the whistler mode is restricted) only the B-cyclotron wave on the beam has the same circular polarization as a whistler mode E.M. wave propagating in the same direction, therefore opposite to the beam velocity u_o.

Energy of Space-Charge Waves. Waves of Positive and Negative Energies.

For the study of the coupling with electromagnetic waves, it is important to have a deeper understanding of the structure of beam waves. Besides the dispersion law the partition of energy in the wave should also be considered. The remarkable concept of waves of negative energy will emerge from this study: the total energy of the beam is diminished by the presence of that type of space-charge wave. The beam (N_O, u_O) with a superposed space-charge wave is represented by the equations:

$$\begin{cases} N = N_O + \delta N\, e^{\,i(\omega t - kz)} \\[2mm] u_z = u_O + \delta u\, e^{\,i(\omega t - kz)} \end{cases}$$

The relationship between the amplitudes of the density and of the velocity perturbations δN and δu is obtained by applying the continuity equation:

$$\frac{\partial N}{\partial t} + \nabla \cdot Nu = 0$$

along the beam, or

$$\frac{\partial N}{\partial t} + u_z \frac{\partial N}{\partial z} + N \frac{\partial u}{\partial z} = 0$$

This furnishes in first approximation:

$$\delta N\,(\omega - ku_O) = kN\delta u$$

or

$$\frac{\delta N}{N} = \frac{k}{\omega - ku_O}\,\delta u \ .$$

This expression can be transformed in other more convenient expressions since (see Appendix A) $\omega - ku_O = \Omega_p$ and $k = \dfrac{\Omega_p}{v - u_O}$

$$\frac{\delta N}{N} = \frac{k}{\Omega}\,\delta u \quad \text{and} \quad \boxed{\frac{\delta N}{N} \qquad \frac{\delta u}{v - u_O}}$$

In this last form it is easy to see that

   —when $v < u_O$ (algebraically), S and the B waves, $\frac{\delta N}{N}$ is of opposite sign to δu: the density and velocity fluctuations are of opposite phases. The density is larger where the velocity is minimum; this organization will produce a decrease of the kinetic energy of the beam.

   —when $v > u_O$, F wave, $\frac{\delta N}{N}$ is of the same sign as δu. The density is larger where the velocity is maximum.

Space-charge waves, therefore, have two fundamentally distinct types of organization. The situation is the same as for the waves, giving the density of presence of free electrons moving in a periodic potential in the quantum theory of metals. The electrons are not trapped in the potential troughs of the wave, since relative to the wave (velocity v) their average velocity is $v - u_O$. But, there is an important difference: in solid state theory the potential is imposed at least in first approximation by the lattice structure. In the case of electron beams the potential results from

184

the distribution of the electrons themselves: this is a <u>self-consistent</u> type of problem, and a quite complicated one. In particular, the potential and the velocity fluctuations cannot be strictly sinusoidal, giving rise to harmonics even in first approximation.

The total energy of each wave is obtained by subtracting from the energy of the perturbed beam, averaged out over many wave lengths at a given instant, the energy of the unperturbed beam. The energy is the sum of the potential electrostatic energy and of the kinetic energy. One needs the relation between the electric field and the density fluctuation. Using Gauss' law:

$$\nabla D = 4\pi Q \qquad D = \epsilon_0 E$$

and the wave of the electric field:

$$E_z = \delta E_z \; e^{i(\omega t - kz)}$$

one obtains $\boxed{4\pi \; (-e\delta N) = -ik\delta E_z}$

The potential energy is $P = \dfrac{\overline{E.D}}{8\pi}$

$$P = \pi e^2 \; (\delta N)^2 \; \frac{1}{k^2}, \text{ always a positive quantity}$$

Eliminate k and δN as functions of δu, Ω, v and u_0

$$P = \tfrac{1}{4} \; mN_0(\delta u)^2$$

The kinetic energy is obtained in averaging $\tfrac{1}{2} \; mNu_z^2$. One finds:

$$\overline{Nu_z^2} - N_0 u_0^2 = \tfrac{1}{2} \; N_0(\delta u)^2 + u_0 \; \delta u \delta N$$

The first term is always positive, but the second, which expresses the correlation between δu and δN discussed above, may be of either sign. In this second term eliminate δN in function of δu:

$$K = \tfrac{1}{4} \; mN_0(\delta u)^2 + \tfrac{1}{2} \; mN_0(\delta u)^2 \; \frac{u_0}{v-u_0}$$

The total energy of the wave is now:

$$P + K = \tfrac{1}{2} \; mN_0(\delta u)^2 \left[1 + \frac{u_0}{v-u_0} \right] = \tfrac{1}{2} \; mN_0(\delta u)^2 \; \frac{v}{v-u_0}$$

From the dispersion law:

$$\frac{v}{v-u_0} = \mp \; \frac{\omega}{\Omega} \qquad \begin{cases} \text{upper sign for S wave} \\ \text{lower sign for F+B waves} \end{cases}$$

Therefore, the <u>total</u> energy of the wave is

$$\boxed{\mp \; \tfrac{1}{2} \; mN_0(\delta u)^2 \; \frac{\omega}{\Omega}}$$

<u>negative for the S wave</u> and positive for the F+B wave.

Consequently, a dissipative process which extracts energy from the initially unperturbed beam, such as for example in producing an electromagnetic wave, tends to increase the amplitude of the S wave.

In summary, it is also interesting to indicate the following properties:

-the <u>potential</u> energies are positive and equal for the two waves. In units of energy density $\frac{1}{2} mN_0(\delta u)^2$ the contribution is $+ \frac{1}{2}$.

-the <u>kinetic</u> energies, for S and F+B waves of equal amplitude, are far from being equal and can be negative as well as positive. In the same unit $\frac{1}{2} mN_0(\delta u)^2$ the kinetic energy is:

$$\frac{1}{2} + \frac{u_0}{v-u_0} = \frac{1}{2} \frac{v+u_0}{v-u_0}$$

In terms of frequency, using the dispersion law to express v,

for S wave   K.E. $= -\left(\frac{\omega}{\Omega} + \frac{1}{2}\right)$ <u>always negative</u>

for F-B wave   K.E. $= \frac{\omega}{\Omega} - \frac{1}{2}$

negative for $0 < \omega < \frac{\Omega}{2}$ (in the B wave range)

positive for $\omega > \frac{\Omega}{2}$

zero for $\omega = \frac{\Omega}{2}$

The ratio of the two kinetic energies is

$$\frac{S}{F} = \frac{1 + 2\frac{\omega}{\Omega}}{1 - 2\frac{\omega}{\Omega}}$$

Contrary to ordinary elastic longitudinal waves, the potential energies and the kinetic energies are not equal. This illustrates the particular and somewhat surprising properties of a plasma beam, resulting from the peculiar role of the plasma oscillations Ω_P.

<u>Remark:</u> When the two waves S and F+B are simultaneously excited with the same amplitude at the same frequency ω the total wave energy added to the beam is not zero. It is not correct to only add the two opposite energies of each wave, independently excited, as a naive usage of the energy formula would lead one to believe. The two waves propagating at two different phase velocities still have definite phase relationships and therefore a correlation of the electron densities and motions exists in such a way that the total energy required for the simultaneous excitation is positive.

FREQUENCIES GENERATED IN THE COUPLING BETWEEN THE BEAM WAVES AND THE E.M. WAVE, PROPAGATING IN THE WHISTLER MODE

The production of E.M. waves now results from the coupling between the beam waves and the E.M. waves traveling together at the same phase velocity for the same wave number, therefore, at a common frequency $\omega = k v_{phase}$. The source of energy of the radiation is the initial kinetic energy of the unperturbed beam. The exact mechanism of transfer of energy from the beam to the waves (both beam waves and E.M. waves) is quite complicated. The general principle has been indicated in the fifth section of this text. Only the coupling conditions will be discussed, but a more complete theory should examine in detail the mechanism of extraction of energy from the beam. The question of essential interest here is the fact that for given beam properties (velocity, density) several discrete frequencies can be emitted. Since the dispersion laws for the beam waves and for the E.M. whistler mode are different, it is only for special values of the frequency that the wave lengths are also the same. In such a case, the electric field structure is common to both types of waves and the two waves stay in phase and interact over an appreciable distance, until the slow gradients in the ambient plasma along the magnetic line of force modify the relative phase velocities to such an extent that the two waves become decoupled at that frequency. One sees that for practical applications of the theory to the true situation of a slowly varying medium one should introduce the concept of a "coherence length" for discussing such questions as the intensity of the radiation.

Coupling With Space-Charge Waves: The Generalized Traveling Wave Tube Mechanism.

The coupling results from the longitudinal electric field common to the two types of waves. This supposes that the wave front of the whistler mode is of finite transverse extension, as of course it is in general, and as it has already been discussed previously. The coupling conditions are obtained by expressing the equality of the two phase velocities.

$$V \text{ (whistler)} = v \text{ (beam wave)}$$

or
$$\frac{1}{n^2} = \frac{v^2}{c^2}$$

1. With the S-waves: $\left(\text{remember } \sigma = \frac{\omega}{\omega_H} \ ; \ b = \frac{\omega_p}{\omega_H} \right)$

$$\frac{\sigma(1-\sigma)}{b^2} = \left(\frac{u_o}{c}\right)^2 \frac{1}{\left(1 + \frac{\Omega}{\omega}\right)^2}$$

or
$$b \, \frac{u_o}{c} = \left[\sigma(1-\sigma)\right]^{\frac{1}{2}} \left(1 + \frac{\Omega}{\omega_H} \, \frac{1}{\sigma}\right)$$

2. With the F+B waves:

$$\frac{\sigma(1-\sigma)}{b^2} = \left(\frac{u_o}{c}\right)^2 \frac{1}{\left(1 - \frac{\Omega}{\omega}\right)^2}$$

or
$$b \, \frac{u_o}{c} = \left[\sigma(1-\sigma)\right]^{\frac{1}{2}} \left(1 - \frac{\Omega}{\omega_H} \, \frac{1}{\sigma}\right)$$

It is seen that the important parameter $b\frac{u_0}{c}$, which characterizes the combination of the properties of the ambient plasma and of the beam, appears again as in the elementary theory. In addition, the finite density of the beam appears in the ratio of the beam plasma frequency Ω to the gyrofrequency ω_H. In the limit of $\frac{\Omega}{\omega_H} = 0$ both formulas recover the result of the old elementary theory, which is entirely contained in the new development. However, for typical situations in the Earth's magnetosphere, using the values given above for the densities of high energy fluxes $\frac{\Omega}{\omega_H}$ could have any value from 10^{-3} to 0.5 or even perhaps somewhat larger, depending on the position along the line of force.

It is convenient to represent the frequencies emitted by a diagram similar to the diagram used in the elementary theory: (see Fig. 5) (For numerical applications, it

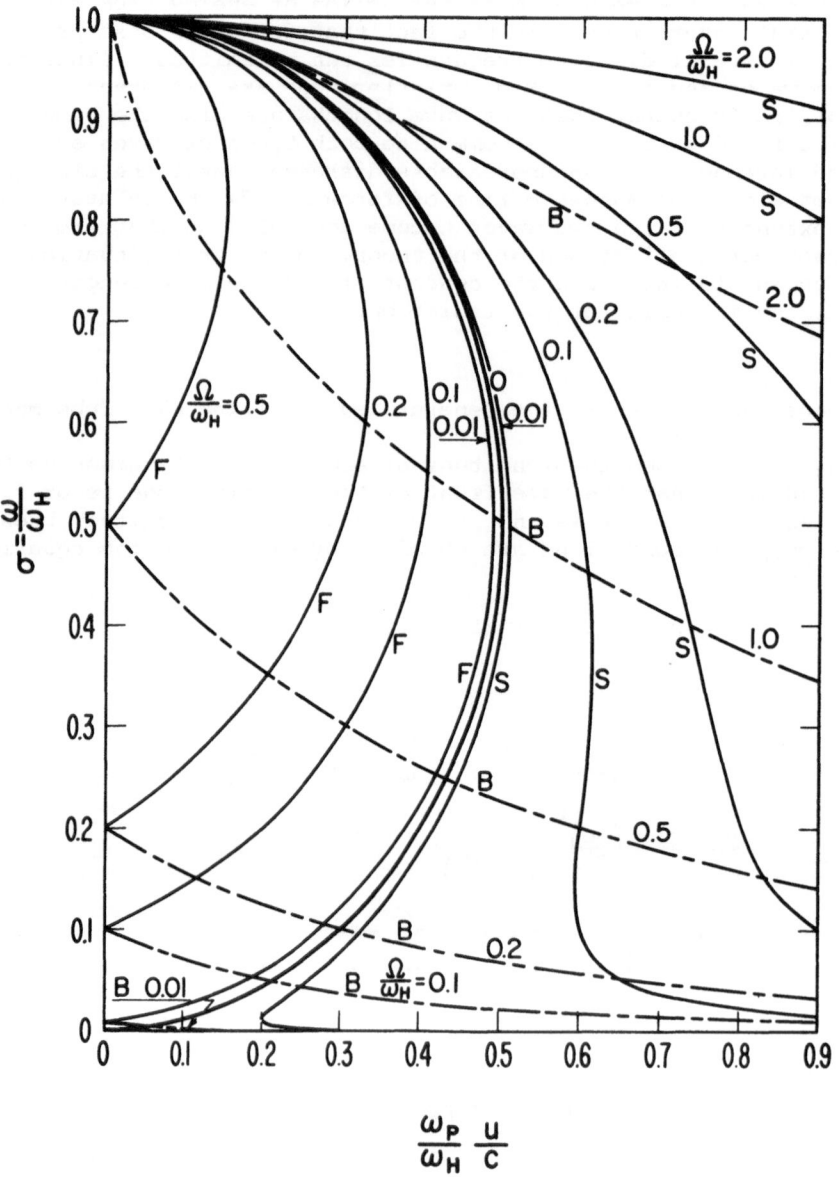

Fig. 5.   Frequencies emitted in the whistler mode by "Traveling wave tube" mechanism.

188

is easier to tabulate the values of $b\frac{u_0}{c}$ obtained as a function of σ, and to solve for the emitted frequencies σ by graphs and successive approximations.) The comparison with the results of the elementary theory (case of vanishingly small beam density) indicates the following important differences:

- The new theory is much "richer" in possibilities. While the old theory indicated only two emitted frequencies only when $b\frac{u_0}{c}<\frac{1}{2}$, the new theory furnishes always some emitted frequencies whatever the value of $b\frac{u_0}{c}$. Each type of space charge wave may furnish as many as three frequencies, therefore altogether six different discrete frequencies may be emitted in favorable circumstances. Consider the case of small values of $\frac{\Omega}{\omega_H}$. Each of the two frequencies predicted by the old theory (for $b\frac{u_0}{c}<\frac{1}{2}$), and symmetrically placed relative to $\sigma=\frac{1}{2}$ ($\omega=\frac{\omega_H}{2}$), is now split in a "doublet" separated by a small frequency interval. Consider the old "low" frequency σ_1: it is now replaced by a doublet in which the lower is due to coupling with an S-wave, and the upper is due to coupling with the F-wave. This order is reversed for the doublet replacing the old "high" frequency σ_2. The point of great physical interest here is that the multiplicity of solutions, near each other when $\frac{\Omega}{\omega_H}$ is small, gives the hope to understand, in a natural way and without special hypotheses, the "fine structure" of the VLF emissions which often display a frequency "split" in their dynamic spectra.

- A completely new phenomenon appears: emission of a VLF wave in a <u>direction opposite to the beam velocity</u> by coupling with the B wave on the beam. This phenomenon will be called "emission in the backward direction," however, one should be very careful to avoid confusion, the E.M. wave emitted is still a <u>forward</u> wave (its phase and group velocity are in the same direction) but it is excited in a direction opposite to the beam velocity by a backward beam wave. The phenomenon is similar to the functioning of certain "backward traveling wave tubes," however, in certain categories of these tubes it is the E.M. wave which is backward by means of special circuits producing a dispersion such that $\frac{d\omega}{dk}$ is negative, and the coupling in that case is with a forward beam wave (for example, with an S wave at a frequency ω much larger than the beam plasma frequency Ω_P: low density beams for amplifiers). Too superficial analogies could be misleading for an understanding of the true state of affairs in the Earth's magnetosphere.

The emission in the backward direction is always present, whatever the values of $b\frac{u_0}{c}$ and $\frac{\Omega}{\omega_H}$ are. This new possibility is of great interest for the natural VLF emissions in the magnetosphere. From the study of observations at magnetically conjugate points the present writer has found a class of VLF emissions triggered by natural whistlers and emitted in a direction opposite to the direction of the propagation of the whistler ("backward interactions"), probably the mechanism, not yet completely explicated, is in two steps: the whistler "organizes" beam waves, when and if a stream of high energy is present; in turn, the beam waves produce an emission in the backward direction.

- even in the case $b\frac{u_0}{c}>\frac{1}{2}$, for which the elementary theory indicated no emission, there are always at least two frequencies emitted: one in the backward direction (coupling with the B wave, just discussed above), and the other in the forward direction due to the S-beam wave. For practical applications, most of the time the value of $b\frac{u_0}{c}$ is quite small in the magnetosphere (case of slow beams), but it may be of interest for streams of high energy electrons near the equator on a line of force where $b\frac{u_0}{c}$ may take large values.

Remark that even small values of σ are important since with the observed VLF emissions between roughly 1 and 10 Kc/sec the value of σ near the Earth is as low as 10^{-3} to 10^{-2}.

It is difficult to predict a priori which frequency will produce the strongest emission. This depends not only on the rate of growth and the type of coupling (with

S, F or B wave), but also on the coherence length for each emitted frequency, therefore, on the exact distribution of electron density in the magnetosphere and the
location at which each one of these frequencies are emitted. In addition, the growth
of the electromagnetic wave is probably limited by <u>non-linear</u> effects when the
amplitudes become appreciable. Since in a space-charge wave the amplitude of the
longitudinal electric field is related to the amplitude of the density modulation on
the beam

$$4\pi(-e\delta N) = - ik\delta E_z$$

and since the beam density modulation is limited by

$$\left(\frac{\delta N}{N}\right)_{Beam} \ll 1$$

considering the low values of the peak density of the electron streams observed in
the magnetosphere (N_{Beam} 10^{-2} to 1 e/cm^3, as indicated in a preceding section), one
obtains here a quantitative limitation on the longitudinal electric field, and
therefore on the intensity of the VLF emission which can be produced. When the beam
modulation $\frac{\delta N}{N}$ is not small, new, more complicated phenomena appear, particularly the
production of space-harmonics, and therefore the possibility to excite, under certain
specific conditions, <u>harmonics and subharmonics</u> for the VLF emissions. These remarks
will be sufficient here to indicate in which direction new developments of the theory
are taking place.

It is perhaps not without interest to underline the role of the coupling with
the S wave, producing an E.M. wave in the forward direction. Since the total energy
density of the S wave is negative

$$- \tfrac{1}{2}\, mN_0(\delta u)^2\, \frac{\omega}{\Omega_P} \qquad (ergs/cm^3)$$

relative to the energy of the unperturbed beam, the coupled E.M. wave (which has
always a positive energy density) tends to increase the amplitude δu of the S-wave
by extracting energy from the beam wave. In this instance the mechanism which tends
to increase the amplitudes of both waves is easy to understand and could be symbolized
in the following diagram.

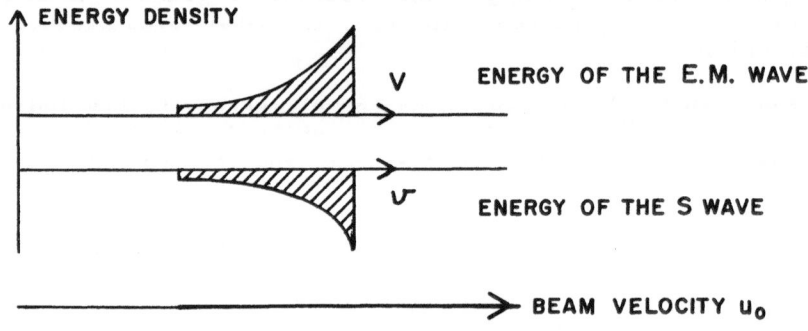

Fig. 6.

The energy transfer for the other couplings is more subtle and will not be explained in detail here. They could be symbolized by the following diagrams.

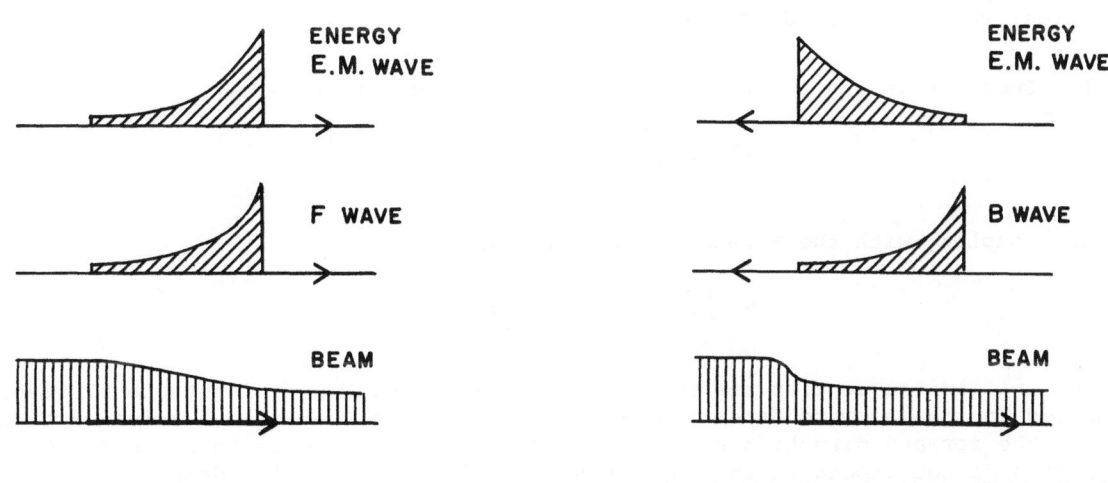

COUPLING WITH F WAVE ($\omega < \Omega_P$)

ENERGY
E.M. WAVE

F WAVE

BEAM

Fig. 7.

COUPLING WITH B WAVE ($\omega < \Omega_P$)

ENERGY
E.M. WAVE

B WAVE

BEAM

Fig. 8.

Coupling with Cyclotron Waves.

The coupling with cyclotron waves on the electron beam represents a new possibility not considered in the elementary theory. The proper frequency of the beam is now the gyrofrequency ω_H. Since the E.M. waves propagating in the whistler mode are restricted to the frequency range $0 < \omega < \omega_H$, no coupling is possible with the F cyclotron wave. This is a first difference with the case of space-charge waves. A further restriction comes from a consideration of the respective circular polarization of the two kinds of waves.

Comparing the equations of the cyclotron waves with those describing the whistler mode, one sees that the identity in form, including the polarization, is obtained only for the B-case of cyclotron waves. More precisely, in that case ($\omega < \omega_H$) k has a sign opposite to the sign of u_o, and consequently identity in form is obtained when u_o is positive between the B cyclotron wave and the whistler mode propagating in the direction opposite to the magnetic field direction. Both the beam wave and the E.M. wave propagate in the same direction and with the same polarization: these two waves could couple. When u_o is negative the B-wave now propagates in the direction of the magnetic field and again has the same polarization as the whistler mode propagating in that direction.

It can be seen that for the S-beam wave and the whistler mode propagating in the same direction the polarizations are opposed. Therefore, in general, the coupling cannot take place, or at least it will be very much weaker; a weak coupling can probably take place in the case of nonuniform waves in the x, y plane, via the longitudinal electric field of the whistler mode which modifies the phase of the cyclotron wave.

It is concluded that the principal effect with the cyclotron waves will be the production of an "emission in the backward direction" (the remarks made in the case of space-charge waves apply also here), but a much weaker emission due to the S wave should not be completely excluded.

The coupling conditions are again obtained by expressing the equality of phase velocities, which furnish directly:

1. Emission in the Backward Direction (coupling with B wave)

$$b^2 \left(\frac{u_O}{c}\right)^2 = \frac{(1-\sigma)^3}{\sigma}$$

2. Coupling with the S wave (weak or absent)

$$b^2 \left(\frac{u_O}{c}\right)^2 = \frac{(1-\sigma)(1+\sigma)^2}{\sigma}$$

The emitted frequencies are represented on the Fig. 9. At least one frequency is emitted in the backward direction, and it is possible that a weak emission takes place in the forward direction also. An important point here is that the emitted frequencies do not depend on the beam density, only the intensity does.

During the last few years several authors, and in particular Dowden, (1962) have considered that certain classes of VLF emissions are produced by Doppler-shifted cyclotron radiation from electrons. The mechanism has even been proposed as a substitution for the general theory of the VLF emissions, rejecting the elementary traveling wave tube mechanism of Gallet and Helliwell (1959). Since the radiated E.M. wave can propagate only at frequencies smaller than the gyrofrequency, only the radiation from electrons moving away from the observer was considered (emission in the backward direction). A formula identical to the B-case above was obtained for the frequency emitted. The reason for the identity is, of course, clear since the observed frequency is in all cases one of the Doppler-shifted proper frequency of the beam (see Appendix A, and the section on Beam Wave Properties). But the explanation from Dowden suffers many difficulties. Dowden himself soon realized that the intensity obtained from the incoherent radiation of the electrons will be many orders of magnitude too small, or will require impossibly large values of the electron fluxes. He then concluded (1963) that the process should be a coherent one. Two other points, however, should be emphasized: the emission process is a coupling between two kinds of waves; the coupling in the forward direction with an S cyclotron wave is certainly possible (and in fact is used in certain types of traveling wave tubes, but, of course, the external circuit guiding the E.M. wave is specially designed to produce an appreciable longitudinal electric field). This is not predicted in the viewpoint of Dowden since the Doppler-shifted frequency in that direction is larger than ω_H and cannot propagate. Dowden does not use the concept of cyclotron waves on the electron beam, which propagate at speeds respectively larger and smaller than the beam velocity, and he considers only the particle velocity.

The second point is of great interest for further development of the theory, which will not be developed here. The E.M. wave cannot propagate at frequencies $\omega > \omega_H$, but the beam wave can. Therefore, energy which propagates in the whistler mode could not pass the equator on a magnetic line of force because the gyrofrequency at the equator is too low, in the presence of an electron beam can be converted in a beam wave (both cyclotron waves and space-charge waves are possible) and be transported at the velocity u_O in form of electromechanical energy on the beam across the

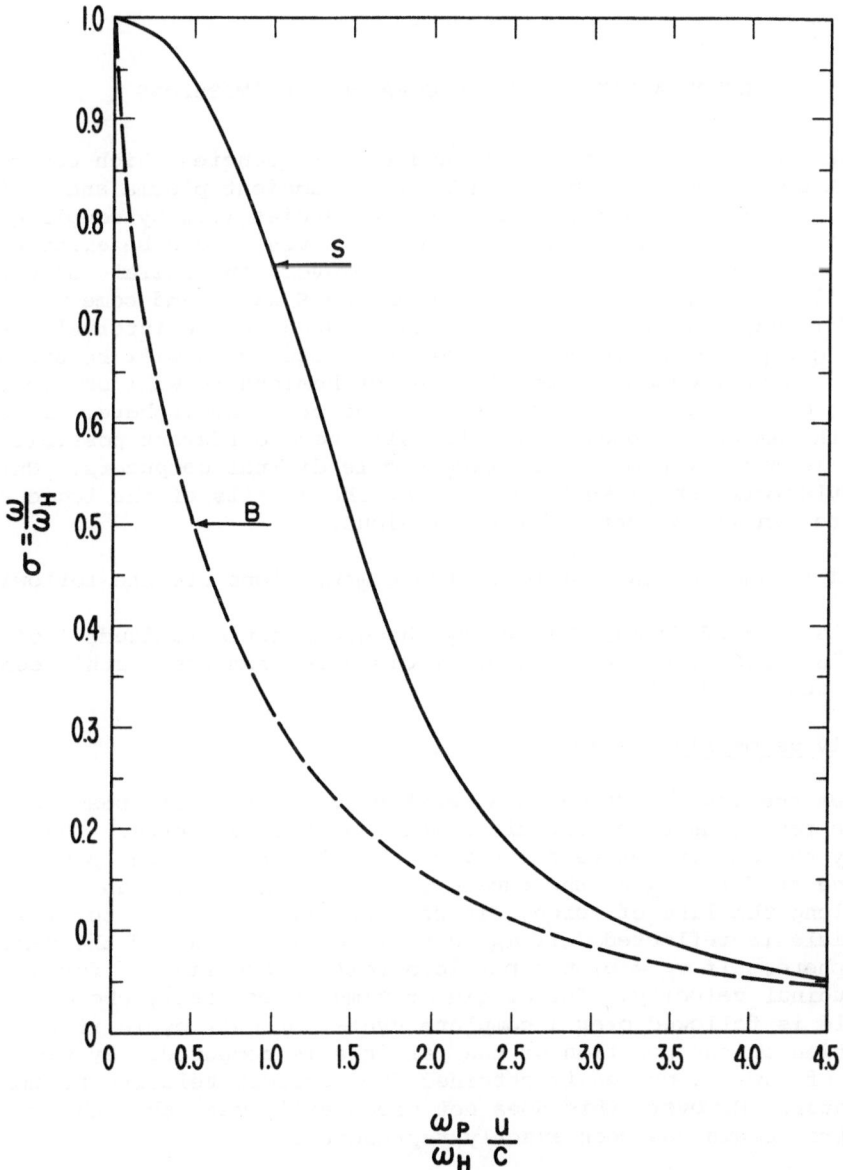

Fig. 9.   Frequencies emitted in the whistler mode by "cyclotron" interaction mechanism.

"forbidden region." At the other end the beam wave excites again the E.M. wave: the total time of propagation is now a composite of two modes of propagation in different regions.

The E.M. wave resulting from coupling with the S cyclotron wave, propagating at a phase velocity inferior to the particle velocity, in a medium of refractive index superior to unity is related to the "anomalous Doppler effect," discussed for the first time by the Russian scientist I. M. Franck (1942). The theory presented in these lectures (both cyclotron waves and space-charge waves) has been obtained by the present author several years ago and was already developed when the papers from Dowden did appear. The two different points of view are quite different, in spite of a superficial similarity for one of the frequencies emitted, and their comparison is presented here for the first time.

COMPUTATION OF THE SPECTRA OF VLF EMISSIONS

In the preceding sections the mechanisms and the frequencies which can be emitted <u>at a given place</u> knowing the local properties of the ambient plasma and of the beam have been discussed. As many as eight different frequencies (six by coupling with space-charge waves, and two by coupling with cyclotron waves) could be emitted, of which two are emissions in the backward direction. However, in general, several of these frequencies will be missing (particularly from the S wave) and some will be relatively weaker. In addition, the frequencies are distributed in the interval $o < \omega < \omega_H$ and ω_H varies along the line of force: some of the frequencies will be unobservably large or small. Also, some frequencies emitted in one hemisphere will be too low to propagate across the equator for being observed in the other hemisphere. In the numerical applications, the amount of bookkeeping for all these different possibilities is considerable, and is made by programs on large-scale digital computers. Only all the essential formulas will be given here, but not the details of the logic necessary to compute complete dynamic spectra of VLF emissions.

The principle and the main steps of the computations are the following:

a) Choose a line of force, defined by the geomagnetic latitude λ of intersection with the Earth's surface, or the equatorial distance from the Earth's center (in units of the Earth's radius).

A purely <u>geometrical</u> step.

b) Consider the law of motion of a relatively small plasma beam as specified by the motion of a representative particle along that line of force. The initial state is specified by the energy, or velocity v of the electron and the pitch angle relative to the line of force α_0 at the equator. Follow the motion as a function of φ the latitude along the line of force. In particular, there is a turning point φ_R at which the particle is reflected. If α_0 is very small $\varphi_R > \lambda$ and the particle dumps into the atmosphere. If $\alpha_0 = o$, the particle follows the line of force without changing its longitudinal velocity. The origin of time is generally chosen at the equator and the particle is followed over a complete cycle, or half cycle if it dumps. The time $\tau(\varphi)$ function of the position of the particle is computed. By varying α_0 and v a large number of laws of motion is obtained from uniform velocity to small bounces around the equator. However, this does not necessarily mean that the true law of motion of electron beams has been exactly represented.

This is a purely <u>dynamical</u> step.

c) Have a physically realistic model of the ambient electron density distribution in the exosphere and the ionosphere.

Some <u>physical</u> assumptions, based on <u>empirical</u> knowledge.

Then at any given position of the particle the component of the velocity parallel to the line of force $v_{\parallel}$ is known and the parameter $b \frac{v_{\parallel}}{c}$ is also calculated at each point.

d) At each value φ, the frequencies emitted are determined for each mode of coupling. Then the time of propagation of the radio wave θ_ω from φ to the Earth's surface

$$\theta_\omega = \frac{1}{c} \int_\varphi^\lambda n' \ ds = \frac{1}{c} \int_\varphi^\lambda n' \left(\frac{ds}{d\varphi}\right) d\varphi$$

at the group velocity $U_{group} = \frac{c}{n'}$ is calculated.

The total time $\quad t_\omega = \tau(\varphi) + \theta$

elapsing for the observation of the frequency ω is determined and tabulated versus ω for each mode of coupling. The calculations are made sequentially for a series of values φ.

Note that for the emissions in the backward direction

$$t_\omega = \tau(\varphi) + \frac{1}{c} \int_\varphi^{-\lambda} n' \left(\frac{ds}{d\varphi}\right) d\varphi$$

with the additional condition that $\omega > \omega_H$ at the equator, otherwise this frequency cannot propagate across the equator and is rejected.

In the following, the formulas corresponding to each main step are given in sequences, with the relevant figures.

FORMULAS FOR COMPUTING FREQUENCY-TIME SPECTRA

A. Formulas Defining a Line of Force; Dipole Magnetic Field.

1. $\quad r = \dfrac{R}{a} = \dfrac{R}{6371} = \dfrac{\cos^2\varphi}{\cos^2\lambda}$ Distance from the earth's center to point on line of force – in earth's radii a = 6371 km.

2. $\quad ds = ad\varphi \dfrac{\cos\varphi}{\cos^2\lambda} (4 - 3\cos^2\varphi)^{\frac{1}{2}}$ Geometrical definition of line of force.

3. $\quad H = 0.315 \dfrac{\cos^6\lambda}{\cos^6\varphi} (4 - 3\cos^2\varphi)^{\frac{1}{2}}$ Field strength (in gauss).

4. $\quad H_{eq} = 0.315 \cos^6\lambda$ Field strength at equator (in gauss).

5. $\quad \dfrac{H}{H_{eq}} = \dfrac{(4 - 3\cos^2\varphi)^{\frac{1}{2}}}{\cos^6\varphi}$

6. $\quad f_H = 2.800 \times 10^6\, H$ Gyrofrequency.

See Fig. 10.

PATH OF PARTICLE ALONG A LINE OF FORCE

PARTICLE OSCILLATES ALONG A SPIRAL TRAJECTORY BETWEEN $-\phi_R$ AND ϕ_R

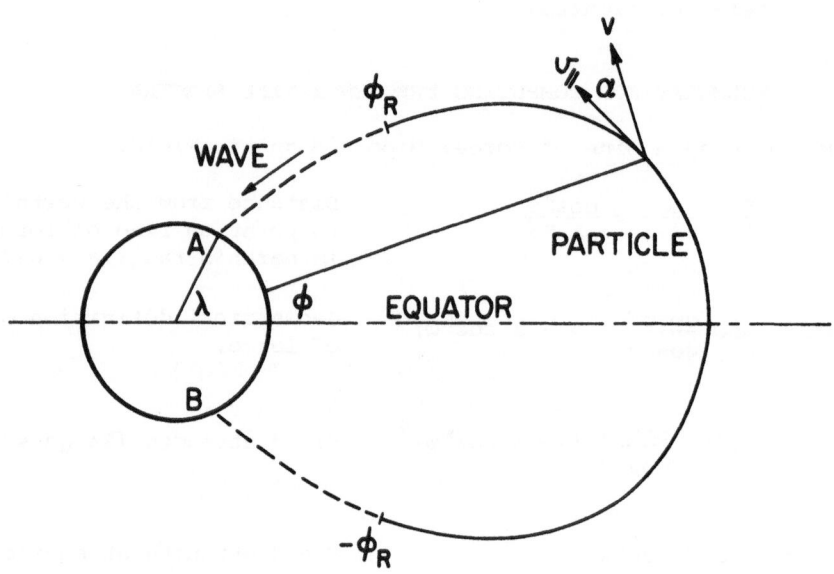

WHEN MOTION IS FROM $-\phi_R$ TO ϕ_R,
  FORWARD WAVE IS RECEIVED AT A;
  BACKWARD WAVE IS RECEIVED AT B.

Fig. 10.

B. Formulas for the Motion of Particles.

1. $K = \frac{1}{2} mv^2$
    kinetic energy of particle, invariant along trajectory.

2. $K = \frac{1}{2} m(v_\perp^2 + v_\parallel^2) = \frac{1}{2} mv_\perp^2 + \frac{1}{2} mv_\parallel^2 = K_\perp + K_\parallel$.

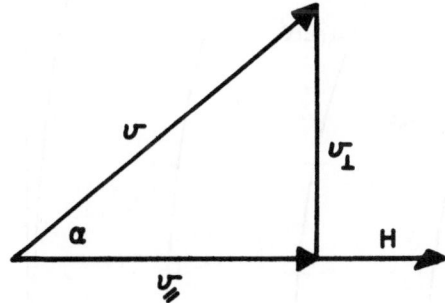

v = velocity of particle.

$v_\perp$ = the component of the velocity normal to the direction of the magnetic field.

$v_\parallel$ = the velocity in the magnetic field direction.

3. $\mu = \dfrac{\frac{1}{2} mv_\perp^2}{H}$
    the magnetic moment of the particle, invariant along the trajectory.

$$(0 \le \mu < \frac{K}{H})$$

4. $\sin\alpha = \dfrac{v_\perp}{v}$; $\tan \alpha = \dfrac{v_\perp}{v_\parallel}$
    α = the pitch angle of the velocity vector.

5. $\mu = \dfrac{\frac{1}{2} mv^2 \sin^2\alpha}{H} = \dfrac{K}{H} \sin^2\alpha$
    invariant along trajectory.

Since K and $\frac{K}{H} \sin^2\alpha$ are invariants of the motion,

$$\frac{\sin^2\alpha}{H} \text{ is constant.}$$

6. $\mu = \dfrac{K}{H_{eq}} \sin^2\alpha_0$
    where α_0 is the pitch angle of the velocity vector at the equator.

7. Then $\sin^2\alpha = \dfrac{H}{H_{eq}} \sin^2\alpha_0$
    which gives α as a function of H knowing α_0.

8. $H_{refl} = H_{eq} \dfrac{1}{\sin^2\alpha_0}$

9. $v = v \cos\alpha = v \sqrt{1-\sin^2\alpha} = v \sqrt{1- \dfrac{H}{H_{eq}} \sin^2\alpha_0} = \sqrt{1 - \dfrac{(4-3 \cos^2\varphi)^{\frac{1}{2}}\sin^2\alpha_0}{\cos^6\varphi}}$

Some curves of this function, with α_0 as parameter, are given in Fig. 11.

197

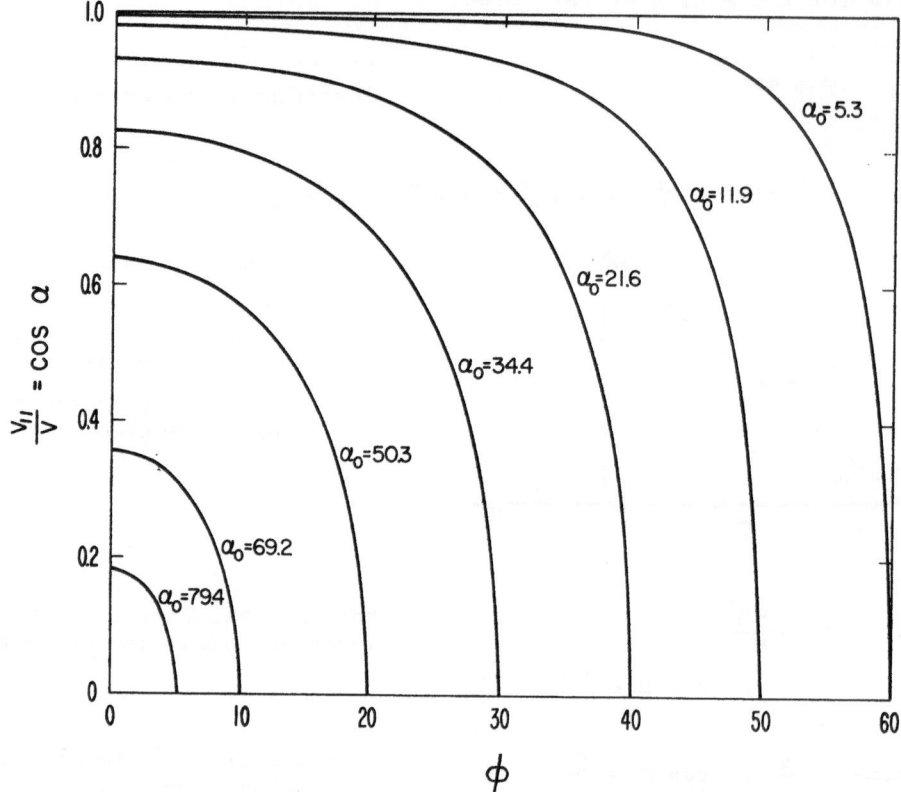

Fig. 11.   Longitudinal Particle Motion Along A Magnetic Line of Force.

10. $d\tau = \dfrac{ds}{v_{//}} \quad \dfrac{ds}{v\sqrt{1-\sin^2\alpha}}$ for particle, along a line of force.

11. $\dfrac{d\tau}{ds} \dfrac{ds}{d\varphi} = \dfrac{d\tau}{d\varphi} = \dfrac{a\ \cos\varphi(4-3\cos^2\varphi)^{\frac{1}{2}}}{v\ \cos^2\lambda\ \sqrt{1-\sin^2\alpha_o}} = \dfrac{a\ \cos\varphi(4-3\ \cos^2\varphi)^{\frac{1}{2}}}{v\ \cos^2\lambda\ \sqrt{1 - \dfrac{H}{H_{eq}}\ \sin^2\alpha_o}}$

12. $\dfrac{d\tau}{d\varphi} \quad \dfrac{R_o}{v}\ \cos\varphi \sqrt{\dfrac{4-3\ \cos^2\varphi}{1 - \dfrac{(4-3\ \cos^2\varphi)^{\frac{1}{2}}\ \sin^2\alpha_o}{\cos^6\varphi}}}$

in which $R_o = \dfrac{a}{\cos^2\lambda}$ distance in kilometers from the earth's
center to the equator of the line of force.

Define the time τ for the motion of the particle from the equator to φ:

$$\tau = \int_0^\varphi \frac{d\tau}{d\varphi} = \frac{R_0}{v} \int_0^\varphi \cos\varphi \sqrt{\frac{(4-3\cos^2\varphi)}{\frac{1-(4-3\cos^2\varphi)^{\frac{1}{2}}\sin^2\alpha_0}{\cos^6\varphi}}}$$

or $\tau = \dfrac{R_0}{v} G(\varphi_1, \alpha_0)$

in which the function $G(\varphi_1 \alpha_0)$ is given by the above integral.

The time is not a function of λ, therefore, one can define the particle motion for any λ by adjusting the scaling factor $\dfrac{R_0}{v}$.

One needs to know the maximum value of φ, turning point φ_R

$$\frac{H \text{ refl}}{Heq} = \frac{1}{\sin^2\alpha_0} \quad , \qquad \qquad \text{from the motion of particles}$$

$$\frac{H \text{ refl}}{Heq} = \frac{(4-3\cos^2\varphi_R)^{\frac{1}{2}}}{\cos^6\varphi_R} \qquad \qquad \text{from the geometry}$$

therefore,

$$\frac{1}{\sin^2\alpha_0} = \frac{(4-3\cos^2\varphi_R)^{\frac{1}{2}}}{\cos^3\varphi_R}$$

The "bounce period," the time for the particle to go from φ_R to $-\varphi_R$ and return to φ_R, is equal to $T = 4 \dfrac{R_0}{v} G(\alpha_0)$.

C.   Electron Density Model "M" For VLF Computations.

As an example, a particular model is given with which many of the computations have been performed.

At any point the electron density is expressed by the sum:

$$N_{e_{total}} = N_{e_{exosphere}} + N_{e_{ionosphere}}$$

where $N_{e_{exosphere}} = H \times .03472 \times F \times \left[1 - e^{-(h/h_{max})^2} \right]$

equivalent to $f_p^2 = F \times f_H$ Take F about 10^6 cps or less to satisfy the whistler observations.

$$N_{e_{\text{ionosphere}}} = N_{max} \; e^{\frac{1}{2}(1-z-e^{-z})}$$

$$N_{max} = (foF_2)^2 \text{x} \; 1.24049 \text{ x } 10^{-8} e/cm^3$$

$$= \text{maximum density in ionosphere}$$

$$foF_2 = \text{critical ionosphere frequency in Mc/sec}$$

$$z = \frac{h - h_{max}}{H}$$

$$h = \text{height in Km. above earth's surface}$$

$$h_{max} = \text{height of maximum density in ionosphere } (\sim 350 \text{ km})$$

$$H = \text{scale height in Km. (of the order of 100 km)}$$

Using the above model the typical behavior of the parameter $b = \frac{\omega_P}{\omega_H}$ along the line of force starting at $\lambda = 60°$ at the earth's surface (or parameter $L = 4$) is illustrated in Fig. 12. The height, in kilometers above the earth's surface, is

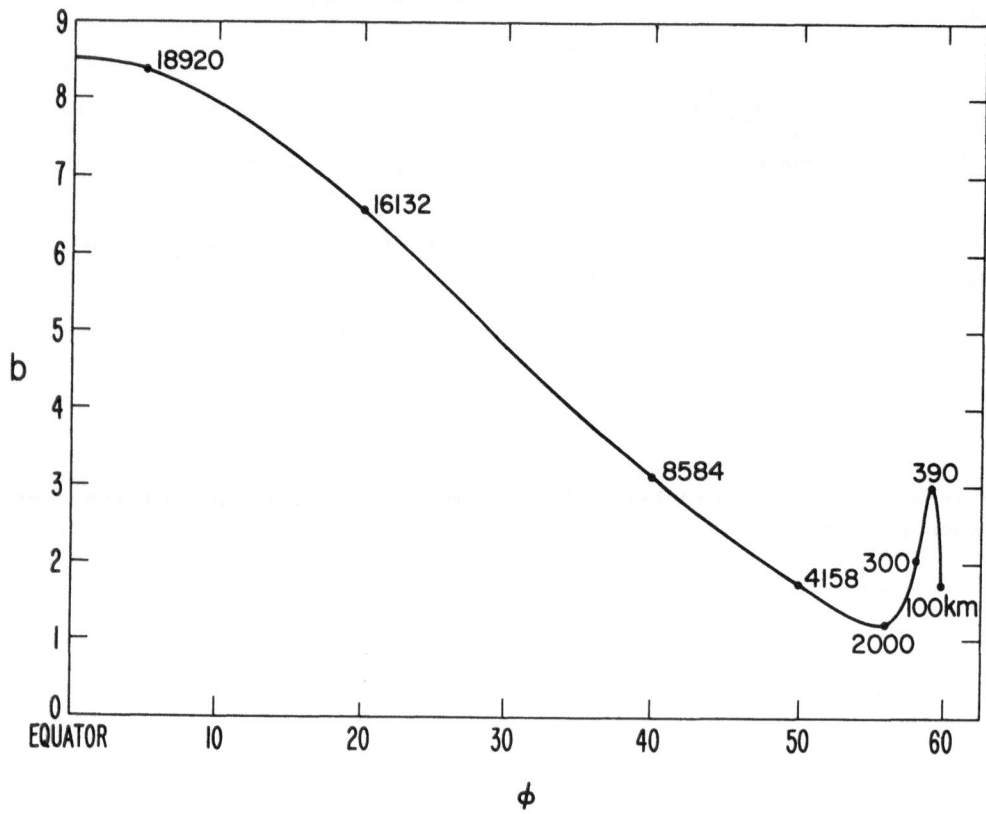

Fig. 12. Distribution of $b = \omega_P/\omega_H$ Along Line of Force
$\lambda = 60°$ For the Adopted Exosphere Model.

also indicated for a selected number of points along this line.

Next, the variation of the important dimensionless parameter $b\,\frac{v_{\parallel}}{c}$ along the same line of force is illustrated for a few particular cases in Fig. 13. In this case the electrons have all the same turning point at the latitude $\varphi_R = 40°$, and the total velocities considered vary from 10,000 to 100,000 km/sec^{-1}.

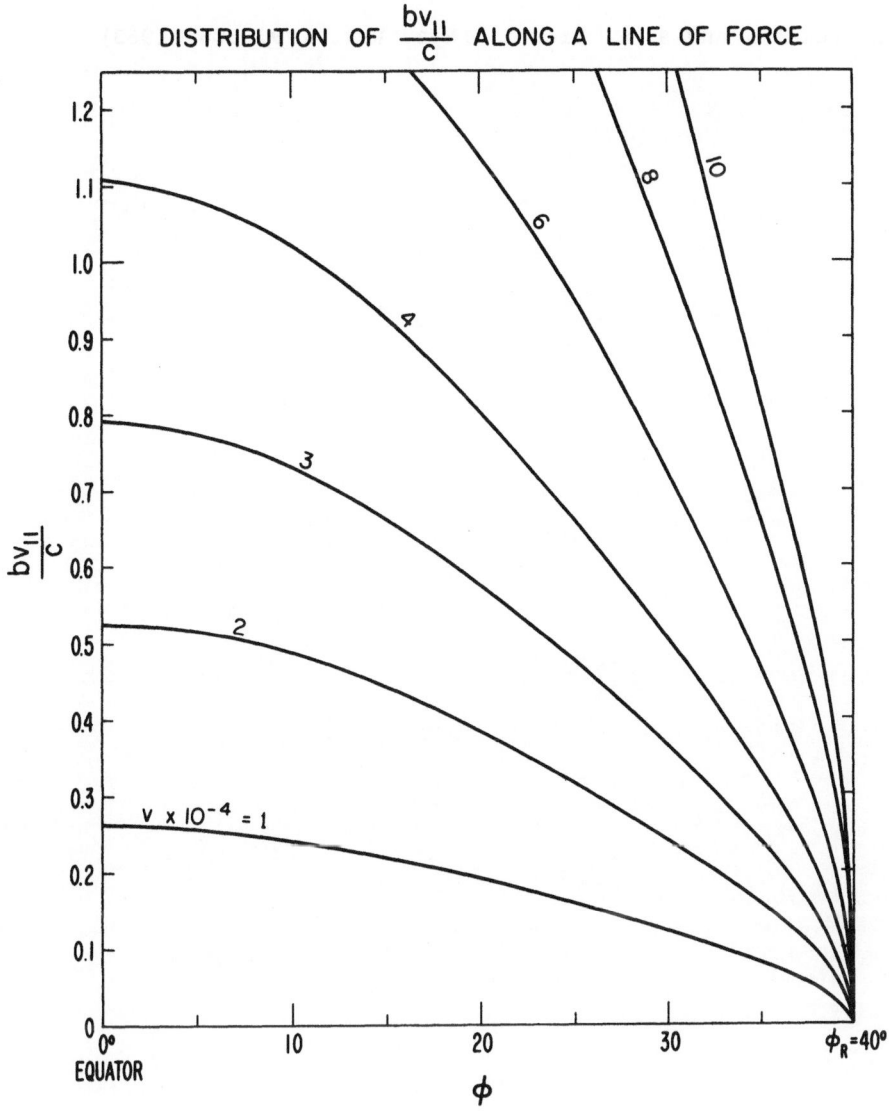

DISTRIBUTION OF $\frac{bv_{\parallel}}{c}$ ALONG A LINE OF FORCE

Fig. 13.

REFERENCES

Dowden, R. L.: J. Geophys. Res. <u>67</u>, 1745 (1962)

       Planet. Space Sci. <u>11</u>, 361 (1963)

Gallet, R. M.: Proc. Inst. Radio. Engrs. <u>47</u>, 211 (1959)

Gallet, R. M., and R. A. Helliwell: J. Res. Nat. Bur. Std. <u>63D</u>, 21 (1959)

Kato, Y., and T. Saito: J. Phys. Soc. Japan <u>17</u>, Suppl. A-D, 34 (1962)

Fields, H., G. Bekefi, and S. C. Brown: Phys. Rev. <u>129</u>, 506 (1963)

APPENDIX A

DEMONSTRATION OF THE DISPERSION LAW FOR BEAM WAVES
USING ONLY THE DOPPLER EFFECT

Equivalent Forms for the Doppler Effect (Non-Relativistic).

A wave observed in the laboratory frame $A\, e^{j(\omega t - kz)}$ in which $k = \frac{2\pi}{\lambda}$ becomes when it is observed by an observer moving with the velocity v (algebraic)

$$A\, e^{j(\omega t - kz^*)}$$

in which $\qquad z^* = z + vt$

But $\qquad A\, e^{j[\omega t - k(z+vt)]} \equiv A\, e^{j[(\omega - kv)t - kz]}$

or $\qquad A\, e^{j[\omega^* t - kz]}$ in which $\boxed{\omega^* = \omega - kv}$

In other terms: charged particles moving through an oscillating field do not "feel" the oscillations at the field frequency ω. Because of the Doppler effect, they feel the oscillations at the frequency ω^*.

In the laboratory frame $V_{phase} = \frac{\omega}{k}$, but in the frame of the moving observer $V^*_{phase} = \frac{\omega^*}{k}$. This is equivalent to say that when one changes from one frame of reference to another moving frame the wave number <u>k is an invariant</u> (it is only related to the <u>counting of the number</u> of crests per unit of length), but both ω and V change simultaneously.

From $\qquad \omega^* = \omega - kv = \omega - \frac{\omega}{V}v$

or $\qquad \boxed{\omega^* = \omega\left(1 - \frac{v}{V}\right)}$ classical form of the Doppler formula.

This is still equivalent to: $\frac{\omega^*}{k} = \frac{\omega}{k}\left(1 - \frac{v}{V}\right)$

or $\qquad V^* = V\left(1 - \frac{v}{V}\right) = V - v \qquad$ or $\boxed{V = v + V^*}$

The phase velocity in one system is equal to the phase velocity in the other system plus the relative velocity of the second system relative to the first one (non relativistic formula).

Dispersion Law for Beam Waves.

When the electrons of the beam have zero <u>relative</u> velocities (zero beam temperature), only oscillations at the proper frequency Ω may exist. They are stationary (do not propagate energy) and <u>any</u> wavelength is possible. A stationary wave is equivalent to the superposition of two traveling waves propagating in opposite directions and of equal amplitude. In the system of reference attached to the beams the phase velocities u of these two opposite waves are such that $\pm k = \frac{\Omega}{u}$, where k is any possible value and therefore u is positive or negative.

It is desired to find the phase velocities v of these two waves as observed by the laboratory observer. Now: $v = u_0 + u$.

Since k is invariant: $k = \frac{\Omega}{u} = \frac{\omega^*}{v} \qquad$ or $\qquad \frac{\omega^*}{v} = \frac{\Omega}{v - u_0}$.

But the value of k is arbitrary; therefore, the value of ω^* is arbitrary: any ω^* can be observed (or excited) in the laboratory coordinate. A wave, which in the laboratory frame is observed with the frequency ω^*, has a (laboratory) phase velocity:

$$1 - \frac{u_0}{v} = \frac{\Omega}{\omega^*} \quad\longrightarrow\quad \boxed{v = u_0 \frac{1}{1-\frac{\Omega}{\omega^*}}}$$

A discussion of the relative values of u and u_0, and of their sign shows that this formula covers both the cases of the S-wave (slow) and of the F-wave (fast, both ranges for which the F-wave is a "forward" or a "backward" wave).

The velocity $u = v - u_0$ could be either positive or negative. Since by definition Ω is positive, it is necessary to take k positive or negative to satisfy $\frac{\Omega}{u}$.

1) Then, if u is positive, v is a fortiori positive and ω^* is positive.

2) If u is negative, v could be positive (as long as $|u| < u_0$), and in that case, it is necessary to choose ω^* negative to satisfy $k = \frac{\omega^*}{v}$ since k is negative (like u):

$$\text{Then:} \quad v = u_0 \frac{1}{1+\frac{\Omega}{|\omega^*|}} \quad\quad\quad \text{S wave (u negative)}$$

3) If <u>v is negative</u> (when $|u| > u_0$), while u is also negative, ω^* is positive in order to satisfy $k = \frac{\omega^*}{v}$ since k is negative (like u):

$$v = u_0 \frac{1}{1-\frac{\Omega}{\omega^*}} \quad ; \quad\quad \text{v can be negative only for } \omega^* < \Omega.$$

Returning to the first case u positive, $v = u_0 + u > u_0$, therefore, this is the case $\omega^* > \Omega$ in $v = u_0 \dfrac{1}{1-\frac{\Omega}{\omega^*}}$ F wave

Consequently, there are <u>two</u> waves for a given excitation frequency ω^* and one given proper frequency Ω.

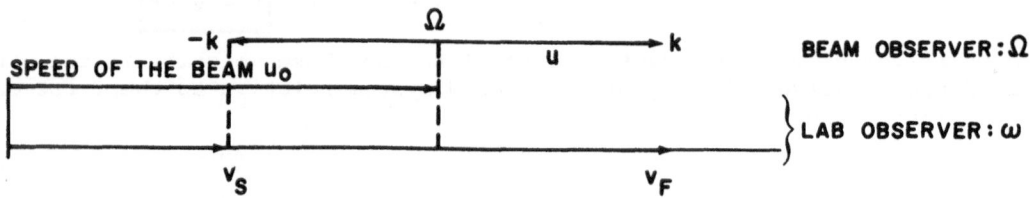

One could as well obtain these results in the following way:

$$\omega^* = \Omega - k u_0 \quad \text{in which k is invariant } k = \frac{\Omega}{v-u_0}$$

$$= \Omega + \Omega \frac{u_0}{v-u_0} = \Omega \frac{v}{v-u_0}$$

or $\dfrac{\omega^*}{v} = \dfrac{\Omega}{v-u_0}$ identical to the expression from which v has been obtained above.

In order to study v_{phase} for the beam wave the value of k was eliminated. One could as well eliminate v_{phase} to obtain the values for k

$$k = \frac{\omega}{v_{phase}} = \frac{\omega \pm \Omega}{u_0}$$

204

TERRESTRIAL EXTREMELY-LOW-FREQUENCY PROPAGATION

J. Galejs
Applied Research Laboratory
Sylvania Electronic Systems
A division of Sylvania Electric Products, Inc.
40 Sylvan Road, Waltham 54, Massachusetts

INTRODUCTION

Numerous papers have been dealing with extremely low frequency (ELF) phenomena recently and a comprehensive bibliography with over one thousand entries and abstracts has been prepared by Brock-Nannestad (1962). A chapter of a recent book by Wait (1962a) discusses ELF waves and it also contains a number of references to published papers. It will not be attempted to relist here all the references of the above publications. Instead, after a brief historical review of the work on ELF propagation and a summary of the fundamental concepts, much of the discussion will be devoted to recent analytical and experimental results.

The wave propagation in a spherical shell between the earth and the ionosphere has been originally analyzed by Watson (1919), and Sommerfeld (1949) considers in detail the propagation around the sphere. Based on this work, Schumann (1952a, 1954a, 1954b and 1957), Budden (1953) and Wait (1957) have developed the mathematical formulations currently used in the analysis of steady state and transient propagation. Also the current work on models involving more refined boundaries of the spherical shell is to a large extent based on this work. Schumann (1952b, 1952c and 1957) originated the concept of earth-to-ionosphere cavity resonances. Wait (1962a) has also provided many results principally in connection with sharply bounded isotropic or anisotropic ionosphere models, and he has clarified the limitations due to various flat earth approximations and has also established the near field behavior of ELF waves. By observing atmospherics, Chapman and Macario (1956) have estimated attenuation rates of ELF waves. König (1959 and 1961) obtained the initial experimental substantiation of Schumann resonances by observing waveforms in the output of a narrow band amplifier. Detailed frequency spectra of this noise suitable for estimating the cavity Q were first obtained by Balser and Wagner (1960). Following Raemer (1961b and 1961c), these noise spectra can be considered as the response of the earth-to-ionosphere cavity due to lightning flashes all over the world, but the sharply bounded ionosphere model of Raemer introduced excessive losses, and he does not succeed in reproducing the measurements of Balser and Wagner (1960). Galejs (1961b and 1961c) has introduced an isotropic ionosphere model of exponentially increasing conductivity, which is based on measured or calculated characteristics of the lower ionosphere. The attenuation rates of ELF waves provided by this model are in agreement with measurements by Jean et al (1961). The same model also permits a close reproduction of the noise spectra measurements in the resonance region. This work has been extended to anisotropic inhomogeneous ionosphere models with a horizontal magnetic field component by Galejs and Row (1963). The earth-to-ionosphere cavity resonances have been analyzed for a nonhomogeneous ionosphere in the presence of a radial (vertical) magnetic field by Thompson (1963). Further references to experimental and theoretical work will be given in the subsequent sections of the text.

The principal amount of ELF wave energy remains within the spherical shell between the earth and the ionosphere. The propagation of the waves depends on the ground characteristics and on the ionosphere properties. These boundaries of the spherical shell will be discussed in the section on boundary characteristics. The excitation of ELF waves by thunderstorms is reviewed in the following section.

Experimental data on propagation characteristics and noise spectra are described in the section devoted to thunderstorm excitation. The following five sections discuss several propagation models. This discussion considers various field representations, the sharply bounded ionosphere and the isotropic and anisotropic inhomogeneous ionosphere models.

BOUNDARY CHARACTERISTICS

Ground Surface.

The electrical ground characteristics have been recently summarized by Watt, Mathews and Maxwell (1963). Measured variations of ground conductivity with changing water content (Vilbig, 1960) can be approximated by the empirical formula of Albrecht (1963)

$$\sigma_g = 7.7 \times 10^{-5} \ (0.73 \ w^2 + 1) \ (1 + 0.03t) \qquad (1)$$

where w is the water content in per cent and t is the temperature in centigrade. It applies to soil, clay or sand specimens with dry conductivity of approximately 10^{-4} mho/m, w < 35, and to the temperature range of $0°C < t < 40°C$.

The ground is usually considered to be a homogeneous isotropic conductor of a given conductivity σ_g, and the displacement currents are negligible in the ELF range. The effective ground conductivity σ_g of 10^{-3} to 1 mhos/m is much higher than the effective ionospheric conductivity σ_i which is estimated to be of the order of 10^{-7} to 10^{-6} mho/m. The ionosphere contributes more than the ground surface to the wave attenuation and the ground conductivity may be assumed therefore infinite in the first approximation.

Multilayer ground models are significant in interpreting micropulsations or in electromagnetic prospecting (Wait, 1962a), but they will not be discussed here in more detail.

It has been suggested that high-frequency values of σ_g are not applicable at ELF, and that in specifying σ_g, one should account for induced polarization (Wait, 1959) due to underground mineral deposits, known to decrease substantially the magnitude of σ_g and to make it strongly dependent on frequency. This could conceivably have an effect on ELF propagation. Raemer (1961a) has considered this possibility and shown that the greatest possible effect of Underground Induced Polarization is decreasing the effective ground conductivity σ_g by a factor of 4. This possible decrease applies to frequencies of the order of 50 cps or less.

In later work, it is found that $\sigma + i\epsilon\omega$ is almost proportional to $A + iB \log \omega + iC$ where A, B, and C depend only weakly on ω. Thus, the apparent dielectric constants are enormous at ELF (Wait, 1963c).

Daytime Ionosphere.

A single composite daytime (and a single nighttime) model will be selected for the lower ionosphere based on various theoretical considerations and experimental data that are available in the literature. More detailed discussions of the D region have recently been presented by Belrose (1962).

In Fig. 1 the solid curve and short dash curve depict the composite electron number density versus altitude to be used for the geomagnetic latitudes 60°N and 0°, respectively. The long dashed curves represent the effective collision frequency.

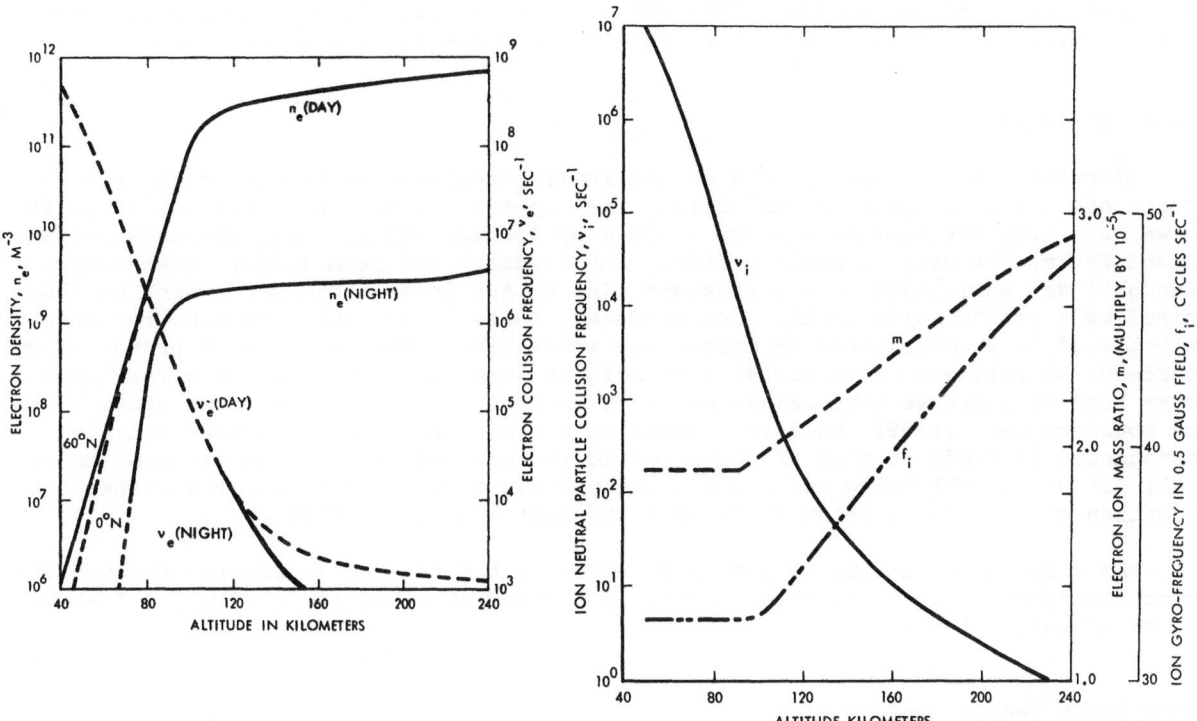

Fig. 1. Daytime and nighttime electron density and collision frequency profiles.

Fig. 2. Daytime and nighttime ion collision frequency mass ratio and gyro frequency.

From an altitude of 40 to 100 km the solid curve is a straight line approximation to the data points taken from the theoretical model of Nicolet and Aikin (1960), and experiments by Pfister, Ulwick and Vancour (1961), and Barrington and Thrane (1962). The data used from Nicolet and Aikin's paper are for the quiet sun directly overhead (0° zenith angle) at a geomagnetic latitude of 60°N. The dashed curve labeled 0° is obtained by reducing the cosmic ray ionization rate for altitudes below 75 km according to data supplied by Van Allen and quoted by Moler (1960). The curve above 100 km conforms to the ionosonde data reported by Pfister, Ulwick and Vancour. The curve for the effective collision frequency of electrons ν_e is arrived at in the following way: It is the sum of the values ν_{en} given by Nicolet (1959) model for the electron neutral particle collision rate, and ν_{ei} the electron ion collision rate calculated from the following formula quoted by Nicolet.

$$\nu_{ei} = \left[59.1 + 4.18 \log_{10}\left(\frac{T^3}{n}\right) \right]\frac{n}{T^{3/2}} \cdot 10^{-6} \qquad (2)$$

where n is the electron density (per meter^3) and T is the electron and ion temperature (thermodynamic equilibrium assumed). The data on temperature at altitudes above
100 km is that quoted by Ramanathan, Shonsle and Dagaonkar (1961). It refers to the
mid-afternoon ionosphere September 7, 1957, above Ahmedabad, India.

The ion neutral-particle collision rate and electron ion mass ratio shown in Fig.
2 is taken from Johnson's data (1961) (for the sunspot minimum). For altitudes below
100 km Johnson does not give the ion-neutral particle collision frequency. This has
been computed using ion mobility data quoted by Brown (1959), and the ion species
concentration (for the daytime D-region) quoted by Nicolet and Aikin (1960).

Nighttime Ionosphere.

Electron and ion density data on the lower ionosphere at nighttime is very
scarce and indicates greater variability than daytime data. For altitudes above 80
km we have used the sunspot minimum profile by Johnson (1961). For altitudes between
72 and 84 km, the data of Mechtly (1962) and Landmark and Leid (1962) (taken near
midnight) are approximated by a straight line on the logarithmic plot which has been
joined by a smooth curve to the Johnson data. Below 70 km nighttime electron density
is believed to be maintained by cosmic ray ionization. The rocket data presented by
Bourdeau, Whipple and Clark (1959) (for heights less than 70 km) shows a very much
lower rate of increase with height below 70 km than Fig. 1 indicates for above 70 km.
The Bourdeau data (1959), however, refers to a time near sunspot maximum and high
geomagnetic latitude so that if corrected to the equator and sunspot minimum (Moler,
1960), it would fall below the lower density limit shown in Fig. 1. The dashed continuation of n_e (night) below 72 km is a straight line extrapolation.

The electron collision frequency at night is the same as in daytime through the
lower ionosphere, but it is significantly decreased at night for altitudes of above
120 km (Johnson, 1961).

Ionospheric Tensor Conductivity.

The usual Appleton-Hartree formulas for the electrical conductivity of a lightly
ionized gas of different ionized species are used. Assuming a static magnetic field
B_0 to act in the z direction the Cartesian form of the tensor conductivity for a
harmonic field with time dependence exp($-i\omega t$) is,

$$[\sigma] = \begin{bmatrix} \sigma_1 & -\sigma_2 & 0 \\ \sigma_2 & \sigma_1 & 0 \\ 0 & 0 & \sigma_0 \end{bmatrix} \tag{3}$$

where

$$\sigma_0 = \epsilon_0 \sum_k \frac{\omega_k^2}{\nu_k - i\omega} \tag{4}$$

$$\sigma_1 = \epsilon_0 \sum_k \frac{\omega_k^2 (\nu_k - i\omega)}{(\nu_k - i\omega)^2 + G_k^2} \tag{5}$$

$$\sigma_2 = -\epsilon_0 \sum_k \frac{\omega_k^2 \, G_k}{(\nu_k - i\omega)^2 + G_k^2} \tag{6}$$

$$\omega_k = \left(\frac{n_k q_k}{\epsilon_0 m_k}\right)^{2\frac{1}{2}} \tag{7}$$

is the plasma frequency of the kth charges species and n_k, q_k, m_k are the number density, charge and mass of the kth specie particle.

$$G_k = \frac{q_k B_0}{m_k} \tag{8}$$

is the kth species gyrofrequency, ν_k is the effective collision frequency between k^{th} species particle and all other constituents of the plasma. For electrons ($k = 1$) in a static field of 0.5 Oersted ($B_0 = 0.5 \times 10^{-4}$ MKS units), $\omega_1^2 = 3.185 \, n_e \times 10^3$, and $G_1 = -1.04 \times 10^7$.

Numerical computations of σ_0, σ_1 and σ_2 have been made using the data shown in Fig. 1 and 2 for frequencies between 0 and 3000 c/s, but space does not permit recording all these results here. The tensor components of the electronic conductivity (the $k = 1$ terms of Eq. (7) to (9)) and/or the electronic plus ionic conductivity have been plotted as solid and dashed curves in Fig. 3 through 8 for f = 0, 20, and 300 cps.

For frequencies above the ion gyrofrequency, which lies in the range 30 to 50 cps, the numerical computations show that the electron contribution to the conductivity tends to dominate; the more so as the frequency increases. Furthermore, the electronic conductivity computed by setting f = 0 is a good approximation to the real part of the total conductivity provided the frequency of interest is lower than the electron collision frequency and the electron gyrofrequency. For frequencies below the ion gyrofrequency the electronic plus ionic conductivity computed by setting f = 0 approximates to the real part of the total conductivity.

In the absence of magnetic field $\sigma_2 = 0$ and $\sigma_0 = \sigma_1$. For $\omega = 0$ the conductivity is real and is given with neglected ions by

$$\sigma_{e\ell} = \frac{\epsilon_0 \omega_1^2}{\nu_1} = 2.83 \times 10^{-2} \, \frac{n_1}{\nu_1} \tag{9}$$

with n_1 defined per cm^3. Through the lower ionosphere of h < 70 km this electronic conductivity is given approximately by

$$\sigma_m(z) = \sigma_m \exp[\beta_m(z - z_0)] \tag{10}$$

where the subscript m allows to differentiate between different regions with different exponential parameters. For daytime m = 1 or 2 and the ionosphere is bounded at h = 45 km. For nighttime m = 3 or 4 and the ionosphere is bounded at h = 60 km. For $z_0 = 60$ km, σ_m and β_m is considered to assume the values

$$\sigma_1 = 6.35 \times 10^{-9} \qquad \beta_1 = 0.382$$

$$\sigma_2 = 4.63 \times 10^{-8} \qquad \beta_2 = 0.308$$

$$\sigma_3 = 6.5 \times 10^{-10} \qquad \beta_3 = 0.44 \qquad (11)$$

$$\sigma_4 = 10^{-9}, \qquad \beta_4 = 0.4$$

where σ is in mho m^{-1} and β is in km^{-1}.

The cosmic rays represent an ionization source effective even at altitudes below the D-layer. The calculated and measured atmospheric conductivity (Bourdeau, Whipple and Clark, 1959) may be approximated by Eq. (10) with $\sigma_0 = 10^{-10}$ mho/m, $z_0 = 50$ km, and $\beta = 0.138$ km^{-1}.

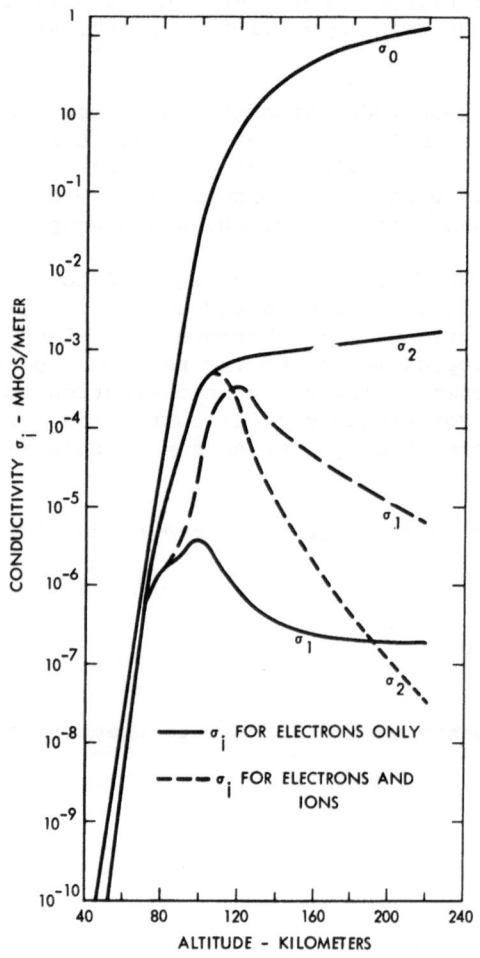

Fig. 3. Components of the daytime tensor conductivity. F = 0

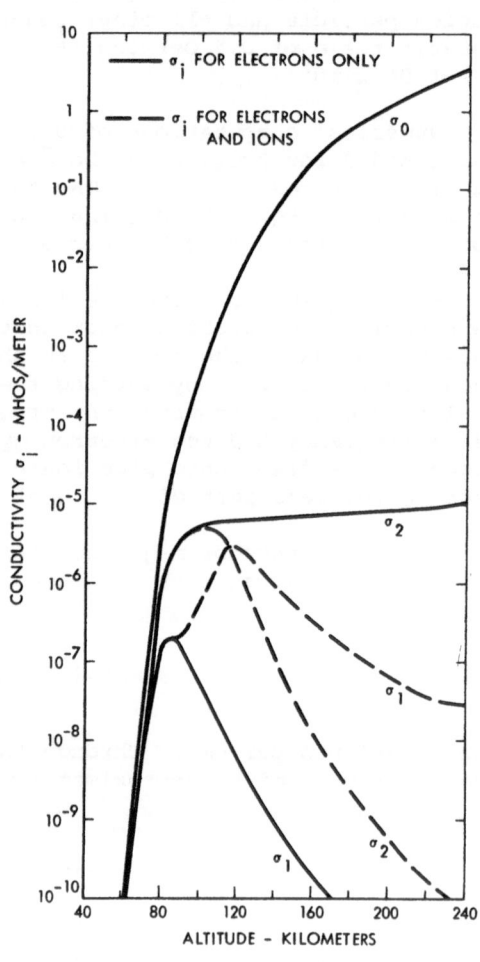

Fig. 4. Components of the nighttime tensor conductivity. F = 0

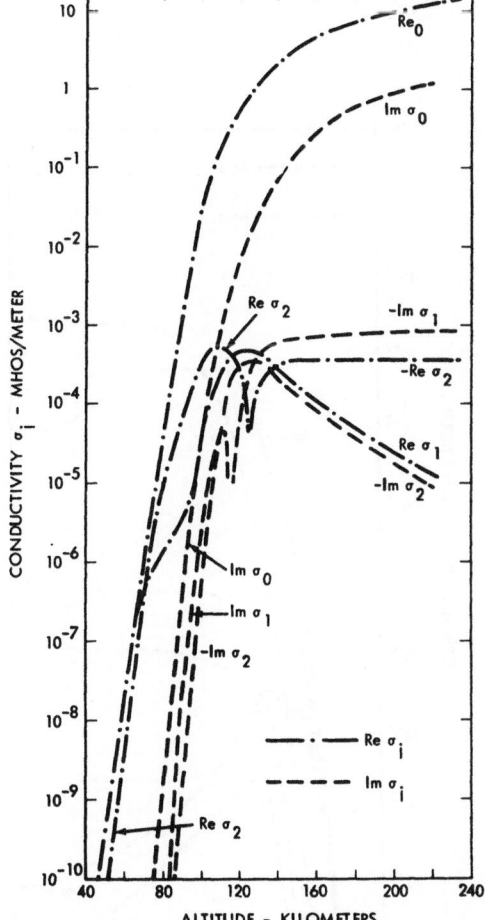

Fig. 5.
Components of the daytime tensor
conductivity F = 20 cps.

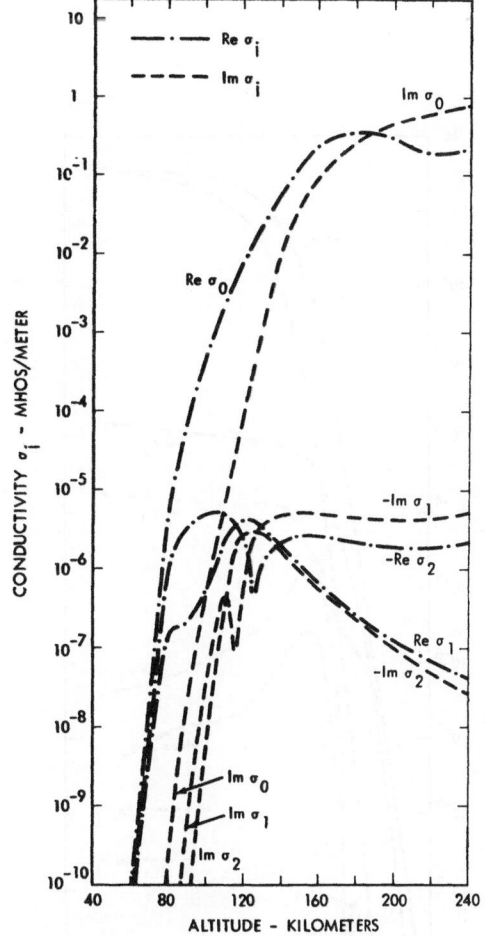

Fig. 6.
Components of the nighttime tensor
conductivity F = 20 cps.

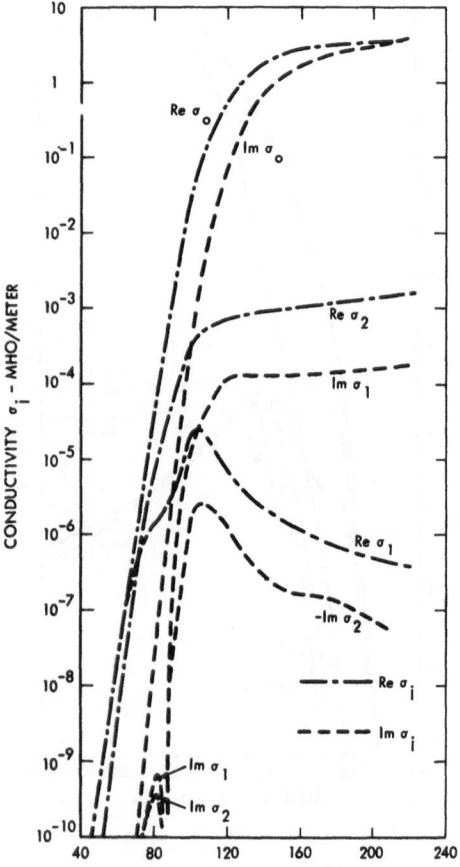

Fig. 7.
Components of the daytime tensor
conductivity F = 300 cps.

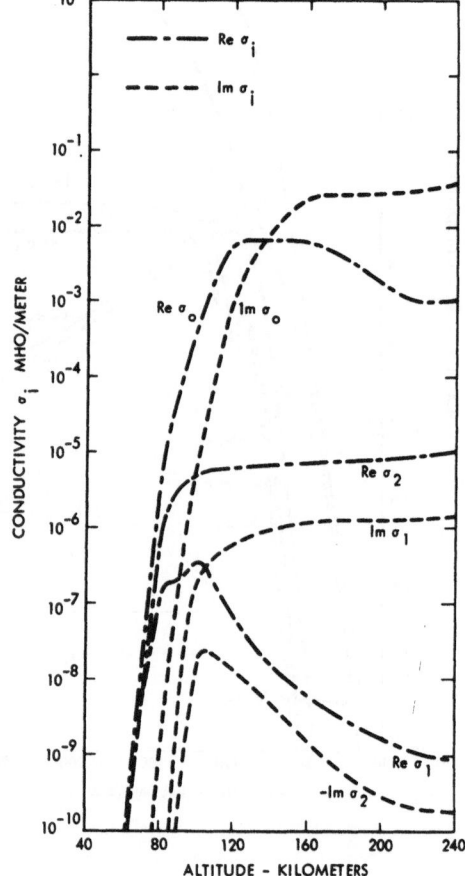

Fig. 8.
Components of the nighttime tensor
conductivity F = 300 cps.

EXCITATION OF ELF WAVES BY THUNDERSTORMS

With the exception of a few relatively rare events (Egeland, 1963), most of the observed ELF energy appears to be due to thunderstorm activities. The "slow-tail"
atmospheric waveforms observed by Liebermann (1956a, 1956b), Hepburn (Hepburn and
Pierce, 1953; Hepburn, 1957, 1958, 1960; Wait 1962a) or Tepley (1959) can be interpreted as caused by lightning discharges. Watt (1960) has examined the varying electromagnetic fields produced by thunderstorms and associated lightning discharges. He
shows that the field variations produced by a typical cloud-to-ground discharge model
agree well with observed fields. Jean et al (1961) have observed individual atmospherics as they travel from storm centers in the Atlantic westward. The ELF noise
spectra in the frequency range of 5 to 30 cps and their changes can be also correlated
with lightning activities (Balser and Wagner, 1962a; Galejs, 1962a). ELF noise
spectra have been calculated by Raemer (1961b, 1961c) as a superposition of responses
from lightning flashes all over the world. In this analysis the power spectra of the
dipole moment in a median lightning flash are computed based on return stroke statistics compiled by Williams (1959). However, Pierce (1963) indicates that other parts
of the lightning flash contribute also to the ELF noise and hence to the dipole
moment, in particular near 10 cps. The spectrum calculations of the dipole moments
by Raemer (1961b, 1961c) are outlined in this section. The resulting noise spectra
will be discussed in later sections.

Williams (1959) has concluded from his studies of thunderstorms that the over-
whelming cause of VLF noise is the presence of vertical currents in the return strokes
associated with cloud-to-ground lightning flashes. These flashes consist of from 2
to 12 strokes of similar shape. He has compiled statistics of flash parameters
including the shape of the median return strokes, interstroke interval, and number of
strokes per flash. The current moment of Williams' median return stroke is proportional to

$$\underset{V}{i}(t-t_k) = \left(\int_0^t \hat{v}(t')dt' \right) i(t-t_k) \tag{12}$$

where $\hat{v}(t)$ is the velocity of the lightning charge along the cloud-ground channel.
The normalized power spectrum of Eq. (12) is designated as $\left| \underset{V}{I}(\omega) \right|^2$.

The actual current moment of a flash is approximated for M identical strokes by

$$\xi(t) = \sum_{\ell=1}^{M} \underset{V}{i}(t-t_\ell), \tag{13}$$

where $\underset{V}{i}(t-t_\ell)$ is given by (1). Its normalized power spectrum is

$$\rho_M(\omega) \equiv \frac{\langle |\tilde{\xi}(\omega)|^2 \rangle}{|\underset{V}{I}(\omega)|^2} = M + \sum_{\substack{\ell,k=1 \\ \ell \neq k}}^{M} e^{i\omega(\ell-k)T} \langle e^{i\omega\Delta T_{\ell k}} \rangle \tag{14}$$

where T is the average interstroke interval and $\Delta T_{\ell k}$ is the deviation of the interval
$(t_\ell-t_k)$ from $(\ell-k)T$. The probability density function for $\Delta T_{\ell k}$ may be approximated
by a Gaussian distribution with standard deviation $\sigma_{\ell k}$, the latter being somewhat

smaller than T. When averaging over all possible values of M, the normalized power spectrum of the current moment follows as

$$\rho(\omega) = \sum_{M=0}^{\infty} \rho_M(\omega) \ p(M) \tag{15}$$

where p(M) is the probability of M strokes in a flash. Evaluation of Eq. (15) gives

$$\rho(\omega) \simeq \langle M \rangle \ \left(1 - e^{-\omega^2 \sigma_t^2 / 2}\right) + e^{-\omega^2 \sigma_t^2 / 2} \sum_{n=0}^{\infty} A_n \cos n \ \omega T \tag{16}$$

where

$$A_k = \sum_{M=K}^{\infty} (M-K) \ p(M). \tag{17}$$

The parameters T and σ_t have been estimated to be about 33 and 7 msec, respectively. The probabilities p(1) through p(10) can be approximated by the numbers 0.15, 0.22, 0.23, 0.15, 0.09, 0.06, 0.04, 0.03, 0.01, and 0; $\langle M \rangle$ is about 2.5 msec.

The power spectrum of the median lightning flash $\rho(\omega)$ has been calculated from Eq. (16) and is shown as a solid curve in Fig. 9. It is seen to exhibit an average

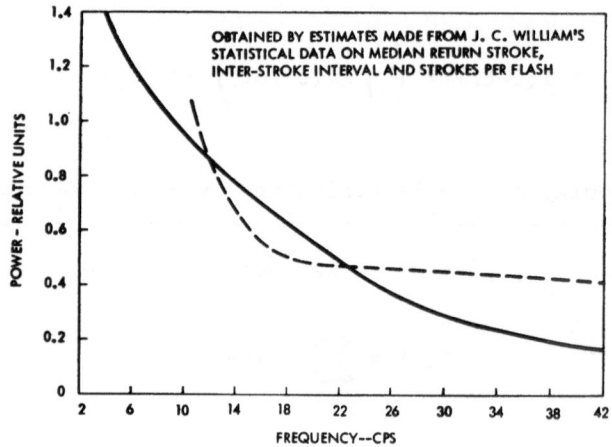

Fig. 9. Estimated Lightning Flash Spectrum.

decay of 3 to 4 db per octave of frequency. Subsequent work of Raemer (1963) has led to a revised value of $\rho(\omega)$ which is indicated dashed. It exhibits a more gradual decay in the frequency range of 20 to 40 cps.

An equivalent power spectrum for the dipole moment of terrestrial noise sources has been deduced by Harris and Tanner (1962) from the noise spectrum measurements of Balser and Wagner (1960) by inferring the complex propagation constant within this

frequency range in a "cut and try" approach. Harris and Tanner (1962) do not require
an ionosphere model or any knowledge of lightning waveforms and their statistics for
obtaining this power spectrum.

The data of Raemer (1961a, 1961b) provide no information for frequencies above
40 cps. The power spectrum of the source dipole moment can be represented as constant in this frequency range in the first approximation.

An estimate of the geographical distributions of thunderstorm activities will be
required for the subsequent calculations of noise spectra. It is known that the
thunderstorms are generally more frequent near the equator and that there are three
equatorial regions of abnormally high average thunderstorm frequency (Handbook of
Geophysics, 1960). These regions are located in (1) South America, (2) Africa, and
(3) the Southwest Pacific area. The regions center approximately at $(50°W, - 10°S)$,
$(15°E, + 10°N)$, and $(110°E, + 10°N)$, respectively, and during all seasons of the year
they show an average thunderstorm frequency considerably higher than is observed elsewhere in the world. The spatial probability density in those regions is from 100 to
1,000 times higher than that throughout most of the earth's surface. There is another
near-equatorial region in Southeast Asia $(\sim 90°E, + 20°N)$, showing substantial year-
round thunderstorm activity, but not as high as in the above three regions. Certain
nonequatorial land regions show intense activity during their summers, e.g., the
Southeastern United States (centered about $75°W, + 30°N$).

EXPERIMENTAL DATA

In experimental observations it is most convenient to record the time variation (or
transient) of the atmospheric signal. An ELF component or slow tail follows the
initial VLF component of the transient. Such transient or slow tail signals have
been reported by Hepburn and Pierce (1953), Liebermann (1956a, 1956b), and Hepburn
(1957, 1958, 1960). A large number of transient signals has been recently recorded
by Tepley (1959). Such transients have been also reviewed by Pierce (1960a, 1960b).
The transient waveform depends on the ionosphere parameters (height of the effective
reflecting layer and its conductivity) and on the distance to the source. The effective ionospheric conductivity has been deduced by transient analysis using a model
of a homogeneous sharply bounded ionosphere. The day and night conductivities (in
mho/m) are estimated as 0.55×10^{-6} and 1.2×10^{-6} by Hepburn and Pierce (1953), as
1.1×10^{-6} and 0.44×10^{-6} by Hepburn (1957) and as 10^{-6} and 3×10^{-6} by Wait (1962a).
However, the limitations of the sharply bounded ionosphere model, outlined later, will
restrict the applicability of the above conductivity figures. Detailed noise spectrum
measurements have been reported by Maxwell and Stone (1963).

From the number of papers dealing with ELF noise observations only relatively
few provide data that lead to signal attenuation rates in the ELF range. With
observations made at a single station such information may be inferred only by
comparing atmospherics that originate at different ranges. The distances between the
lightning discharge, which is the source of the atmospheric, and the observer may be
determined by a direction finding system which operates in a different frequency
range. The signal attenuation is estimated from differences between the waveform
spectra received from different lightning sources which are at different distances
from the observer. This technique is not very reliable because there are no assur-
ances that the original spectra of several different lightning discharges are the
same. The data of Chapman and Macario (1956) observed at a single station must be

interpreted in view of the above uncertainty. Still, the data provide valuable information about ELF day- and nighttime attenuation rates and will be quoted in connection with the theoretical considerations.

The difficulties of a one-station approach have been avoided in a three-observer system described by Jean et al (1961), who observed a single atmospheric waveform as it passed in an approximately great circle path over two of these stations. Also, the observed lightning discharges were at distances where the near field effects are negligible and a single propagating mode characterizes the field variations. The data were taken with the sunset (or an ionospheric discontinuity between the day-night boundary) approaching their eastern station. These attenuation rates are also referred to later in this paper. However, the measured increase of the attenuation rates below 50 cps is contrary to the theoretical considerations, and it cannot be substantiated by Q measurements of the earth-to-ionosphere resonances.

The first experimental indication of Schumann resonances was provided by König (1958, 1959). He recorded noise waveforms passed through a band pass filter of 12 cps upper cut-off frequency and noted oscillatory waveforms with an estimated frequency of 9 cps. A similar observation technique has also been reported by Polk and Fitchen (1962). Waveforms of about 50 cps upper cut-off frequency have been recorded by Balser and Wagner (1960) and then processed through a narrow band digital filter. The resulting power spectra shown later exhibit the first sharp peak near 8 cps. Further measurements by Balser and Wagner (1962a) have correlated diurnal variations of measured noise spectra with world-wide thunderstorm activities. Their measurements of the diurnal variations of the power level at the first three resonance frequencies are shown in Fig. 10, together with similar measurements by Gendrin and Stefant (1962a), that have also been correlated with spatial and time variations of thunderstorms. Gendrin and Stefant (1962a) record the signals on a magnetic tape and by playing them back at 160 times increased speed are able to use a 0 to 8 kc Sonagraph for display purposes. Diurnal variations of the fields near the two lower resonance frequencies have been measured at Kingston, Rhode Island, by Polk, Huck and Yu (1963). The maximum signal intensity occurs between 1500 and 1800 UT. In an amplitude versus time plot the signal peak is broader during summer than winter. Seasonal and diurnal variations of the noise power in the ELF frequency band below 35 cps have been described by Wright (1963) without singling out the individual resonance modes. The measurements are made at Byrd Station, Antarctica (80°S, 120°W) and show peak intensities near 1000 UT during spring and summer and near 2300 UT during most of the year. Spectra similar to those of Balser and Wagner (1962a) have been obtained recently by Benoit and Houri (1962). Other noise spectra measurements which indicate earth-to-ionosphere cavity resonances have been reported by Fournier (1960), Benoit and Houri (1961), and Lokken et al (1961 and 1962). Lokken (1963) has also reported noise recordings made simultaneously in Antarctica and in Canada. As seen from Fig. 11, there is good correlation between the high intensity ELF noise bursts observed at the various sites. Diurnal frequency variations of the earth-ionosphere cavity modes have been reported by Balser and Wagner (1962b). The resonance frequencies vary by about ± 0.2 to 0.3 cps, but the changes are not simultaneous for the different modes; a low resonance frequency of one mode occurs simultaneously with a high frequency of a different mode (Gendrin and Stefant, 1963).

A high altitude nuclear explosion has been shown to affect the earth-to-ionosphere cavity resonances by simultaneously lowering all the resonance frequencies (Gendrin and Stefant, 1962a and 1962b). This lowering of resonance frequencies seen in Fig. 12 has been attributed to world-wide changes of the lower ionosphere layers (Gendrin and Stefant, 1962b). Phase and amplitude observations of VLF transmissions over paths that are at large distances from the nuclear explosions tend to substantiate such changes of the ionosphere (Decaux et al, 1963; Zmuda, Shaw and Haave, 1963) although the ionosphere perturbations may be expected to vary with distance. The lowering of the first resonance mode after a nuclear explosion has been also reported by Balser and Wagner (1963). Ward (1963) has noted changes of polarization in the magnetic field fluctuations after a nuclear explosion.

216

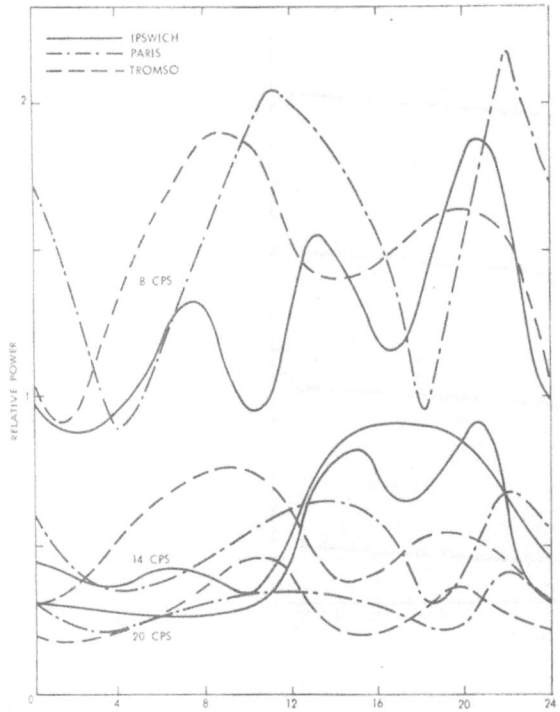

Fig. 10.
Diurnal variation of the resonance
modes at Ipswich, Massachusetts,
February 1961 (vertical electric
field); Chambon-la-Foret (near Paris)
July 1962, and at Tromso, Norway,
April 1962 (horizontal /magnetic
field).

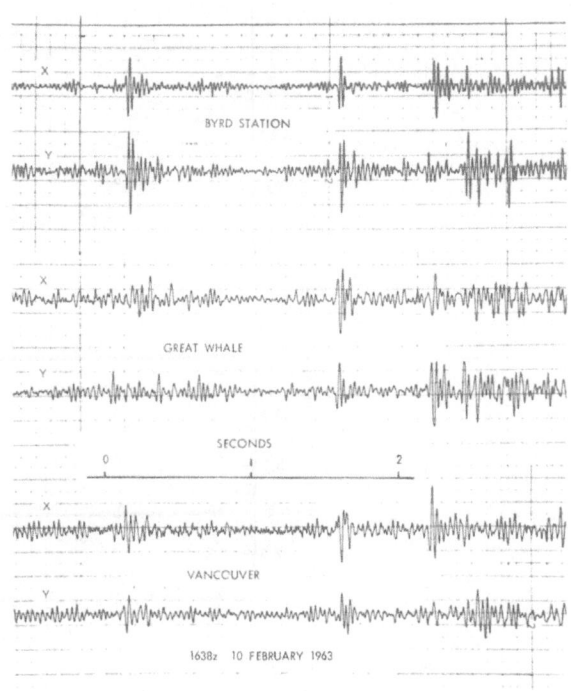

Fig. 11.
Simultaneous Noise Recordings.

The earth-to-ionosphere cavity resonances provide also signals useful in electromagnetic prospecting; the surface impedance of the earth may be determined from
measuring the horizontal electric and magnetic field components in two mutually perpendicular directions (Fournier, 1963). Such a recording of approximately 8 cps
signals is shown in Fig. 13.

It may be noted that waveform observations provide higher apparent resonant
frequencies than spectral measurements, which is the consequence of the significant
high-frequency components contained by the composite noise waveform. For given power
spectrum of the noise the apparent average frequency $\bar{f}$ of the noise waveform may be
estimated from the mean time interval between its axis crossings. This average
frequency is given by Rice (1944, 1945) as

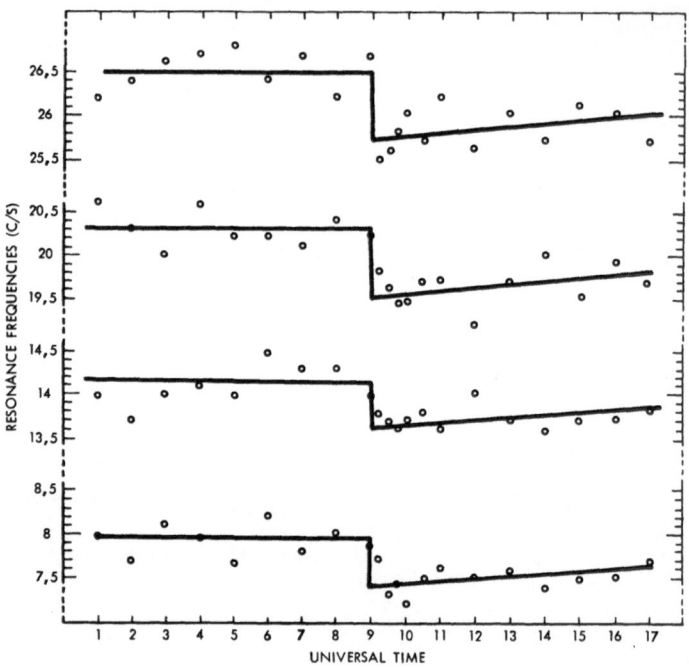

Fig. 12.   Nuclear Effects on Earth-Ionosphere Cavity Resonances.

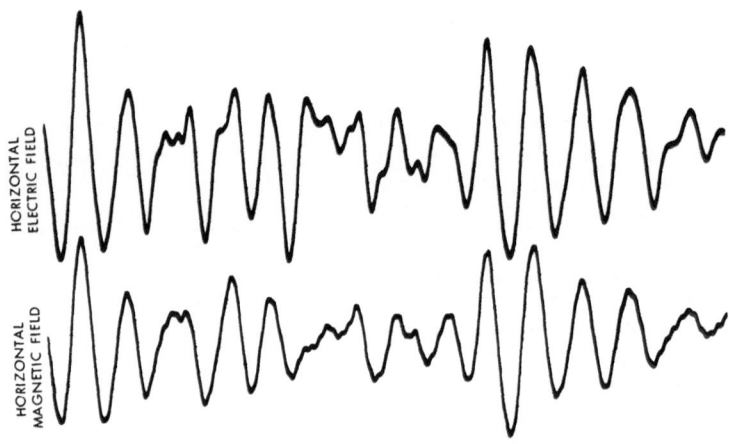

Fig. 13.   Surface Impedance Measurements at 8 cps.

$$\overline{f} = \frac{\displaystyle\int_0^\infty g(f)\,f^2\,df}{\displaystyle\int_0^\infty g(f)\,df} \qquad\qquad (18)$$

where $g(f)$ is the power spectrum of the noise waveform. Noise of spectrum similar to that recorded by Balser and Wagner (1960, 1962a, 1962b) is assumed to be passed through idealized filters with bandwidths of 5 to 12 and 5 to 20 cps. For these two bandwidths the average frequency is calculated from Eq. (18) as $\overline{f}$ = 8.7 and 12.9, respectively. This average frequency $\overline{f}$ depends on the bandwidth of the waveform, and is generally higher than the frequency of the first spectral peak.

FIELD REPRESENTATIONS IN THE SPHERICAL SHELL
BETWEEN THE EARTH AND THE IONOSPHERE

The propagation of the terrestrial ELF waves is confined principally within the spherical shell between earth and ionosphere. The field penetration below ground surface is negligible and the ground may be approximated by a perfect conductor in the analysis. The fields penetrate into the ionosphere and the exact structure of its lower edge (altitudes below 100 to 130 km) must be considered. The spherical shell between the earth and the ionosphere is assumed to be bounded by an equivalent surface impedance Z_s which depends on the ionosphere structure beyond the boundary, operating frequency, magnetic field effects, and also on the propagation constant of the fields below the ionosphere. The surface impedance representation is satisfactory at distances from sources that are large relative to the ionosphere height (1000 km and more) and where only a single propagation mode exists.

An example of field representations will be developed for an assumed vertical electric dipole excitation. This vertical electric dipole excites transverse magnetic (TM) field components which are not coupled to the transverse electric (TE) component in the absence of the magnetic field. In the presence of the magnetic field the coupling may be made small by selecting the ionospheric height at a sufficiently low altitude. Following Schumann (1952a, b, c) or Wait (1962a), the TM vertical components can be derived from a Hertz potential which has only a single radial component Π.

Within a spherical shell of inner radius a, and outer radius c = a + h the field components are given by

$$E_r = -\frac{1}{rb}\,\frac{1}{\sin\theta}\,\frac{1}{\partial\theta}\left(\sin\theta\,\frac{\partial\Pi}{\partial\theta}\right) \qquad\qquad (19)$$

$$E_\theta = \frac{1}{rb}\,\frac{\partial^2}{\partial r\partial\theta}\,(r\Pi) \qquad\qquad (20)$$

$$H_\phi = -\frac{k^2}{i\omega\mu b}\,\frac{\partial\Pi}{\partial\theta} \qquad\qquad (21)$$

where b is the radius of the source ($a \le b \le c$) and where

$$k^2 = \sqrt{\epsilon\mu\omega^2 + i\omega\mu\sigma}\tag{22}$$

An exp$(-i\omega t)$ time variation is assumed for the field components. The primary excitation is given by

$$\Pi_o = \frac{Ids}{4\pi i\omega\epsilon R}\, e^{ikR}\tag{23}$$

where R is the distance from the source. This primary excitation can be expressed in terms of spherical wave functions as

$$\Pi_o = \frac{-kIds}{\omega\epsilon 4\pi}\sum_{n=0}^{\infty}(2n+1)\,P_n(\cos\theta)\begin{cases}j_n(kr)h_n^{(1)}(kb) & \text{for } r \leq b\\[2mm] h_n^{(1)}(kr)\,j_n(kb) & \text{for } r \geq b\end{cases}\tag{24}$$

The secondary Hertz vector components satisfy the homogeneous wave equation and they can be expressed as

$$\Pi_s = \sum_{n=0}^{\infty}\left[A_n h_n^{(1)}(kr) + B_n j_n(kr)\right]P_n(\cos\theta)\tag{25}$$

where $P_n(\cos\theta)$ is the Legendre polynomial of order n. The spherical Bessel functions are defined as

$$j_n(x) = \sqrt{\frac{\pi}{2x}}\, J_{n+\frac{1}{2}}(x)\tag{26}$$

$$h_n^{(1)}(x) = \sqrt{\frac{\pi}{2x}}\, H_{n+\frac{1}{2}}^{(1)}(x)\tag{27}$$

where $J_n(x)$ is the Bessel function, $H_n^{(1)}(x)$ is the Hankel function of the first kind. The net potential

$$\Pi = \Pi_o + \Pi_s\tag{28}$$

must satisfy the impedance boundary conditions at the earth $(r = a)$ and at the lower edge of the ionosphere $(r = c)$. For $r \leq b$ the potential is expressed as

$$\Pi = \sum_{n}(2n+1)\,\frac{G_n}{D_n}\left[j_n(kr) + R_{gn}h_n^{(1)}(kr)\right]P_n(\cos\theta)\tag{29}$$

For a perfectly conducting ground $E_\theta\big|_{r=a} = 0$ leads to

$$\frac{\partial}{\partial x}(x\Pi) = 0\bigg|_{x = ka}\tag{30}$$

For $r \geq b$ the potential is expressed as

$$\Pi = \sum_n (2n+1) \frac{F_n}{D_n} \left[h_n^{(1)} (kr) + R_{in} \; j_n(kr) \right] P_n(\cos \theta) \qquad (31)$$

The impedance boundary condition

$$E_\theta = Z_s H_\phi \qquad (32)$$

for $r = c$ leads from (2) and (3) to

$$\frac{\partial}{\partial x} (x\Pi) = i\zeta x\Pi \Big|_{x = kc} \qquad (33)$$

where $\zeta = \sqrt{\frac{\epsilon_o}{\mu_o}} \; Z_s$. For an homogeneous sharply bounded ionosphere

$$\zeta = \sqrt{\frac{\epsilon_o}{\mu_o}} \; Z_s \approx (n_i)^{-1} = \left[1 + \frac{i\sigma_i}{\omega\epsilon_o} \right]^{-\frac{1}{2}} \qquad (34)$$

where n_i is the refractive index of the ionosphere. The boundary conditions Eq. (30) and Eq. (33) determine R_{gn} and R_{in}, and the Eq. (24), (25), (28), (29) and (31) can be solved for A_n, B_n, D_n, F_n and G_n. This completes the formal determination of Π.

The series representation Eq. (29) or Eq. (31) is very slowly converging and the number of modes required for an accurate field representation is estimated as 2 to 10 ka. Although this series can be used in the frequency range of earth-to-ionosphere cavity resonances (Madden, 1961), Row (1962), Thompson, (1963) computations at higher frequencies are feasible only with the aid of high-speed computers (Johler and Berry, 1962). It is generally advantageous to use the Watson transform to change the series Eq. (29) into a more rapidly converging representation. The Watson transformation has been discussed in detail by Sommerfeld (1949), Bremmer (1949) or Wait (1962a). It changes the summation over n into a complex integral, the residues of which are evaluated at the poles of the integrand ν, which are the zeroes of the denominator D_n. This results in

$$\Pi = - \pi \sum_\nu \frac{2\nu+1}{\sin \nu\pi} \; \frac{G_\nu \left[j_\nu(kr) + R_{g\nu} h_\nu^{(1)}(kr) \right]}{\partial D_n / \partial n \Big|_{n=\nu}} \; P_\nu (-\cos \theta) \qquad (35)$$

The further considerations are restricted to geometries with the source and observation point on the ground surface (a=b=r). Introducing the notation

$$\Psi(x) = x j_\nu(x) \qquad (36)$$

$$\phi(x) = x h_\nu^{(1)}(x) \qquad (37)$$

221

and applying the Wronskian

$$\Psi(x)\emptyset'(x) - \emptyset(x)\Psi'(x) = i \tag{38}$$

it follows that

$$\Pi = \frac{iIds}{4} \sqrt{\frac{\mu_o}{\epsilon_o}} \frac{1}{(ka)^2} \sum_{\nu} \frac{2\nu+1}{\sin \nu\pi} \frac{g_\nu}{f'_\nu} P_\nu (-\cos \theta) \tag{39}$$

with

$$g_\nu = \Psi(x) \emptyset'(y) - \emptyset(x)\Psi'(y) - i\zeta \left[\Psi(x) \emptyset(y) - \emptyset(x)\Psi(y)\right] \tag{40}$$

$$f_\nu = \Psi'(x)\emptyset'(y) - \emptyset'(x)\Psi'(y) - i\zeta \left[\Psi'(x)\emptyset(y) - \emptyset'(x)\Psi(y)\right] \tag{41}$$

where $x = ka$, $y = kc = k(a+h) = x+\Delta$ and $f'_\nu = \partial f/\partial \nu$. The eigenvalues ν are determined from the solution $f_\nu = 0$. The functions $\Psi(y)$ and $\emptyset(y)$ of Eq. (40) and Eq. (41) are expanded in a Taylor series about $x = ka$ following Schumann (1952a, b, c). This Taylor series is written as

$$u(y) = u(x+\Delta) = u + u'\Delta + \frac{u''}{2} \Delta^2 + \frac{u'''}{6}\Delta^3 + \frac{u'^V}{24} \Delta^4 \tag{42}$$

The functions $\Psi(y)$ and $\emptyset(y)$ satisfy the differential equation

$$u''(y) + \left[1 - \frac{\nu(\nu+1)}{y^2}\right] u(y) = 0 \tag{43}$$

u'' is related to u by Eq. (43) and u''', u'^V are related to u and u' by differentiating Eq. (43) as

$$u''' = - \frac{2\nu(\nu+1)}{x^3} u - \left[1 - \frac{\nu(\nu+1)}{x^2}\right]u' \tag{44}$$

$$u'^V = \left\{\frac{6\nu(\nu+1)}{x^4} + \left[1 - \frac{\nu(\nu+1)}{x^2}\right]^2\right\} u - \frac{4\nu(\nu+1)}{x^3} u' \tag{45}$$

After applying the Wronskian Eq. (38) it follows that

$$f_\nu = i\Delta \left\{1 - \frac{\nu(\nu+1)}{x^2} \left[1 - \frac{\Delta}{x} + (\frac{\Delta}{x})^2\right] - \frac{\Delta^2}{6} \left[1 - \frac{\nu(\nu+1)}{x^2}\right]^2\right\} - \tag{46}$$

$$\zeta \left\{1 - \frac{1}{2} \Delta^2 + \frac{1}{2} \Delta^2 \frac{\nu(\nu+1)}{x^2} (1 - \frac{2}{3}\frac{\Delta}{x})\right\} + \cdots$$

222

In the above approximation f_ν is quadratic with respect to $\nu(\nu+1)/x^2$.
After neglecting terms $\frac{\Delta}{x}$ relative to 1, it follows that

$$\frac{\nu(\nu+1)}{x^2} = 1 + i\frac{\zeta}{\Delta} - \frac{1}{3}\zeta^2 \approx 1 + i\frac{\zeta}{\Delta} \tag{47}$$

Solving Eq. (47) as a quadratic equation for ν

$$\nu = -\frac{1}{2} + \sqrt{\frac{1}{4} + \left(1 + i\frac{\zeta}{\Delta} - \frac{1}{3}\zeta^2\right)x^2} \tag{48}$$

or

$$S = \frac{\nu + \frac{1}{2}}{x} = \sqrt{1 + i\frac{\zeta}{\Delta} - \frac{1}{3}\zeta^2 + \frac{1}{4x^2}} \approx \sqrt{1 + i\frac{\zeta}{\Delta}} \tag{49}$$

The derivative f'_ν is computed from Eq (46) as

$$f'_\nu = -i\frac{2\nu+1}{x^2}\Delta\left\{1 - \frac{\Delta}{x} + \frac{\Delta^2}{x^2} - \frac{1}{3}\left[1 - \frac{\nu(\nu+1)}{x^2}\right]\Delta^2 + \frac{1}{2i}\zeta\Delta\left(1 - \frac{2}{3}\frac{\Delta}{x}\right)\right\} \approx -i\frac{2\nu+1}{x^2}\Delta \tag{50}$$

The function g_ν follows as

$$g_\nu = i\left\{1 - \frac{\Delta^2}{2}\left[1 - \frac{\nu(\nu+1)}{x^2}\right] - \frac{2}{3}\frac{\Delta^3}{x}\frac{\nu(\nu+1)}{x^2}\right\} + \Delta\zeta\left[1 - \frac{\Delta^2}{6}\left(1 - \frac{\nu(\nu+1)}{x^2}\right)\right] \approx i \tag{51}$$

The above expansions can be expected to be accurate only in the ELF range where $h < \lambda$. Over these frequencies only a single mode propagates and the summation sign may be omitted from Eq. (39). Substituting the approximate values of f'_ν and g_ν in Eq. (39) gives

$$\Pi = -\frac{iIds}{4\omega\epsilon h}\frac{1}{\sin\nu\pi}P_\nu(-\cos\theta) \tag{52}$$

The radial electric field component follows from (1) and (34) as

$$E_r = \frac{iIds}{4\omega\epsilon h a^2 \sin\nu\pi}\frac{1}{\sin\theta}\frac{1}{\partial\theta}\left[\sin\theta\frac{\partial}{\partial\theta}P_\nu(-\cos\theta)\right] \tag{53}$$

$$= -\frac{iIds}{4\omega\epsilon h a^2}\frac{\nu(\nu+1)}{\sin\nu\pi}P_\nu(-\cos\theta)$$

which is a commonly used expression. However, its validity depends on the

223

approximations Eq. (50) and Eq. (51). The Legendre function of Eq. (53) can be approximated by its asymptotic expansion. It follows from 3.9.1.2 of Erdelyi (1953) that

$$P_\nu \ (-\cos \ \theta) \approx \sqrt{\frac{2}{\pi (\nu + \frac{1}{2}) \ \sin \ \theta}} \ \cos \ \left[(\nu + \frac{1}{2}) \ (\pi \ - \ \theta) \ - \ \frac{\pi}{4} \right] \tag{54}$$

For $\nu \gg 1$ with Im $\nu > 0$ (36) simplifies further to

$$P_\nu \ (-\cos \ \theta) \approx \sqrt{\frac{1}{2\pi (\nu + \frac{1}{2}) \ \sin \ \theta}} \ e^{-i} \ \left[(\nu + \frac{1}{2}) \ (\pi \ - \ \theta) \ - \ \frac{\pi}{4} \right] \tag{55}$$

The distance along the surface of the earth is designated as D = aθ. Hence

$$i (\nu + \frac{1}{2}) \ \theta = \frac{\nu + \frac{1}{2}}{ka} \ ikD \tag{56}$$

The factor

$$\frac{\nu + \frac{1}{2}}{ka} \ = \ \frac{\nu + \frac{1}{2}}{x} \ = S \tag{57}$$

of Eq. (56) can be interpreted as the ratio between the propagation constant along the surface of the earth k_θ and the propagation constant of the free space k. The Legendre function $P_\nu(-\cos \theta)$ of Eq. (53) may be also expanded in a series of Legendre polynomials $P_n(\cos \theta)$ following Magnus and Oberhettinger (1949) as

$$P_\nu (-x) \ = \ - \ \frac{\sin \ \nu \pi}{\pi} \ \sum_{n=0}^{\infty} \ P_n (x) \ \frac{2n+1}{n(n+1) \ - \ \nu(\nu+1)} \tag{58}$$

Substituting Eq. (58) in Eq. (53) gives

$$E_r \ = \ \frac{iIds \ \nu (\nu+1)}{4\pi \omega \epsilon h a^2} \ \sum_{n=0}^{\infty} \ P_n (\cos \ \theta) \ \frac{2n+1}{n(n+1) \ - \ \nu(\nu+1)} \tag{59}$$

The change from Eq. (53) to Eq. (59) can be interpreted as an inversion of the Watson transform between Eq. (29) and Eq. (35). Therefore, Eq. (59) should be also obtainable directly from Eq. (29). After introducing the notations Eq. (40) and Eq. (41), Π of Eq. (29) is expressed for r = a = b as

$$\Pi = \ - \ \frac{iIds}{4\pi} \ \sqrt{\frac{\mu_o}{\epsilon_o}} \ \frac{1}{(ka)^2} \ \sum_{n=0}^{\infty} \ \frac{(2n+1) g_n}{f_n} \ P_n (\cos \ \theta) \tag{60}$$

Applying Eq. (46) and (47) f_n is approximated as

$$f_n \approx \frac{ikh}{(ka)^2} \left[\nu(\nu+1) - n(n+1) \right] \tag{61}$$

Substituting Eq. (61) and g_n of Eq. (51) in Eq. (60) gives

$$\Pi = \frac{iIds}{4\pi\omega\epsilon_0 h} \sum_{n=0}^{\infty} \frac{2n+1}{n(n+1) - \nu(\nu+1)} \, P_n(\cos\theta) \tag{62}$$

Applying Eq. (19) and the differential equation for $P_n(\cos\theta)$ gives

$$E_r = \frac{iIds}{4\pi\omega\epsilon h a^2} \sum_{n=0}^{\infty} \frac{n(n+1)}{\nu(\nu+1)} \frac{(2n+1)}{-n(n+1)} \, P_n(\cos\theta) \tag{63}$$

which may be obtained from Eq. (59) by multiplying each term of it by a factor $n(n+1)/[\nu(\nu+1)]$. The resonant terms of E_r where $\nu(\nu+1) \approx n(n+1)$ will be approximately the same in Eq. (59) and Eq. (63), but differences between corresponding terms will occur as n deviates from this resonance condition. The field representations Eq. (59) and Eq. (63) both involve approximations, although Eq. (59) is used more frequently.

The above derivation was based on the most commonly considered vertical dipole excitation with an assumed perfect ground conductivity. A generalization to finite ground conductivity can be seen from the work of Wait (1962a). Multilayer ionosphere models have been considered by Wait (1958, 1962a) and Madden (1961). Other excitations have been also worked out. The excitation by a vertical magnetic and horizontal electronic dipole have been considered by Schumann (1954a, b, c) and Wait (1962a), but some of the earlier results (Schumann, 1954c, p. 268) are in disagreement with later work (Wait, 1962a). The horizontal magnetic dipole has been discussed by Galejs (1961a). Magnetic field effects have been treated by Schumann (1955, 1956). A quasi-longitudinal and transverse approximation has been discussed by Wait (1962a). The anisotropic ionosphere has been treated in the spherical geometry by Wait (1963a) and Thompson (1963).

There are several simplifications permissible in the ELF range.

(1) The surface impedance may be calculated as for an ionosphere bounded by a flat plane because the fields penetrate in the ionosphere only over a distance that is small relative to earth radius. Also numerical work that is based on such a flat ionosphere model appears to agree with available experimental data.

(2) Simple approximations of the radial wave functions yield fields that do not depend on the absolute radius of the ionosphere c = a + h or earth a, but only on the difference h. The subsequent model equation for determining the propagation constant is also dependent only on h and one has a plane geometry as far as radial wave functions are concerned.

(3) The θ functions may be represented by a series of Hankel functions with $\rho = a\,\theta$ in argument. This approximation has been considered by Wait (1962a). He shows that the leading term of this series is a sufficient approximation when $\sqrt{\theta/\sin\theta}$ is nearly unity over the propagation distance. No use of this approximation will be made in the following discussion.

SHARPLY-BOUNDED IONOSPHERE MODELS

A model of a sharply-bounded homogeneous ionosphere is simplest to analyze and is commonly used in investigations of VLF and ELF propagation (Budden, 1953; Schumann, 1952a, b, c; Wait, 1957). This model is particularly useful near the source, and at distances of the order of 30 km the ionosphere can be neglected altogether (Wait, 1962a). Generally, a homogeneous ionosphere does not account for measured attenuation characteristics. An ionosphere model which consists of several homogeneous layers has been described by Wait (1958, 1960a, 1960b, 1962a). It has been possible to construct two-layer models that provide reasonable agreement with measure ELF attenuation rates over 100 to 400 cps frequency range (Jean et al, 1961). The ionosphere of this model is bounded at h = 90 km. The lower layer has a thickness of 20 km and a conductivity of $\sigma \approx 0.9 \times 10^{-6}$ mho/m. The semi-infinite upper layer has a conductivity of $\sigma \approx 9 \times 10^{-6}$ mho/m.

The ionospheric model of exponentially increasing conductivity has been introduced in mode theory by Shmoys (1956) in an analysis that is based on the differential equations for horizontally-polarized waves. The exponential model of Wait (1958, 1960a, 1960b, 1960c, 1962a) where the refractive index n changes exponentially after a sudden transition from n = 1 in the atmosphere to n >> 1 at the lower ionosphere edge, accounts for attenuation rates measured by Chapman and Macario (1956) only by using more gradual changes of the refractive index than those obtained from ionospheric profile data. In an alternate approach, the attenuation rates calculated with a homogeneous ionosphere may be brought into agreement with measured data by defining a frequency-dependent ionosphere altitude, which is tailored to fit measured attenuation rates (Pierce, 1960a, 1960b).

The model of a sharply-bounded homogeneous ionosphere has been also used in investigations of Schumann resonances (1957). Substituting Eq. (35) and Eq. (54) in Eq. (53) and letting $\omega = ip$ it follows that

$$E_r = - \frac{I \; ds \; (p + \alpha \sqrt{p})}{4\pi \epsilon h a^2} \sum_{n=0}^{\infty} P_n(\cos \theta) \; \frac{2n+1}{p^2 + \alpha p^{1.5} + \omega_n^2} \tag{64}$$

where

$$\omega_n^2 = \left(\frac{c}{a}\right)^2 n(n+1) \tag{65}$$

$$\alpha = \frac{1}{h\sqrt{\mu_0 \sigma_i}} \tag{66}$$

and where the ionospheric displacement currents are neglected. The complex resonant frequencies of the Schumann modes are given by the solutions of

$$p^2 + \alpha p^{1.5} + \omega_n^2 = 0 \tag{67}$$

For $\sigma_i = \infty$, $\alpha = 0$ and resonances occur for

$$\omega = ip = \omega_n = 15\pi \sqrt{n(n+1)} \tag{68}$$

which gives the sequence of resonances as 10.6, 18.3, 25.9, 33.5 cps. Finite iono-spheric conductivity σ_i decreases the resonant frequencies, which is in agreement with observations.

The validity of Eq. (67) has been examined by Row (1962) who considered the natural modes of oscillations in the earth-to-ionosphere cavity by neglecting the source terms Eq. (23) or Eq. (24), and who obtained field components that are similar to those derivable from Eq. (60). It follows from the work of Row that the approxi-mation Eq. (67) is valid for

$$\left(\frac{\omega_n}{\omega}\right)^2 \ll 2 \frac{i\sigma_i}{\omega\epsilon_o} \approx 2 n_i^2 = \frac{2}{\zeta} \tag{69}$$

which seems to hold for most ionospheric parameters of interest.

Madden (1961) has considered the earth to ionosphere electromagnetic resonances in a multi-layer ionosphere formulation. For a uniform homogeneous ionosphere of $\sigma_i = 1.8 \times 10^{-6}$ mho/m and h = 80 km he obtains approximately correct resonance fre-quencies, but the cavity Q is 2 to 2.5 times too low. The agreement with measurements of Balser and Wagner (1960) is improved by introducing a perfectly reflecting inter-face at the height of 149 km and by changing σ_i to 2.1×10^{-6} mho/m.

Raemer (1961b, 1961c) computes terrestrial noise spectra by integrating the power which is radiated from a distribution of lightning sources. The radiated power is proportional to the squared magnitude of Eq. (64) and to the power spectrum of the dipole moments Eq. (16). The loss parameter α is adjusted to make the computed spectral peaks to agree with the measured ones. For h = const this gives the spectral curves of Fig. 14, while for σ_i = const, this gives the curves of Fig. 15. It is seen that the calculated spectrum has more damped spectral peaks than the measured spec-trum. Excessive ionosphere losses are therefore introduced if α is made so large as to obtain the correct resonance frequencies. This appears to be a consequence of the assumed homogeneous isotropic model ionosphere.

The sharply-bounded ionosphere models seem to provide an agreement with measure-ments only over a limited frequency range; a two-layer model which is suited for frequencies of 100 to 400 cps (Jean et al, 1961) is different from a model for frequencies of 8 to 30 cps (Madden, 1961). The optimum model is usually established by a cut and try procedure without attempting to correlate the model with measured or calculated profiles of ionospheric refractive index or conductivity, such as shown in Fig. 3 through 8.

The effects of a longitudinal and transverse magnetic field have been considered by Wait (1960b, 1962a) for a homogeneous sharply-bounded ionosphere. The longitudinal magnetic field tends to increase the attenuation, particularly if the longitudinal component of gyro-frequency exceeds the electron collision frequency. A transverse magnetic field decreases the attenuation, but very slightly. The phase velocity is increased for East to West propagation and decreased for West to East propagation.

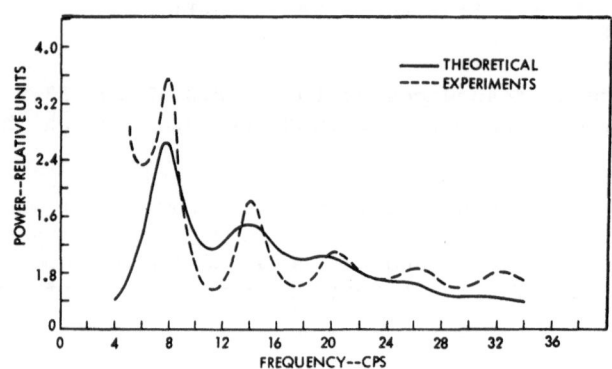

Fig. 14.
ELF Noise Spectrum -- Effective Ionospheric Height Assumed Constant.

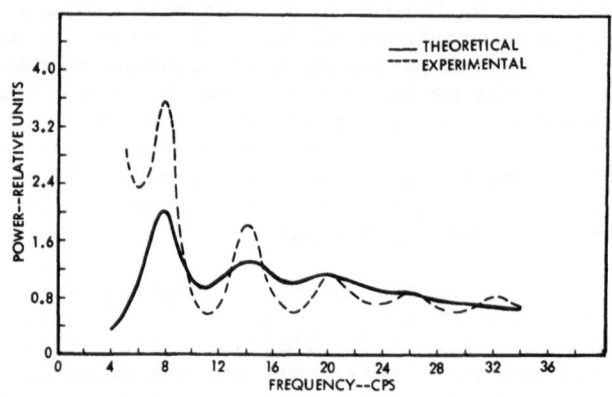

Fig. 15.
ELF Noise Spectrum -- Effective Ionospheric Height Assumed Frequency Dependent.

ISOTROPIC IONOSPHERE MODELS WITH A DIFFUSE LOWER BOUNDARY

The discussion of the preceding section dealt with models where the refractive index n, was assumed to change suddenly from n = 1 in the atmosphere to n >> 1 at the ionospheric boundary. However, the ionospheric conductivity plots, Fig. 3 through 8, indicate a gradual change of ionospheric conductivity with a similarly gradual change of the refractive index near the lower edge of the ionosphere. Wait (1962b) shows that a gradual boundary tends to increase the attenuation for frequencies in the ELF range. Harris and Tanner (1962) develop an iterative integral equation method for determining the complex propagation constant in the space between the earth and ionosphere as a function of the gradually changing ionospheric conductivity profile, but no examples are given to indicate the accuracy or the validity of this procedure.

It is also possible to consider such approximations to the ionospheric conductivity profile which make it possible to obtain exact solutions of the required differential equations. The exponential model provides a good representation of the conductivity σ through the lower edge of the D and E regions. As long as the ELF signals do not penetrate beyond the E region (this does not take into consideration possible whistler-mode signals), the exponential increase of σ that is assumed in this model above the E layer would not affect the propagation below the ionosphere. Galejs (1961c) has obtained solutions for such an exponential model. The propagation characteristics of the fields below this ionosphere will be discussed after determining the surface impedance at the assumed ionospheric boundary. The boundary may be at a height where the conduction currents are smaller than the displacement currents ($\sigma < \omega \epsilon$).

Propagation Constant.

It follows from Maxwell's equations and from the equation of continuity that the vertical electric-field component E_z in a vertically-stratified lossy medium of complex dielectric constant

$$\epsilon = \epsilon_0 + \frac{i\sigma}{\omega} \tag{70}$$

is of the form

$$E_z = Z(z) \, e^{ik_x x - i\omega t} \tag{71}$$

where x is the direction of propagation of the vertically-polarized fields below the dielectric medium. $Z(z)$ satisfies the differential equation

$$Z'' + \frac{\epsilon'}{\epsilon} Z' + \left[\mu_0 \epsilon \omega^2 + \frac{\epsilon''}{\epsilon} - \left(\frac{\epsilon'}{\epsilon} \right)^2 - k_x^2 \right] Z = 0. \tag{72}$$

With E_z determined, the field components E_x and H_y follow from Maxwell's equations. The field components will be examined, first, for an exponentially increasing refractive index n = ϵ/ϵ_0, and second, for exponentially increasing conductivity σ.

For exponential variation of the refractive index n, the variable in Eq. (72) is changed from z to

$$u = n^2 = \frac{\epsilon}{\epsilon_o} = iBe^{\beta z} \tag{73}$$

The surface impedance may be computed as

$$\frac{E_x}{H_y} = \sqrt{\frac{\mu_o}{\epsilon_o}} \; \frac{1}{n} \; \frac{K_{\nu-1}(-iw)}{K_\nu(-iw)} + \frac{\beta}{2i\omega\epsilon_o} \; \frac{1 - \nu}{n^2} \tag{74}$$

where K_ν is the modified Bessel function of the second kind of order

$$\nu = \sqrt{1 + 4 \; \frac{k_x^2}{\beta^2}} \tag{75}$$

and where

$$w = 2k_o n / \beta \tag{76}$$

with $k_o = \omega \sqrt{\mu_o \epsilon_o}$. For $|w| \ll 1$ Eq. (74) may be approximated by

$$\frac{E_x}{H_y} = \sqrt{\frac{\mu_o}{\epsilon_o}} \; \frac{k_o}{\beta} \left[-2i \left(0.116 - \ln \frac{2k_o|n|}{\beta} \right) + \frac{\pi}{2} + \frac{ik_x^2}{k_o^2 n^2} \right] . \tag{77}$$

The variation of n^2 in Eq. (73) applies only to those regions of the ionosphere where $|n| \gg 1$. It cannot be used in ionosphere models where $\sigma_i < \omega \epsilon_o$.

For an exponential variation of the ionospheric conductivity σ the variable in Eq. (72) is changed from z to

$$v = \frac{\sigma}{\omega \epsilon_o} = Be^{\beta z} \tag{78}$$

and the surface impedance follows from Galejs (1961c) as

$$\frac{E_x}{H_y} = \sqrt{\frac{\mu_o}{\epsilon_o}} \; \frac{k_o}{\beta} \left\{ + i \ln v + i \frac{k_x^2}{k_o^2} \left[\ln(v - i) - \ln v \right] + \frac{K\beta}{k_o^2} \right\} . \tag{79}$$

The models of exponential refractive index n and conductivity σ differ insignificantly for $n \gg 1$. Hence, the constant K of Eq. (79) may be determined by equating Eq. (74) and Eq. (79) for $|n| \gg 1$. This gives

230

$$\frac{E_x}{H_y} = \sqrt{\frac{\mu_o}{\epsilon_o}} \frac{k_o}{\beta} \left[-2i \left(0.116 - \ln \frac{2k_o}{\beta} \right) + \frac{\pi}{2} \frac{k_o^2 + k_x^2}{k_o^2} \right.$$

$$\left. - i \frac{k_x^2 - k_o^2}{k_o^2} \ln v - i \frac{k_x^2}{k_o^2} \sum_{m=1}^{\infty} \frac{(-iv)^m}{m} \right] . \tag{80}$$

The validity of Eq. (79) and Eq. (80) is restricted by the inequality.

$$\frac{|k_x^2 - k_o^2|}{\beta^2} < \frac{k_o^2}{\beta^2} \ll \frac{\beta}{k_o} \tag{81}$$

For $\beta = 1/3.25$ km^{-1} and $f = 1000$ cps, v should satisfy $0.46 \times 10^{-2} \ll v \ll 15$.

The propagation constant in the space between earth and ionosphere is

$$ik_x = ik_o S = -\alpha + i\beta \tag{82}$$

For a homogeneous ionosphere of refractive index n_i that is much smaller than the refractive index of the ground n_g, the first-order perturbation solution of the model equation gives for the zero-order mode

$$S = \sqrt{1 + \frac{i}{k_o h n_i}} \sqrt{1 - \left(\frac{S}{n_i} \right)^2} \tag{83}$$

where h is the height of the ionosphere. In the presence of a non-homogeneous ionosphere, one may define an equivalent homogeneous ionosphere of refractive index n_i^e which exhibits the same surface impedance Z_i as the non-homogeneous ionosphere. It follows from Galejs (1961c) that for $n_i^e \gg 1$,

$$\left(n_i^e \right)^2 \approx \frac{\mu_o}{\epsilon_o Z_i^2} - S^2 , \tag{84}$$

Examination of Eq. (79) and Eq. (80) shows that Z_i is also dependent on $S = \frac{k_x}{k_o}$ and Eq. (83) can be solved by an iteration procedure.

The attenuation constants α_i are calculated from Eq. (80), (82), and (84) and are shown in Fig. 16 for an ionosphere which is sharply bounded at an altitude h where $v = \sigma(h)/(\omega \epsilon_o) = $ const. The curves in Fig. 16 show that significant contributions to α are made by ionospheric layers where $v < 1$. The curve indicated by dots and dashes is obtained using Eq. (74) and Eq. (77) for the surface impedance with $n^2 = 10/i$. It differs from the $v = 10$ curve at the higher frequencies where the inequality Eq. (81) is not satisfied strictly.

The cosmic-ray ionization (Bourdeau, Whipple and Clark, 1959) will also contribute to the attenuation of ELF radio waves. The atmospheric conductivity that is due to cosmic-ray ionization can be approximated by the exponential curve of a gradient which is smaller than that for ionospheric conductivity. Provided that the atmospheric losses are small, the ionospheric fields can be assumed to be locally unperturbed by the former losses. The additional attenuation of the fields due to atmospheric losses (attenuation constant α_a) can be calculated as in a waveguide filled with a lossy dielectric. Such calculations will result in a correction of the attenuation constant $\alpha = k$ Im S, but will not change the phase constant $\beta = k$ Re S.

Roots S of the model equation have been plotted in Fig. 17 for the four ionosphere models of Eq. (11). The attenuation constant α of Eq. (82) is shown in Fig. 18 and 19 for the same conductivity profiles $\sigma_m(z)$, which are bounded at a height h = constant. To show the effects of a shifted conductivity profile, calculations

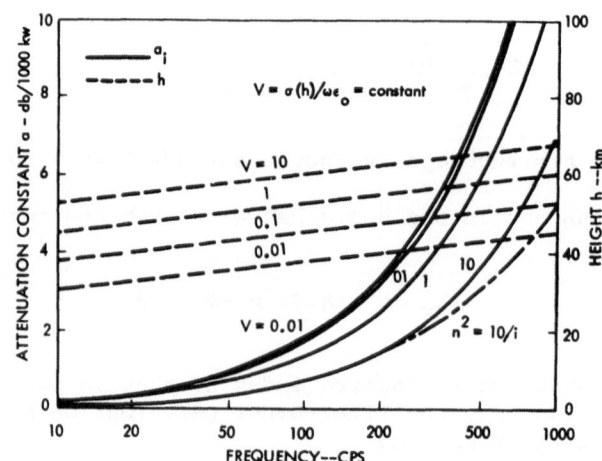

Fig. 16.
Daytime attenuation constants for an ionosphere of exponential conductivity profile which is sharply bounded at h where $V = \sigma(h)/\omega\epsilon_o$ = Constant.

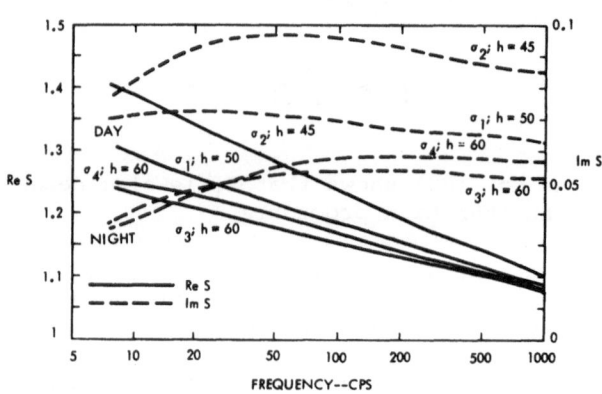

Fig. 17.
Roots of the model equation.

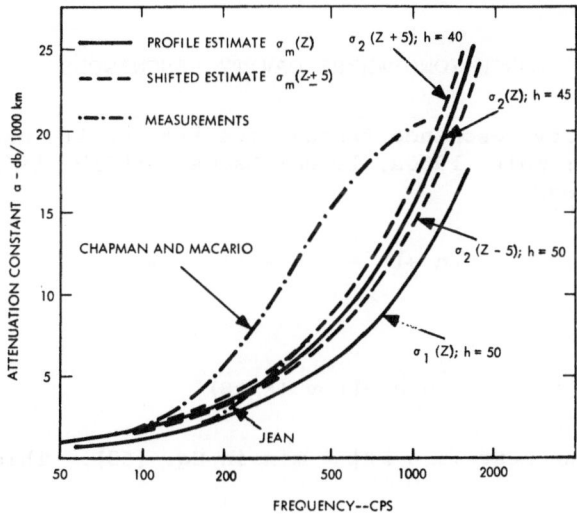

Fig. 18. Daytime attenuation constant.

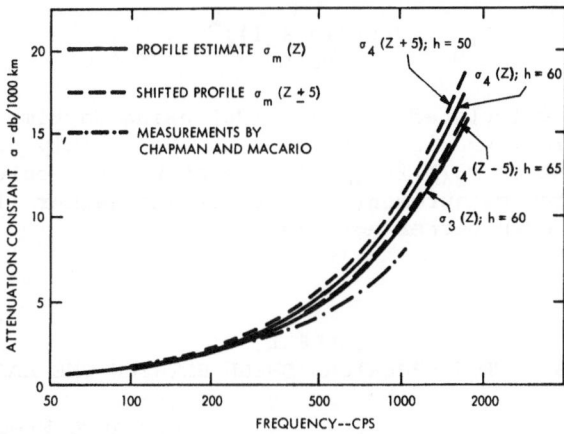

Fig. 19. Nighttime attenuation constant.

are made for $\sigma_m (z \pm \Delta)$, the results of which are shown by dashed curves in Fig. 18 and 19. The daytime ionosphere of conductivity σ_2, which is bounded at h = 45 km, provides fair agreement with the measurements of Jean et al (1961). The attenuation is lower than was indicated by the earlier measurements of Chapman and Macario (1956). The conductivity of the profile σ_1 increases more rapidly than σ_2 and provides less attenuation. The nighttime ionosphere of conductivity σ_3, which is bounded at h = 60 km, provides a somewhat higher attenuation than that measured by Chapman and Macario. The lower conductivity gradient of σ_4 provides more attenuation than σ_3. An upward shift of the ionosphere profile $\sigma_m (z - \Delta)$ decreases the attenuation; a downward shift $\sigma_m (z + \Delta)$ increases it. The atmospheric conductivity (Bourdeau, Whipple, and Clark, 1959) makes a negligible contribution to the signal attenuation in the frequency range shown in Fig. 18 and 19.

233

EARTH-IONOSPHERE CAVITY RESONANCES

The earth-ionosphere cavity resonance frequencies are usually examined by study of Eq. (59) (Schumann, 1957; Wait, 1960a, 1960b; Raemer, 1961b, 1961c). It follows that E_r will exhibit maxima near

$$|n(n+1) - \nu(\nu + 1)| = \min \tag{85}$$

or

$$n(n+1) \approx (k_o aS)^2 \tag{86}$$

The same maxima occur near $|\sin \nu\pi| = \min$ in Eq. (53). This condition is equivalent to

$$\text{Re } \nu = k_o a \text{ Re } S - \tfrac{1}{2} = m \tag{87}$$

where m is an integer. The resonance frequency f_n of the nth mode is obtained from Eq. (86) as

$$f_n \approx \frac{7.5\{n(n+1)\}^{\frac{1}{2}}}{\text{Re } S} \quad \text{(cps)} \tag{88}$$

The resonance frequencies calculated from Eq. (88) using daytime S and the average of day and night S figures are summarized in Table 1. Using only daytime data with either h = 45 km = constant or $\sigma_2(h)/\omega\epsilon_o = 1$ = constant, the calculated resonances differ by 1 cps or less from measurements by Balser and Wagner (1960). Combined day and night data provide an even better agreement.

TABLE 1
RESONANCE FREQUENCIES OF THE SPHERICAL SHELL BETWEEN THE EARTH AND IONOSPHERE

DATA	Resonance Frequency f_n, cps			
	n=1	n=2	n=3	n=4
Lossless cavity (Schumann, 1957; Wait, 1960b)	10.6	18.3	25.9	33.5
Day with σ_2 and h = 45 km	7.6	13.5	19.4	25.4
Average data of day with σ_2 and h = 45 km and of night with σ_3 and h = 60 km	8.05	14.25	20.5	26.8
Day with $\sigma_2(h)/\omega\epsilon_o = 1$ = constant	8	13.8	19.5	25.4
Measurements by Balser and Wagner (1960)	8	14.1	20.3	26.4

The ionospheric and atmospheric losses are accounted for by defining separate Q factors in analogy with microwave cavities (Slater, 1950). The Q factor due to ionospheric losses is

$$\frac{1}{Q_i} = \frac{c}{\omega} \cdot \frac{\left.\dfrac{Re\, n_i}{|n_i|^2}\right|_{day} + \left.\dfrac{Re\, n_i}{|n_i|^2}\right|_{night}}{h_{day} + h_{night}} .$$ (89)

where n_i is the equivalent refractive index n as in Eq. (84). This factor considers only the losses due to the fundamental cavity modes. The losses due to the higher modes which are excited by the height discontinuity between day and night hemispheres can be shown to be negligible. The Q factor due to atmospheric losses Q_a is given by Galejs (1961c).

Numerical values for the combined Q factor

$$\frac{1}{Q} = \frac{1}{Q_a} + \frac{1}{Q_i}$$ (90)

are shown in Table 2 for daytime and combined daytime and nighttime data.

TABLE 2
Q FACTORS OF EARTH-IONOSPHERE CAVITY RESONANCES

DATA	Q FACTORS Frequency		
	10 cps	30 cps	100 cps
Day with σ_2 and h = 45 km, night with σ_3 and h = 60 km	6.75	5.7	5.9
Day only with σ_2 and h = 45 km	4.6	4	4.3
Day with σ_2 and h = 45 km, night with σ_3 and h = 60 km, cosmic-ray ionization	4.45	4.7	5.5
Day only with $\sigma_2(h)/\omega\epsilon_o = 1$ = constant	3.8	4.8	5.7
Measurements by Balser and Wagner (1960)	4	6	–

Combined day and night data result in Q figures that are too high at the lower frequencies relative to the measurements of Balser and Wagner (1960). The Q figures agree more closely with the measurements when considering the atmospheric conductivity due to cosmic-ray ionization ($Q_a \neq 0$). The discontinuity in conductivity between the lower edge of the exponential ionosphere $\sigma_m(h)$ and the atmospheric conductivity (Bourdeau, Whipple, and Clark, 1959) at the same height is suppressed by assuming an exponential variation of the atmospheric conductivity between $\sigma_a = 10^{-13}$ mho/m at ground level to $\sigma_a = \sigma_m(h)$ at height h.

Noise Spectra.

The power spectrum received by a vertical stub antenna that is due to a distribution of vertical electric dipoles is computed starting with Eq. (53). It follows (Galejs, 1961b) that

$$G(i\omega) = \int_0^{2\pi} \int_0^{\pi} g(i\omega, \theta, \emptyset) a^2 \sin \theta$$

$$\cdot \left| \frac{\nu(\nu + 1)}{4k_oha^2} \frac{P_\nu(-\cos \theta)}{\sin \nu\pi} \right|^2 d\theta \, d\emptyset \tag{91}$$

where $g(i\omega, \theta, \emptyset)$ is proportional to the squared dipole moment of the sources per unit area. Assuming that

$$g(i\omega, \theta, \emptyset) = g(i\omega) \tag{92}$$

for $\theta_1 < \theta < \theta_2$ and is equal to zero elsewhere, Eq. (91) may be evaluated with the aid of 3.12.3 of Erdelyi (1953). The integral Eq. (91) may also be evaluated after replacing the Legendre functions by their asymptotic expansions. Applying Eq. (53) gives

$$G(i\omega) = \frac{g(i\omega) \, |\nu(\nu + 1)|^2}{8(ak_oh)^2 \, | \, \nu + \frac{1}{2} \, | \, [\cosh (2\pi \, \text{Im} \, \nu) - \cos (2\pi \, \text{Re} \, \nu)]}$$

$$\tag{93}$$

$$\cdot \left\{ \frac{\sinh[2 \, \text{Im} \, \nu \, (\pi-\theta)]}{(-\text{Im} \, \nu)} - \frac{\sin[(2 \, \text{Re} \, \nu + 1)(\pi-\theta) - (\pi/2)]}{\text{Re} \, \nu + \frac{1}{2}} \right\}_{\theta_1}^{\theta_2}$$

The parameter ν is defined by Eq. (49). To account for the losses below the ionosphere edge, Im S is related to Q by

$$\text{Im} \, S = \frac{1}{2Q \, \text{Re} \, S} \tag{94}$$

Eq. (94) is established by squaring Eq. (83) and by considering the imaginary parts of the resulting equation. For $n_i \gg 1$ the right hand side of this equality is 1/Q according to Eq. (89).

The power spectrum of the dipole moment in a median lightning flash in Fig. 9 may be approximated by

236

$$g(i\omega) = 1.69 \exp (-9.1 \times 10^{-3}\omega) \tag{95}$$

The sources are assumed to be uniformly distributed over the polar angle interval $(90° - \Delta) \le \theta \le (90° + \Delta)$ and typical calculations are shown in Fig. 20. In more recent measurements Balser and Wagner (1962a) have correlated the diurnal variations of the ELF noise spectra with world-wide thunderstorm activities. They indicate that the noise sources of their winter measurements appear to be concentrated in South America and Central Africa. The noise spectrum plots of Fig. 4a and 4b of Balser and Wagner (1962a) are compared with computations in Fig. 21 and 22.

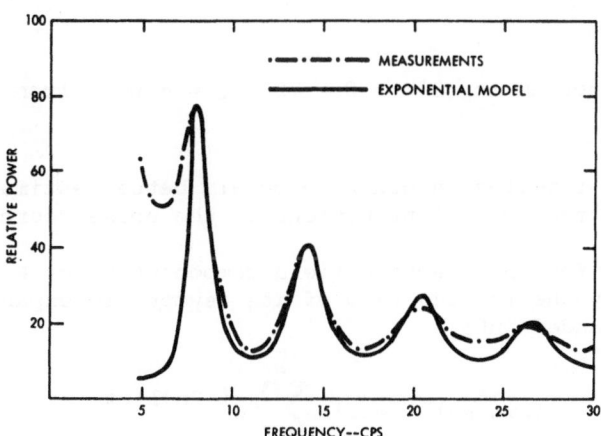

Fig. 20. Noise spectrum; Q = 4 to 6 from f = 10 to 30 cps, $\Delta = 60°$.

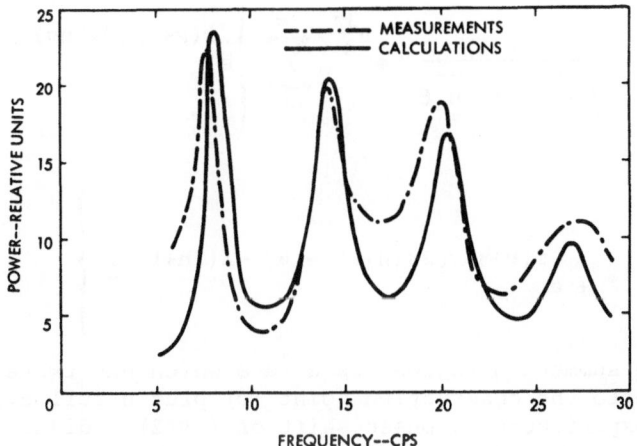

Fig. 21. Noise spectrum for $47° \le \theta \le 128°$; Q = 4 to 6 from f = 10 to 30 cps.

The above equations are also applicable to noise spectrum calculations at higher frequencies and $g(i\omega)$ of Eq. (92) may be assumed to be constant with frequency in the first approximation. Such calculations have been carried out by Abraham (1963) who

237

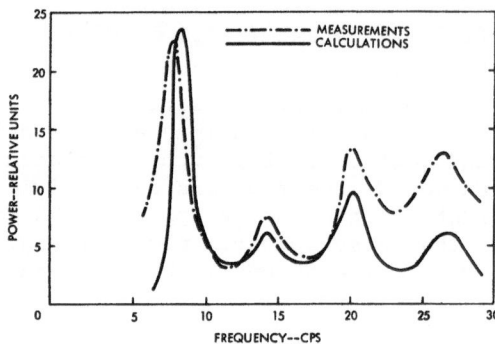

Fig. 22. Noise spectrum for $38° \leq \theta \leq 72°$; Q = 4 to 6 from f = 10 to 30 cps.

modified the preceding formulas in order to obtain better estimates of the day-night effects and of the geographical distributions of the noise sources.

In the expression for the electric field component E_r of Eq. (53) the Legendre function is replaced by the leading term of its asymptotic expansion Eq. (54). The function sin $\nu\pi$ is expanded into

$$(\sin \nu\pi)^{-1} = 2i \sum_{n=0}^{\infty} e^{i(2n+1)\nu\pi} \tag{96}$$

which is valid for Im $\nu > 0$. The cosine function of Eq. (54) is expressed as a sum of two exponentials, which results in

$$E_r = \frac{Ids}{2\omega\epsilon ha^2} \frac{\nu(\nu+1)}{\sqrt{2\pi(\nu+\frac{1}{2})\sin \theta}} e^{i\frac{3\pi}{4}} \sum_{n=0}^{\infty} \left\{ e^{i(\nu+\frac{1}{2})(\theta+2n\pi) - i2n\frac{\pi}{2}} \right. $$

$$\left. + e^{i(\nu+\frac{1}{2})[2\pi(n+1) - \theta] - i(2n+1)\frac{\pi}{2}} \right\} \tag{97}$$

The first term of the summation represents a wave which has traveled the direct distance from the source to the observation point (θ) plus n full circles around the sphere ($2n\pi$), while experiencing a phase shift of ($-\pi/2$) radians at each traversal of the source point or its antipode. The other term of the summation represents a wave which has traveled toward the observation point across the antipode of the source ($2\pi - \theta$) and has made furthermore n circles around the sphere ($2n\pi$) and has also experienced a phase shift of ($-\pi/2$) radians at each traversal of the source point or its antipode. This physical interpretation of Eq. (97), which is equivalent to Eq. (53) for ν = cost., permits a heuristic consideration of the ν differences between the day and night hemispheres. The multiplicative ν factors in front of the n summation are assumed to be constant and equal to $\bar{\nu}$ which is intermediate between the day and night ν values. The θ factors of the exponentials are replaced by integrals in order to account for the ν variation seen by the circular wavefront as it propagates in the θ direction with respect to the source.   Thus

$$E_r \approx \frac{Ids}{2\omega\epsilon ha^2}\sqrt{\frac{\overline{\nu}(\overline{\nu}+1)}{2\pi(\overline{\nu}+\frac{1}{2})\sin\theta}}\ e^{i\frac{3\pi}{4}}\sum_{n=0}^{\infty}\left\{\exp\left[i\int_0^{\theta+2n\pi}(\nu+\frac{1}{2})d\theta\right.\right.$$

(98)

$$\left.\left.-\ i2n\frac{\pi}{2}\right]+\exp\left[i\int_0^{2\pi(n+1)-\theta}(\nu+\frac{1}{2})d\theta-i(2n+1)\frac{\pi}{2}\right]\right\}$$

The amplitude changes at the day and night boundary can be also considered for each traveling wave (Wait, 1962c).

To allow for wide time and space variations of the surface density of noise sources, the surface of the earth was subdivided into 10° cells. Each cell had a constant surface density of noise sources but a variation was allowed from cell to cell. The plots of thunderstorm days over the world were used to set relative values of lightning activity for each of the cells, by assuming that the thunderstorm days plots were directly proportional to the distribution of squared dipole moment per unit area according to the Handbook of Geophysics (1960). The diurnal variations were accounted for by a multiplicative factor in terms of local time at the source (Williams, 1959).

After evaluating the θ-integrals of Eq. (98) the noise contribution from a single cell is computed as in Eq. (91) to Eq. (92). The contributions of the individual cells are then added to get the total noise at the receiver. The final equation is similar to Eq. (93), although it is algebraically more involved. An example of spectrum calculations is shown in Fig. 23. It indicates a decreased noise

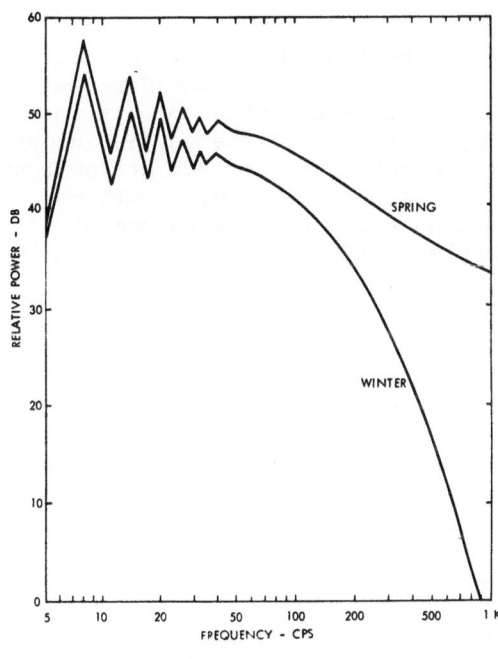

Fig. 23.
Calculated noise spectrum for local noon in Northern Hemisphere.

power near the higher frequencies and/or during winter months. The diurnal variation of the noise power at the first three Schumann resonances have been also computed and are shown in Fig. 24. The calculations have disregarded the noise sources near the

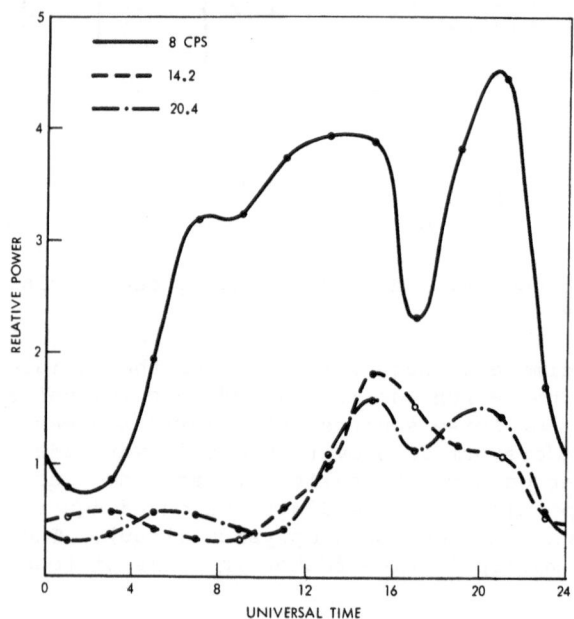

Fig. 24. Calculated diurnal variation of the resonance
modes. Wintertime for Boston, Massachusetts.

antipode where the asymptotic approximation of the Legendre functions Eq. (54) is inaccurate. Fig. 24 shows a qualitative agreement with the measurements of Balser and Wagner (1962a), shown in Fig. 10.

In the above analysis, the propagation seems to be considered more accurately than the noise source distribution. Still, it seems desirable to consider the variations of ionospheric parameters more rigorously. The present method appears to provide a reasonable approximation to the day and night effects if the day and night boundary is located symmetrically with respect to source (i.e., source at local noon or midnight). For nonsymmetrical day and night boundaries, it is difficult to estimate the effects of wavefront distortion, particularly in the presence of isolated ionosphere perturbations (like due to nuclear effects).

The ionospheric anisotropy is considered in this section with a horizontal magnetic field either for transverse (East-West or West-East) or for longitudinal (South-North) propagation following Galejs and Row (1963) and Galejs (1963). For transverse propagation in a vertically stratified medium, the differential equations of the various field components are uncoupled and relatively simple solutions are obtainable. For longitudinal propagation, the differential equations of the various field components are coupled, with the coupling being particularly strong above the D-region. However, the differential equations are considerably simplified by assuming no coupling in the lower ionosphere and strong coupling above a pre-selected altitude.

Transverse Propagation.

For an assumed $\exp(-i\omega t)$ time dependence of the fields the relative tensor permitivity is of the form

$$
[\epsilon] = \begin{bmatrix} \epsilon_1 & -\epsilon_2 & 0 \\ \epsilon_2 & \epsilon_1 & 0 \\ 0 & 0 & \epsilon_3 \end{bmatrix} = \begin{bmatrix} 1 + \dfrac{i\sigma_1}{\omega\epsilon_o} & -\dfrac{i\sigma_2}{\omega\epsilon_o} & 0 \\ \dfrac{i\sigma_2}{\omega\epsilon_o} & 1 + \dfrac{i\sigma_1}{\omega\epsilon_o} & 0 \\ 0 & 0 & 1 + \dfrac{i\sigma_o}{\omega\epsilon_o} \end{bmatrix} \tag{99}
$$

if the static magnetic field is in the z direction. It is assumed that the field components exhibit no variation in the z direction and that $[\epsilon]$ is changing only in the y direction, which would be the South-North and vertical directions in terrestrial propagation problems. It follows from the Maxwell equations that H_z and E_z satisfy the differential equations

$$
\frac{\partial^2}{\partial x^2} H_z + \frac{\partial^2}{\partial y^2} H_z + k_o^2 \frac{\epsilon_1^2 + \epsilon_2^2}{\epsilon_1} H_z + \frac{\epsilon_1^2 + \epsilon_2^2}{\epsilon_1} \left[\frac{\partial}{\partial y} \frac{\epsilon_1}{\epsilon_1^2 + \epsilon_2^2} \frac{\partial}{\partial y} H_z \right.
$$

$$
\left. - \frac{\partial}{\partial y} \frac{\epsilon_2}{\epsilon_1^2 + \epsilon_2^2} \frac{\partial}{\partial x} H_z \right] = 0 \tag{100}
$$

$$
\frac{\partial^2}{\partial x^2} E_z + \frac{\partial^2}{\partial y^2} E_z + k_o^2 \epsilon_3 = 0 \tag{101}
$$

where $k_o = \omega \sqrt{\mu_o \epsilon_o}$. The derivation of the above equations is given by Galejs and Row (1963). There is no coupling by the anisotropic medium between the field

241

components E_x, E_y, H_z and E_z, H_x, H_y. If the excitation is such that the latter components are not excited, (e.g., far fields of a vertical electric dipole of dipole moment I dy) the surface impedance in the anisotropic medium is simply

$$Z_s = - \frac{E_x}{H_z} = i\omega\epsilon_o \left[(\epsilon_1^2 + \epsilon_2^2)\, H_z \right]^{-1} \cdot \left[\epsilon_1 \frac{\partial}{\partial y} H_z - \epsilon_2 \frac{\partial}{\partial x} H_z \right] \tag{102}$$

This surface impedance applies for East to West propagation along the magnetic equator. The sign of ϵ_2 is reversed for West to East propagation, which causes an obvious non-reciprocity. For propagation in the x direction, the magnetic field component H_z will exhibit the x and y dependence of

$$H_z = \exp(ik_x x)\, Y(y). \tag{103}$$

Substituting Eq. (103) in Eq. (100) gives a second order differential equation for the y-dependent variable $Y(y)$.

An exact closed form solution of this differential equation is possible in terms of Bessel functions for identical exponential height variations of the two components ϵ_1 and ϵ_2, which determines the surface impedance at altitudes y where $\epsilon_1 \approx \pm\, \epsilon_2$. Lower layers of the ionosphere will also affect the propagation characteristics of fields below it. These effects can be considered most simply by neglecting the anisotropy of the lower layers in a two-layer model, where the lower layer of $y < y_1$ is isotropic and the upper layer of $y > y_1$ is anisotropic. The isotropic layer is characterized by $\sigma(y)$, while the anisotropic layer by $\sigma_1(y)$, $\sigma_2(y)$ and $\sigma_o(y)$ as indicated in Eq. (99). The isotropic layer is assumed to exhibit an exponential conductivity variation. The surface impedance below this isotropic layer is obtained from Eq. (79) with K determined from the known value of the surface impedance $Z_s(y_1)$ at the edge of the anisotropic layer, and with $Z_s(y_1)$ given by Eq. (102).

The exact shape of the conductivity-height profiles of Fig. 3 through 8 may be considered by numerical techniques. The surface impedance at the bottom of an anisotropic ionosphere is determined from the solution of a Riccati-type differential equation, which is well suited to numerical integration. The surface impedance Eq. (102) can be expressed as

$$Z_s = \frac{1}{i\omega\epsilon_o} \left\{ \left(\frac{\epsilon_1}{\epsilon_1^2 + \epsilon_2^2} \right) \left[V(y) - \tfrac{1}{2}P(y) \right] - \frac{\epsilon_2}{\epsilon_1^2 + \epsilon_2^2}\, ik_x \right\} \tag{104}$$

$$P(y) = \frac{d}{dy} \log \left(\frac{\epsilon_1}{\epsilon_1^2 + \epsilon_2^2} \right) \tag{105}$$

The new variable $V(y)$ satisfies the Riccati-type differential equation

$$\frac{dV}{dy} = - (V^2 + K) \tag{106}$$

with

$$K(y) = Q - \tfrac{1}{2} \frac{dP}{dy} - \tfrac{1}{4} P^2 \tag{107}$$

and

$$Q(y) = \frac{\epsilon_1^2 + \epsilon_2^2}{\epsilon_1} k_o^2 - k_x^2 - ik_x \frac{\epsilon_2}{\epsilon_1} \frac{d}{dy} \log \left(\frac{\epsilon_2}{\epsilon_1^2 + \epsilon_2^2}\right) \tag{108}$$

which can be integrated numerically if $V = V(y_u)$ is specified at a given altitude $y = y_u$. $V(y_u)$ may be determined from the surface impedance Z_s at the upper edge $y = y_u$ of the ionosphere model, after suitably simplifying the differential equations. The surface impedance Eq. (104) may be also expressed in terms of a new variable

$$U(y) = V(y) - \tfrac{1}{2} P(y) \tag{109}$$

which satisfies the differential equation

$$\frac{dU}{dy} = - U^2 (y) - P(y) \; U(y) , - Q(y) \tag{110}$$

Eq. (110) can be integrated numerically similarly as Eq. (106). However, Eq. (110) gives more accurate solutions for the lower frequencies (f < 100 cps), where $V(y) \approx \tfrac{1}{2} P(y)$ near the lower boundary of the ionosphere.

Longitudinal Propagation.

Near the magnetic equator the static earth's magnetic field is in the South-to-North direction, which is the z-direction in the definition of the permitivity tensor Eq. (99). The tensor $[\epsilon]$ changes only in the vertical y direction and the field components exhibit no variation in the x-direction (West-to-East direction). It follows from the Maxwell's equations that H_x and E_x satisfy the differential equations

$$\epsilon_1 \frac{\partial}{\partial y} \left[\frac{1}{\epsilon_3} \cdot \frac{\partial}{\partial y} H_x \right] + (\epsilon_1 k_o^2 - k_z^2) \; H_x = \omega \, \epsilon_o \epsilon_2 k_z E_x \tag{111}$$

$$\epsilon_1 \frac{\partial^2}{\partial y^2} E_x + \left[k_o^2 \; (\epsilon_2^2 + \epsilon_1^2) - k_z^2 \; \epsilon_1 \right] E_x = -\epsilon_2 k_z \mu_o H_x \tag{112}$$

where the field components exhibit an $\exp(ik_z z)$ z-variation. For vertically polarized fields and for weak coupling the principal field components will be the set E_y, E_z and H_x. The magnitude of the coupling effects can be readily examined after defining

$$\frac{\partial^2}{\partial y^2} E_x = \gamma_y^2 E_x \tag{113}$$

where γ_y is complex and y-dependent. After applying Eq. (113), Eq. (112) is solved for E_x, which is substituted in Eq. (111). This gives

$$\epsilon_1 \frac{\partial}{\partial y} \left[\frac{1}{\epsilon_3} \frac{\partial}{\partial y} H_x \right] + (\epsilon_1 k_o^2 - q k_z^2) H_x = 0 \qquad (114)$$

where

$$q = \frac{\gamma_y^2 - k_z^2 + k_o^2 \epsilon_1}{\gamma_y^2 - k_z^2 + k_o^2 (\epsilon_2^2 + \epsilon_1^2)/\epsilon_1} . \qquad (115)$$

In the lower ionosphere $|\epsilon_2| \approx |\epsilon_1|$, $\gamma_y^2 \gg (k_o^2 \epsilon_1)$ and q approaches unity. Eq. (114) is now the same as Eq. (111) with the right hand side set equal to zero, which implies negligible coupling between E_x and H_x components. At higher altitudes $|\epsilon_2| \gg |\epsilon_1|$ and q < 1. A simple approximation can be worked out by assuming

$$q = \begin{cases} 1 \text{ for } y < y_1 \\ \\ 0 \text{ for } y \geq y_1 \end{cases} \qquad (116)$$

which implies negligible coupling between E_x and H_x for $y < y_1$, and strong coupling for $y \geq y_1$, where $y_1 < y_u$. This artificial boundary at $y = y_1$ will introduce an impedance discontinuity, but at ELF this discontinuity appears at a height, where it does not affect the fields below the ionosphere to any significant extent. With neglected coupling between H_x and E_x, the surface impedance for vertically polarized fields is simply

$$Z_s = \frac{E_z}{H_x} = \frac{1}{i\omega\epsilon_o\epsilon_3} \frac{1}{H_x} \frac{\partial}{\partial y} H_x = \frac{1}{i\omega\epsilon_o} \frac{1}{\epsilon_3} \left[V(y) - \frac{1}{2} P(y) \right] \qquad (117)$$

where

$$P(y) = - \frac{d}{dy} (\log \epsilon_3) \qquad (118)$$

V(y) is determined from the solution of Eq. (106) or Eq. (110), where K(y) is given by Eq. (107) with

$$Q(y) = \epsilon_3 k_o^2 - \frac{\epsilon_3}{\epsilon_1} q k_z^2 . \qquad (119)$$

244

Numerical Results.

For ELF waves the propagation constant ik_x of the section on Transverse Propagation, or ik_z of the section on Longitudinal Propagation is given by $ik_0 S$ where S is computed from an iteration solution.

The daytime conductivity-height profiles of Fig. 3 are approximated first by an exponential

$$\sigma_i = 10^{-6} \exp\left[0.291 \ (y - 74.1)\right] \tag{120}$$

where $i = 0$, 1 or 2 and where the altitude y is given in kilometers. The real and imaginary parts of S figures of a two-layer exponential ionosphere are depicted in Fig. 25. The upper layer of $y > 70$ km is anisotropic ($\epsilon_1 = \pm \ \epsilon_2$), while the lower layer (70 km $> y > 50$ km) is isotropic ($\epsilon_2 = 0$). East-West propagation exhibits a

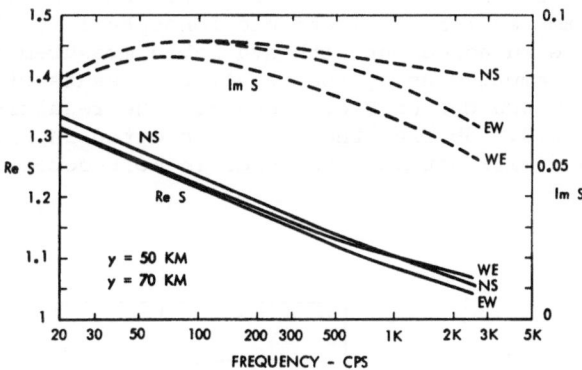

Fig. 25. Roots of the model equation. Closed form solutions for an exponential two-layer model. $\sigma_2 = 0$ in lower layer. $\sigma_1 = \sigma_2$ in upper layer.

lower Re S and a higher Im S, although a lower Im S can be expected for a model of a homogeneous sharply bounded anisotropic ionosphere (Wait, 1962a). These differences between the S figures of various directions of propagation can be explained with the aid of the closed form solution which is valid for a lower ionosphere boundary at y = 70 km and at the lower frequencies.

The conductivity-height profiles of Fig. 3 through 8 have been approximated by polynomials for a more accurate numerical integration.

The S figures of Fig. 26 are computed using such a polynomial approximation to the daytime conductivity curves of Fig. 3 for f = 0 and neglected ions. They exhibit approximately the same frequency dependence as the S figures for a two-layer model approximation in Fig. 25, except for a rapid increase of Im S for West-East propagation with f < 1000 cps. This is due to the absorption in the lower ionosphere layers where $\epsilon_1 \approx 1$ and $\epsilon_2 \ll 1$. This increase of Im S_{we} can be accounted for after simplifying the differential Eq. (108) for the above values of ϵ_1 and ϵ_2 and after examining its closed form solutions. The South-to-North direction (SN) gives results which are intermediate relative to East-West and West-East data. The numerical integration was started with $y_u = 100$ km in Fig. 26. Increasing the value of y_u has been shown to

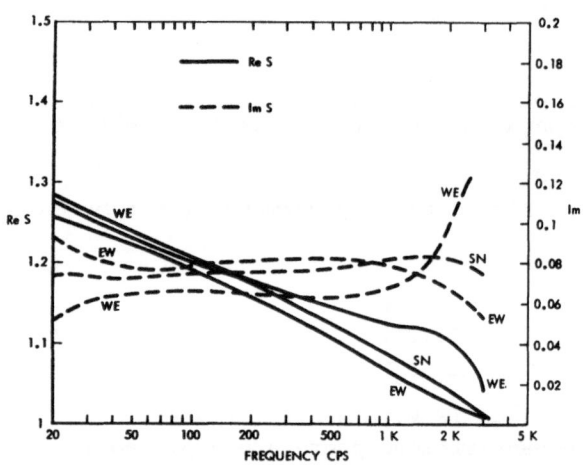

Fig. 26.
Roots of the model equation. Electronic daytime conductivity for f = 0.

have no effect on the computed S figures. The nighttime S figures are generally lower than the corresponding daytime figures in Fig. 26. The attenuation constants $\alpha = k_o \, \mathrm{Im} \, S$ have been plotted in Fig. 27. It is significant to note that WE direction provides the lower attenuation for f < 1000 cps. The magnitude of the calculated attenuation constants is about the same as for the isotropic ionosphere model in Fig. 18 and 19, which has been shown to be in agreement with available measurements. The attenuation constants have been also computed using the tensor components of the electronic plus ionic conductivity for f = 300 cps of Fig. 7 and 8. The resulting attenuation constants differ by less than 0.5 db from the figures plotted in Fig. 27 for electronic conductivity at f = 0, and will not be considered in more detail.

Fig. 27.
Attenuation constants.

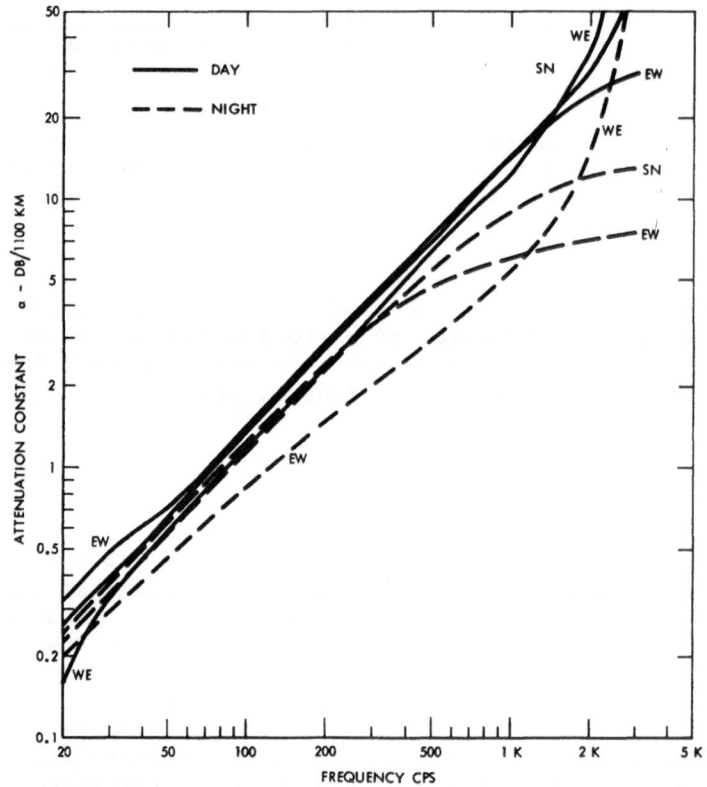

246

The S figures have been computed also for the electronic plus ionic conductivity at f = 20 cps of Fig. 5 and 6, which is applicable to frequencies below the ion gyro-frequency (30 to 50 cps). There is a significant decrease of the West-to-East attenuation near 8 cps. The attenuation constant is α_{ew} = 0.16 db/1000 km and α_{we} = 0.022 db/1000 km at 8 cps.

EFFECTS OF THE RADIAL MAGNETIC FIELD

A formalism that may be used for treating the multilayer spherically stratified aniso-tropic ionosphere has been presented by Wait (1963b) in the presence of a superimposed radial magnetic field. This problem is formulated for frequencies in the VLF range and hence it employs the Legendre functions $P_{\nu}(-\cos\ \theta)$ for the azimuthal field repre-sentation. For an M layer ionosphere the boundary conditions lead to 4M + 2 algebraic equations connecting 4M + 2 coefficients. Setting the determinant of these co-efficients equal to zero gives a complicated transcendental equation for determining the eigenvalues ν of the modes. When restricting the considerations to the lower ELF range, the fields may be treated as a superposition of earth-to-ionosphere cavity modes, and the azimuthal field variation can be represented in terms of Legendre poly-nomials $P_n(\cos\ \theta)$. In this formulation there is no need to solve a large number of simultaneous transcendental equations which is a computational simplification. Detailed solutions in such a geometry have been worked out by Thompson (1963) by assuming the radial magnetic field to be constant all over the globe. This represents an extension of the multilayer matrix formalism of Madden (1961) that was originally applied only to a two-layer ionosphere. Similar techniques for treating multi-slab approximations have been discussed also by Ferraro and Gibbons (1959) and Ferraro (1962).

In the following discussion, some of the Thompson (1963) results will be general-ized to allow also for ionospheric displacement currents which are not negligible near the lower edge of a diffuse ionospheric boundary.

Fields in the Anisotropic Ionosphere.

The wave solutions for the fields in the anisotropic ionosphere can be derived from Maxwell's equations

$$\triangledown \times \underline{E} = i\omega\mu_o\ \underline{H} \tag{121}$$

and

$$\triangledown \times \underline{H} = i\omega\epsilon_o[\epsilon]\ \underline{E} \tag{122}$$

The permitivity tensor $[\epsilon]$ is defined by Eq. (99). The field components E_r and H_r are proportional to $f_r(\theta) = P_n(\cos\ \theta)$, while E_θ, H_θ, E_ϕ and H_ϕ are proportional to $f_\theta(\theta) = f_\phi(\theta) = dP_n(\cos\ \theta)/d\theta$. After writing out the 2 vector Eq. (121) and (122) in their 6 scaler components, the θ functions can be cancelled leaving 6 scaler equations in the r-dependent components $E_j^O(r) = E_j/f_j(\theta)$, $H_j^O(r) = H_j/f_j(\theta)$, where $j = r$, θ or $\emptyset$. This gives

$$- n(n+1) \; \frac{E_\emptyset^O}{r} \;\; = \;\; i\omega\mu_o H_r^O \tag{123}$$

$$- \frac{1}{r} \; \frac{\partial}{\partial r} \; (rE_\emptyset^O) \;\; = \;\; i\omega\mu_o H_\theta^O \tag{124}$$

$$\frac{1}{r} \; \frac{\partial}{\partial r} \; (rE_\theta^O) \; - \; \frac{E_r^O}{r} \;\; = \;\; i\omega\mu_o H_\emptyset^O \tag{125}$$

$$n(n+1) \; \frac{H_\emptyset^O}{r} \;\; = \;\; i\omega\epsilon_o \, \epsilon_3 E_r^O \tag{126}$$

$$\frac{1}{r} \; \frac{\partial}{\partial r} \; (rH_\emptyset^O) \;\; = \;\; i\omega\epsilon_o \left[\epsilon_1 E_\theta^O \; + \; \epsilon_2 E_\emptyset^O \right] \tag{127}$$

$$\frac{1}{r} \; \frac{\partial}{\partial r} \; (rH_\theta^O) \; - \; \frac{H_r^O}{r} \;\; = \;\; i\omega\epsilon_o \left[\epsilon_2 E_\theta^O \; - \; \epsilon_1 E_\emptyset^O \right] \tag{128}$$

After introducing the notation

$$k_j^2 \;\; = \;\; \omega^2 \, \mu_o \epsilon_o \epsilon_j \tag{129}$$

($j = 1, 2, 3$) a manipulation of Eq. (123) to Eq. (128) results in

$$\frac{\partial^3}{\partial r^3} \; (rE_\emptyset^O) \; - \; n(n+1) \; \frac{\partial}{\partial r} \; (\frac{E_\emptyset^O}{r}) \; + \; k_1^2 \; \frac{\partial}{\partial r} \; (rE_\emptyset^O) \; - \; k_2^2 \, E_r^O \left[1 - \frac{(k_3 r)^2}{n(n+1)} \right] = 0 \tag{130}$$

and

$$\frac{\partial}{\partial r} \; (rE_\emptyset^O) = \frac{1}{k_2^2} \left\{ \frac{\partial^2}{\partial r^2} \left[\frac{(k_3 r)^2}{n(n+1)} \; E_r^O \right] - k_1^2 \, E_r^O \left[1 - \frac{(k_3 r)^2}{n(n+1)} \right] \right\} \tag{131}$$

248

Two kinds of approximations are considered at this point. Negligible displacement currents of the ionosphere imply that $k_j \gg k_o = \omega \sqrt{\mu_o \epsilon_o}$. It may be also noted that $n(n+1)$ is of the order of $(k_o r)^2$ for ω near the resonances (this follows from Eq. (65) and (68)). As a consequence

$$\frac{k_j^2 r^2}{n(n+1)} \gg 1 \tag{132}$$

Applying Eq. (132) to simplify Eq. (130) and (131) gives

$$\frac{\partial^4}{\partial r^4} (r^2 E_r^o) + 2 k_1^2 \frac{\partial^2}{\partial r^2} (r^2 E_r^o) + (k_1^4 + k_2^4) \, r^2 \, E_r^o \approx 0 \tag{133}$$

Eq. (133) has solutions

$$r^2 E_r^o = e^{ikr} \tag{134}$$

where

$$k = \pm \sqrt{k_1^2 \pm i \, k_2^2} \tag{135}$$

The four possible sign combinations represent four waves: two of them are outgoing and two are incoming.

In the other approximation the r factors are considered to change only gradually relative to E_r^o and $E_\emptyset^o$ in Eq. (130) and (131). The r factors are replaced by constants r_m and Eq. (130) and (131) can be shown to give

$$\frac{\partial^4}{\partial r^4} E_r^o + 2 k_1^2 \left\{ 1 - \frac{n(n+1)}{2} \left[\frac{1}{(k_3 r_m)^2} + \frac{1}{(k_1 r_m)^2} \right] \right\} \frac{\partial^2}{\partial r^2} E_r^o$$

$$+ \left\{ k_1^4 \left[1 - \frac{n(n+1)}{(k_1 r_m)^2} - \frac{n(n+1)}{(k_3 r_m)^2} + \frac{[n(n+1)]^2}{(k_1 k_3 r_m^2)^2} \right] + k_2^4 \left[1 - \frac{n(n+1)}{(k_3 r_m)^2} \right] \right\} E_r^o = 0 \tag{136}$$

Solutions of Eq. (136) are of the form

$$E_r^o = e^{ikr} \tag{137}$$

where

$$k^2 = k_1^2 \left\{ 1 - \frac{n(n+1)}{2} \left[\frac{1}{(k_3 r_m)^2} + \frac{1}{(k_1 r_m)^2} \right] \right\}$$

$$\pm ik_2^2 \sqrt{ 1 - \frac{n(n+1)}{(k_3 r_m)^2} - (\frac{k_1}{k_2})^4 \left[\frac{n(n+1)}{2} \right]^2 \left[\frac{1}{(k_3 r_m)^2} - \frac{1}{(k_1 r_m)^2} \right]^2 } \qquad (138)$$

$$\approx (k_1^2 \pm ik_2^2) \left[1 - \frac{n(n+1)}{2(k_3 r_m)^2} \right] - \frac{n(n+1)}{2 r_m^2}$$

The solution Eq. (134) and (135) neglects displacement currents in a strictly spherical geometry, while solutions Eq. (137) and (138) are plane wave approximations which consider displacement currents.

The remaining field components are obtained by substituting Eq. (134) or (137) in Eq. (123) to (128). It follows that the waves are circularly polarized, the same way as in the quasi-longitudinal approximation of Booker.

The Matrix Solution.

The space above the earth surface ($r = r_0 = a$) is subdivided in a number of concentric spherical shells. A source is located above the ground surface at $r = r_1 = b$, the lower boundary of the ionosphere is at $r = r_2 = c = a + h$, and the boundaries between the various ionospheric layers are located at $r = r_m (m > 2)$. The fields within the nth layer and on its boundaries can be represented in matrix form as

$$\left[S_n(r) \right] = \left[a_n(r) \right] \left[C_n \right] \qquad (139)$$

where $\left[S_n(r) \right]$ is a column matrix of the field components, $\left[C_n \right]$ is a column matrix of coefficients, $\left[a_n(r) \right]$ is a matrix of the functions or the solution matrix. Considering the two boundaries r_{n-1} and r_n of the nth layer, it follows that $\left[S_n(r_{n-1}) \right]$ is related to $\left[S_n(r_n) \right]$ as

$$\left[S_n(r_{n-1}) \right] = \left[a_n(r_{n-1}) \right] \left[a_n^{-1}(r_n) \right] \left[S_n(r_n) \right] \qquad (140)$$

The boundary condition at r_n between the nth and $(n+1)^{st}$ layer is given as

$$\left[S_n(r_n) \right] = \left[S_{n+1}(r_n) \right] + \left[Q \right] \qquad (141)$$

where the source [Q] is assumed to be on the interface and causes a discontinuity in one or several field components. When combining Eq. (140) and (141), the field components on the surface of the earth r_o are related to the source $[Q]$ and to the field components at the upper boundary r_m by

$$\left[s_1(r_o)\right] = \left[a_1(r_o)\right] \left[a_1^{-1}(r_1)\right] \left[Q\right]$$

$$+ \left[a_1(r_o)\right] \left[a_1^{-1}(r_1)\right] \left[a_2(r_1)\right] \left[a_2^{-1}(r_2)\right] \cdots \left[a_m(r_{m-1})\right] \left[a_m^{-1}(r_m)\right] \left[s_m(r_m)\right]$$

(142)

The column matrix is assumed to represent the tangential field components in the form

$$[s_n(r)] = \begin{bmatrix} E_\theta^o(r) \\ H_\phi^o(r) \\ E_\phi^o(r) \\ H_\theta^o(r) \end{bmatrix}$$

(143)

For excitation by a vertical electric dipole the source matrix $[Q]$ is given by

$$[Q] = K_n \begin{bmatrix} 1 \\ 0 \\ 0 \\ 0 \end{bmatrix}$$

(144)

K_n is computed by applying the boundary condition Eq. (141) to the E_θ component of the dipole excitation, which is determined from Eq. (20) and (24) with the aid of Eq. (37). It follows that

$$K_n = \frac{i(2n+1)}{4\pi\omega\epsilon_o b^3} Ids$$

(145)

where n designates the order of the Legendre polynomial. K_n may be set equal to unity when considering a single mode at a time.

The semi-infinite homogeneous anisotropic ionosphere provides a simple example of the matrix solution. It follows that

$$
\begin{bmatrix} E_\theta^o(r_o) \\ H_\phi^o(r_o) \\ E_\phi^o(r_o) \\ H_\theta^o(r_o) \end{bmatrix}
=
\begin{bmatrix} 0 \\ H_\phi^o(a) \\ 0 \\ H_\theta^o(a) \end{bmatrix}
= [a_1(a)][a_2^{-1}(c)][a_3(c)]
\begin{bmatrix} A \\ 0 \\ B \\ 0 \end{bmatrix}
+
\begin{bmatrix} 1 \\ 0 \\ 0 \\ 0 \end{bmatrix}
\tag{146}
$$

where $[a_1(a)][a_1^{-1}(b)]$ and $[a_1^{-1}(b)][a_2(b)]$ have been set equal to the unit matrix I.

There are only outgoing waves in the ionosphere which give two zeros in the co-efficient matrix. The solution to Eq. (146) has been discussed by Thompson (1963). In the absence of the magnetic field, it leads to the resonance condition Eq. (67).

The calculations of Thompson (1963) have been made for the spherically stratified ionosphere with up to sixty layers, with ionospheric conductivity components similar to those discussed in the section on Boundary Conditions. The earth is considered to be a perfect conductor, and the daytime or nighttime ionospheric parameters are assumed to change only radially. Increasing the earth's magnetic field from zero to about 0.4 Gauss decreases somewhat the resonant frequencies of the day and night ionosphere models. The average of the computed average day and night frequencies differs by only 0.1 to 0.4 cps from the measurements by Balser and Wagner (1960). The magnetic field decreases the nighttime cavity Q approximately twice, while the daytime Q is decreased only in the first mode. This can be explained qualitatively by considering the dissipation and field penetration effects. At daytime, most of the energy is dissipated below 85 km, and very little of it escapes to higher altitudes even in the presence of a magnetic field. At nighttime, there is no dissipation at these altitudes, but there are also no reflections, and one of the up-going waves suffers little attenuation between 85 to 200 km in the presence of a magnetic field. This wave gets into the lossless region above 200 km and escapes from the cavity. The field penetrations to various altitudes have been checked in the model of Thompson (1963) by placing a perfect boundary at different heights. Such a boundary does not affect the solution if it is placed at h ≈ 100 km at daytime, and at h = 120 km at nighttime for B = 0. For B = 0.4 Gauss the boundary may be placed at h = 150 km at daytime. At nighttime, the solution is still affected by a boundary at h = 200 km. Therefore, there are significant fields near this boundary level, and also above it during nighttime. The computed Q figures are between 15 to 20 with the exception of the nighttime anisotropic solution where Q is from 7 to 10. These are higher than the measured values of Q = 4 to 6 (Balser and Wagner, 1960). Q of the daytime isotropic case of Thompson (1963) should be comparable with the computation of Galejs (1962a) or Galejs and Row (1963). However, the Q figures computed by Eq. (94) from the data of Fig. 17 and 26 for h = 50 km are 5.5 to 6.5. This discrepancy is probably caused by the neglect of displacement current in Thompson's analysis. According to Fig. 4 of Galejs (1961c), a sizeable dissipation takes place in the lower ionospheric layers where displacement currents cannot be neglected. The difference between measured and computed Q figures may be caused to some extent by "line spreading" due to the ϕ - unsymmetry of the actual earth-to-ionosphere geometry (Madden, 1963). In this case associated Legendre function $P_n^m(\cos \theta)$ with $m \neq 0$ will be required in the field representations in addition to the Legendre polynominals $P_n^o(\cos \theta) = P_n(\cos \theta)$. The terms proportional to $P_n^m(\cos \theta)$ may exhibit peaks conceivably at slightly different frequencies, thus causing a broadening of the computed resonance peaks.

ACKNOWLEDGEMENT

The author wishes to acknowledge support by the Office of Naval Research under Contract Nonr 3185(00).

REFERENCES

Abraham, J.: Sylvania Electronic Systems, Waltham, Mass. (Private communication)(1963)

Albrecht, H. J.: Fluggeraetewerk Bodensee, Ueberlingen, W. Germany (Private communication) (1963)

Balser, M., and C. A. Wagner: "Observations of Earth-Ionosphere Cavity Resonances," Nature 188, 638 (1960)

"Diurnal Power Variations of the Earth-Ionosphere Cavity Modes and Their Relationship to World-Wide Thunderstorm Activity, " J. Geophys. Res. 67, 619 (1962a)

"On Frequency Variations of the Earth-Ionosphere Cavity Modes," J. Geophys. Res. 67, 4081 (1962b)

"Effect of a High-Altitude Nuclear Detonation on the Earth-Ionosphere Cavity," J. Geophys. Res. 68, 4115 (1963)

Barrington, R. E., and E. Thrane: "The Determination of D-Region Electron Densities from Observations of Cross-Modulation," J. Atmos. Terr. Phys. 24, 31 (1962)

Belrose, J. S.: "Present Knowledge of the Lowest Ionosphere," Proc. of Ionospheric Research Committee of AGARD, Munich, in publication, Pergamon Press (1962)

Benoit, R., and A. Houri: "Propagation of Very Low Frequencies in the Earth-Ionosphere System," Ann. Geophys. (France) 17, 370 (1961)

"Powerspectrum Measurements of Geophysical Noise. Applications to Earth-to-Ionosphere Cavity," C. R. Acad. Sci. (France) 255, 2496 (1962)

Bourdeau, R.E., E.C. Whipple, and J.F. Clark: "Analytic and Experimental Conductivity Between the Stratosphere and the Ionosphere," J. Geophys. Res. 64, 1363 (1959)

Bremmer, H.: "Terrestrial Radio Waves," Elsvier, New York and Amsterdam (1949)

Brock-Nannestad, L.: "Bibliography Electromagnetic Phenomena With Special Reference to ELF (1-3000 cps)," TR 10 Saclant, ASW Res. Center, La Spezia, Italy (1962)

Brown, S. C.: "Basic Data of Plasma Physics," John Wiley and Sons, New York, 73 (1959)

Budden, K. G.: "The Propagation of Very Low Frequency Radio Waves to Great Distances" Phil. Mag. 44, 504 (1953)

Chapman, F. W., and R. C. Macario: "Propagation of Audio Frequency Radio Waves to Great Distances," Nature 177, 930 (1956)

Decaux, B., A. Frances, A. Gabry, and M. Reysat: "Perturbations Produced by High Altitude Thermonuclear Explosions (particularly by that of 9 July 1962) on transmission delay and amplitude of VLF waves," C. R. Acad. Sci. 256, 481 (1963)

Egeland, A.: Kiruna Geophysical Observatory (Private communication) (1963)

Erdelyi, A.: "Higher Transcendental Functions" 1, McGraw-Hill Book Co., New York (1953)

Ferraro, A. J., and J. J. Gibbons: "Polarization Computations by Means of the Multi-Slab Approximation," J. Atmos. and Terrest. Phys. 16, 136 (1959)

Ferraro, A. J.: "Multi-Slab Concept Applied to Radio-Wave Propagation in the Ionosphere and its Limit to the Continuous Ionosphere," J. Geophys. Res. 67, 3817 (1962)

Fournier, H.: "Some Aspects of the First High-Frequency Geomagnetic Recordings Obtained at Garchy," C. R. Acad. Sci. (France) 251, 962 (1960)

Geophysical Station of Nivernais, France (Private Communication) (1963)

Galejs, J.: "Excitation of VLF and ELF Radio Waves by a Horizontal Magnetic Dipole," NBS J. Res. 65D, 305 (1961a)

"Terrestrial Extremely Low Frequency Noise Spectrum in the Presence of Exponential Ionospheric Conductivity Profiles," J. Geophys. Res. 66, 2787 (1961b)

"ELF Waves in Presence of Exponential Ionospheric Conductivity Profiles," IRE Trans. on Antennas and Propagation, AP-9, 554 (1961c)

"A Further Note on Terrestrial Extremely Low Frequency Propagation in the Presence of an Isotropic Ionosphere with an Exponential Conductivity-Height Profile," J. Geophys. Res. 67, 2715 (1962a)

"Terrestrial Extremely Low Frequency Propagation in the Presence of an Isotropic Ionosphere with an Exponential Conductivity Height Profile," Proc. of the International Conference on the Ionosphere, London, Chapman and Hall, 467 (1962b)

ELF and VLF Waves Below an Inhomogeneous Anisotropic Ionosphere, Research Report 350, Applied Research Laboratory, Sylvania Electronic Systems, Waltham 54, Mass. (1963)

Galejs, J., and R. V. Row: "Propagation of ELF Waves Below an Inhomogeneous Anisotropic Ionosphere," Research Report 334, Applied Research Laboratory, Sylvania Electronic Systems, Waltham 54, Mass. (1963)

Gendrin, R., and R. Stefant: "Magnetic Records Between 0.2 and 30 cps" AGARD Conference on Propagation of radio frequencies below 300 kc, Munich, Germany (1962a)

"Effects of the Very High Altitude Nuclear Explosion of the July 9, 1962 on the Earth to Ionosphere Cavity Resonances, Experimental Results and Interpretation," C. R. Acad. Sci. 255, 2273, 2493 (1962b)

CNET, Issy-les-Moulineaux, France (Private Communication) (1963)

Handbook of Geophysics: USAF Air Res. and Dev. Command, AF Cambridge Res. Center, MacMillan Co., New York, N. Y. (1960)

Harris, F. B., and R. L. Tanner: "A Method for Determination of Lower Ionosphere Properties by Means of Field Measurements on Sferics," J. Res. Nat. Bur. Stand. 66D, 463 (1962)

254

Hepburn, F.: "Atmospheric Waveforms with Very Low Frequency Components Below 1 kc Known as Slow Tails," J. Atmos. and Terr. Phys. 10, 266 (1957)

"Classification of Atmospheric Waveforms," J. Atmos. and Terr. Phys. 12, 1 (1958)

"Analysis of Smooth Type Atmospheric Waveforms," J. Atmos. and Terr. Phys. 19, 37 (1960)

Hepburn, F., and E. T. Pierce: "Atmospherics with Very Low Frequency Components," Nature 171, 837 (1953)

Jean, A. G., Jr., A. C. Murphy, J. R. Wait, and D. F. Wasmundt: "Propagation Attenuation Rates at ELF," J. Res. NBS 65D, (1961)

Johnson, F. S.: Satellite Environment Handbook, Stanford University Press (1961)

Johler, J. R., and L. A. Berry: "Propagation of Terrestrial Radio Waves of Long Wavelength--Theory of Zonal Harmonics with Improved Summation Techniques," J. Res. NBS 66D, 737 (1962)

Kamke, E.: "Differential Gleichunqen," Chelsea Publishing Co., New York (1948)

König, von, H. L.: "Atmospherics geringster Frequenzen," Dissertation an der Technischen Hochschule München, (1958)

"Atmospherics geringster Frequenzen," Z. Angew Phys. 11, 264 (1959)

König, von, H. L., E. Haine, and C. H. Antoniadis: "Messung von 'Atmospherics' geringster Frequenzen in Bonn," Z. Angew. Phys. 13, 364 (1961)

Landmark, B., and F. Leid: "Radiowave Absorption in the Ionosphere," AGARDograph 53, Pergamon Press, 92 (1962)

Liebermann, L.: "Extremely Low Frequency Electromagnetic Waves: I. Reception from Lightning," J. Appl. Phys. 27, 1473 (1956a)

"Extremely Low Frequency Electromagnetic Waves: II. Propagation Properties," J. Appl. Phys. 27, 1477 (1956b)

Lokken, J. E., J. A. Shand, C. S. Wright, L. H. Martin, N. M. Brice and R. A. Helliwell: Stanford Pacific Naval Laboratory Conjugate Point Experiment, Nature 192, 319 (1961)

Lokken, J. E., J. A. Shand, and C. S. Wright: A Note on the Classification of Geomagnetic Signals Below 30 Cycles per Second, Can. J. Phys. 40, 1000 (1962)

Lokken, J. E.: Pacific Naval Laboratory, Esquimalt, B. C. (Private Communication) (1963)

Madden, T. R.: "An Analysis of the Earth-Ionosphere Electromagnetic Resonances," unpublished class notes for MIT Course 12.88, April 24, 1961

Massachusetts Institute of Technology, (Private Communication) (1963)

Magnus, F., and F. Oberhettinger: "Special Functions of Mathematical Physics," Chelsea Publishing Co., New York (1949)

Maxwell, E. L., and D. L. Stone: "Natural Noise Fields from 1 cps to 100 kc," Trans. IEEE PGAP, AP-11, 339 (1963)

Mechtly, E. A.: "Nighttime Lower Ionosphere Electron Density Measurements by Radio Wave Propagation to Ascending Rockets," Sci. Rept. 60, Pennsylvania State University (1962)

Moler, W. F.: "VLF Propagation Effects of a D-Region Layer Produced by Cosmic Rays," J. Geophys. Res. 65, 1459 (1960)

Nicolet, M.: "The Collision Frequency of Electrons in the Terrestrial Atmosphere," Phys. of Fluids 2, 95 (1959)

Nicolet, M. and A. C. Aikin: "The Formation of the D-Region of the Ionosphere," J. Geophys. Res. 65, 1464 (1960)

Pfister, W., J. C. Ulwick, and R. P. Vancour: "Some Results of Direct Probing of the Ionosphere," J. Geophys. Res. 66, 1290 (1961)

Pierce, E. T.: "The Propagation of Radio Waves of Frequency Less than 1 kc," Proc. IRE, 48, 329 (1960a)

"Some ELF (Extremely Low Frequency) Phenomena," J. Res. NBS, 64D, 383 (1960b)

"Excitation of Earth-Ionosphere Cavity Resonances by Lightning Flashes," J. Geophys. Res. 68, 4125 (1963)

Polk, C., and F. Fitchen: "Schumann Resonances of the Earth-Ionosphere Cavity -- Extremely Low Frequency Reception at Kingston, Rhode Island," J. Res. NBS, 66D, 313 (1962)

Polk, C., F. Huck, and I. P. Yu: University of Rhode Island (Private Communication) (1963)

Raemer, H.: "Effect of Underground Induced Polarization on ELF Propagation," J. Geophys. Res. 66, 1596 (1961a)

"On the Extremely Low Frequency Spectrum of Earth-Ionosphere Cavity Response to Electric Storms," J. Geophys. Res. 66, 1580 (1961b)

"On the Spectrum of Terrestrial Radio Noise at Extremely Low Frequencies," J. Res. NBS, 65D, 581 (1961c)

Sylvania Electronic Systems, Waltham, Mass. (Private Communication) (1963)

Ramanathan, K. R., R. V. Shonsle, and S. S. Dagaonkar: "Effect of the Ionosphere on the Absorption of Cosmic Radio Noise at 25 Mc/s. at Ahmedabad," J. Geophys. Res. 66, 2763 (1961)

Rice, S. O.: "Mathematical Analysis of Random Noise," Bell System Tech. J. 23, 282 (1944); 24, 46, Eq. 3.3-11 (1945)

Row, R. V.: "On the Electromagnetic Resonant Frequencies of the Earth Ionosphere Cavity," Trans. IRE PGAP, AP-10, 766 (1962)

Schumann, W. O.: " ber die Ausbreitung sehr langer elektrischer Wellen und der Blitzentladung um die Erde," Z. Angew Phys. $\underline{4}$, 474 (1952a)

"Über die strahlungslosen Eigenschwingungen einer leitenden Kugel die von einer Luftschicht und einer Ionosphärenhülle umgeben ist," Z. Naturforsch $\underline{7a}$, 149 (1952b)

"Über die Dämpfung der elektromagnetischen Eigenschwingungen des Systems Erde-Luft-Ionosphäre," Z. Naturforsch $\underline{72}$, 250 (1952c)

"Über die Oberfelder bei der Ausbreitung langer, elektrischer Wellen im System Erde-Luft-Ionosphäre und 2 Anwendungen (horizontaler und senkrechter Dipol)," Z. Angew. Phys. $\underline{6}$, 35 (1954a)

"Über die Strahlung langer Wellen des horizontalen Dipols in dem Lufthohlraum zwischen Erde und Ionosphäre," Z. Angew. Phys. $\underline{6}$, 225 (1954b)

"Über die Strahlung langer Wellen des horizontalen Dipols in dem Lufthohlraum zwischen Erde und Ionosphäre," Z. Angew. Phys. $\underline{6}$, 267 (1954c)

"Der Einfluss des Erdmagnetfeldes auf die Ausbreitung elektrischer Längstwellen," Z. Angew. Phys. $\underline{7}$, 284 (1955)

"Der Einfluss des horizontalen Erdmagnetfeldes auf elektrische Wellen zwischen Erde und Ionosphäre, die schräg zum magnetischen Meridian verlaufen," Z. Angew. Phys. $\underline{8}$, 126 (1956)

"Elektrische Eigenschwingungen des Hohlraumes Erde-Luft-Ionosphäre," Z. Angew. Phys. $\underline{9}$, 373 (1957)

Seigel, H. O.: "A Theory of Induced Polarization Effects," Overvoltage Research and Geophysical Applications, J. R. Wait, ed., Pergamon Press, London (1959)

Shmoys, J.: "Long-Range Propagation of Low-Frequency Radio Waves Between the Earth and Ionosphere," Proc. IRE, $\underline{44}$, 163 (1956)

Sommerfeld, A. N.: "Partial Differential Equations," Academic Press, New York (1949)

Slater, J. C.: "Microwave Electronics," Van Nostrand Co., Inc., New York (1950)

Tepley, L. R.: "A Comparison of Sferics as Observed in the Very-Low-Frequency and Extremely-Low-Frequency Bands," J. Geophys. Res. $\underline{64}$, 2315 (1959)

Thompson, W. B.: "A Layered Model Approach to the Earth-Ionosphere Cavity Resonance Problem," Ph.D. thesis, Dept. of Geology and Geophysics, MIT, (1963)

Vilbig, F.: "Lehrbuch der Hochfrequenztechnik," $\underline{I}$, Akademische Verlagsgesellschaft, Frankfurt am Main, Table 4-1 (1960)

Wait, J. R.: "The Mode Theory of VLF Ionosphere Propagation for Finite Ground Conductivity," Proc. IRE, $\underline{45}$, 760 (1957)

"Extension to Mode Theory of VLF Propagation," J. Geophys. Res. $\underline{63}$, 125 (1958)

"The Variable Frequency Method," Overvoltage Research and Geophysical Applications, J. R. Wait, ed., Pergamon Press, London (1959)

"Terrestrial Propagation of VLF Radio Waves -- A Theoretical Investigation," J. Res. NBS, 64D, 153 (1960a)

"Mode Theory and the Propagation of ELF Radio Waves," J. Res. NBS, 64D, 387 (1960b)

"On the Propagation of ELF Radio Waves and the Influence of a Non-homogeneous Ionosphere," J. Geophys. Res. 65, 597 (1960c)

"Electromagnetic Waves in Stratified Media," MacMillan Co., New York (1962a)

"On the Propagation of VLF and ELF Radio Waves When the Ionosphere is Not Sharply Bounded," J. Res. NBS, 66D, 53 (1962b)

"An analysis of VLF Mode Propagation for a Variable Ionosphere Height," J. Res. NBS, 66D, 453 (1962c)

"The Mode Theory of VLF Radio Propagation for a Spherical Earth and a Concentric Anisotropic Ionosphere," Can. J. Phys. 41, 299 (1963a)

"Concerning Solutions of the VLF Mode Problem for an Anisotropic Curved Ionosphere," J. Res. NBS, 67D, 297 (1963b)

National Bureau of Standards, (Private Communication) (1963c)

Ward, S. H.: University of California, Berkeley, (Private Communication) (1963)

Watson, G. N.: "The Transmission of Electric Waves Round the Earth," Proc. Roy. Soc. A95, 546 (1919)

Watt, A. D.: "ELF Electric Fields from Thunderstorms," J. Res. NBS, 64D, 425 (1960)

Watt, A. D., F. S. Mathews, and E. L. Maxwell: "Some Electrical Characteristics of the Earth's Crust," Proc. IEEE, 51, 897 (1963)

Williams, J. C.: "Thunderstorms and VLF Radio Noise," Ph.D. thesis, Harvard Univ., Div. of Engineering and Applied Physics (1959)

Wright, C. S.: "Conjugate Relations of Micropulsations and Other Ionospheric Phenomena," NATO Advanced Study Institute on Low Frequency Electromagnetic Radiation, Bad Homburg, Germany (1963)

Zmuda, A. J., B. W. Shaw, and C. R. Haave: "VLF Disturbances and the High-Altitude Nuclear Explosion of July 9, 1962," J. Geophys. Res. 68, 745 (1963)

DISCUSSION

S. H. Ward

Polarization of Sferics. Sferic activity at a given site may be represented by
a series of delta functions of magnetic field intensity, whose amplitude and time distributions are essentially random. The azimuthal distribution on the other hand is
somewhat ordered, giving rise to an equivalent "ellipse of polarization" or "integration ellipse" provided sferic bursts are studied over a sufficient period of time.
For distant sources and a perfectly and uniformly conducting earth, the plane of the
ellipse is horizontal. Local sources and local conductivity inhomogeneities tilt the
plane of the ellipse (Jewell and Ward, 1963). Of course, each sferic burst may be
elliptically polarized in the conventional sense, either at the source or as a result
of local conductivity inhomogeneities to produce a further complexity.

Some sferic energy is always measurable in any three orthogonal directions so
that in the sense considered herein, an "integration ellipsoid of polarization" is
required to provide complete definition of the mean polarization of sferic energy.
The eccentricity of the integration ellipsoid of polarization in the horizontal plane
is low, often approaching unity, while the eccentricity in the plane through the
major and minor axes is high, sometimes as large as five.

Both of these eccentricities are dependent upon the secondary fields from local
conductivity inhomogeneities as well as the primary field character.

The azimuth of the major axis will change hourly, sometimes by as much as 90°
during a 24-hour interval. The tilt of the minor axis away from the vertical is remarkably stable (Ward, 1959), seldom varying by more than 2° from a mean position.
The azimuth tends to undergo a fairly regular diurnal variation with short period
fluctuations superimposed. The intensity of any component undergoes a regular diurnal
and seasonal variation as has been reported previously (Ward, 1959).

Local conductivity inhomogeneities can cause the minor axis to tilt from the
vertical by as much as 50 or 60 degrees, and commonly varies from, say, 20° in one
direction to 20° in the opposite direction over a distance of several hundreds of
feet. The azimuth may similarly be rotated by induced secondary fields arising in
subsurface conductivity inhomogeneities.

Generally, the integration polarization ellipse in the vertical plane is remarkably constant in time, but extremely variable in space, while the reverse is true for
the integration polarization ellipse in the horizontal plane. It is important in
studying polarization of sferics to recognize both the space and time dependencies.

Interestingly enough, the November 1, 1962, nuclear detonation in the lower
ionosphere produced a rotation of the major axis of the integration ellipsoid of
polarization by about 10° in azimuth at our observatory near San Francisco. This
effect seems to be a result of increased D layer ionization and a lowering of the D
layer, over a local area perhaps two or three thousands of kilometers in diameter.
The propagation path for sferics from the East Indian source to San Francisco would
thus be affected by the change in reflectivity of the lower surface of the ionosphere.
The field due to this source might then predominate over others which, for a normal
D layer, would predominate. A rotation of the integration ellipsoid of polarization
would then result.

References

Jewell, T. R., and S. H. Ward: "The Influence of Conductivity Inhomogeneities Upon Audio-Frequency Magnetic Fields," Geophys. <u>XXVIII</u>, 201 (1963)

Ward, S. H.: AFMAG - Airborne and Ground, Geophys. <u>24</u>, 761 (1959)

"Polarization of Sferics," (manuscript in preparation)

WHISTLERS AND VLF EMISSIONS

J. Katsufrakis
Radioscience Laboratory
Stanford Electronics Laboratory
Stanford University
Stanford, California

INTRODUCTION

The purpose of this paper is to review briefly some of the research on whistlers and VLF (very low frequency) emissions, and to present the spectra of VLF phenomena observed at geomagnetically conjugate points.

Before reviewing the results of some of the research, a brief description of the phenomena should be helpful. Whistlers are produced by the electromagnetic impulses from lightning and are observed mainly in the range of frequencies from 300 cps to 30,000 cps. Much of the energy of the impulse is reflected from the ionosphere back to the ground where it is the main source of interference to radio communication from very low to high frequencies. Some of the energy enters the ionosphere, where it is guided by the earth's magnetic field into the opposite hemisphere. Dispersion in the ionosphere stretches the original impulse into a gliding tone (order of one second in duration), which is called a whistler. A typical whistler path is shown in Fig. 1, and the associated spectra which would be observed at the two ends of the path are sketched in the figure inserts. An actual recording is shown in Fig. 2. Whistler paths appear to be fixed in the ionosphere at any given time and often two or more such paths can be observed simultaneously, giving rise to what is known as the multiple path whistler. The shape of the frequency vs. time curve of the whistler is determined by the dispersion, which is controlled by the electron density and the earth's magnetic field. Details of the dispersion and guiding of whistlers are treated elsewhere (Storey, 1953; and Helliwell, 1964).

A second class of phenomena, known as VLF emissions, are observed at whistler frequencies. These phenomena are less well understood than whistlers, but are believed to originate in the ionosphere.

VLF emissions can be detected and recorded with the same equipment employed in the study of whistlers. In fact, several types of emissions are often observed in close association with whistlers. Emissions appear in a wide variety of spectral forms; an emission may be relatively steady over a period of minutes, or even hours, or occur in discrete bursts as short as a fraction of a second. The amplitude of emissions frequently fluctuate in a periodic or quasi-periodic manner, with periods ranging from less than a second to more than a minute.

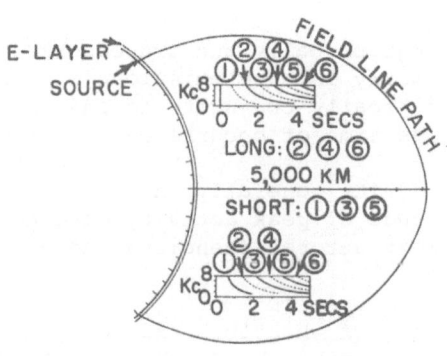

Fig. 1.   Dipole field line model with inserts of even and odd whistler echo trains.

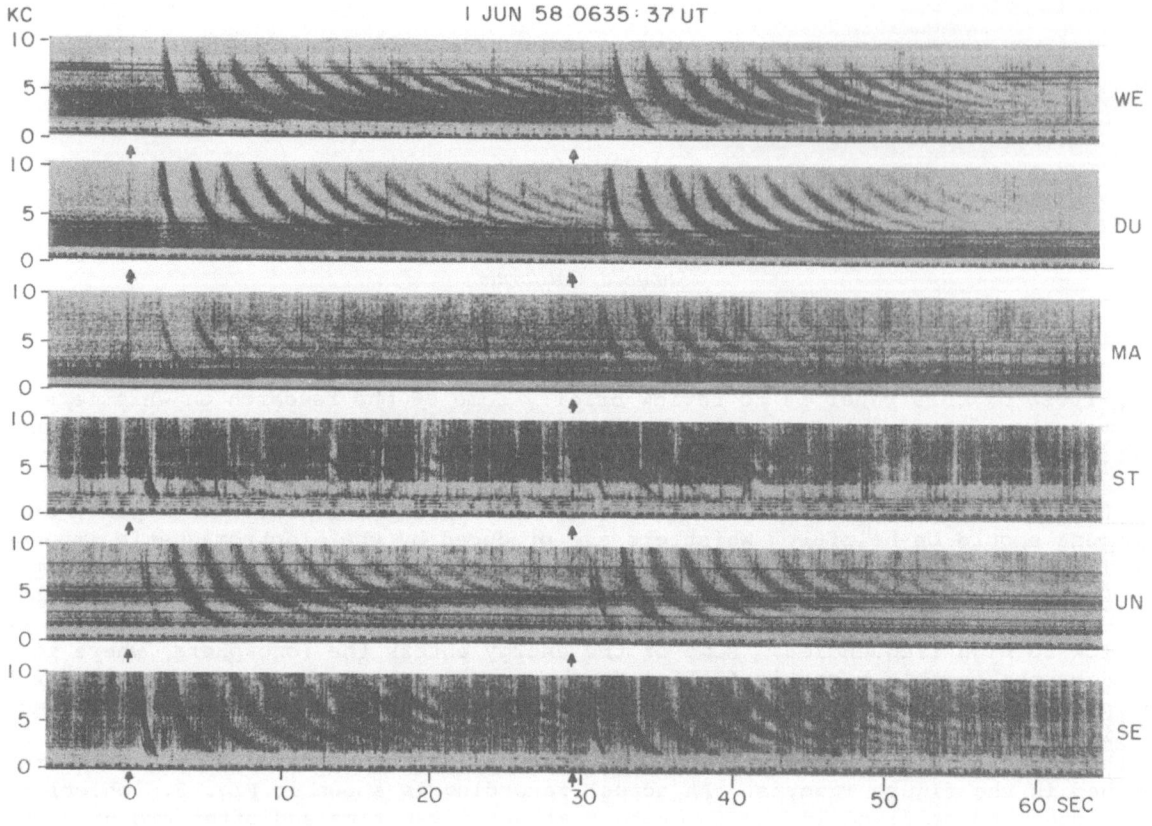

Fig. 2.   Long enduring echo trains recorded at Wellington, New Zealand;
          Dunedin, New Zealand; MacQuarie Island; Stanford, California;
          Unalaska, Alaska; and Seattle.

     VLF emissions, like whistlers, are localized geographically and are observed
most commonly at middle and high latitudes.   Particularly intense emissions are found
in the auroral zone.   At locations which are geomagnetically conjugate to one another,
emissions are often closely related in form as well as time of occurrence.

     As a function of magnetic activity, emission activity generally increases at low
latitudes and decreases at high latitudes.   The latitude of peak activity descends
with increasing K_p.   A close association exists between auroral phenomena and certain
types of emission.

     Before the beginning of the IGY, the understanding of whistlers and related phe-
nomena was relatively limited.   Whistlers were first described by Barkhausen (1919),
who heard them while eavesdropping on Allied telephone communications at the front
lines during World War I.   In 1928, Eckersley (1928) in England reported an association between whistlers and lightning and also between whistlers and solar activity.
The first evidence of VLF emissions was reported by Marconi workers, who heard and
described the so-called "dawn chorus" in 1931.   Measurements of the dispersion of a
whistler were reported by Burton and Boardman (1933) of the U.S.A., following which
Eckersley published an approximate form of the dispersion law which provided an explanation for the observed dispersion.   Further research on whistlers was interrupted
by World War II, following which, Storey began his investigations (1953) at Cambridge,
England.   He published his well-known paper in which the field aligned whistler path

262

was proposed and a calculation of the outer ionosphere electron density was made. Storey's results were first presented to the scientific community by J. A. Ratcliffe at the URSI General Assembly held in Sydney, Australia, in 1952, and stimulated a number of groups in various countries to begin studies of this fascinating topic. Several tests of Storey's theory of the field-line path were carried out including the observation by Morgan and Allcock (1956) of whistlers at conjugate points showing approximately the relationship sketched in Fig. 1. Extension of observations to high latitudes revealed a new type of whistler which showed in addition to the falling tone a rising component that was joined smoothly to the descending component at the beginning of the whistler. This strange shape was called the "nose" whistler and was explained by the magneto-ionic theory after the limiting restrictions of Storey and Eckersley were removed. Later, nose whistlers were discovered at middle latitudes in accord with predictions of the theory. A multiple-path nose whistler is illustrated in Fig. 3. The nose whistler provides a means for determining the latitude of the associated field-line path, called the path latitude, necessary in using dispersion data for the study of the electron density in the magnetosphere.

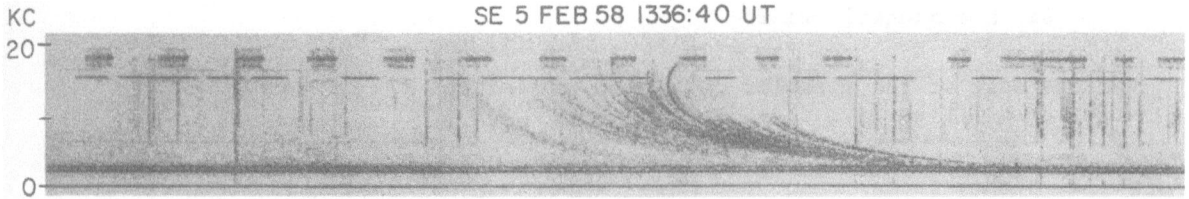

Fig. 3. A multi-path nose whistler recorded at Seattle, Washington.

The plan of the paper is to present introductory remarks on whistler theory, and follow this with three main sections, the first devoted to some recently obtained experimental results on the shape and variations of the equatorial profile of magnetospheric electron density, the second describing periodic emissions and their probable generating mechanisms, and the third an Atlas of whistlers and emissions recorded at geomagnetically conjugate points.

WHISTLER PROPAGATION THEORY

Whistler dispersion is essentially the result of an interaction between the wave, free electrons, and the geomagnetic field. Since the number density of electrons with low thermal energies ($T \sim 1500°K$) is usually far greater than the density of energetic electrons, a cold plasma theory is sufficient to predict the dependence of dispersion on electron concentration.

The principal features of whistler propagation can be deduced from the magneto-ionic theory, which shows that in the ionosphere, a very low frequency wave can propagate freely in the circularly polarized ordinary mode. At very low frequencies the refractive index of the ionosphere can be represented by the quasi-longitudinal approximation of the magneto-ionic theory (Budden, 1961). This is given by

$$n^2 = 1 - \frac{X}{1-iZ \pm |Y_L|} \tag{1}$$

where

n = complex refractive index = $\mu - i\chi$

$X = f_O^2/f^2$ $\qquad\qquad$ f_O = plasma frequency

$Z = \nu/2\pi f$ $\qquad\qquad$ ν = electron collision frequency

$Y_L = f_L/f$ $\qquad\qquad$ f_H = electron gyrofrequency

f = wave frequency $\qquad\qquad$ $f_L = f_H \cos\theta$

θ = angle between wave normal and earth's magnetic field

Because most of the whistler path is characterized by the conditions

$$Z \ll Y_L \quad \text{and} \quad X/|Y_L| \gg 1$$

it is convenient to neglect the effect of collisions as well as the first term on the right side of Eq. (1). When this is done, the expression for the refractive index is

$$n^2 = \mu^2 = \frac{X}{Y_L - 1} = \frac{f_O^2}{f(f_L - f)} \tag{2}$$

The expression for the group refractive index is obtained from the relation

$$\mu_g = \frac{d}{df} (\mu f) \tag{3}$$

which gives

$$\mu_g = \frac{f_O f_L}{2f^{\frac{1}{2}}(f_L - f)^{3/2}} \tag{4}$$

In order to make ray-path calculations using Eq. (4), it is necessary to make an additional assumption about the distribution of f_O (or N) in the magnetosphere. If a "smooth" model is used, with electron density decreasing smoothly with height, the calculations predict that whistler ray paths will depart significantly from the direction of the geomagnetic field. On the other hand, if the distribution is assumed to involve field-aligned irregularities, then it is possible to cast the ray path problem in terms of trapping and guiding along field-aligned ducts. Fortunately, the question of a choice between the two assumptions is easily resolved in favor of the latter, since experimental data provide an unambiguous indication that whistlers propagate on discrete, field-aligned paths.

264

A study of guiding along ducts (Smith, 1961a) has shown that the actual travel time on a duct may be closely approximated by assuming strictly longitudinal field-line propagation. Use of this important simplifying assumption means that: (1) all frequencies are considered to travel along the same path; (2) the complications of a twisting, only approximately field-aligned ray path, are avoided. In Eq. (4) we set $f_H + f_L$, and for the travel time at a given frequency obtain the expression

$$t = \frac{1}{c} \int_{path} \mu_g ds = \frac{1}{2c} \int_{path} \frac{f_o f_H ds}{f^{\frac{1}{2}}(f_H-f)^{\frac{3}{2}}} \tag{5}$$

where c is the velocity of light and ds is an element of the field-line path.

The justification of the assumption of strictly longitudinal propagation is obviously of crucial importance, and it is discussed in detail in Helliwell (1964). For the present, we shall take the assumption to be justified, and will remark briefly on the properties of Eq. (5).

This equation shows that whistler travel time is proportional to a weighted integral of plasma frequency ($f_o \propto N^{\frac{1}{2}}$, where N is the number of density electrons). The incremental travel time of a whistler varies approximately as $f_H^{-\frac{1}{2}}$, which means that the observed frequency vs. time, or dispersion curve of a whistler will be particularly sensitive to electron density and magnetic field strength along the outer, equatorial portion of the path. It is this sensitivity which is used to advantage in deducing equatorial profiles of electron density from whistler data (Smith, 1961b; Helliwell, 1961; Carpenter, 1962a, 1963a).

THE EQUATORIAL PROFILE OF MAGNETOSPHERIC ELECTRON DENSITY

Studies of electron density distribution using whistlers made rapid progress following the IGY. An initial survey of temporal variations in the equatorial profile has been provided by: (1) low-frequency dispersion measurements, which essentially follow variations in the profile at a given height; (2) nose-frequency measurements, which can follow temporal variations over significant segments of the profile. These initial results will be briefly summarized in this section. Limitations of space prevent detailed remarks on regions of applicability. [Present methods of studying temporal density variations assume that the field-line model of f_o used in the calculations does not change between successive measurements. Removal of this assumption will make it possible to present a more physically detailed picture, particularly in the case of the diurnal and solar cycle variations.]

The Diurnal Variation.

This variation is one of the most difficult to study because of the amount of data and analysis required. As reported, the diurnal variation involves an afternoon density maximum and a post-midnight minimum, the amount of the reduction being about 30 per cent in many cases. (Iwai and Outsu, 1958; Rivault and Corcuff, 1960; Allcock, 1960). The diurnal variation may exhibit important changes in amplitude with season and solar epoch (Kimpara, 1962 and Corcuff, 1962). As yet, no detailed diurnal study based on nose-frequency information has been reported.

The Annual Variation.

The annual variation in electron density in the magnetosphere shows a maximum near the December solstice and a minimum near June (Smith, 1961c and Helliwell, 1961). This variation has been studied over a relatively wide range of geomagnetic longitudes and is found to show a declining amplitude with decreasing solar activity, (Carpenter, 1962a and Gomez et al, 1962). From nose-frequency data at Stanford and Seattle for the range 2 to 4 R_E (R_E is the radius of the earth), Carpenter (1962a) observed a reduction from January to June in 1958 by about 35 per cent and a corresponding reduction in 1961 by about 20 per cent. It is interesting that the phase of the annual variation agrees with the corresponding observations made by satellite drag measurements, which show a December maximum and June minimum in air density in the height range 200 to 600 km (Paetzold and Zschörner, 1961a).

The Semi-Annual Variation.

The semi-annual variation of magnetospheric electron density has recently been studied by Corcuff (1962), who reports finding density maximums in the October-November period and in the April-May period. For dispersion data taken at Poitiers, France (49.5°N geomagnetic) in the period June 1957 to December 1958, Corcuff finds a density ratio of minimum to maximum of about 0.75. As in the case of the annual variation, the semi-annual variation is approximately in phase with the air-density variations deduced from satellite drag (Paetzold and Zschörner, 1961b; Martin et al, 1961; Jacchia and Slowey, 1962).

The Solar Cycle Variation.

Beyond 2 R_E (geocentric distance), the change in electron density over the solar cycle appears to be relatively small. From Stanford and Seattle nose-frequency data, Carpenter (1962a) estimated the density reduction from early 1958 to early 1961 to be 20 to 25 per cent, in contrast to a corresponding reduction in f_oF2, (at Washington, D. C.) of about 60 per cent. The density reduction in the magnetosphere from 1958 to 1961 was found to be most pronounced near the December solstice (Carpenter, 1962a, and Gomez et al, 1962). More detailed studies will probably reveal a relatively large solar-cycle variation in the profile between 3000 and 7000 km altitude. If the plasma frequency below about 2 R_E (geocentric) drops significantly, this would probably be evident in the dispersion of Alouette satellite whistlers (Barrington and Belrose, 1963) recorded at low latitude. The drop in f_o might also be manifested by anomalous behavior of the extraordinary trace on topside ionograms, an effect discussed by Thomas et al (1963).

The Magnetic Storm Variation.

During the main phase and recovery phase of magnetic storms, electron-density levels in the magnetosphere are often depressed. The effect has been observed at relatively low latitudes following severe storms (Kimpara, 1962, and Outsu and Iwai, 1962) and it is regularly observed at latitudes corresponding to the equatorial range 2 R_E to 7 to 8 R_E (Carpenter, 1962a, 1962b, 1963b, and Corcuff, 1962). In the geocentric range 2 to 4 R_E, storm-period density levels are often depressed by 15 to 20 per cent, occasionally being reduced by a factor of as much as 10. The "knee" in the profile has a strong dependence on geomagnetic activity and appears to account for a substantial number of the observations of deep density depressions.

Conclusions.

An initial survey of the temporal variations in magnetospheric density has been made. Future investigations will present a more detailed picture of this subject through considerations of variations with time in the form of the field-line distribution of ionization.

The whistler technique has proven relatively easy to use. As a passive technique, it is relatively inexpensive, and is logistically flexible. The method appears capable of increasing theoretical and experimental refinement.

PERIODIC VLF EMISSIONS

Among the many remarkable types of VLF emissions is one which consists of short bursts of noise repeated at a regular interval, the length of which is of the order of a few seconds. These events we shall call periodic VLF emissions, examples of which have been reported by a number of investigators.

The first spectra of periodic emissions were presented at a meeting of the U.S.A. National Committee of the URSI, by Dinger (1957). Other examples have been published by Gallet (1959), Pope and Campbell (1960), Lokken et al (1961), and Brice (1962).

Periodic VLF emissions can be classified as either "dispersive" or "non-dispersive." In the dispersive type the period between bursts varies in a systematic way with frequency. Often this variation is exactly the same as that found in associated echoing whistlers, from which it is deduced that each burst is the result of whistler-mode echoing of the previous burst. In non-dispersive periodic emissions there is little or no observable systematic change in period with frequency.

The dispersive type of periodic emission is illustrated in Fig. 4, which shows

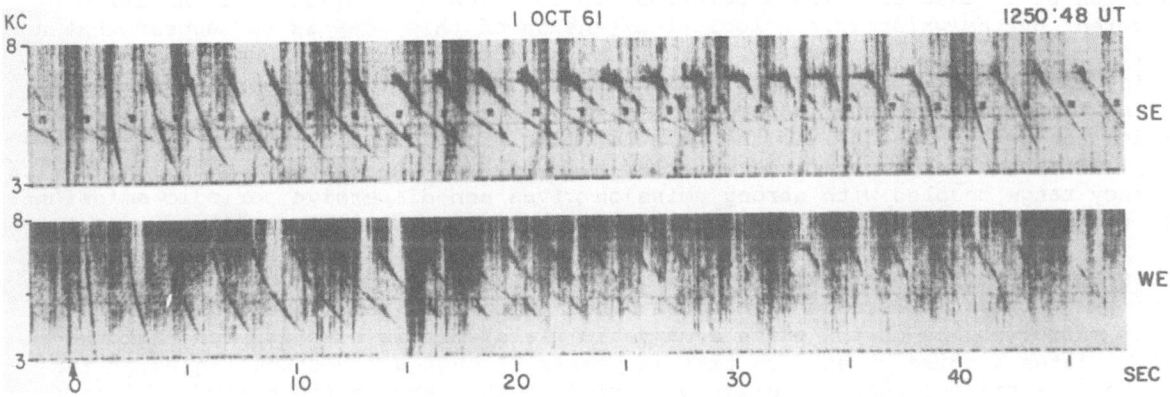

Fig. 4. Whistler-echo trains and associated VLF emissions recorded simultaneously at Seattle, Washington, and Wellington, New Zealand.

the spectra made from tapes recorded simultaneously at Seattle, Washington (North geomagnetic latitude 53°), and Wellington, New Zealand (South geomagnetic latitude 45°), on 1 October 1961, 1250 UT (Helliwell et al, 1962, and Helliwell, 1963). The period of observation was preceded by a sudden commencement magnetic storm beginning at 2109 UT, September 30, 1961. A magnetic K_p index of 8 was observed during the associated three-hour period. The echo train is of the two-hop, or "long" type at Seattle, and of the one-hop, or "short" type at Wellington.

The same whistler-echo train and related noise bursts were observed at several other stations (spectra not shown), including Stanford, California; Logan, Utah; Lauder, New Zealand; and Byrd Station, Antarctica. Since the details of the spectral shapes recorded at these other stations are identical to those presented in Fig. 4 from Seattle and Wellington, only the latter two recordings are presented. It is concluded that the whistlers and associated noise traveled over the same magnetic field-line path and, after exiting from the ionosphere, reached the receivers by propagating in the earth-ionosphere wave guide.

The echoing of the whistler can clearly be detected up to about the 26th hop on the Seattle record. After the first one or two hops, an emission appears in the form of a slight irregular broadening on the trailing edge of the whistler trace, mostly in the frequency range 6 to 7 kcps and to a lesser extent in the range 4 to 5 kcps. With each successive echo the intensity of the emission in the upper frequency range increases while the echo of the original whistler becomes weaker.

The difference between the echoing emission and the echoing whistler is especially obvious at and following the 16th hop, after which the emission appears to be stronger than the associated whistler echo. On the 39th hop, at Wellington, a marked downward extension of the noise burst (a falling tone) appears for the first time and then echoes in the whistler mode to the end of the record. On the 40th hop at Seattle a similar downward extension occurs, preceding the echo of the first falling tone at Wellington, and it too echoes in the whistler mode. The echoing periods of both of the falling tones are exactly the same as those of the parent whistler at corresponding frequencies.

It will be recalled that an association between emissions and whistlers was reported by Storey (1953), who observed that if a whistler occurred during a period of general riser activity, it would usually be followed by a riser. He cited spectrograms of such events which were published by Potter (1951). The relationship between whistlers and "triggered" emissions is covered by Helliwell (1963). There is a clear suggestion of a cause and effect relationship in which the emission is initiated or "triggered" by the whistler. A logical extension of this idea is to suggest that an emission may also be triggered by another emission propagating in the whistler mode. If this triggered emission echoes in the whistler mode at the appropriate frequency, then the triggering process is repeated, giving rise to a set of periodic emissions in which the period is equal to the whistler mode group delay at the triggering frequency. Strong absorption of the whistler-mode echo of the emission over most of its frequency range coupled with strong emission gives non-dispersive periodic emissions, while the reverse situation gives the dispersive type (Helliwell, 1943).

In all attempts to explain the non-dispersive periodic emissions, the authors assume the a priori existence of a small bunch of charged particles that oscillates between mirror points in the earth's magnetic field. It is also assumed that the bunch radiates a noise burst each time it passes through a favorable region of the ionosphere plasma, so that the emission period is the same as the mirror period of the bunch. We shall refer to this explanation as the "particle-bunch" theory of non-dispersive periodic VLF emissions.

For the particle-bunch theory of non-dispersive periodic emissions to be acceptable, the particle-bunch would have to be related to the whistler producing atmospheric and the mirror period for the particle-bunches would have to be the same as the whistler-mode echo period at the triggering frequency. Furthermore, the nearly identical periods for several sets of periodic emissions would require that the several corresponding particle-bunches have almost exactly the same mirror period. Since these conditions appear most unlikely, the particle-bunch theory is ruled out as an explanation of non-dispersive periodic emissions.

Acceptance of the triggering hypothesis, on the other hand, leads to explanations of other emission phenomena not previously understood. Under conditions similar to those which yield non-dispersive emissions (Helliwell, 1963), a parent whistler might be undetectable because of absorption, so that only the triggered emissions would be observed (Fig. 5 could be placed in this category).

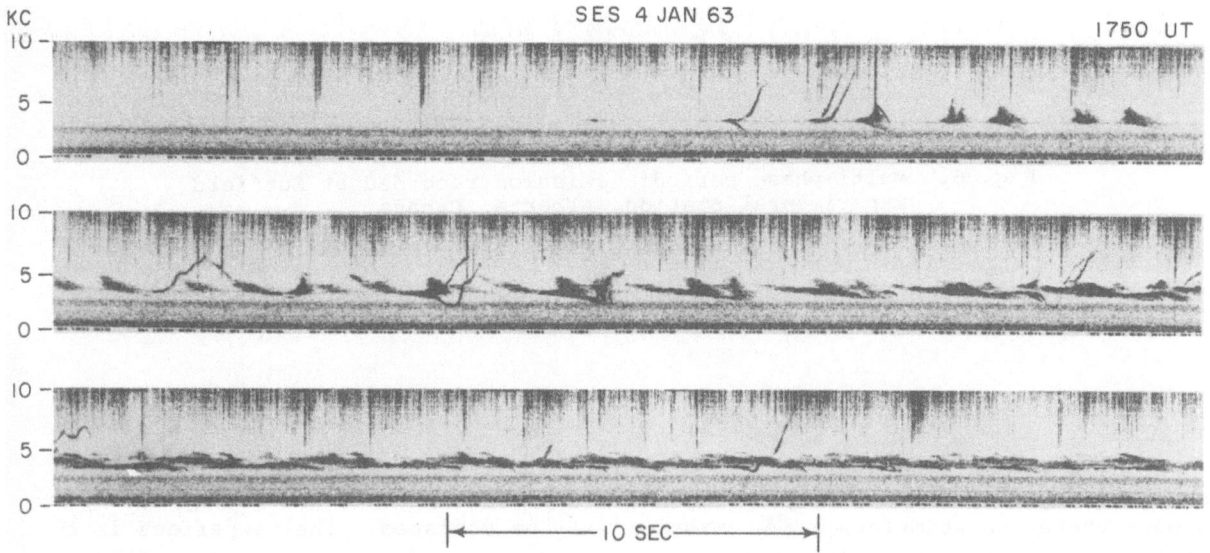

Fig. 5.   Multi-phase periodic emission recorded at Suffield
Experimental Station, Alberta, Canada.

Multi-phase periodic emissions are a common occurrence. An example of three phases is shown in Fig. 6. The sets appear in a limited frequency band and are identified by the letters A, B, and C. An example of twelve phases is shown in Fig. 7. Also in Fig. 7 can be seen an emission, which then echoes, triggered by $W_{4,T}$ of whistler W_O. A multi-phase non-dispersive periodic emission which goes from a three, to a two, to a one phase emission twice in thirty minutes is shown in Fig. 8. A superposition of multi-phase emissions is shown in Fig. 9 and Fig. 10.

The cause of the equal spacing between sets of emissions in Fig. 8 is not known but it may be related to a "recovery" time in the medium. Let us suppose that the

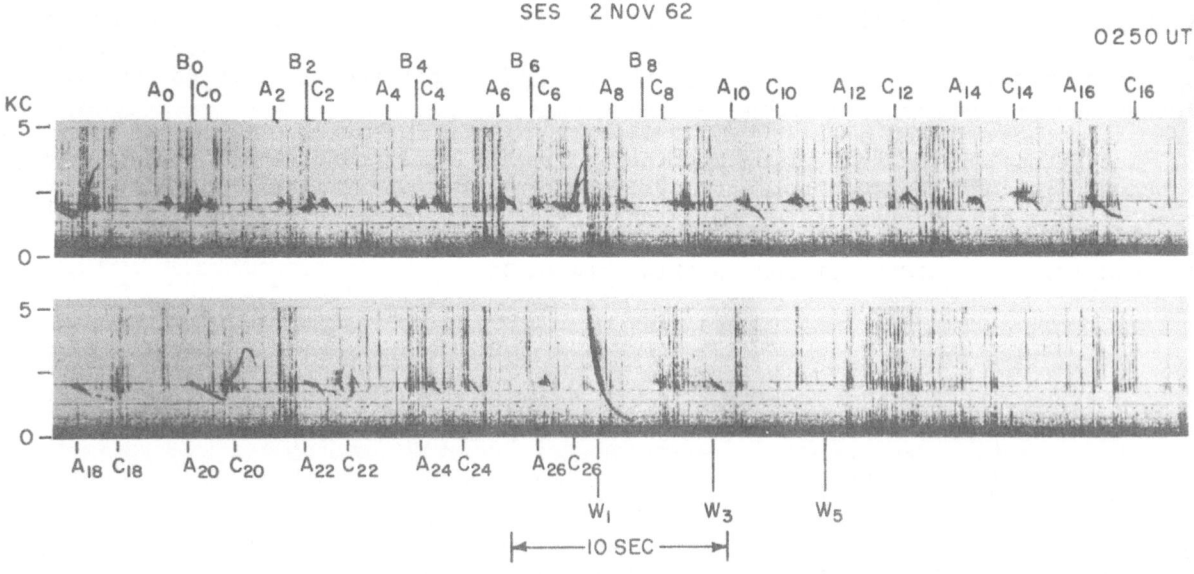

Fig. 6. Multi-phase periodic emission recorded at Suffield
         Experimental Station, Alberta, Canada.

emitting region is temporarily disorganized after each burst, and so any attempt by
a wave packet to stimulate a new emission will be resisted. The new effect is to
delay very slightly the generation of a burst with respect to the preceding burst
(or a different set, of course). After many periods this tendency would produce
roughly equal spacings, as observed.

Of particular interest in multi-phase emissions are those cases in which the
number of phases is even and the spacings between phases the same. The conjugate
point records would then show an in-phase relationship if the basic period were
taken (incorrectly) to be the <u>true period divided by an even integer equal to or
less than the number of phases</u>. On three-phase emissions one must be sure that he
is not misinterpreting the basic period, taking the true period to be one-third of
the basic period.

270

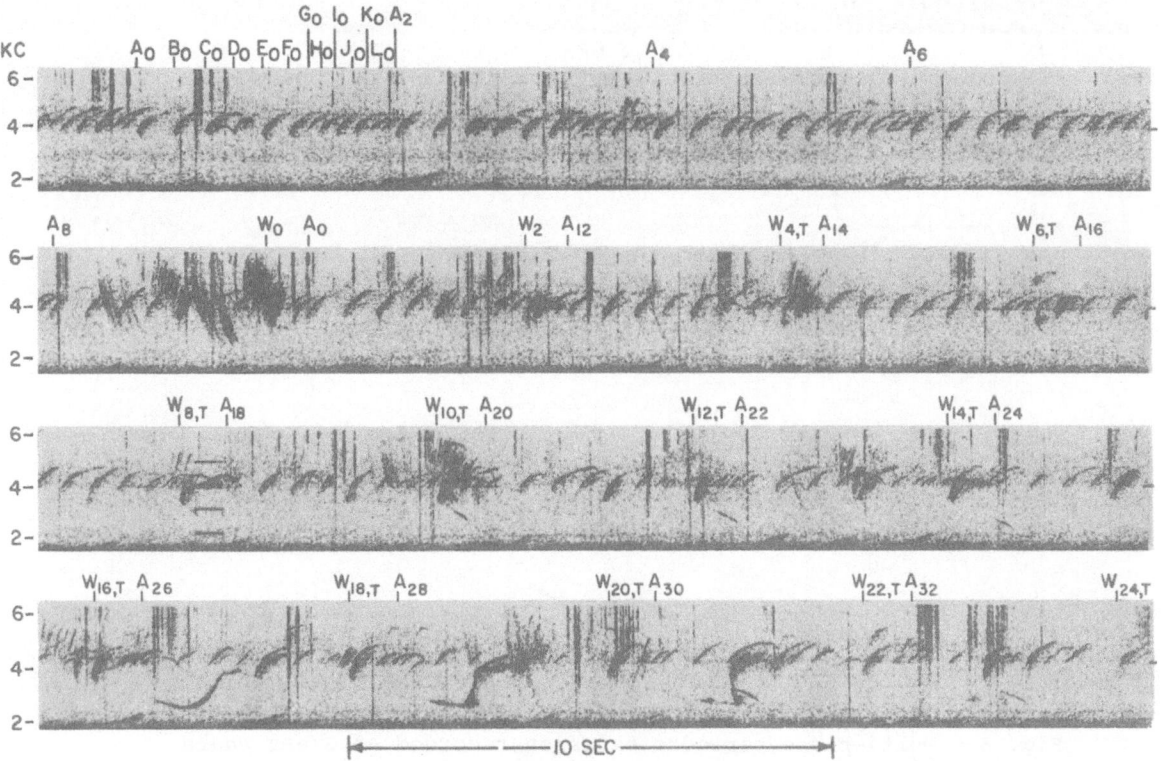

Fig. 7.   Multi-phase periodic emission and triggered emission
recorded at Byrd Station, Antarctica.

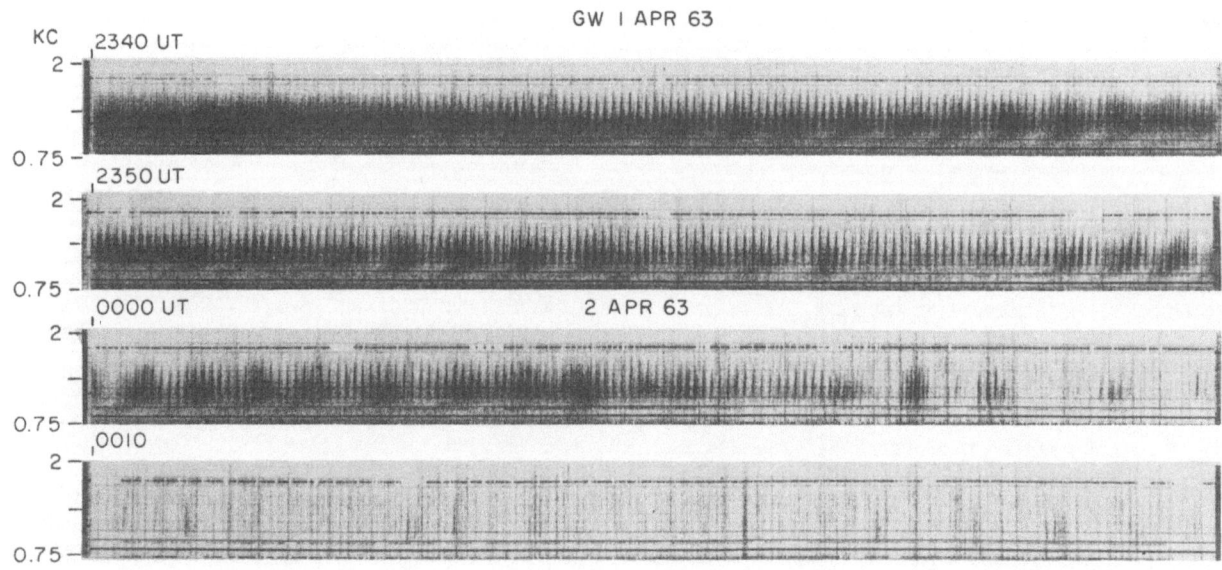

Fig. 8. Multi-phase periodic emission recorded at Great Whale
River, Canada.

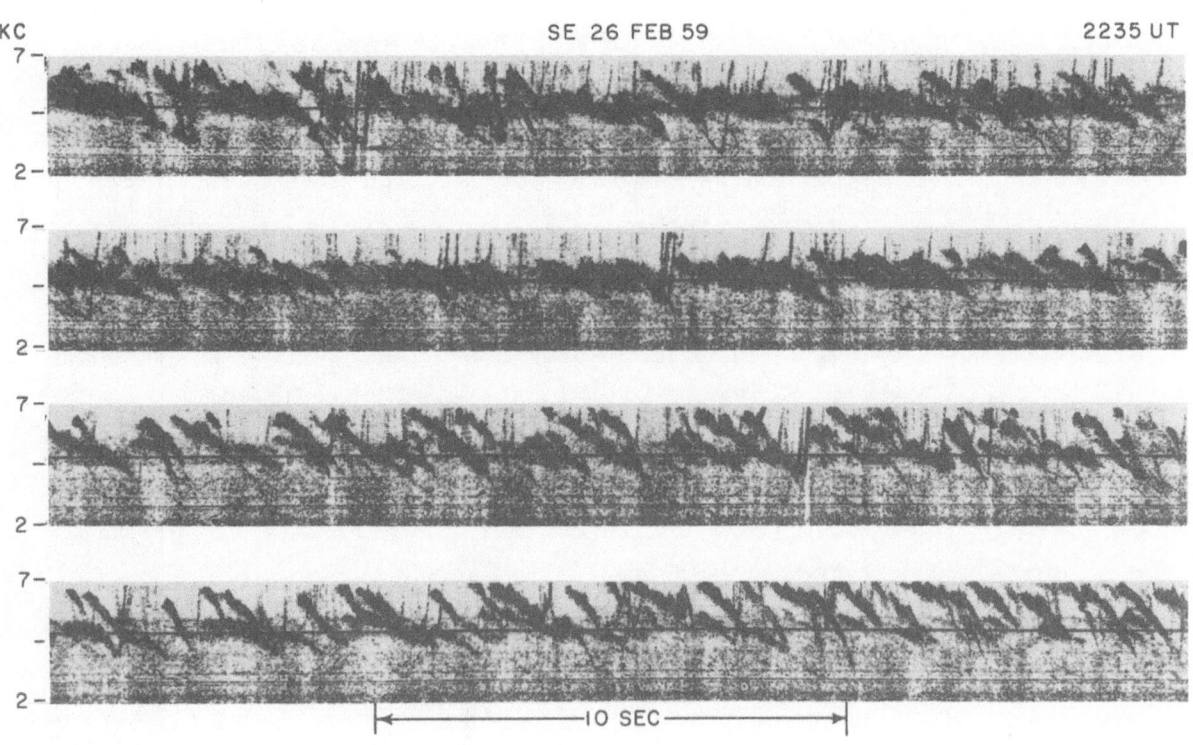

Fig. 9. Multi-phase periodic emissions recorded
at Seattle, Washington.

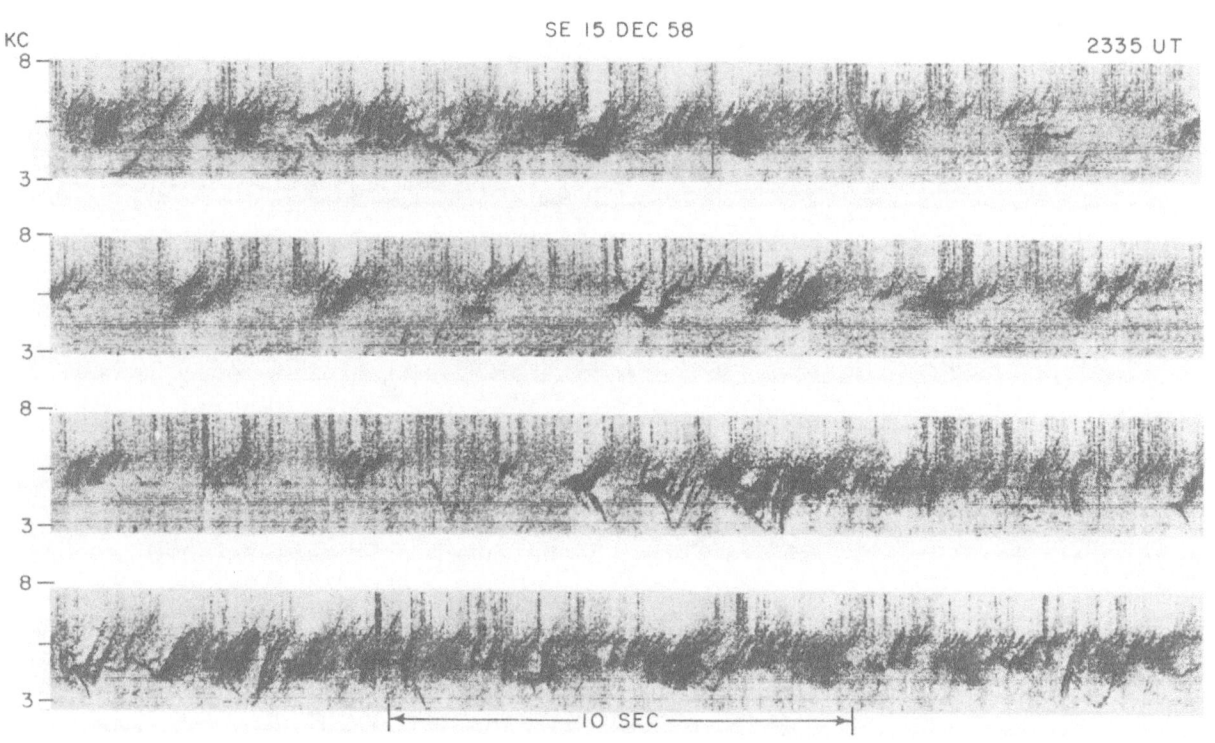

Fig. 10. Multi-phase periodic emissions recorded
at Seattle, Washington.

Emission Mechanisms.

Although no detailed theory of VLF emissions has yet been developed, there is general agreement that their sources must be situated in the ionosphere. Extra-terrestrial sources have not been considered because the magnetosphere is thought to be opaque to VLF energy from outside and because the noise is often closely related to whistlers. In current theories the electromagnetic energy of the emission is derived primarily from the kinetic energy streams of charged particles trapped on the lines of force of the earth's magnetic field. Possible sources of power for the streams and mechanisms for conversion of this power are considered by Sturrock (1962).

Conversion mechanisms that have been considered can be divided into two main categories, depending on whether the longitudinal motion or the transverse motion of the charged particles is the controlling factor. Mechanisms depending on longitudinal motion include Cerenkov radiation and a kind of amplification somewhat analogous to that observed in a laboratory traveling wave tube. The transverse motion of charged particles in the earth's magnetic field is very nearly circular and the associated radiation is of the cyclotron type.

A mechanism for "organizing" the transverse radiation from the particles has been proposed by Brice (1963). An instability in the transverse interaction has been demonstrated by Bell and Buneman (to be published), an important step in establishing the validity of this mechanism.

AN ATLAS OF VLF PHENOMENA RECORDED AT
GEOMAGNETICALLY CONJUGATE POINTS

The whistlers and emissions presented here were recorded at three pairs of conjugate stations as shown in Fig. 11. The three are:

(1)   Unalaska - Dunedin

(2)   Great Whale River - Byrd Station

(3)   Carde (near Quebec City) - Eights Station (also known
       as Sky-Hi during the Austral Summer of 1961-62)

The spectra were prepared at Stanford University with the aid of the Rayspan spectrum analyzer (Helliwell et al, 1961). The phenomena spectrum analyzed were recorded in real time on magnetic tape.

Simultaneous spectrograms have been accurately synchronized by comparing the VLF transmitter signals that are included in the broadband information recorded on the tapes. Frequency conversion of these signals places them at the bottom of the spectrogram. The same technique has been used in some illustrations to provide absolute time reference using transmitter NBA, which transmits 300 ms pulses whose leading edges mark the beginning of each second, except for the 29th, 56th, 57th, 58th, and 59th seconds, at which the pulses are omitted.

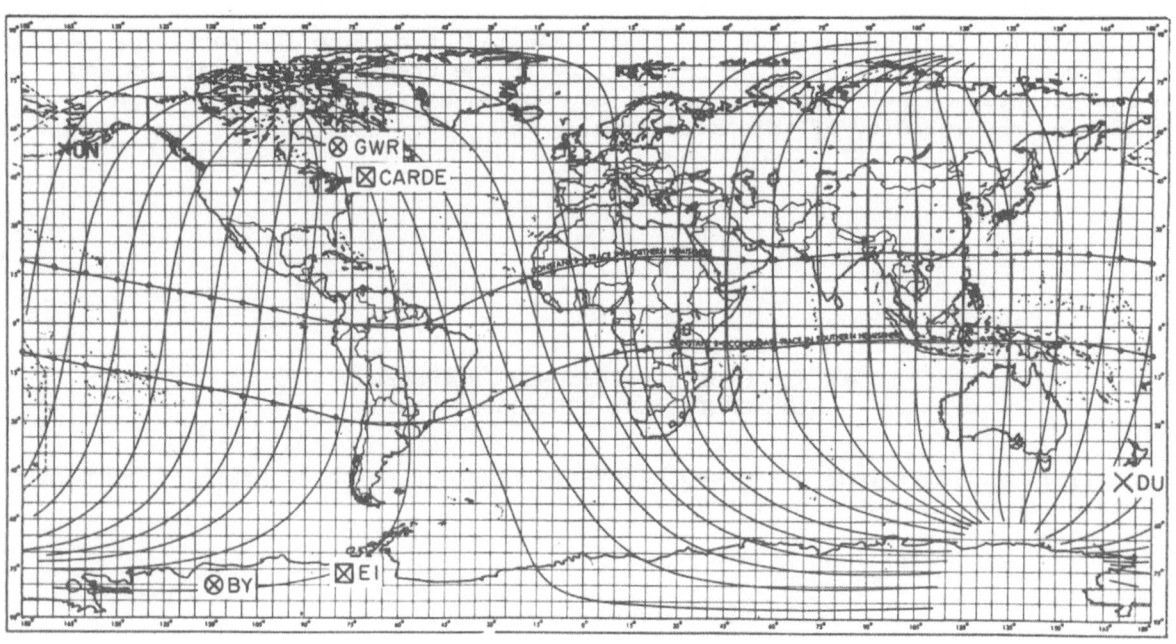

Fig. 11. Map showing the location of Byrd Station (BY), Carde, Dunedin (DU), Eights Station (EI), Great Whale River (GWR) and Unalaska (UN).

276

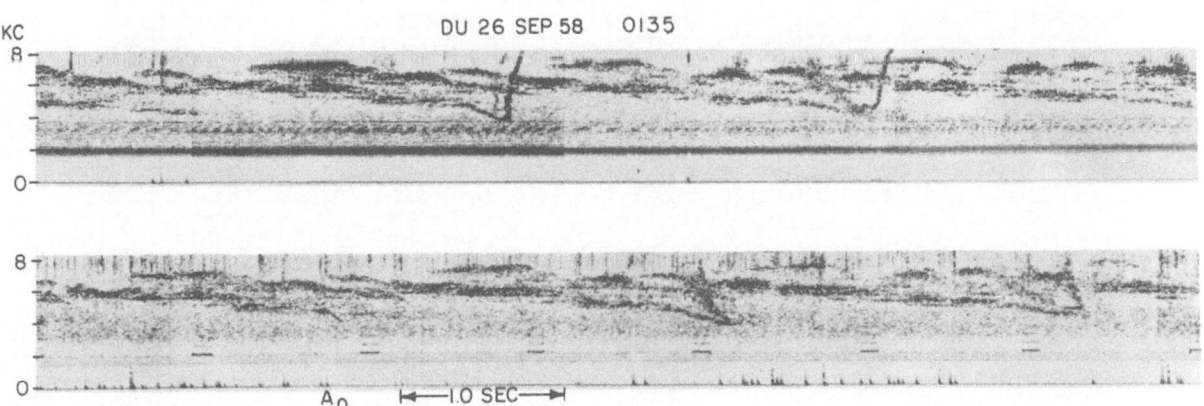

Fig. 12. Echoing hooks and quasi-constant periodic VLF noise recorded at Dunedin, New Zealand and Unalaska, Alaska, on 26 September 1958.

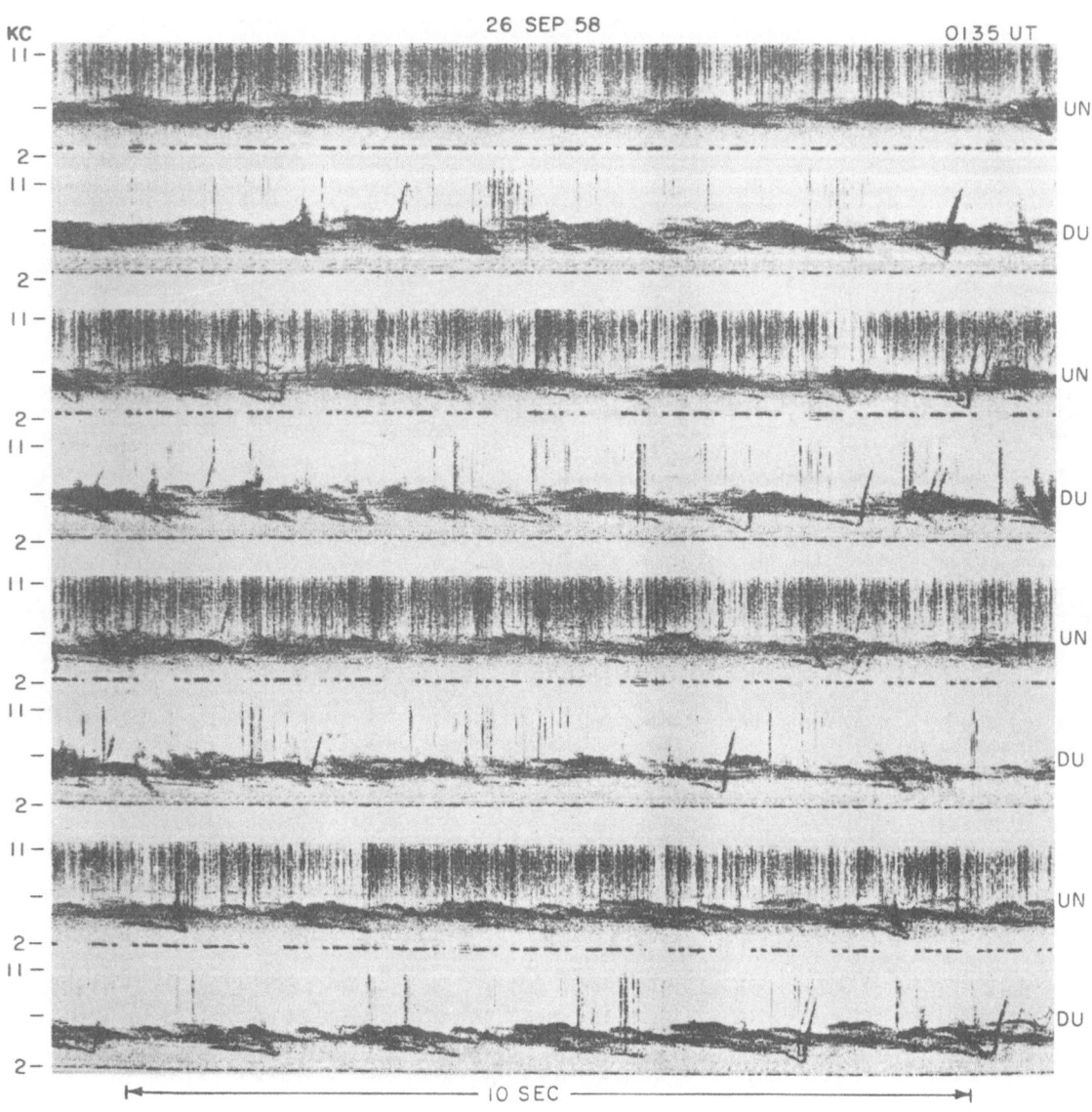

Fig. 13. Echoing hooks and quasi-constant periodic VLF noise recorded at Dunedin, New Zealand and Unalaska, Alaska, on 26 September 1958.

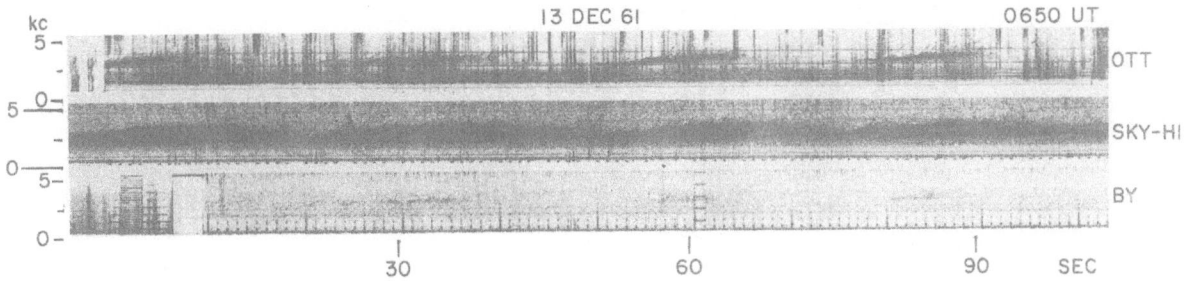

Fig. 14.   Periodic emissions with a period of 30 seconds recorded at Ottawa,
Canada; Sky-Hi, Antarctica; and Byrd Station, Antarctica.   Periodic
emissions with this long a period are always observed to be in
phase in opposite hemispheres.

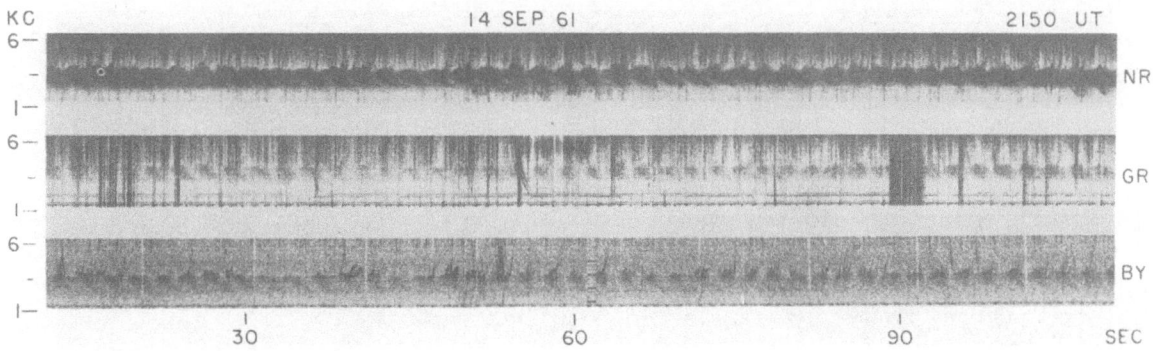

Fig. 15. Periodic emissions recorded at Norwich, Vermont; Greenbank, West
Virginia; and Byrd Station, Antarctica.   The emissions are in
anti-phase in opposite hemispheres.

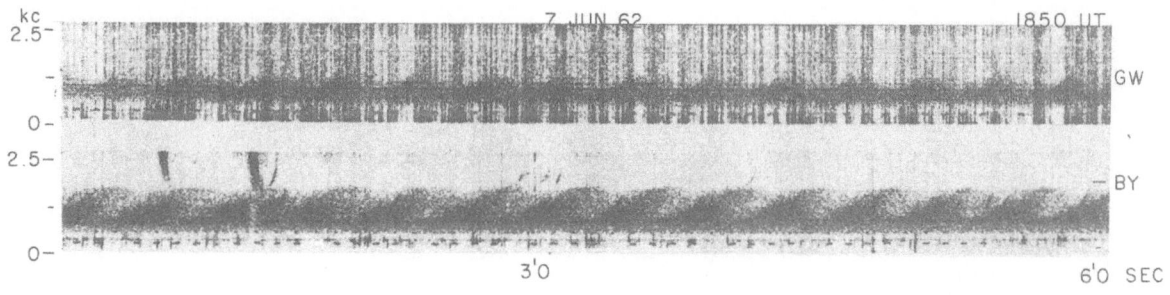

Fig. 16.   Periodic emissions recorded at Great Whale River, Canada, and
Byrd Station, Antarctica.   The emissions are in anti-phase in
opposite hemispheres.

279

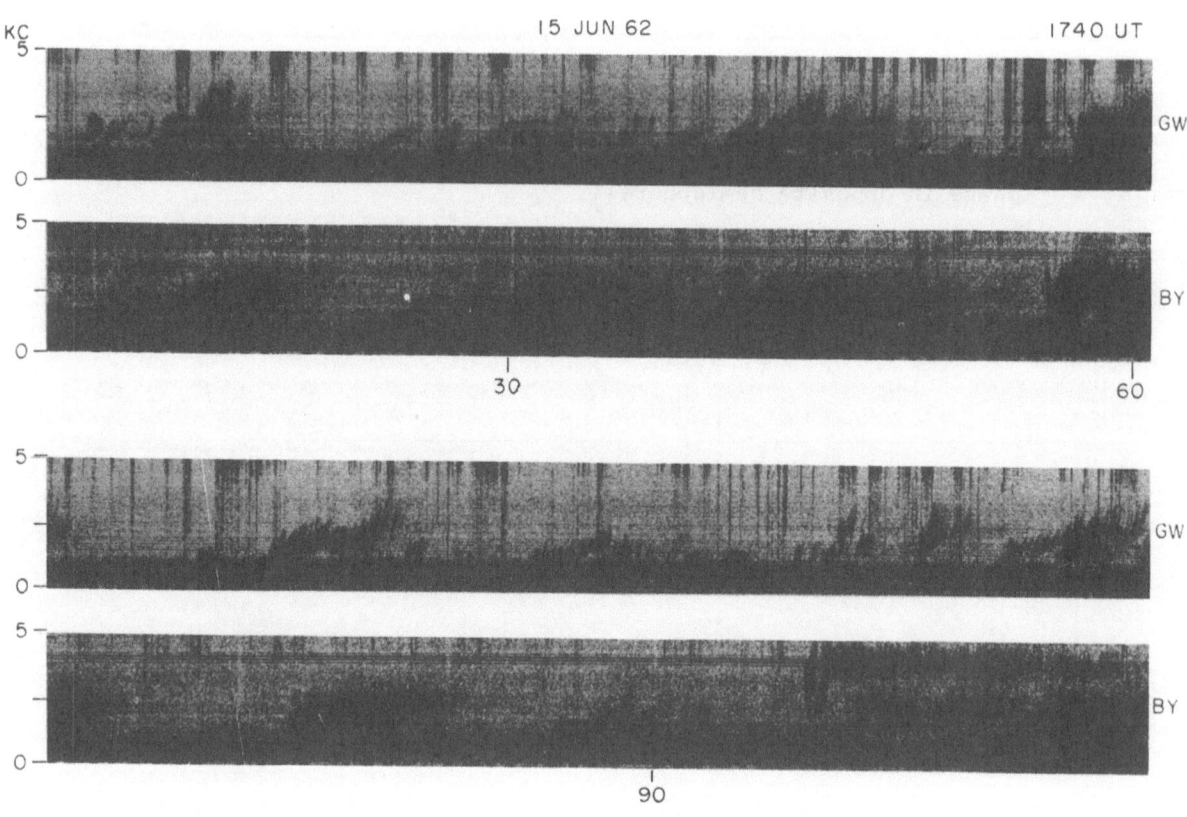

Fig. 17. Chorus bursts and triggered risers observed at Great Whale River and Byrd Station.

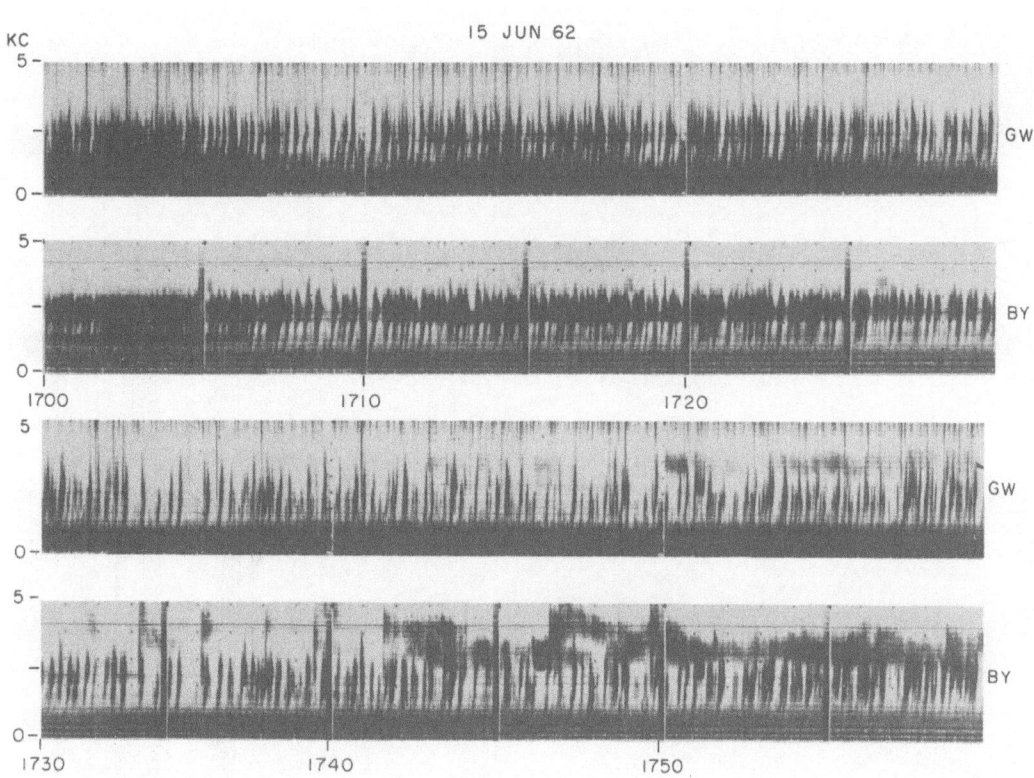

Fig. 18.
A low time resolution spectrogram of the phenomena like that in Fig. 17.

281

20 JUN 62

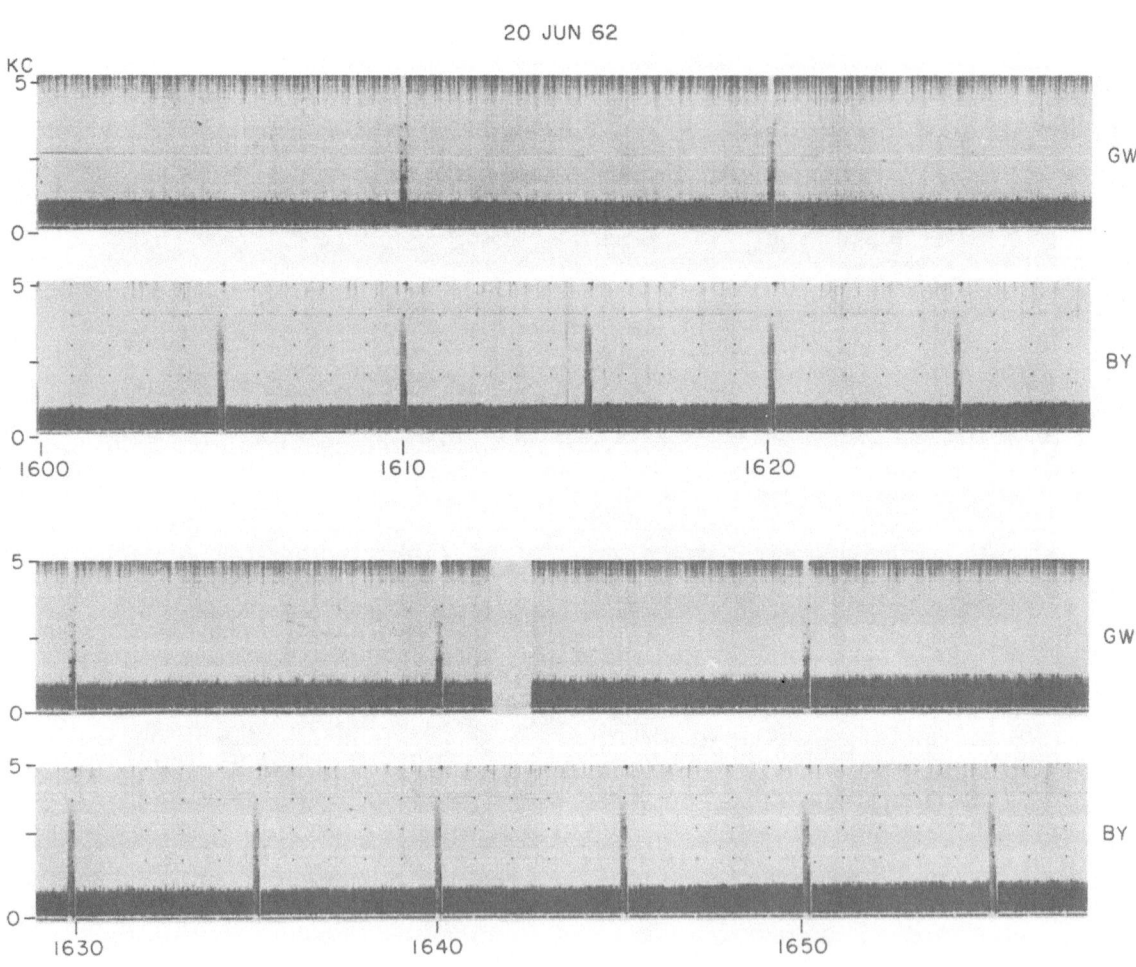

Fig. 19. A low time resolution spectrogram of polar chorus
observed at Great Whale River and Byrd Station.

282

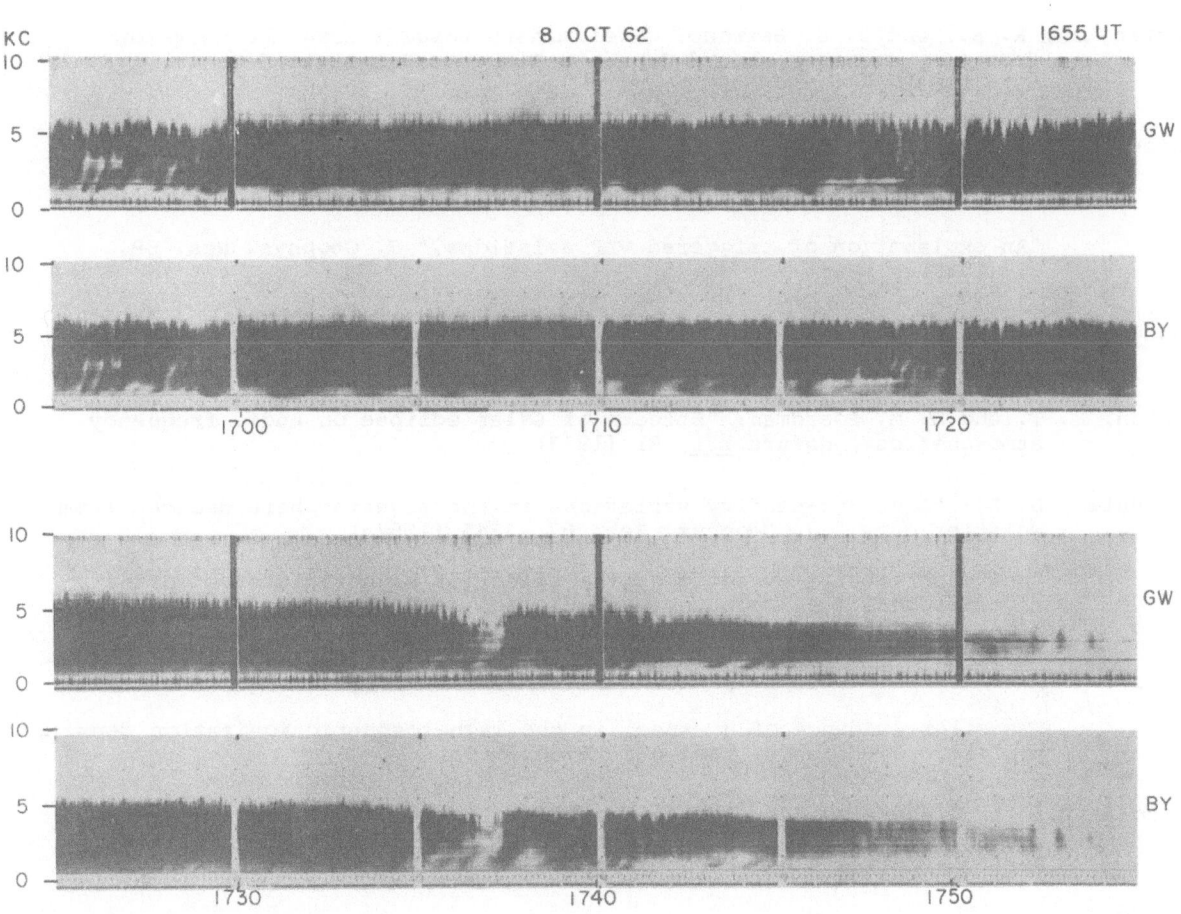

Fig. 20.   A low time resolution spectrogram of hiss and chorus
recorded at Great Whale River and Byrd Station.

REFERENCES

Allcock, G. McK.: "IGY Whistler Results," Proc. of Commission IV on Radio Noise of Terrestrial Origin During XIIIth General Assembly of URSI, London, 1960, Elsevier, Amsterdam-New York, p. 1934 (1962)

Barkhausen, H.: "Zwei mit Hilfe der Neuen Verstäker Entdeckte Erscheinungen," Physik. Zeitschr. 20, 401 (1919)

Barrington, R. E., and J. S. Belrose: "Preliminary results from the very-low-frequency receiver aboard Canada's Alouette Satellite," Nature 198, 651 (1963)

Brice, N. M.: "Discussion of paper by R. L. Dowden, doppler-shifted cyclotron radiation from electrons: a theory of very low frequency emissions from the exosphere," J. Geophys. Res. 67, 4897 (1962)

"An explanation of triggered VLF emissions," J. Geophys. Res. 68, 15 (1963)

Budden, K. G.: "Radio Waves in the Ionosphere," The University Press, Cambridge, England (1961)

Burton, E. T. and E. M. Boardman: "Effects of solar eclipse on audio frequency atmospherics," Nature 131, 81 (1933)

Carpenter, D. L.: "Electron-density variations in the magnetosphere deduced from whistler data," J. Geophys. Res. 67, 3345 (1962a)

"New experimental evidence of the effect of magnetic storms on the magnetosphere," J. Geophys. Res. 67, 135 (1962b)

"Whistler Measurements of electron-density and magnetic field strength in the remote magnetosphere," J. Geophys. Res. 68 (1963a)

"Whistler evidence of a 'knee' in the magnetospheric ionization density profile," J. Geophys. Res. 68, 1675 (1963b)

Corcuff, Y.: "La Dispersion des Sifflements Radioelectriques au cours des Orages Magnetiques; ses Variations Nocturne, Annuelle, Et Semiannuelle en Periodes Calmes," Ann. Geophys. 18, 334 (1962)

Dinger, H. E.: "Periodicity in dawn chorus," paper presented at IRE - URSI Symposium, May 22-25, 1957, Washington, D. C.

Eckersley, T. L.: Letter to the Editor, Nature 122, 768 (1928)

Gallet, R. M.: "The very low frequency emissions generated in the earth's exosphere," Proc. IRE 47 (2), 211 (1959)

Gomez, D., M. G. Morgan, and T. Laaspere: "The diurnal and long-term variations of whistler dispersion," (1962) (in preparation)

Helliwell, R. A., J. Katsufrakis, M. Trimpi, and N. Dunckel: "Periodic noise associated with whistler echo trains observed at spaced stations in opposite hemispheres (abstract)," IRE Trans. Antennas Propagation 10 (4), 492 (1962)

Helliwell, R. A.:"Whistler-triggered periodic very-low-frequency emissions," J. Geophys. Res. 68 (19), 5387 (1963)

"Whistlers and Related Ionospheric Phenomena," Stanford University Press, Stanford, California (in press), (1964)

Iwai, A., and J. Outsu: "On the characteristic phenomena for short whistlers observed at Toyokawa in winter," Proc. Res. Inst. Atmos. Nagoya Univ. 5, 53 (1958)

Jacchia, L. G. and J. Slowey: "Accurate Drag Determinations for Eight Artificial Satellites: Atmospheric Densities and Temperatures," Smithsonian Institution Astrophysical Observatory, Spec. Rept. No. 100 (1962)

Kimpara, A.: "Some characteristics of the dispersion of whistlers," Proc. Res. Inst. Atmos. Nagoya Univ. 9, 5 (1962)

Lokken, J. E., J. A. Shand, S. C. Wright, L. H. Martin, N. M. Brice, and R. A. Helliwell: "Stanford-Pacific Naval Laboratory conjugate point experiment," Nature 192 (4800), 319 (1961)

Martin, H. A., W. Neveling, W. Riester and M. Roemer: "Model of the Upper Atmosphere from 130 km Through 1600 km, Derived from Satellite Orbits," Proc. 2nd Int. Space Science Symp., Florence, 1961, North-Holland, Amsterdam, 1961, p. 903

Morgan, M. G. and G. McK. Allcock: "Observations of whistling atmospherics at geomagnetically conjugate points," Nature 177, 30 (1956)

Outsu, J. and A. Iwai: "Some correlations between occurrence rate and dispersions of whistlers at lower latitudes and magnetic K index," Proc. Res. Inst. Atmos. Nagoya Univ. 9, 19 (1962)

Paetzold, H. K. and H. Zschörner: "An annual and a semi-annual variation of the upper air density," Geofis. Pura e Appl. 48, 85 (1961a)

"The Structure of the Upper Atmosphere and its Variations After Satellite Observations," Proc. 2nd Int. Space Sci. Symp., Florence, 1961, North-Holland, Amsterdam,(1961b) p. 958

Pope, J. H. and W. H. Campbell: "Observations of a unique VLF emission," J. Geophys. Res. 65 (8), 2543 (1960)

Potter, R. K.: "Analysis of audio-frequency atmospherics," Proc. IRE 39 (9), 1067 (1951)

Rivault, R. and Y. Corcuff: "Recherche du Point Conjugue Magnetique de Poitiers-Variation Nocturne de la Dispersion des Sifflements," Ann. Geophys. 16, 530 (1960)

Smith, R. L.: "Propagation characteristics of whistlers trapped in field-aligned columns of enhanced ionization," J. Geophys. Res. 66, 3699 (1961a)

"Properties of the outer ionosphere deduced from nose whistlers," J. Geophys. Res. 66 (11), 3709 (1961b)

"Electron densities in the outer ionosphere deduced from nose whistlers," J. Geophys. Res. 66 (8), 2578 (1961c)

Storey, L. R. O.: "An investigation of whistling atmospherics," Phil. Trans. Roy. Soc. A, 246, 113 (1953)

Sturrock, P.: "Generation of radio noise in the vicinity of the earth," J. Res. NBS 66D (2), 153 (1962)

Thomas, J. O., A. R. Long, and D. Westover: "The Calculation of Electron Density Profiles from 'Topside Sounder' Records," URSI-IEEE Meeting, Washington, D. C., April 29 - May 2, 1963

MICROPULSATIONS AT NEAR CONJUGATE STATIONS IN THE AURORAL ZONES AND THEIR ASSOCIATION WITH OTHER IONOSPHERIC PHENOMENA

Sir Charles Wright
Pacific Naval Laboratory
Defence Research Board of Canada
Esquimalt, British Columbia

ABSTRACT

After a short and incomplete review of previous work, this report deals with the results of the joint operation with Stanford University at Fort Churchill, Great Whale River and its near conjugate at Byrd Station.

On the assumption that data recorded at conjugate points will be very similar and will occur nearly simultaneously, attention has been focused on the micropulsation and magnetograph "events" which occur quite frequently and introduce Birkeland's Polar Elementary Storms. These have adequately sudden commencements to enable them to be timed with an accuracy of about ½ minute for a single micropulsation event. These events are preceded by a quiet interval and followed by bays, or DP's.

There is a pronounced difference between the daytime and the nighttime regimes, records of the former often being of regular (Pc) shapes and the latter nearly always of irregular shape (Pt) with components of much higher frequency (riders). The duration of these riders is seldom longer than eight minutes. Simultaneous auroral events and activity also show similar differences between the daytime and nighttime regimes and the evidence suggests the same is also true of ionospheric absorption and hiss.

All these ionospheric events occur simultaneously when measured at the near-conjugate sites of Great Whale and Byrd and sometimes also at Churchill. They occur most frequently in the two hours ending 04 G.M.T. which is not far from geomagnetic midnight at Great Whale. The events can also be recorded on sensitive micropulsation instruments in middle latitudes, presumably as "leakage" currents from the auroral electrojets. Those which include the higher frequency riders, which also appear nearly simultaneously with the negative H events in the auroral zones, have a pronounced maximum of occurrence at Victoria about midnight local mean time. The events can also sometimes be recognized on mid-latitude magnetograms as small perturbations if it is known from other sources when to look for them.

It is suggested that the opportunity exists of examining in more detail the relative importance of local mean and geomagnetic time in the incidence of these events in different longitudes around the globe and the difference in form between daytime and nighttime events.

In the frequency range of 0.005 to 3 cps, with which this report is almost entirely concerned, the balance of evidence is that conjugate points in the auroral zones are of fleeting duration in the sense that points which are conjugate one day are not necessarily conjugate next day or next week.

It seems that the origin of the events lies within the magnetosphere, but a direct relation with fluctuations in the solar wind cannot be excluded.

INTRODUCTION

The examination of geomagnetic micropulsations of the earth's magnetic field is not entirely a modern development and may be said to stem from the study of the variations in earth current - the electric component of the natural electromagnetic background which showed many years ago as disturbances of the telegraph systems during severe magnetic storms.

The early experiments on the magnetic component dates back to the closing years of the last century (Angenheister, 1896; Arendt, 1896; van Bemmelen, 1900). My own direct participation in micropulsations dates back only to 1919. Later, in 1926, a large horizontal loop was set up at Eskdalemuir at my request, through the good offices of Sir George Simpson FRS, then Director of the Meteorological Office in the U.K. The records from this station were examined by E.R.R. Holmberg at a much later date (Holmberg, 1953).

Dr. Troitskaya (1953) was experimenting about this time in Russia. The results of both investigators were roughly in agreement in reporting a pronounced difference between the day and night effects. The inquiry into this natural noise background has been carried on continuously since that time and with increasing tempo and Pacific Naval Laboratory has been actively engaged since 1954. It is only recently that the inquiry has been extended to high and low latitudes. The International Geophysical Year has been largely responsible for the increased interest in micropulsations, but the data from low latitudes is still scanty. However, taking things by and large, the early results of Holmberg and Troitskaya relating to the pronounced difference between the day and the night regimes are confirmed, with one exception noted later. The chief features which distinguish the two classes of background signals from one another are: The pronounced difference in their diurnal occurrence; the included frequency band or bands of the micropulsations; their dynamic range; their continuity; and a preference or lack of preference for appearance in conjugate zones. Thus, stations well spaced in latitude and longitude show differing responses and this is no doubt also true of the corresponding electric components measured as earth currents.

The obvious differences between the day and night background are that the daytime signals (Pc) are regular, smooth and relatively continuous, though they may appear briefly at night in mid-latitudes. On the contrary, the nighttime signals (Pt) are impulsive in character, irregular and rough, due to the inclusion of signals within a wide frequency band. Troitskaya's diagram (Fig. 1) plotted on a universal time basis exhibits very clearly the occurrences of these two regimes during the day and night.

These two regimes seem to hold sway from northern to southern auroral zone with one major exception. Benioff's type A signals of about 1 cps frequency (Benioff, 1960) are shown as regular night occurrences in Southern California. The words daytime and nighttime are used deliberately since the conjugate relationship between Byrd and Great Whale Stations seems to remain even though one of the two is in darkness and the other in light. While there may be other cases confined to restricted areas of the earth, it is believed that there are few other exceptions to the general rule stated above. To render the picture more complete, one may add that both regular and irregular (impulsive) classes appear of much greater amplitude in the two auroral zones and decrease rapidly towards lower magnetic latitudes. The exception to this generalization is that the regular Pc signals of about 20 seconds period are not very greatly enhanced in the auroral zones.

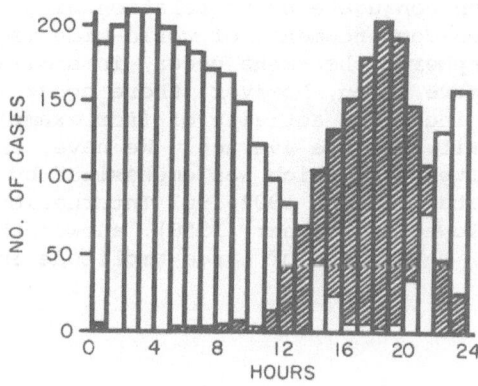

Fig. 1.
The statistical diurnal variation (G.M.T.) of occurrence of the two main
types of micropulsation observed in Russia by V. A. Troitskaya. The Type
I, regular and continuous (Pc), occurs most frequently between 0200 to
0400 G.M.T. and the Type II, irregular and impulsive (Pt), between 18 and
19 hours. Note that 0200 to 0400 is the time of maximum occurrence of
Type II events at Byrd, Great Whale and Churchill, and of Type I events
in Russia. The Type II events are thus nighttime events in Russia as at
our three auroral stations.

In recent years, considerable interest has been shown in the correlation at conjugate stations of various magnetic and ionospheric effects. Much of the work has,
naturally, dealt with the large effects which take place during world-wide magnetic
storms, the results of which are felt over a large part, or the whole, of the earth.
Much of this work has been recorded at stations close to the auroral zones where
calculations of conjugate positions owing to shortage of data is likely to be less
accurate than for positions in lower magnetic latitudes. However, even in and near
the auroral zones it is clear that calculation is able to choose pairs of stations
with a close degree of conjugate relationship, judging by similarity of experimental
data at the same time at the two stations.

While I have seen no definition of what is involved in "conjugacy" other than
the fact that the two stations are at opposite ends of a field line, it is reasonable
to assume that conjugacy involves similarity of micropulsation records (for example)
at the same time at both stations, for those phenomena which are directed to the two
stations along the field line joining them and at such speed that the phenomena take
place at the same time within the limits of accuracy of the timing system. This
immediately raises the question of the need for accurate timing of the recording
system if one is to gain some idea of what is happening along the line of force.
This can be achieved if one can recognize that the event seen at both ends of the
field line is the same event, i.e., has a characteristic shape and shows a change in
the phenomenon which is sufficiently sudden to be timed with the desired accuracy.
Preferably, these events should also occur with sufficient frequency to permit the
data to be handled statistically. The sudden commencements of world-wide magnetic
storms can be timed with some accuracy and have been used lately by many investigators to show the reality of "conjugacy" for various ionospheric phenomena, especially
at sites in or near the auroral zones. World-wide storms are, however, not very
frequent in these years of approach towards the Year of the Quiet Sun.

It is also necessary in such an inquiry into conjugate relationships to question
whether a pair of stations which are conjugate today remain conjugate on other days -
if the relevant field line appropriate to one station has a conjugate point at the
other end which does not vary in position.

Recent published work on conjugate point relationships have naturally dealt chiefly with the large sudden commencements of world-wide magnetic storms and have shown that a number of ionospheric phenomena occur, in similar form, simultaneously at conjugate points. Much more often, however, there occur in the auroral zones sudden events which precede increased activity of (for example) the magnetic elements and these happen at least daily, on the average. We have, therefore, concentrated on this type of event, the existence of which was emphasized by Birkeland (1902-1903), using the magnetic-aurora data from the 1902-1903 International Magnetic Expeditions to high latitudes. He, followed by Heppner (1958), showed the existence of "Polar Elementary Storms" (Fig. 2), which were of large amplitude in the auroral zones and

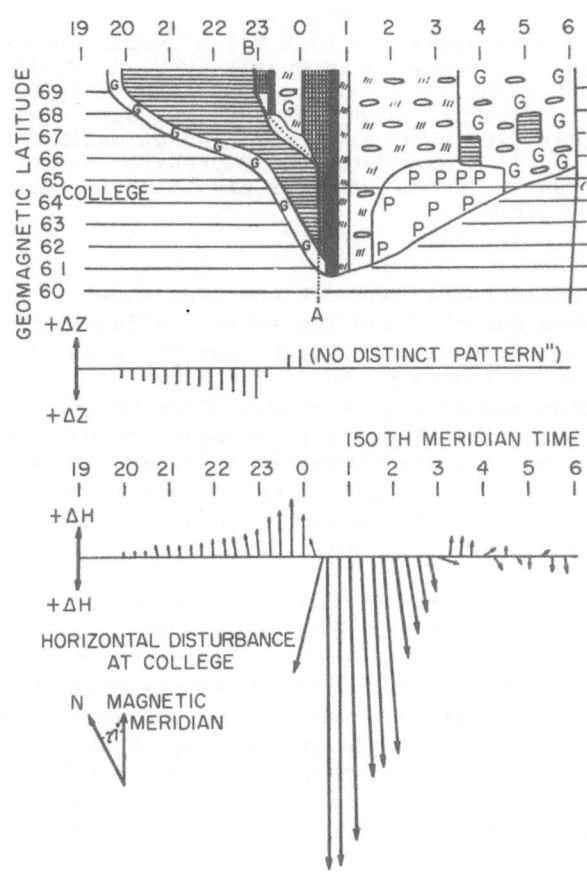

Fig. 2. One example of the coincidence of aurora at College, Alaska, with change of horizontal magnetic field from plus to minus ΔH. The line AB indicates the time of increase in intensity of the auroral arc closely followed by its break-up.   --After J. P. Heppner.

are known to decrease rapidly in intensity toward lower magnetic latitudes. These local magnetic storms which were examined by Heppner at College, Alaska, commenced suddenly during the nighttime. The sudden changes lasted only a few minutes, involved a negative change in H ($-\Delta H$) and were associated with significant changes in the auroral arc overhead. The pioneer work of Birkeland and Heppner shows the auroral arc nearly overhead exhibits a large increase in intensity closely followed by a sudden breakup of the arc. This corresponds to a reversal of the ionospheric current and as Heppner states: "Simultaneously with the transition to active rayed aurora, ΔH changes sign and rapidly reaches large negative values..." At College, the negative disturbance on the H trace occurs often after midnight while the positive ΔH disturbances appeared generally before midnight (Fig. 2). These sudden changes in H and the corresponding nighttime changes in auroral form and activity are followed by bay disturbances, recently renamed DP's by Akasofu and Chapman (1963), so that the whole magnetic disturbance is made up of a series of DP's often prefaced by sudden H changes. This is equally true during world-wide magnetic storms. The negative H changes seem to occur more suddenly than the positive H changes and these sudden commencements may even approach the magnitude of those which occur with large magnetic storms which are known to take place at the same time - within about a minute - at a large number of observatories well spaced around the globe.

RECENT WORK ON THE ASSOCIATION BETWEEN IONOSPHERIC PHENOMENA

Of recent years considerable interest has developed in the simultaneous occurrence at the same observing site of such related ionospheric phenomena as variations of the magnetic elements, of micropulsations, of aurora, of ionospheric absorption (riometer), of bremsstrahlung X-rays, of hiss and chorus. To a considerable extent, the experimental work has been related to the problems of simultaneous and similar activity at conjugate stations and particularly at stations near the auroral zones. These two interests are of course related and cannot usefully be separated. The aurora, for example, is only visible simultaneously at Byrd and Great Whale for a small fraction of the year since it is winter at one station and summer at the other. A strong association both statistical and time-coincident, between aurora and micropulsations at one station and between micropulsations at the two stations can be taken as assurance that aurora at the two stations are also well conjugated. Another example which actually happened with the first radio station setup in 1911-1914 in the Antarctic by Sir Douglas Mawson at Cape Denison could only be analyzed, and that not very satisfactorily, because the incomplete radio and auroral records showed associations with the continuously recorded magnetic data (Wright, 1940).

Sugiura has recently reviewed some, but not all, of the recent reports (Sugiura, 1963). The experimental results have been reported chiefly by workers of the Geophysical Institute at College, Alaska, which is not far from the boreal auroral zone in Alaska and at near-conjugate stations in New Zealand, Campbell Island and Macquarie Island. (Fig. 3 shows the positions of the chief conjugate stations referred to in this report.) The National Bureau of Standards at Denver have worked at Eights Station in the Antarctic and in its conjugate zone (three stations) near Quebec City. The University of California at Berkeley have done much work using balloons, while Pacific Naval Laboratory, jointly with Stanford University, have worked at Fort Churchill, Great Whale River and its near-conjugate at Byrd Station. Three reports have been issued by P.N.L. (Lokken et al, with Martin et al, 1961; Lokken et al, 1962; and Lokken et al, 1963) dealing with the early results. They are supplemented by this paper, using additional data of 1961 and, to some extent, of 1962.

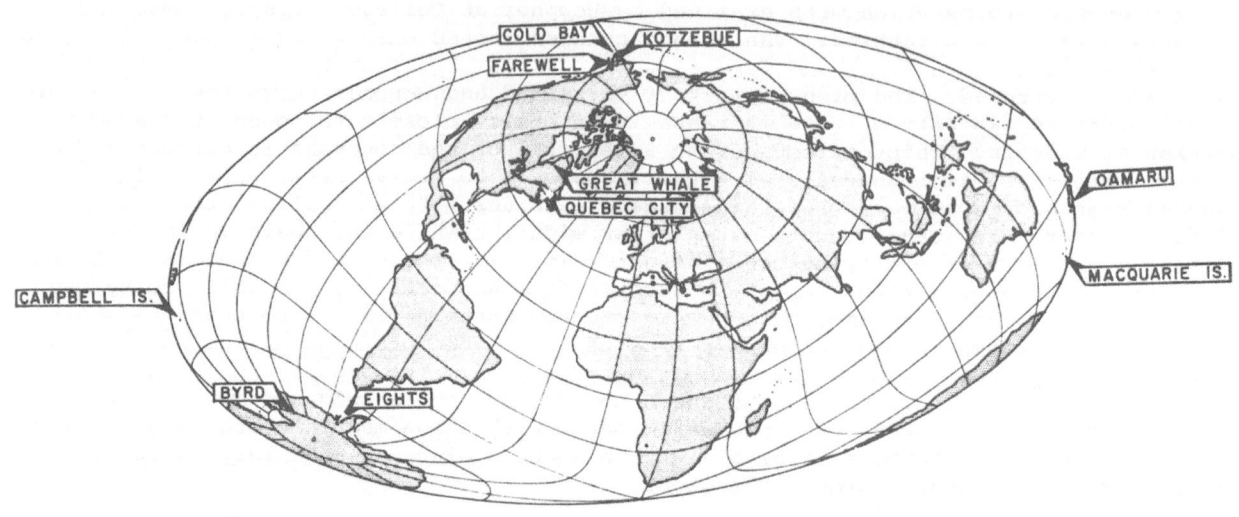

Fig. 3. Map to display positions of some near conjugate
stations referred to in text.

Ionospheric Phenomena Recently Measured At One Station.

Priority of time and pride of position goes to the measurements by Heppner (loc. cit.) at College. An example of one type of the magnetic-aurora association is shown in Fig. 2. Here AB represents the time of sudden auroral change which corresponds to the abrupt change in H from positive to negative near midnight. Campbell (1960) reported on the simultaneous occurrence of λ 3914 auroral pulsations with the X micropulsations measured at another station in Alaska 30 miles away (Fig. 4). A later report (Campbell and Rees, 1961) showed (Fig. 5) that these auroral pulsations at λ 3914 occurred roughly at the same time as an outburst of regular micropulsation activity and of cosmic noise absorption (riometer record). An association on June 17, 1960 was reported by Barcus and Brown (1962) (Fig. 6) between the auroral electrojet current in the N-S direction, ionospheric absorption and X-ray observations at College. On June 28, 1960, records were made (Fig. 7)(Campbell and Matsushita, 1962) of the horizontal component of magnetic intensity at College, of the magnetic N-S micropulsation activity, the cosmic noise absorption on 27.6 m.c. riometer and of the bremsstrahlung counts in balloon-borne equipment. Macquarie Island (Anger et al, 1963) also furnished a riometer record (Fig. 8) which showed the course of ionospheric absorption on March 5, 1962, compared with the counting rates of a Geiger counter and NaI scintillation detectors in a balloon.

Within the limits of timing of the slow traverse records displayed, it looks as if all these ionospheric events started and ended about the same time.

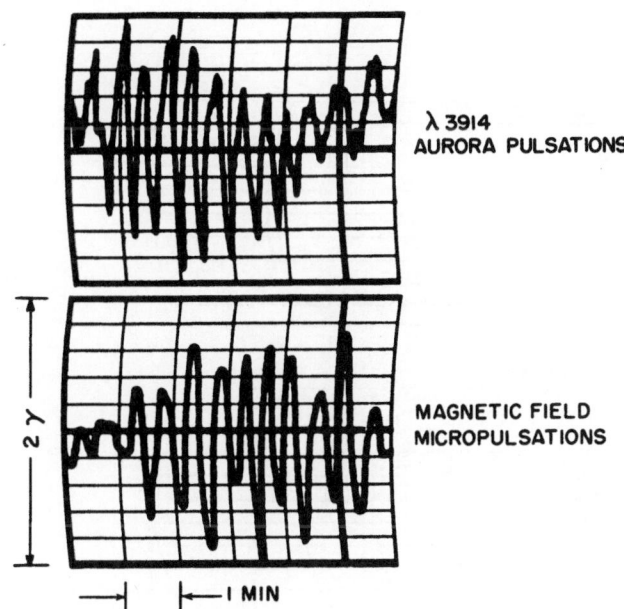

λ 3914
AURORA PULSATIONS

MAGNETIC FIELD
MICROPULSATIONS

2 γ

I MIN

Fig. 4.
Pulsations of λ 3914 auroral
intensity and corresponding fluctuations of micropulsations of the same
period near College.
                    --After W. H. Campbell.

Ionospheric Phenomena Recently Measured at Near-Conjugate Stations in the Auroral
Zones.

     Records of the magnetic variations on March 17, 1958, at Macquarie Island and
Kotzebue (Fig. 9) (Westcott, 1962) showed a quite remarkable agreement in the H variations at the two places, though the D variations did not correlate nearly so well. A
similar comparison (Fig. 10) (Boyd, 1963) between the H component at Eights Station,
Quebec South Station and Ottawa showed equally good correlations for the conjugates
and a surprising similarity with Ottawa which is nearly 400 km distant from the calculated position of the conjugate to Eights.

     Axford and Reid (1963) (Fig. 11) discuss the increases of solar cosmic rays
measured in Explorer XII in advance of sudden commencements of magnetic storms. However, with the possible exception of the three reports (loc. cit.) from P.N.L., the
times on the records are now shown with precision. They also seem to refer chiefly to
occasions of considerable magnetic activity when one might expect disturbances to
occur similarly over a larger area in the two conjugate zones than is the case in
quieter times, magnetically. To settle such points more work is required of the type
carried out with multiple stations near Quebec City by N.B.S. and near Kotzebue by the
Geophysical Institute at College, with particular attention to the size of the area of
good correlation of small events and adequate accuracy of timing of those events which
occur with sufficient suddenness. It is, however, also necessary that there shall be
adequate speed of traverse and of response of the recording systems, if one is to be
sure that the two sets of records refer to the same event.

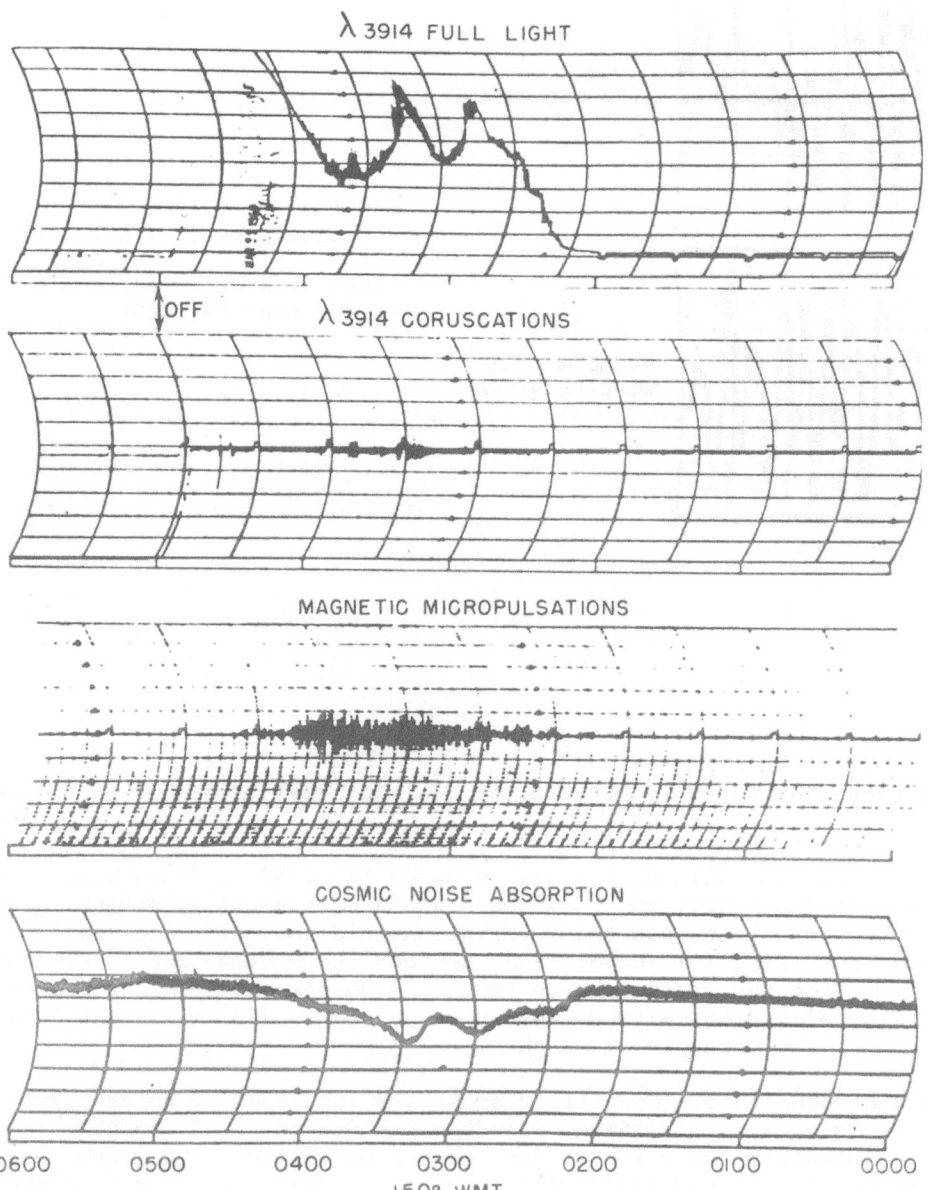

Fig. 5.   Corresponding fluctuations of λ 3914 total light
          intensity, magnetic micropulsations and cosmic
          noise absorption. This event took place during
          the nighttime regime.
                    --After W. H. Campbell and M. H. Rees.

294

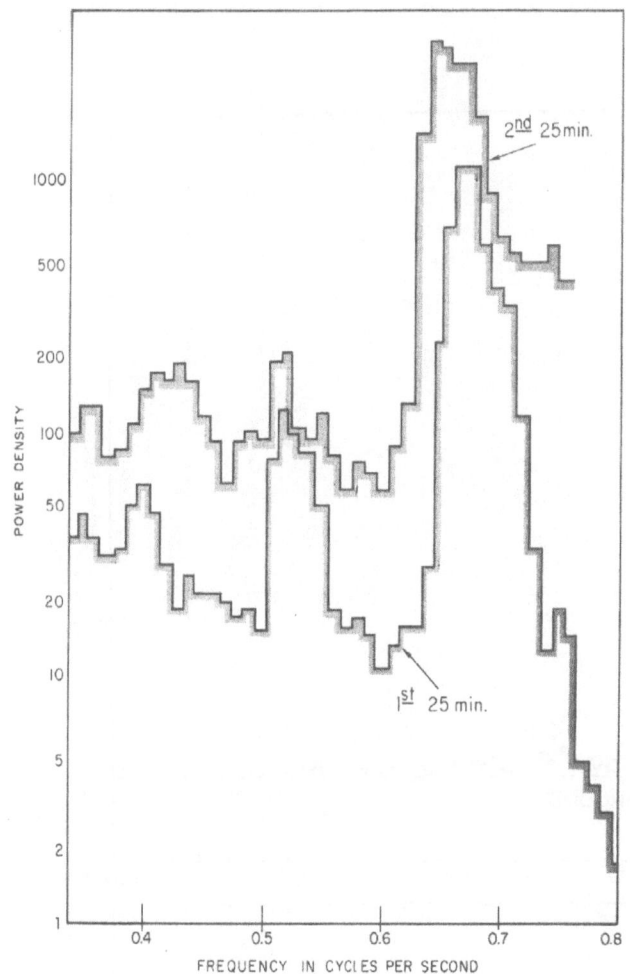

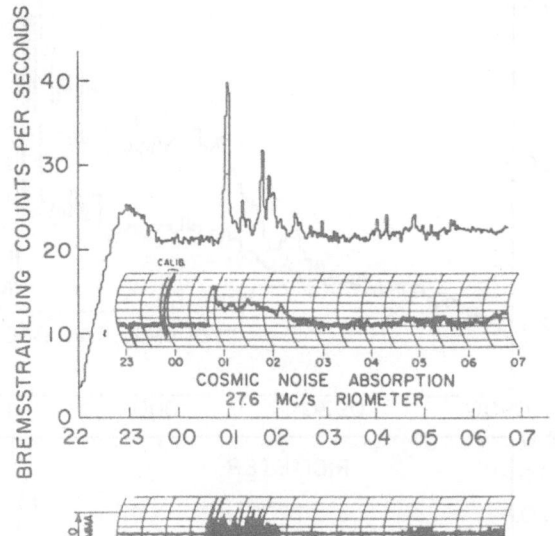

COSMIC NOISE ABSORPTION 27.6 Mc/s RIOMETER

MAGNETIC N-S 5-30 SEC PERIOD MICROPULSATIONS

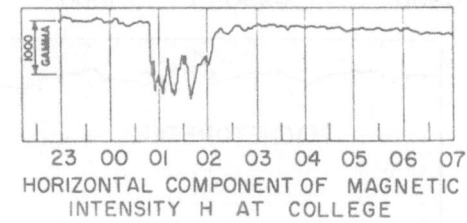

HORIZONTAL COMPONENT OF MAGNETIC INTENSITY H AT COLLEGE

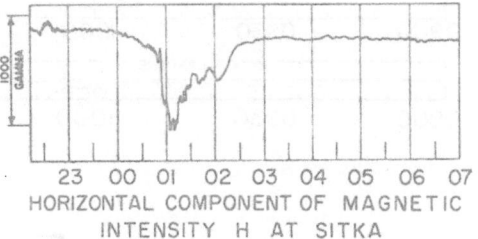

HORIZONTAL COMPONENT OF MAGNETIC INTENSITY H AT SITKA

—TIME: 150° WMT—

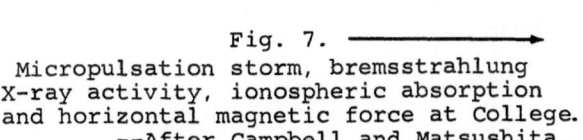

Fig. 6.
Ionospheric current, absorption of cosmic noise and X-ray variations at College.   --After Barcus and Brown.

Fig. 7.
Micropulsation storm, bremsstrahlung X-ray activity, ionospheric absorption and horizontal magnetic force at College. --After Campbell and Matsushita.

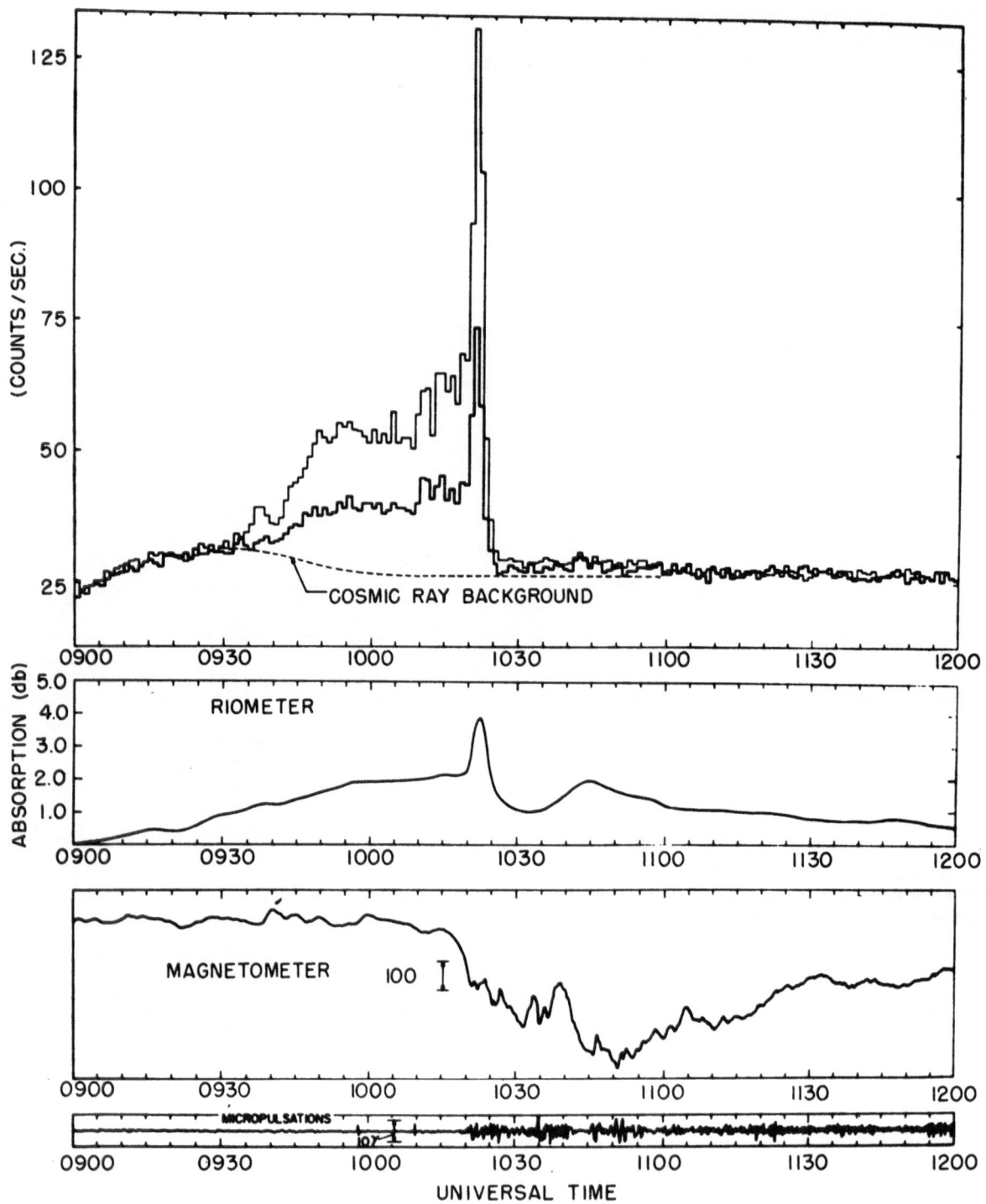

Fig. 8. Records from Macquarie Island of counting rates in balloon
and ionospheric absorption.   --After Anger et al.

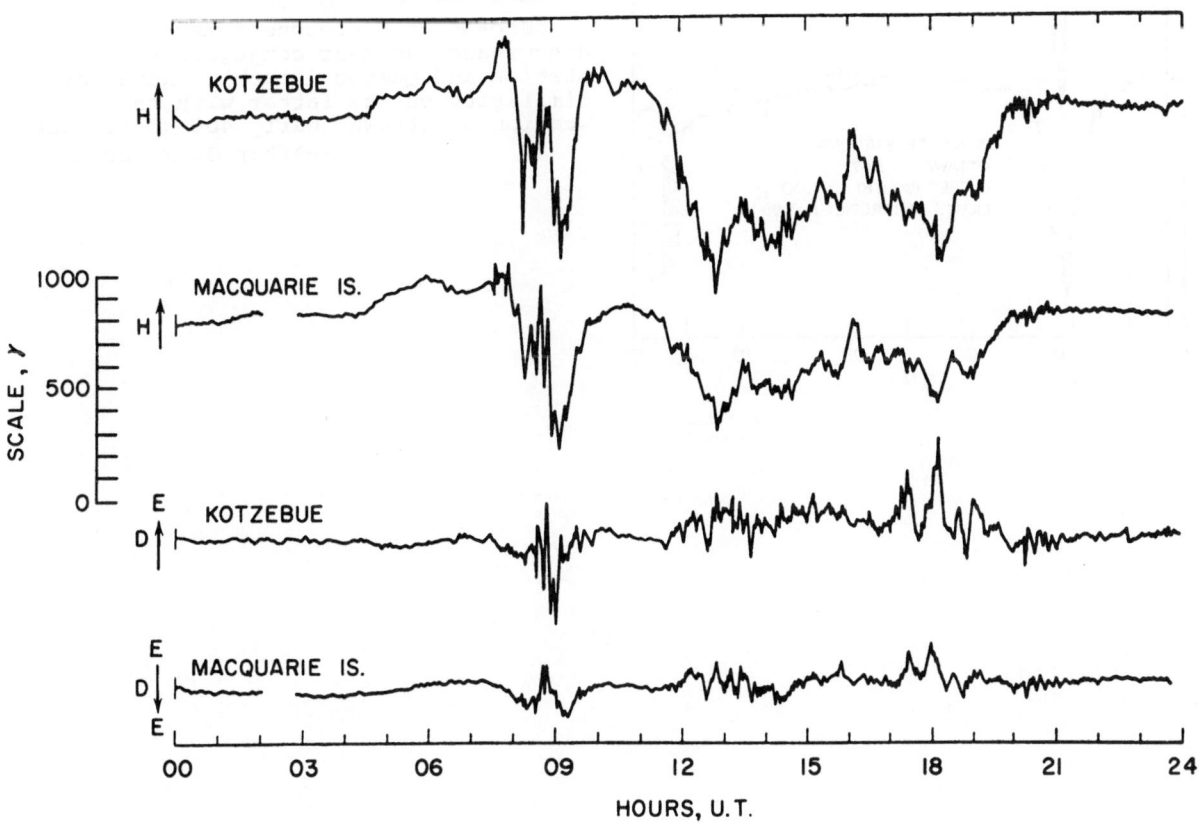

Fig. 9.   Record showing the variation of the magnetic
          elements H and D at the near conjugate stations
          Kotzebue and Macquarie Islands.
                              --After E. M. Westcott.

297

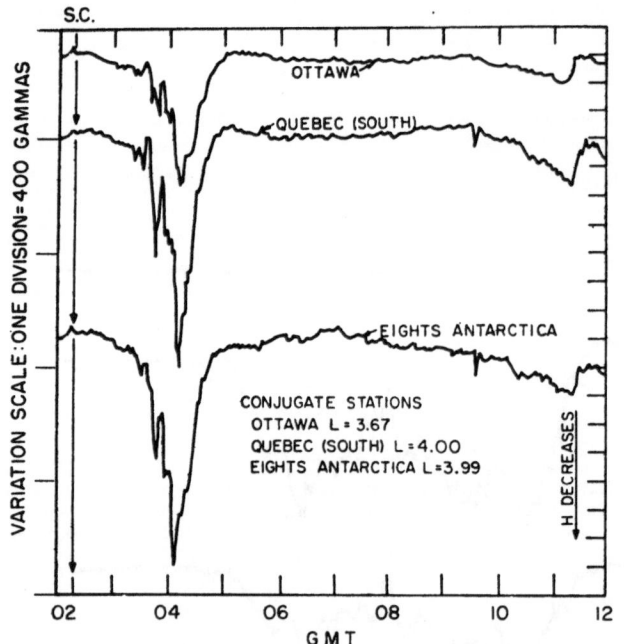

Fig. 10.
Records of H component magnetic disturbance at near conjugate Eights Station and Quebec (south). Note the similarity of the latter with the station at Ottawa nearly 400 km distant.
--After G. M. Boyd.

Fig. 11.
Comparison of cosmic noise absorption at Churchill and square root of the 9 to 31 Mev. proton flux recorded in Explorer XII. SC marks the time of the sudden commencement.
--After Bryant et al.

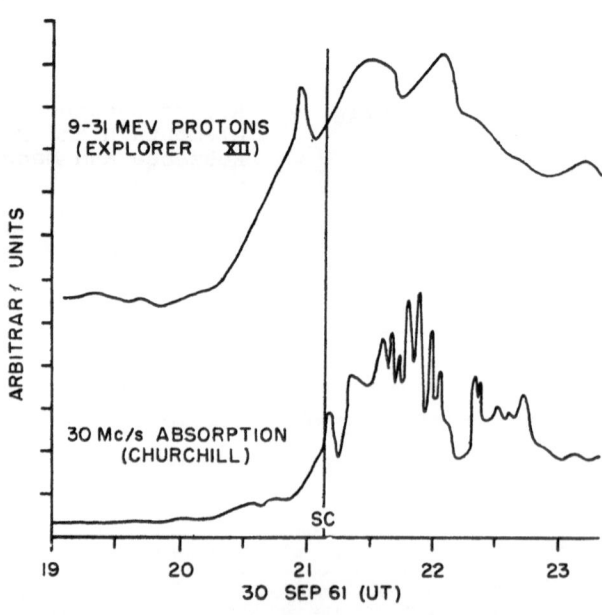

298

EXPERIMENTAL EQUIPMENT AND RESULTS

This inquiry was, and is, a joint operation between Stanford University, the Defense Research Board of Canada and their Pacific Naval Laboratory at Esquimalt, B. C. The original intention was to close the frequency gap between the VLF whistler activities which were known to follow the field lines and the micropulsations which were expected to do the same. "Conjugate" stations were established at Byrd Station (Antarctica) and Great Whale River (Quebec) on the eastern shore of Hudson Bay and, for a short time also, at Fort Churchill on its western shore. All three stations were in, or close to, an auroral zone. Three preliminary reports have been issued (Lokken et al, loc. cit.). The equipments were first set up and records made at all three stations during three weeks in January 1961. Since mid-April 1961, except for January to March 1962, the micropulsation equipment has been operated most efficiently by the over-wintering staff of Stanford University, while Great Whale has been operated fully during specially chosen intervals. This system of measurement will be continued. We are greatly indebted to Keith Marks, to Ward Helms, and now to Henry Morozumi for the efficient operation of Byrd Station equipment during the rest of 1961, in 1962, and in 1963. In the future, at other than the special intervals, data from Great Whale will usually be confined to Esterline-Angus records at slow speed of traverse, chiefly to give statistical data. Accuracy of timing from these records will thus be achieved only during some of the special World Intervals chosen for IQSY.

The micropulsation equipment installed at the three stations each employed three detectors with 35 strips of Telcon 79 Mumetal six feet long, 0.015 inch thick and 3/4 inch wide, lightly buried and joined by separate leads ≥100 feet long, to the chopper amplifier systems at the recording site (English et al, 1961). Two of the detectors were laid geographic NS and EW (X and Y) while the third was buried vertically. This has been our standard procedure for some years. The detectors are wound with 20732 turns of No. 2 Formex type copper wire with windings graduated so that there are fewer layers toward the ends. A relatively new departure was the addition of a large horizontal (V) loop of a few low resistance turns (area x turns variable, but about 5×10^8 sq. cm x turns). This was designed to enable records to be made on magnetic tape to cover the first few modes of the Schumann-ELF bands. Records of X, Y, Z and V were made on FM magnetic tape at 3-3/4 inches/second at all stations at pre-selected times. The X, Y and Z records during a large part of 1961 were also recorded continuously on Esterline-Angus recorders and the V record at the same speed of traverse after rectification and smoothing. The V records on magnetic tape, filtered to accept the nominal range of 2 to 30 cps recorded not only the first few Schumann bands, but also the frequencies below 1 cps which showed the micropulsation events very well on the rectified and smoothed paper records as well as directly on the magnetic tape. The higher frequency components so recorded accompany the sudden -H changes. The differences in the forms of the event which appear on the paper records can be seen in Fig. 12. Fig. 5 of Lokken et al (1962) shows the form assumed on the V paper record.

At 3/4 inch/minute the paper records should enable the time of start of an event, which occurs sufficiently suddenly, to be measured within about two seconds. We have not found that this was possible, the most likely reason being that the record is subject to a noise background, as well as the relatively large variations of longer period, measured in minutes and large fractions of a minute. However, the timing of an event can be improved by filtering out these longer periods and concentrating on the higher frequency components.

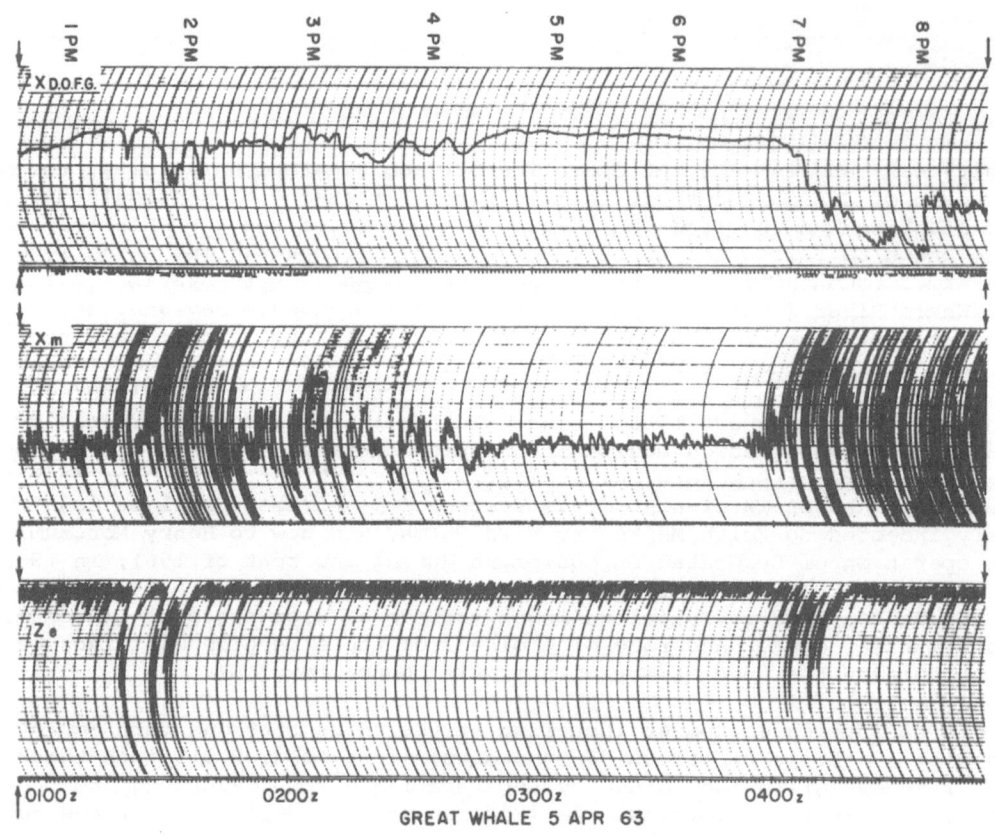

Fig. 12.

Illustrating the Esterline-Angus records of fluxgate magnetometer at Great Whale (top), the broad band X record, and the X and Z records in the nominal frequency range of 2 to 30 cps, the two at bottom of the figure. The onset of the higher frequency components occurs when the fluxgate X record decreases sharply. These components can also be seen on the original broad band X record as well as the slower movements which can be traced also on the fluxgate record.

EXPERIMENTAL DATA

Micropulsation and Magnetic Records.

We have seen from the observations of Birkeland and Heppner that the sudden commencements of magnetic storms are associated with sudden changes of the auroral form and activity and that these occur coincidentally within the limits of accuracy of timing the records. As might be expected, the study of the micropulsation records by the induction method, which records rate of change of the small fluctuations in the X, Y and Z directions in the two ranges of about .005 to 3 cps and, nominally, from 2 to 30 cps in the vertical direction [X, Y and Z in the Schumann-ELF frequency range of 2 to 40 cps since January 1962] shows these sudden changes and enables the time of their occurrence usually to be determined within about half a minute. These correspond to the sudden commencements of Birkeland's "polar elementary storms" which may reach high amplitudes in the auroral zones, but decrease rapidly towards lower magnetic latitudes. As already stated, these storms form a series of bays (or DP's) which are thus usually introduced by sudden changes in micropulsation activity, that is by micropulsation

300

"events."[1] These events occur, on the average, more than once a day so that they can be examined statistically as well as individually. The events show up at the sudden commencement of a magnetic bay (Fig. 13 at 0600), at the commencement of micropulsation activity (Fig. 14 and 15 at about 0352), and of auroral activity with breakup of the enhanced auroral arc (Fig. 16 at about 0628). The duration of the sudden negative change in H is usually less than eight minutes and seems to be the duration of

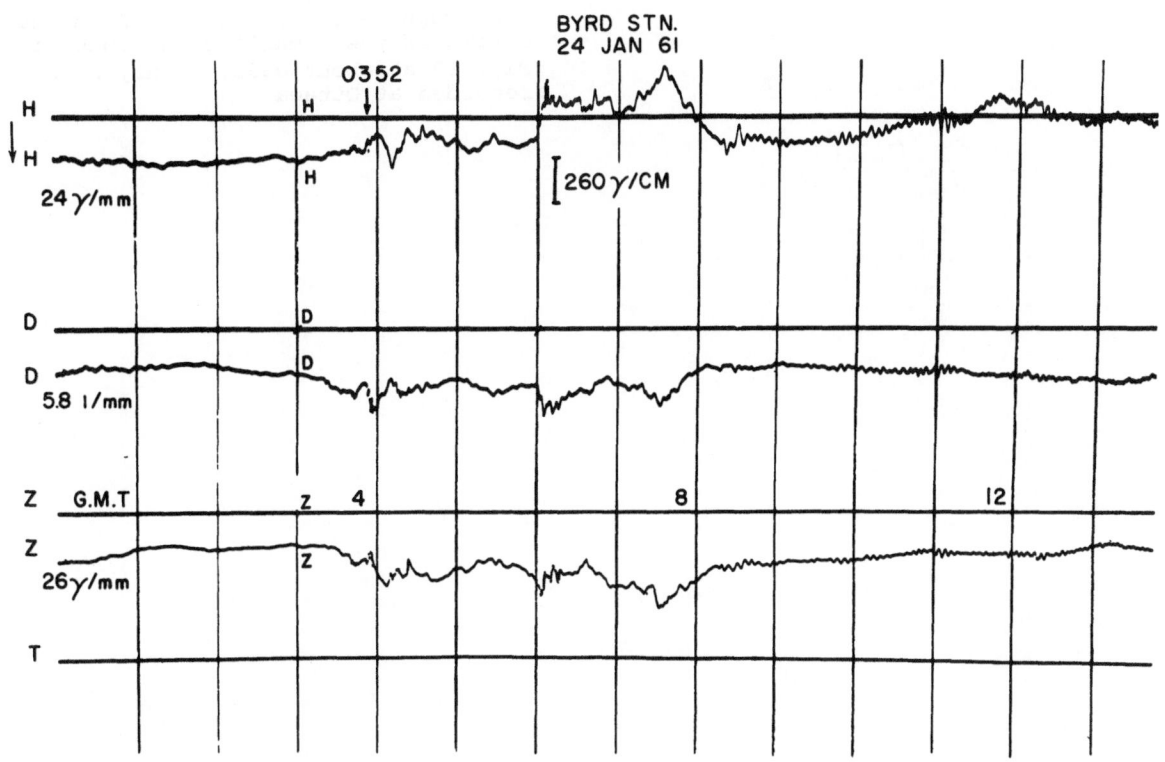

Fig. 13.
     Record of H, D and Z on the slow-speed magnetometer at Byrd Station. There is a large positive H event at 0600 G.M.T. and a much smaller but more sudden negative event at about 0350. This is, however, quite a large event on the micropulsation gear (Fig. 14 and 15) due, presumably, to the fact that micropulsation equipment measures rates of change. Note the relative smoothness of the records for some time prior to 0600 and prior to 0810 when the regular, long period daytime variations take over from the impulsive fluctuations between 0600 and 0800 G.M.T. These quiet periods, introductory to the events, often show more clearly on the micropulsation records. Compare Fig. 14.

the higher frequency components at the beginning of the micropulsation activity. The occurrence of a micropulsation event is easily recognized and the usual time of most frequent occurrence was noted after only three days operation in January 1961. To some extent, this is due to its occurrence after a relatively quiet interval. Fig. 14

---

[1]The word "event" is used as a common description, since other ionospheric phenomena also experience sudden changes at the same time. Thus, we refer to magnetic events, micropulsation events, auroral events, and so on.

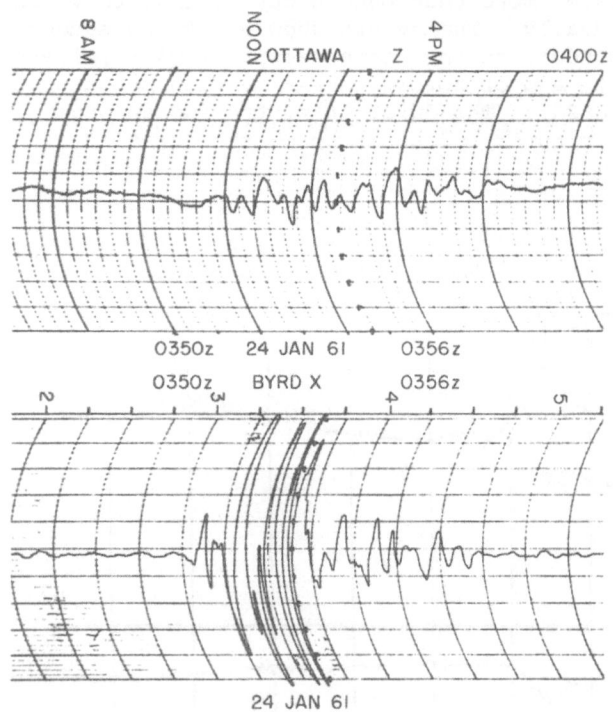

Fig. 14.
Micropulsation records at Byrd and Ottawa of the "small" event shown in Fig. 13 at about 0352. Only Z was recorded at Ottawa.

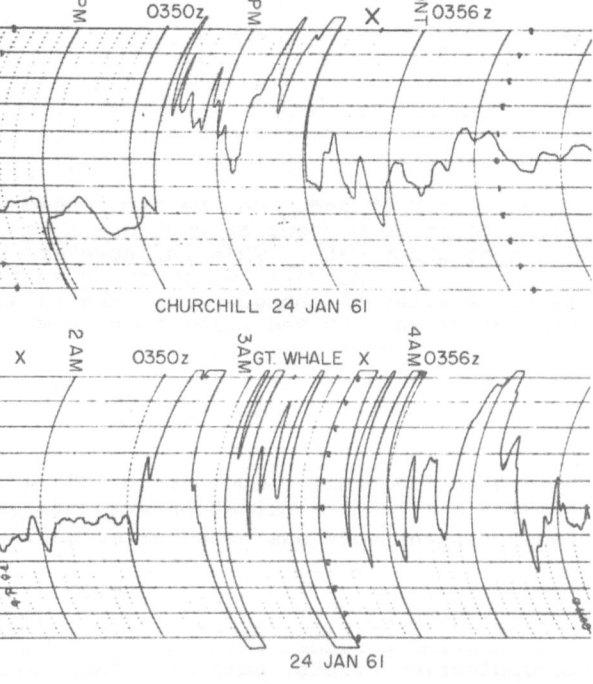

Fig. 15.
Similar X micropulsation records taken at Churchill and Great Whale at the same time. In this and the preceding figure, the reduction in amplitude following the initial burst is due to increased attenuation set by the operator in attendance.

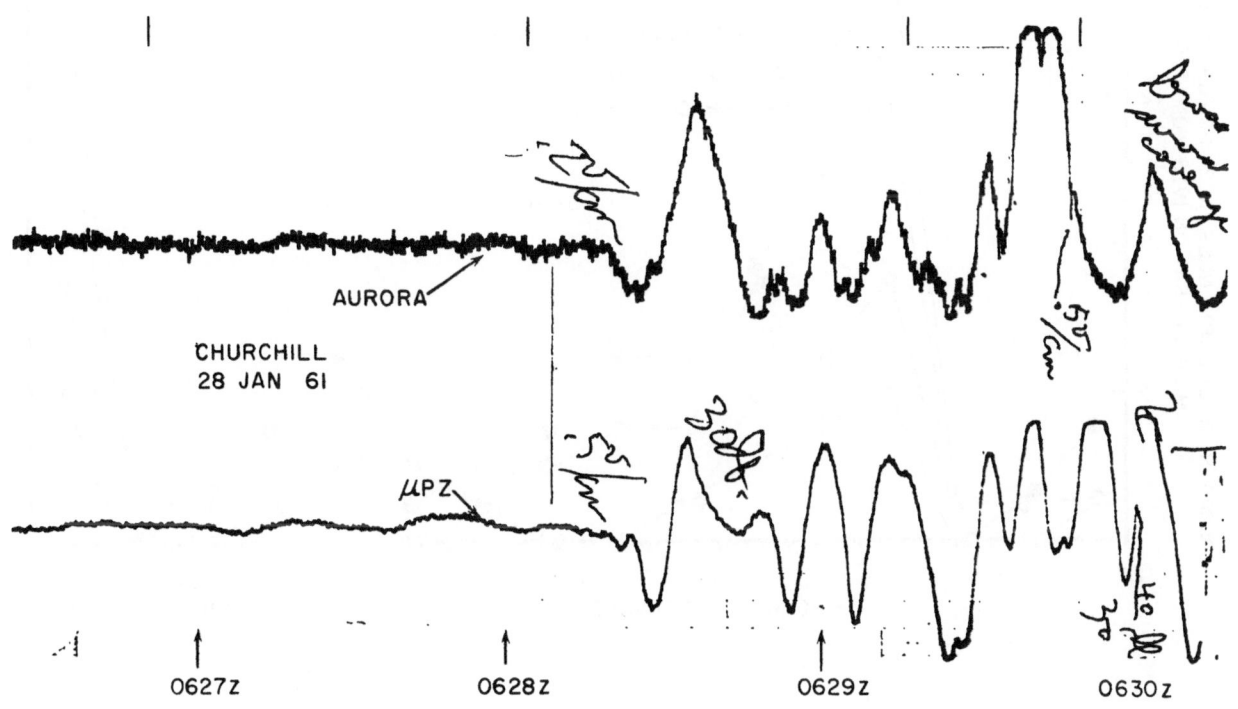

Fig. 16.   An event between 0628 and 0629 on January 28, which
           appeared simultaneously at Churchill on Z micropulsa-
           tion record (below) and narrow band, zenith directed,
           auroral recorder. The fluctuations do not long remain
           in phase.

shows this quiet interval prior to the burst of micropulsation activity which is also
frequently matched by a quiet interval on the magnetograph record, preceding the large
H change at 0600 hours in Fig. 13.

    Fig. 17 shows the statistical occurrence of the micropulsation events during
part of January 1961 when all three auroral stations were operating. This has a
similar characteristic shape at each station, the number of events rising steeply
after a quiet interval of a few hours to a sharp peak in the two hours between 0200
and 0400 hours G.M.T. and a subsequent more gradual reduction in the number of events.
The plot for January 1961 seems to be of unusual shape since Fig. 18 and 19 for most
of the year and for the period from mid-April to mid-May are of different shape at
Byrd. However, the time of maximum occurrence is about the same and is preceded by
the quiet interval. [January 1961 was a quiet month magnetically.] Fig. 18 and 19
show also the number of occurrences of -ΔH events on the magnetograph records at Byrd.
All these events were counted independently, i.e., without reference to other records
and exhibits a clear statistical similarity of the plots of the two phenomena. This
is not, of course, surprising since each -ΔH magnetic event should, and usually does,
have its counterpart in a micropulsation event. The reverse need not be true always,
and judgment plays some part in deciding whether an event is one which is worthy of
inclusion in the count. One should also emphasize that Fig. 13, which shows part of
the Byrd magnetograph for January 24, 1961, has been specially chosen because it

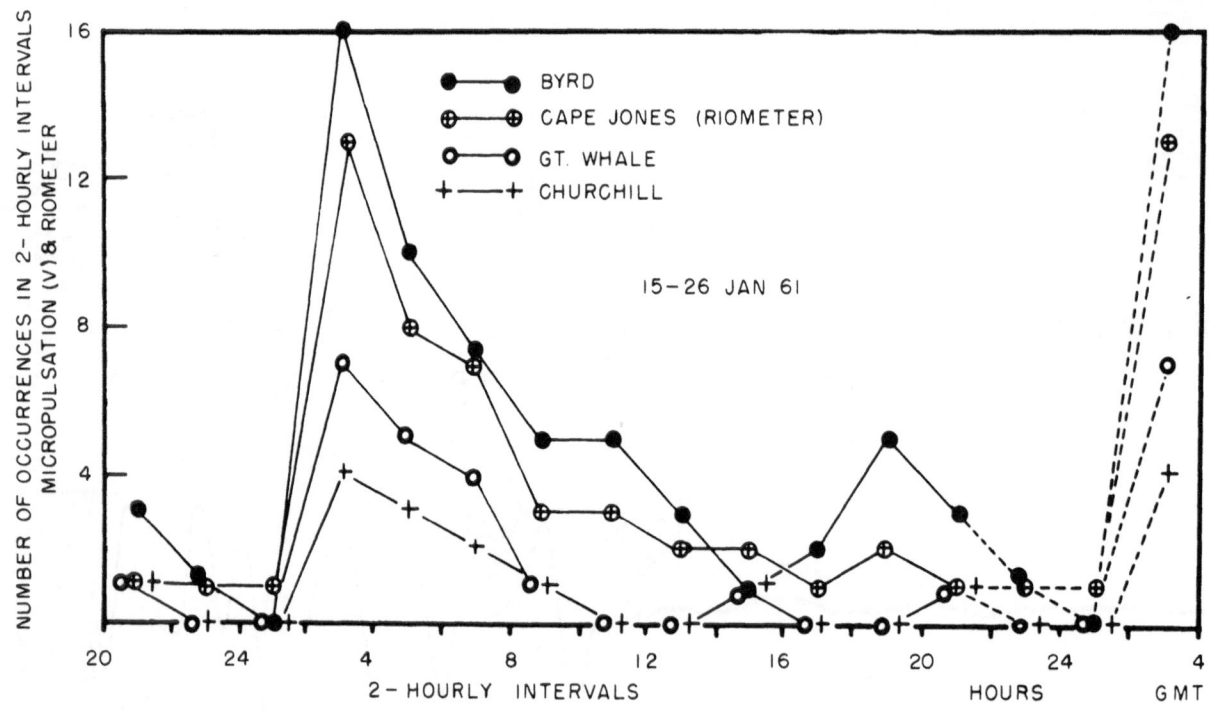

Fig. 17.

Independent count of V micropulsation events shown at Churchill, Great Whales and Byrd, as number of occurrences in two-hourly intervals during the Greenwich day. The similarity of shape at all three stations for this short interval is striking, as is also the quiet interval which precedes the events, on the average. The number of simultaneous riometer events at Cape Jones, for the same interval, is also shown.

displays two events of greatly differing size - small at about 0350 [The amplitude change is small, but the rate of change seems to be large.], and large at about 0600 hours G.M.T. For the small one, Fig. 14 and 15 show the corresponding micropulsation event at Byrd, Great Whale, Churchill and even at Ottawa. The reduced amplitude shown on the first three of these records after the first two or three minutes is the result of increasing the attenuation of the chopper-amplifiers in an attempt to keep the records on the chart.

As to the association between Byrd and its near-conjugate Great Whale, the available coincident data do not lend themselves usefully to a display of the coincidence of individual micropulsation at the two places. They occurred on the average more than once a day.

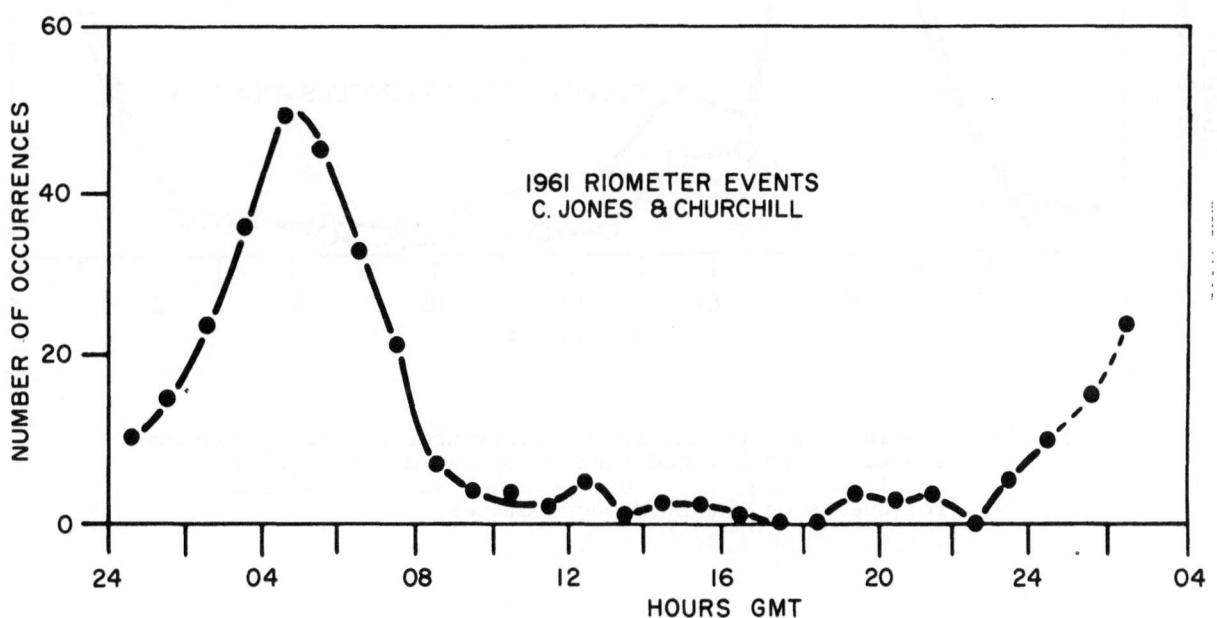

Fig. 18. Similar plot to Fig. 17, but using all available
data taken in 1961 on the V records and on Byrd
magnetograms. Note the persistence of the quiet
period preceding the events at 0400 in this, in
the two preceding Figures, and in Fig. 19.

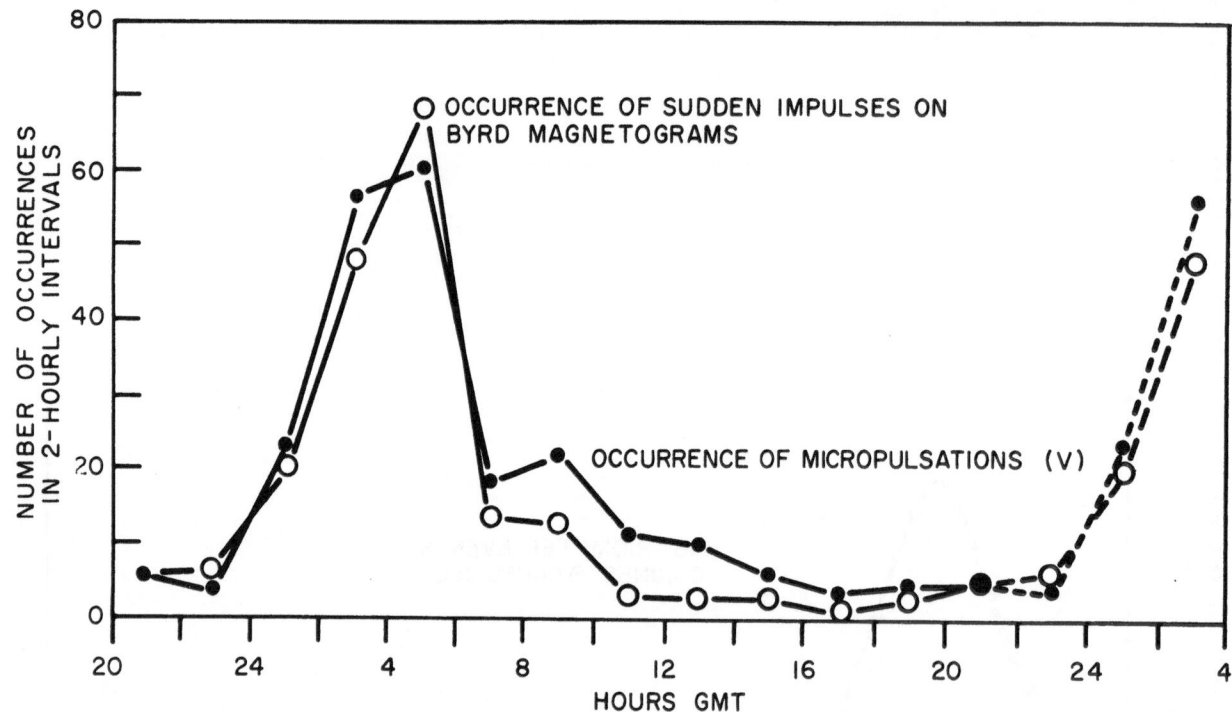

Fig. 19.   Similar plot to Fig. 18 of occurrences of micropulsation
events on the X broad-band equipment during the period
April 14 to May 12, 1961, together with the similar plot
for events on the Byrd magnetometer.

Micropulsations and Auroral Activity.

No provision was made at Great Whale for recording aurora, though Churchill and
Byrd Station were so implemented. Some auroral records were made at Churchill in
January 1961, for comparison side by side with micropulsations on the same record and
it was seen that auroral events showed simultaneously with the micropulsation events.
Fig. 16 showed such a pair of events at about 0628. It will be seen that the times
of occurrence are closely in agreement and that the two records do not long remain in
phase with one another. Table 1 refers to simultaneous auroral and micropulsation
events at the approximate times of events recorded on January 28.

It is somewhat surprising that the magnetometers at Churchill and Byrd showed no
effect at any of these times, due presumably to insufficient sensitivity of the magnetometer equipment. A narrow, $\pm 30°$ beam, photometric equipment and filter for the
3914 line directed to the zenith was used at Churchill.

Other micropulsation and auroral events were recorded at Churchill, including
the small event about 0352 on January 24 which appears on some of the micropulsation
figures relating to other stations (Fig. 14 and 15) and also on the Byrd magnetogram
(Fig. 13). This event occurred during one of the intervals chosen for recording on
magnetic tape, when one would expect a closer agreement in time than can be read from

TABLE 1

Auroral Events Churchill Jan. 28, 1961	Micropulsations Churchill	Micropulsations Great Whale	Micropulsations Byrd
0515 G.M.T.	Yes	Yes	No
0527	Yes	Yes	Yes
0629	Yes	Yes	Yes
0654	Yes	Yes	Yes
0723	Yes	No	Yes
0817	No	No	Yes

The micropulsation stations at Albert Head and Royal Roads, within a few miles of Victoria, were not in operation at these times.

the paper records, the total spread in time on paper records at the different stations being about one minute. To some extent, one can blame clock errors and differences between the recording instruments, but the most likely causes of timing uncertainty are probably the background of noise, effect of ground conductivity and insufficiently-sudden start of the events. Examination of the tape records shows a difference of nearly one-half minute between Byrd and Great Whale, the Byrd event occurring later.

A previous taped event only ten minutes earlier on the same day suggested that the Byrd event was then about 0.3 minute later than at Great Whale, a difference which, if real, is not easily explained. A better timing system is now available and the equipment has been modified better to exhibit the components of higher frequency.

The slow run magnetograms do not permit close timing of magnetic events, but we have found at Great Whale that a fluxgate magnetometer, which is readily displayed on Esterline-Angus, Brush or Sanborn recorders, is useful because it can be recorded at the same speed and side by side with other events restricted by filters to the same frequency band.

These remarks refer to events occurring in January 1961, which took place during the micropulsation night regime. Fig. 20, which refers to the coincident events (V record bursts and auroral flare-ups) at Byrd Station during the winter of 1962, shows clearly that the coincidences occur during the night regime and most frequently just before 0400 hours G.M.T. This figure gives the coincident count of events by Ward Helms (micropulsation) and John Turtle (aurora) at Byrd Station during the winter, to whom I am indebted for permission to use this figure.

The situation is quite different during the day regime of quiet auroral activity, seemingly of quite different type from the active aurora of the night regime which seems to be associated with the influx of electrons whereas there may well be some association during the day regime with an influx of protons. However this may be, the characteristic association in the day regime seems to be of the type shown in Fig. 3. I am informed by John Turtle that evidence of such simultaneous light intensity auroral and micropulsation records of regular type and of essentially the same period, occurred frequently during the winter at Byrd, but that usually the daytime

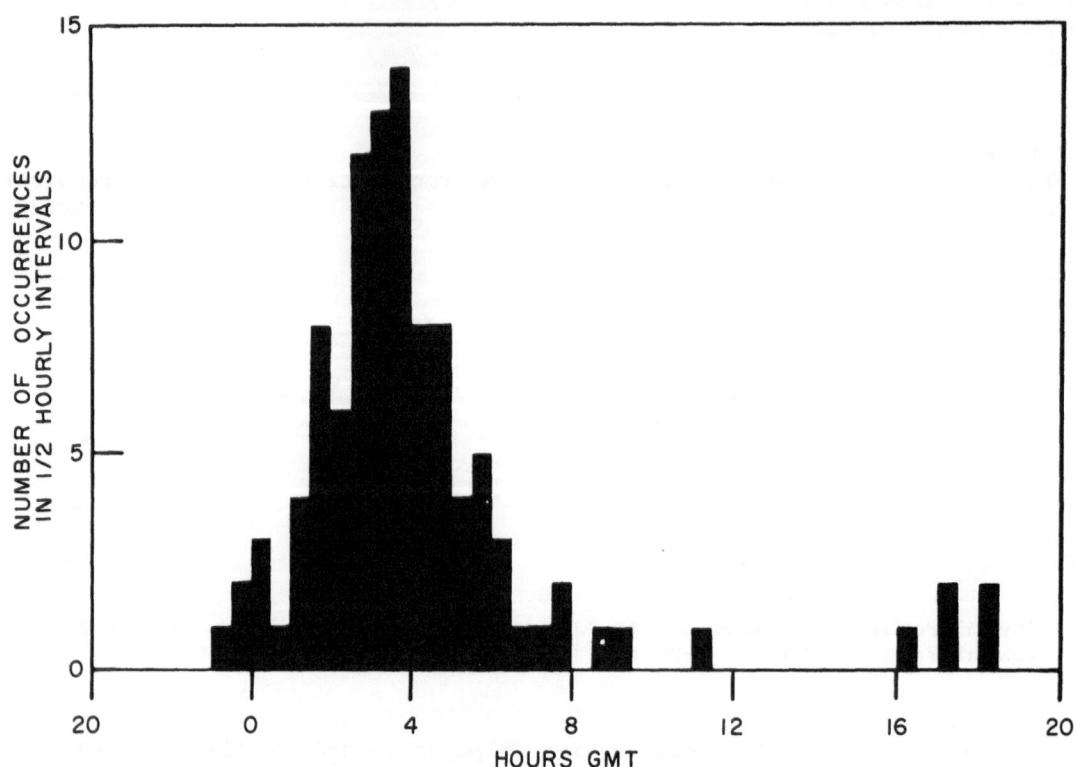

Fig. 20. This figure illustrates the diurnal variations of the
number of occurrences during the dark months at Byrd
of coincident V micropulsation events and auroral
events of the type shown in Fig. 16.

auroral intensity was insufficient to show a satisfactory record. When the amplitude
was sufficiently large, one could readily see that the period remained nearly constant and that the phase difference between auroral intensity and micropulsations
changed quite slowly in contradistinction to the quick changes in phase difference of
the nighttime auroral and micropulsation events. Records made at Byrd by Ward Helms
and John Turtle show that the day regime association could last for many minutes.

Ward Helms has further reported from Byrd Station that a one-to-one correlation
exists between micropulsation (V) and auroral events, but that a strong overhead
aurora is not always accompanied by a micropulsation event.

We have not yet studied intensively the day regime of micropulsations or the magnetic elements. One can only repeat that long period regular micropulsations and
magnetic fluctuations of three- to ten-minute periods are very common in these quiet
(magnetic) years during the day regime and in the auroral zones. Thus, they conform
as regards regularity but not period with the regular daytime micropulsations of
about 15- to 30-second periods (Pc's) which govern the daytime activity in mid-
latitudes. A relevant incident was reported by Harang (1939) of the simultaneous
occurrence of micropulsations and of radar echoes of the same 116-second period,
presumably from aurora at a great height or as variations impressed on the radar beam
reflected from the earth at a great distance.

It seems to be clear that the difference between the day and the night regimes of the auroral are as great as, and similar to, the differences between the micropulsation day and night regimes.

MICROPULSATION AND OTHER IONOSPHERIC PHENOMENA

During 1961, no riometer equipment existed at Byrd or at Great Whale for measurement of ionospheric absorption, but this is being done in 1963 as part of the continuing auroral zone experiments. To get some information about the suitability of Cape Jones, on Hudson Bay, about 180 km S.W. of Great Whale, as a possible conjugate to Byrd, Dr. R. Montalbetti kindly loaned copies of his Cape Jones riometer records for that part of January 1961 when the P.N.L. party was recording at Byrd.

A very quick comparison of the micropulsation (V) and Schumann-ELF records (2 to 30 cps, nominally) at Byrd with D.R.T.E.s rectified and smoothed riometer record at Cape Jones showed that a large number of the sudden brief micropulsation events at Byrd could be recognized almost simultaneously on the Cape Jones riometer records. Fig. 21 shows the excursion at 0358 of January 24, which appeared in Fig. 13, 14 and

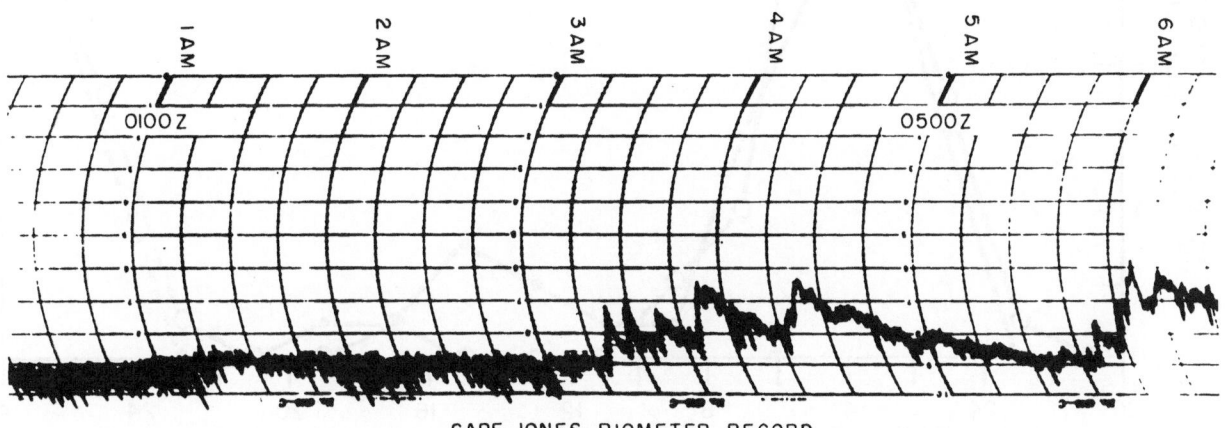

CAPE JONES, RIOMETER RECORD
24 JAN 61

Fig. 21.   Riometer record at Cape Jones showing, among others, the
event at about 0352 on January 24. This event also appears
on Fig. 13, 14 and 15.

15. The diurnal counts of the riometer events at Cape Jones (Fig. 16) is included for the short period of overlapping records in January 1961 and shows a surprisingly good statistical agreement with the micropulsation records. The riometer events were in this case counted <u>during comparison</u> with the micropulsation records and were therefore not chosen independently. More recent records from Byrd show such close individual agreement in time with the riometer records at Cape Jones (Fig. 22) as to justify much closer examination and this will be the subject of a separate inquiry with D.R.T.E. A large number of riometers are already installed and working in high and mid-latitudes so that, if a more open time scale and good time control could be adopted, it should be possible to settle some questions about the relative importance

of local and geomagnetic (or auroral) time and in examining the changing form of an event recorded at a number of well separated stations.

We have no information about the difference between day and night regimes of riometer events, but one may note the remark (Montbriand et al, 19) about the differ ence in duration of the two absorption types which they consider "consistent with the assumption that a different mechanism is responsible."

There is at least the hint of a similar difference in duration of hiss events (date kindly supplied from the South Pole Station in 1960 by Henry Morozumi). Also, the early reports from Churchill and Great Whale made it clear that there was a real and nearly time-coincident association between audible auroral hiss and micropulsation events at the same station. The recorded data of hiss in various frequency bands are

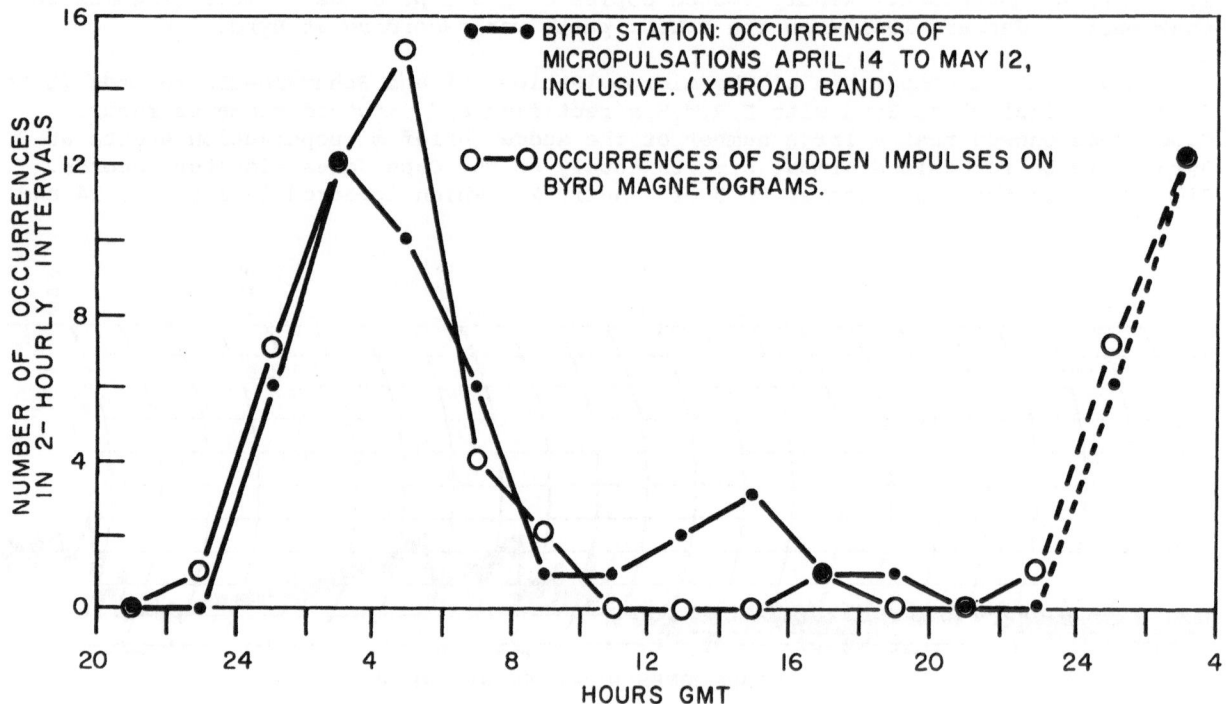

Fig. 22. 1961 Riometer events at Cape Jones or Churchill
corresponding to micropulsation events at Byrd.

being examined in detail at Byrd Station by the team under Henry Morozumi and some information on the association of hiss, aurora and magnetic activity at the Pole Station has now been published (Morozumi, 1963). The data recorded at Byrd Station by Ward Helms and John Turtle on micropulsations, hiss, aurora and magnetic K indices are now being prepared by them for publication. As might be expected, the statistical relationships of hiss activity with aurora and micropulsations are clearly indicated and there is a similarity in the diurnal occurrence of hiss and the occurrence of micropulsations (of the shape shown in Fig. 18, for example). However, there is an indication that the night maximum of hiss activity precedes the maximum of auroral activity. This is shown by examination of individual events which suggests that, sometimes, the hiss activity is suppressed by ionospheric absorption associated with the aurora as suggested by Martin et al (1960).

Clearly, there is scope for much more work on the inter-relationships of ionospheric phenomena in and near the auroral zones.

SUMMARY

Based on the coincidence in time of occurrence of ionospheric "events," the weight of evidence, though scanty, suggests that conjugate points are not, in the auroral zones, maintained from day to day and week to week. On some days, the two near-conjugate points of Byrd and Great Whale seemed to be more nearly conjugate than on other days. This is not surprising when one considers the movement of the auroral arcs during the night and the changes of position of the aurora with magnetic activity. Presumably, we can only speak of conjugate auroral zones within which stations may be conjugate, the criterion of conjugacy being the similarity of coincident magnetic and micropulsation records and of records of auroral, ionospheric absorption and other ionospheric phenomena at the two conjugate stations.

In the auroral zones, the micropulsation and other related events are identified with a magnetic event - the sudden commencements of Birkeland's Polar Elementary Storms or the DP's of Akasofu and Chapman. While they are known to be of relatively great intensity in the auroral zones, the amplitude decreases rapidly toward lower magnetic latitudes. These events appear most frequently and of greatest amplitude at night and correspond to a sudden negative change in the general level of horizontal magnetic force ($-\Delta H$).

At each of the three stations there is a similar statistical variation, plotted against universal time, which is common to the events on magnetographs, micropulsation records, aurora, ionospheric absorption and hiss. These events occur most frequently in the two-hourly interval preceding 0400 hours G.M.T.

This statistical similarity occurs because there is a strong tendency toward coincidental occurrence of individual ionospheric events at the observing stations - for example, coincidence of magnetograph and micropulsation events. Individual events at the two near-conjugate stations of Great Whale and Byrd thus occur usually at the same time, less frequently between Churchill and another station, while coincident events at all three stations rarely occur. It is as if the same event occurs at any pair of stations either simultaneously, or not at all.

Statistically, there is a quiet interval which immediately precedes the magnetic and micropulsation events. Individual events are also usually preceded by a quiet interval also. The reason for this is not clear, but may be of some importance.

These events are detected also in middle latitudes on sensitive micropulsation equipment. They can often be found simultaneously on the normal slow speed magnetographs when the time of occurrence is known from other sources of information. This should be of interest in view of the large number of magnetic observatories which operate in these northern latitudes. Thus, the possibility exists of examining events on a world-wide basis. Evidence already available (Kato and Saito, loc. cit.) and private communications with Saito suggests there is a difference between the frequency content of the simultaneous events on the day and night sides of the earth. This seems to fit in with the Axford-Hines picture of the more turbulent conditions in the earth's tail on the night side of the globe.

The difference between the day and night regimes of the events with which we are dealing is very clear in the auroral zones as also in mid-latitudes. In quiet conditions, it is the difference between the regular and relatively continuous daytime signal background (of Pc type) and the impulsive, irregular, nighttime signals (of Pt type). Both of these can be seen on the magnetograph in Fig. 13, the long period and relatively continuous daytime signals are those to which Jacobs gave the name long Pc's (LPc) which are probably allied to the giant pulsations of the auroral zones.

GENERAL REMARKS

The experimental results which have been reported in the last section are summarized later, but some comment seems to be desirable on the global aspects of the problems.

The "Events" and Their Relationship to Conjugacy.

The concentration upon the occurrence and timing of the sudden short events results from the feeling that conjugate relationships in the auroral zones must be associated with propagation of similar signals along or directed by the field line joining two stations, the effects at the two ends of the line being nearly simultaneous. Such events do occur more than once a day on the average and the event can be timed with reasonable accuracy because of its suddenness. The disturbances (bays or DP's) lasting sometimes for hours, which follow the events have not yet been examined, nor has the day regime, of regular and relatively continuous magnetic LPc's, which does not lend itself to a similar method of examination. Thus, statistical data on the variation of the general activity during 24 hours is lacking at present.

The more powerful events corresponding to the negative H changes of the magnetic field, which occur at night, are larger than those coincident with positive H changes in the daytime. The former include much higher frequencies approaching 1 cps and we can reasonably regard the irregularity of the night regime as associated with the turbulent area of the Axford-Hines model (Axford and Hines, 1961), which appears on the side of the earth remote from the sun, in contrast to the regular and more continuous pattern of activity facing the sun on days of low activity of the solar wind. If this picture is correct, the statistical diurnal variation of (say) the micropulsations could be little more than an expression of the circumstance that there are characteristic differences between the results at stations facing and remote from the sun. But what does require more investigation is why there are simultaneous changes occurring suddenly in both auroral zones, especially in the two-hourly period from 0200 to 0400 hours G.M.T. Why this time, in fact? We can hardly blame this occurrence directly on the sun's activity, as we do in the case of sudden commencements of world-wide magnetic storms, since the sudden commencements of sun storms should show no strong preference for this time of night. It is worth noting that the events shown in all the illustrations of Heppner's report (loc. cit.) are clustered round 2300 to 0200 hours, local time. All the examples, he states, are of type ssc, which denotes a sudden commencement followed by a magnetic storm.

The circumstance that magnetic storm activity is often introduced by a sudden commencement at about the same time over the globe and that this is followed by a series of bays (or DP's) which reach their maximum development in the auroral zones, raises the question whether the smaller events which normally precede DP's in the auroral zones are due directly to the action of the solar wind in the magnetosphere, of if they result from some process of dumping particles from an already overloaded reservoir within the magnetosphere. When there is simultaneous conjugate operation in the two zones, this must mean that a common field line is acting to lead or direct the particles in the reservoir toward the two zones at a relatively high speed. The preference of the events for about 0200 to 0400 hours G.M.T. is an indication that they do not depend directly on the activity of the solar wind. Thus, the events probably originate within the magnetosphere and come to light for reasons which seem obscure and at times which may be associated with the geographic longitudes of the geomagnetic (or auroral) poles, as has often been suggested.

Mid-Latitude Effects.

We have known for some time that the events which appear in the auroral zones are simultaneously seen in middle latitudes on sensitive micropulsation equipments. Fig. 23 shows two examples. Our micropulsation gear was run continuously for 3½ months in

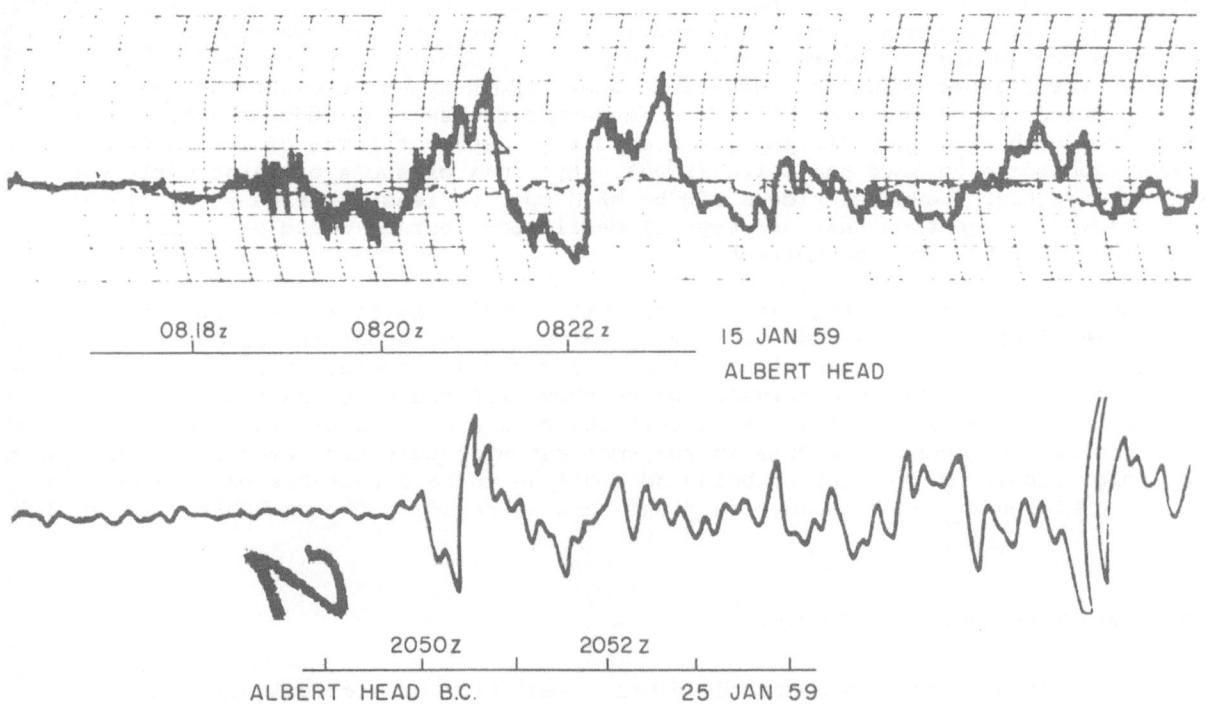

Fig. 23.
Two examples of micropulsation events at Victoria. The top record contains riders of nearly 1 cps. Statistically, it is found that the micropulsation events at Great Whale which contain riders of such high frequency appear also at Victoria with riders also of high frequencies. The maximum of occurrence at the latter station is close to midnight. The bottom record shows the ordinary daytime type astride the long period damped event.

early 1959, for a special purpose, and it was found at that time (Duffus et al, 1959) that the local micropulsation event often included much shorter period components approaching 1 second period (top record of Fig. 23). These occur during the negative ΔH events in the auroral zones which include short period components during the event. The word "simultaneously" means within the accuracy of timing, which is probably no better than ½ minute for a single micropulsation event, but could probably be improved by concentrating on the timing of the shorter period components. The equipment in 1959 was poorly adapted to record these higher frequency components, but the data have been used to display those Victoria micropulsation results which were coincident with ΔH deflections on the magnetographs of Churchill, College, Sitka, Fredericksburg and San Juan. Much more data is needed before drawing firm conclusions but the immediate result of the comparison was to show that the occurrence of events which were (nearly) simultaneous at Victoria and at the other stations were practically confined to the night regime. For the northern observatories these coincident occurrences were usually associated with -ΔH deflections. For Fredericksburg and San Juan, however, they coincided with +ΔH deflections although the events took place also chiefly during the night hours between 0200 and 0700 hours G.M.T. The frequent occurrence of these micropulsation events in Victoria and other mid-latitude stations in North America is considered to be the result of the "leakage" current of the auroral electrojet which fails to close its circuit over the polar cap. The paper by Akasofu and Chapman (loc. cit.) is relevant since it shows that the equatorial electrojet is enhanced during polar sub-storms (DP's) suggesting that a part of the return leakage current spreading to middle and low latitudes may approach the magnetic equator. The link between the auroral and equatorial electrojet should be further examined by more measurements of micropulsations near the magnetic equator, including frequencies at least up to 3 cps. It is possible that the high frequency components occur only with the more impulsive negative ΔH events and are detectable in the northern hemisphere in lower latitudes than the generally less intense positive H events, for reasons which are suggested later. Both types of events are similar damped pulses of period about 1 to 2 minutes and the smooth +ΔH type ("fine-damped"), free from high frequency riders, often occurs in Japan at times of sudden commencements, according to Kato and Saito (1958). At times of ssc's of large world-wide storms, the high frequency riders can be of 3 cps, at least (Duffus et al, 1958). Troitskaya's suggestion that her type II oscillatory bursts would be simultaneous over the world may well be correct.

As pointed out by Duffus et al (loc. cit. 1959) many of the micropulsation events in middle latitudes can also be found on the magnetographs of observatories in North America if it is known when to look for the small perturbations. Study of the magnetographs of the many observatories in these latitudes at the times of the events should be of interest and might show that there is a continuous gradation of activity from the largest magnetic storms to the smallest micropulsation events and the following polar sub-storm, the latter being so small as to be detectable only in or near the auroral zones. Such an inquiry might help in deciding the probable cause of the events.

The Diurnal Variation of Events.

Much has been written about the diurnal variation of micropulsations since Troitskaya's early paper (loc. cit.), which showed the distribution in Russia of her Type I, Continuous, and Type II, Oscillatory Bursts, the latter of which showed simultaneous bursts at widely separated stations and were persistently accompanied by higher frequency oscillations with periods of the order of two to three seconds. Her bursts showed a maximum of occurrence in Russia near midnight, L.M.T., with the maximum of the continuous type about 12 hours away. In this paper, she points out that the largest number of oscillatory bursts (Pt) occur in Russia at the time when the sun is crossing the meridians on which, in the northern hemisphere, the north magnetic pole lies, while continuous (Pc) oscillations are detected when the sun is crossing the meridian of the south magnetic pole. But, of course, each meridian is crossed twice in 24 hours - at magnetic noon and magnetic midnight in the south as well as in the north. Ohchi (1958) examined this Pt type of disturbance at Sitka, Cheltenham and Toledo[2], and found, in agreement with Yanagihara's results on such currents at Kakioka, that the times of occurrences of the maximum were close to 23 to 24 hours L.M.T. Holmberg's records for Eskdalemuir showed a maximum between 1900 and 0200, while Troitskaya's local time was also not far from midnight, as already stated. The paper by Jacobs and Sinno (1960), though it deals with the diurnal occurrence of the continuous Pc type, is of some interest in view of the normal 12-hour difference in time between the maximum occurrence of the two types.

Many authors, including Troitskaya, Heppner, Davis and Nakamura have suggested that geomagnetic time is important to the diurnal variation, while Davis has mentioned the importance of auroral time. The latter view is strongly supported by the Stanford University staff (Neil Brice et al, private communications) in Antarctica, who have stated that the use of auroral time shows a maximum of occurrence at Byrd and South Pole stations of events at the same auroral time. The auroral pole is stated to lie about halfway between the geomagnetic and dip poles. In low and middle latitudes, there is not much to choose between local mean time and geomagnetic (or auroral) time, but the situation is different near and especially within the auroral zones and it is from stations in such areas that one should investigate the factor, or factors, which determine the type of variation with time of day. Comparisons of individual events, not their statistical occurrence, are necessary.

The pronounced difference between the day and night regimes in the auroral zones and the recognition of the more powerful, negative ΔH, micropulsation events at our local station led to an inquiry regarding the higher frequency components, or riders. It was found that there was a pronounced statistical maximum of these frequencies at local midnight, but that there were other events, even at midnight, which in 1959 did not show these components, possibly to some extent because the equipment was not very responsive to these frequencies. This led to a request to Saito at Onagawa Observatory for information about the type of micropulsation records on certain dates and times in January 1959. No definite result has yet come from this inquiry apart from the fact that events were coincident and it is clear that direct comparison of the two sets of records is desirable, as one would like to know with certainty whether the day and night regimes show, for the same events, regular and smooth records on the side facing the sun and irregular (and rough with higher frequencies) on the night side. Both sets of records are, of course, of the Pt type. Possibly the best opportunity for such an inquiry would be afforded by investigating the records of the large number of Russian and European micropulsation and earth current stations which are well distributed in middle latitudes and examining the changes in form of the event with change in longitude[3]. In this connection, Troitskaya seems first to have remarked on the mutually exclusive character (in G.M.T. of occurrence) of the two types Pc and Pt. On a statistical basis these represent our slightly overlapping day and night micropulsation regimes in middle as well as in auroral latitudes. Our positive events are generally less intense than the negative ones in the auroral zones. Thus, if we are measuring the leakage currents in mid-latitudes, we expect the negative events also to be more powerful here, being relatively close to the auroral electrojet at night with which they are associated. The positive daytime events are not only farther, by the diameter of the auroral zone from the auroral electrojet, but also there exists between mid-latitude station and the electrojet the more conducting ionosphere above the polar cap which may be expected to short circuit the local leakage current to an appreciable extent.

---

[2] Earth current records.

[3] The suggestion that ssc's at stations 12 hours removed in the auroral zones should be compared was mentioned in his paper (Heppner, 1955).

FUTURE WORK

While this investigation arose out of the P.N.L. interest in micropulsations and commenced as a joint effort with Stanford University to close the gap between the frequency of whistlers and allied radio activity such as hiss and the frequency of micropulsations and the Schumann bands, with special reference to the results at near-conjugate sites, the inquiry has led us into other interesting bypaths. The work on micropulsations and its association with other ionospheric phenomena must go on, but it seems to be less important and less promising than the study of auroral events and activity. Indeed, the opportunity might be afforded of learning not only the source(s) and energy of the auroral particles appropriate to the day and night regimes but a great deal more about the position of the aurora, at the times of the events, in varying magnetic conditions. This might well involve direct visual observation in auroral zones as a preliminary to the construction of new instruments to display statistically the most important features. The occurrence of Doppler-displaced hydrogen lines during the quiet arcs of the daytime should be investigated more fully, also the daytime micropulsation regime, both from the point of view of the incidence of positive H events and of the association of activity with LPC occurrence and with the quiet daytime auroras.

As usual, the result of recent work has solved fewer old problems than the number of new problems uncovered. These new problems seem to demand examination of ionospheric phenomena on a global basis, and this will need the fullest cooperation between interested parties in all countries.

ACKNOWLEDGEMENTS

I am indebted to the Commanding Officers and Officers-in-Charge at the Stations at Fort Churchill, Great Whale River and Byrd Station for their interest and assistance in this investigation. I also wish to thank Mr. D. J. Evans, who was responsible for the design and manufacture of the micropulsation equipment and for the erection and operation of the gear at Byrd in January 1961, and also Keith Marks, who kept the instruments operating efficiently during the rest of 1961.

During 1962, Ward Helms carried on and supplemented the work of Keith Marks. He and John Turtle contributed the data relative to the incidence of aurora and hiss during the winter at Byrd Station.

The investigation was made possible by the support of the National Science Foundation and with the help of Prof. R. A. Helliwell and his staff from Stanford University. This was supplemented by information supplied by other investigators at Byrd – from Coast and Geodetic Survey and Cambridge Air Force Research Center. The investigation, especially at the stations in the northern auroral zone, was also supported by the Defence Research Board of Canada under Project D 45-95-11-39.

REFERENCES

Akasofu, S., and S. Chapman: J. Geophys. Res. 68, 9 (1963)

Angenheister, M.: Terr. Mag. I, 55 (1896)

Anger et al: J. Geophys. Res. 68, 4 (1963)

Arendt, T.: Das Wetter 12, 276 (1896)

Axford, W. I., and C. O. Hines: Can. J. Phys. 39, 10 (1961)

Axford, W. I., and G. C. Reid: J. Geophys. Res. 68, 7 (1963)

Barcus, J. R., and R. R. Brown: J. Geophys. Res. 67, 7 (1962)

Benioff, H.: J. Geophys. Res. 65, 5 (1960)

Birkeland, K.: The Norwegian Polaris Expedition, 1902-1903, Vol. I, Sects. 1 and 2
          Longmans, Green and Co., London and New York

Boyd, G. M.: J. Geophys. Res. 68, 4 (1963)

Campbell, W. H.: J. Res. Nat. Bur. Stand. 64D, No. 4 (1960)

Campbell, W. H., and S. Matsushita: J. Geophys. Res. 67, 20 (1962)

Campbell, W. H., and M. H. Rees: J. Geophys. Res. 66, 1 (1961)

Davis, T. N.: J. Geophys. Res. 67, 1 (1962) (two papers)

Duffus et al: Nature 181, (1958)

          Nature 183, (1959)

English, W. N.: Inst. Soc. of America (preprint) (June 1961)

Harang, L.: Terr. Mag. 44, 11 (1939)

Heppner, J. P.: J. Geophys. Res. 60, 1 (1955)

          D.R. 135, Defence Research of Canada (1958)

Holmberg, E. R. R.: Mon. Nat. Roy. Ast. Soc. Geophys. Supp. 6, No. 8 (1953)

Jacobs, J. A., and K. Sinno: Geophys. J. Roy. Ast. Soc. 3, 3 (1960)

Kato, Y., and T. Saito: Sci. Rep. Tohôku Univ., Series 5, Geophys. 9, No. 3 (1962)

Lokken, J. E. et al: Nature 192, 4800 (1961)

          Can. J. Phys. 40, (1962)

          J. Geophys. Res. 69, 3 (1963)

Martin, L. H., R. A. Helliwell, and K. R. Marks: Nature 187, 751 (1960)

Montbriand et al: E.R.T.E. <u>1095</u>, 12 (19)

Morozumi, H. M.: I. G. Bull. Nat. Acad. Sci. <u>72</u>, (1963)

Nakamura, J.: T78J Defence Research Board of Canada, Translation by E. R. Hope
          (1962)

Ohchi, K.: T61J Defence Research Board of Canada, Translation by E. R. Hope (1958)

Rodés, L.: T1 Sp. Defence Research Board of Canada, Translation by E. R. Hope
          (1959)

Sugiura, M.: Trans. A. G. U. Geomag. Disturbances <u>44</u>, 2 (1963)

Troitskaya, V. A.: T174R Defence Research Board of Canada, Translation by
          E. R. Hope (1953)

van Bemmelen, W.: Terr. Mag. <u>5</u>, 42 (1900)

Westcott, E. M.: J. Geophys. Res. <u>67</u>, 4 (1962)

Wright, C. S.: Australian Antarctic Expedition, Sci. Rep. Series B. Vol. II,
          Part IV, Government Printer, NSW Australia (1940)

MICROPULSATIONS OF THE EARTH'S ELECTROMAGNETIC
FIELD IN THE FREQUENCY RANGE 0.1-10 CPS

J. A. Jacobs

Institute of Earth Sciences
University of British Columbia
Vancouver, Canada

INTRODUCTION

There are two scales on which one can study geophysical phenomena: Regional and global. Thus, on the one hand, a geologist may map and study in great detail a very small area of the earth, while on the other, a geophysicist may view the area as part of a very much larger tectonic feature. Such a division applies equally well to a study of rapid variations in the earth's electric and magnetic fields - both local and world-wide effects must be studied. We must also consider geomagnetic micropulsations against the whole background of solar terrestrial relationships, and it may well be that the secrets we strive so hard to learn could be more easily obtained from other upper atmospheric phenomena. There are a whole host of such phenomena such as visual and radar aurora, airglow, whistlers, solar flares, cosmic rays, and ionospheric disturbances, and it is not surprising that correlations should exist between many of these events. In many instances we may be observing different manifestations of what was originally some solar disturbance, but it is not easy to distinguish between what are fundamental phenomena and what may turn out to be relatively unimportant side issues. It must not be forgotten that whereas in the laboratory, experiments can be carefully planned and executed with a view to settling definite questions, Nature, on the other hand, is continually making a multitude of "experiments" simultaneously all over the earth. It is not easy to disentangle all these "experiments" and to sort out cause and effect. This may require long periods of observation from many stations well distributed over the earth. With Nature working 24 hours a day, seven days a week, 52 weeks a year, it is difficult to take a representative sample of her work - especially when diurnal, seasonal and annual trends are to be looked for.

Geomagnetic micropulsations are fluctuations of the earth's magnetic field with periods from about 0.1 second to 10 minutes. Most work has been done in the lower frequency end of this range and the literature has recently been extensively reviewed (Jacobs and Westphal, 1963). In recent years, interest has spread to higher frequencies, particularly to oscillations around 1 cps. Around the turn of the century, van Bemmelen noticed micropulsations with about a one-second period and in 1906, Ebert (1906), using an induction magnetometer, reported the existence of pulsations with frequencies from 6 to 40 cps. In 1936, Harang (1936) observed micropulsations with periods less than one second at Tromsö and in the same year Sucksdorff (1936) found them with periods around two to three seconds at Sodankyla. Since the IGY, a large number of investigators in many countries have studied the phenomena using equipment with higher sensitivity and a faster speed of recording. Campbell (1960) found that in California the transition of the dominate received natural signals from slow tail sferics to geomagnetic micropulsations occurred between 2.0 and 0.2 cps, although it has been concluded (Vladimirov and Kleimenova, 1962) that the highest frequency of geomagnetic oscillations is 20 cps. This paper will be limited to micropulsations of the earth's electromagnetic field whose ultimate source lies beyond the ionosphere. Thus, earth-ionosphere cavity resonances (predicted by Schumann in 1952) which are induced by lightning strokes, will not be considered. König (1959) has reviewed our knowledge of atmospherics in the frequency range under discussion.

The whole subject of micropulsations has been plagued by a lack of uniformity in notation and definitions of the different phenomena. Some attempt at rectifying this has been made (Jacobs et al, 1963; Matsushita, 1963). [For a full discussion of the subject, see the report of the IAGA committee meeting at the IUGG General Assembly held at Berkeley in August, 1963.] In this paper the symbolism and notation of the various investigators will be clearly defined, but will not be changed in order to avoid further confusion. Until we know more about the cause of micropulsations and their relationship to other upper atmospheric phenomena some formal classification is necessary, but it is very important that restrictions are not imposed by too rigid definitions. Strict adherence to a narrow classification can but hamper further progress in this field.

The most commonly used notation in the frequency range under consideration is that due to Benioff (1960), Troitskaya (1961), and Tepley (1961a, 1961b). Benioff's Type A oscillations are nearly sinusoidal in shape with periods ranging from about 0.3 to 2.5 seconds. The wave trains usually exhibit beats and have been called "pulsations of the pearl beating type" by Troitskaya (1957). Benioff's Type B oscillations have nearly sinusoidal forms with periods ranging from about three to eight seconds.

Troitskaya (1961) recognizes four characteristic types of pulsations in the period range 1 to 15 seconds:

(1) irregular pulsations with periods from 1 to 15 seconds (SIP)

(2) pulsations of pearl type (PP)

(3) intervals of pulsations diminishing by periods (IPDP)

(4) continuous pulsations with small periods (T<15 seconds)

In an analysis of the data in the frequency range 0.5 to 5 cps obtained from the Lockheed Magnetic Observatory at Palo Alto, Tepley (1961a, 1961b) refers to the signals as "hydromagnetic (hm) emissions" since their properties seem consistent with their generation either above or high in the ionosphere and with their propagation downward through the ionosphere by hydromagnetic waves. Signals of similar appearance but of extremely high amplitude and associated with magnetic storms have been called "solar whistles" by Duffus et al (1958, 1960b). An hm emission is defined as an oscillation at a single frequency which may, however, vary slowly with time. In contrast, pearls or Type A oscillations may include a number of frequencies, i.e., a pearl oscillation train may include a number of distinct hm emissions.

In a later paper, Tepley and Wentworth (1962) extend their classification of micropulsations in the frequency range 0.4 to 7 cps into hydromagnetic emissions and noise bursts. Hydromagnetic emissions are characterized by well-defined frequency bands on a sonagram, and often exhibit a regular and reproducible fine structure. The corresponding waveforms are approximately sinusoidal and exhibit an irregular amplitude modulation pattern. In noise bursts the signals are characterized by a broad spectral energy distribution. Well-defined frequency bands are not observed on sonagrams. The waveforms are jagged, and the signals are frequently superimposed on micropulsations of longer period, in which case they may be interpreted as the high-frequency components of the latter events. It seems likely that both SIP and IPDP (Troitskaya, 1961) consist of a superposition of both hm emissions and noise bursts. Furthermore, different SIP and IPDP events are likely to vary greatly in appearance so that classification of signals into these categories may be highly subjective. Both Type A and B (Benioff, 1960) oscillations fall into the category of hm emissions.

Lokken (1962) divided geomagnetic micropulsations into two classes: impulsive and regular, according to their nature and associations. Impulsive activity normally contains a broad range of frequencies and hence presents an irregular record. It has great dynamic range. A prominent characteristic of impulsive activity is the nature of the signal enhancement above a quiet or steady background. Impulsive activity occurs more frequently and is of greater amplitude in the auroral zones than in lower latitudes. On the other hand, regular signals consist of regular series of pulsations, generally nearly sinusoidal and disposed in one or more discrete frequency bands. By comparison with impulsive activity, regular signals have a limited dynamic range and are often capable of maintaining their identity over regions at least of continental extent. The successive sections of Fig. 1 illustrate in A the background continuum, in B a regular signal emerging above it, and in C the signal approaching its maximum

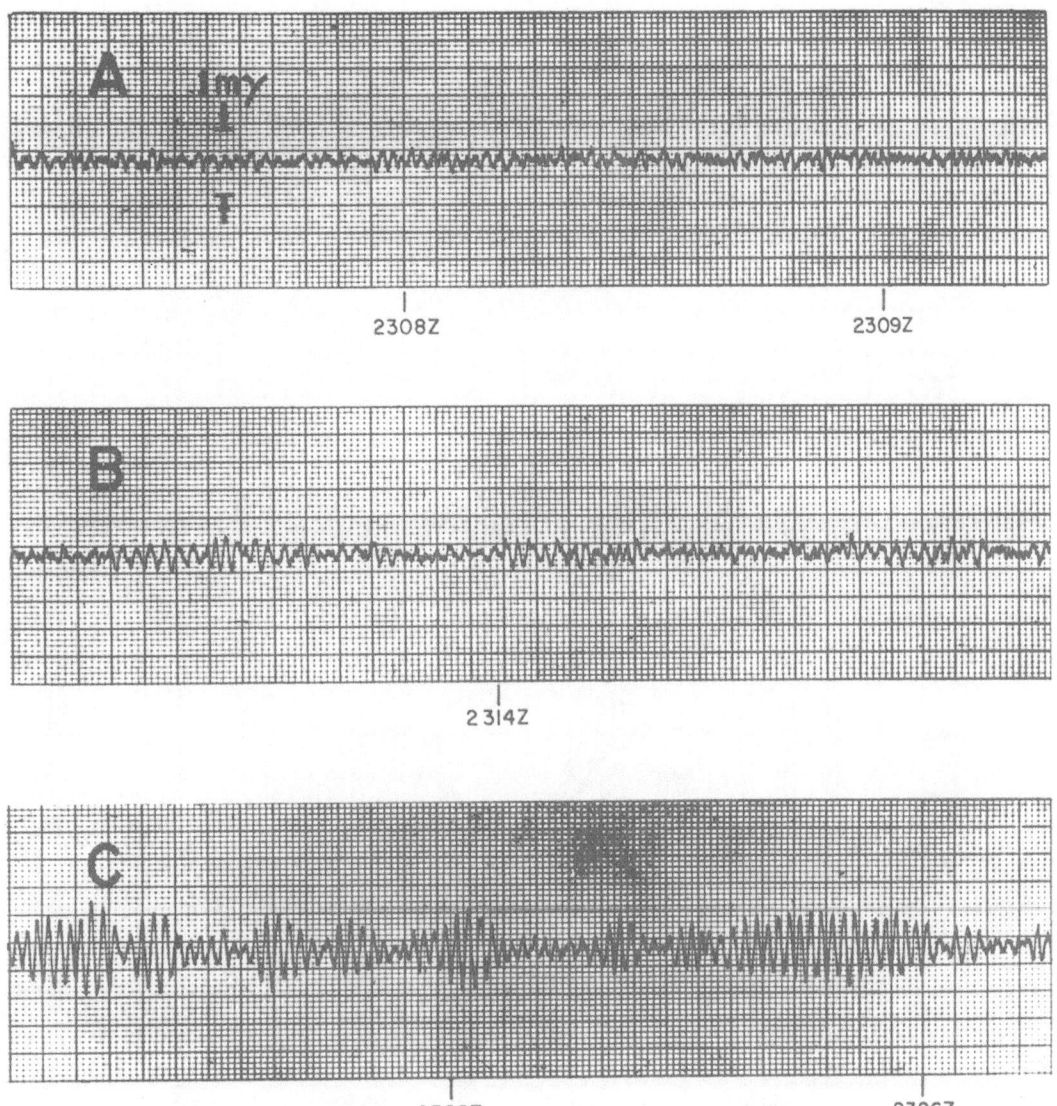

Fig. 1. Three successive sections from a system tuned to 0.81 cps, A showing the background continuum, B showing an emerging regular micropulsation signal of 0.72 cps, and C showing the signal approaching its maximum value, about 1 mγ peak-to-peak (Lokken, Shand and Wright).

value, all within 20 minutes. Such enhanced activity is recognized as belonging to the regular class of micropulsations because of its nearly uniform frequency, and because changes in amplitude tend to be smooth rather than abrupt. The impulsive class of signal is illustrated by the event shown in Fig. 2. Signals of this kind commence and terminate abruptly and occur in bursts which are not smoothly developed.

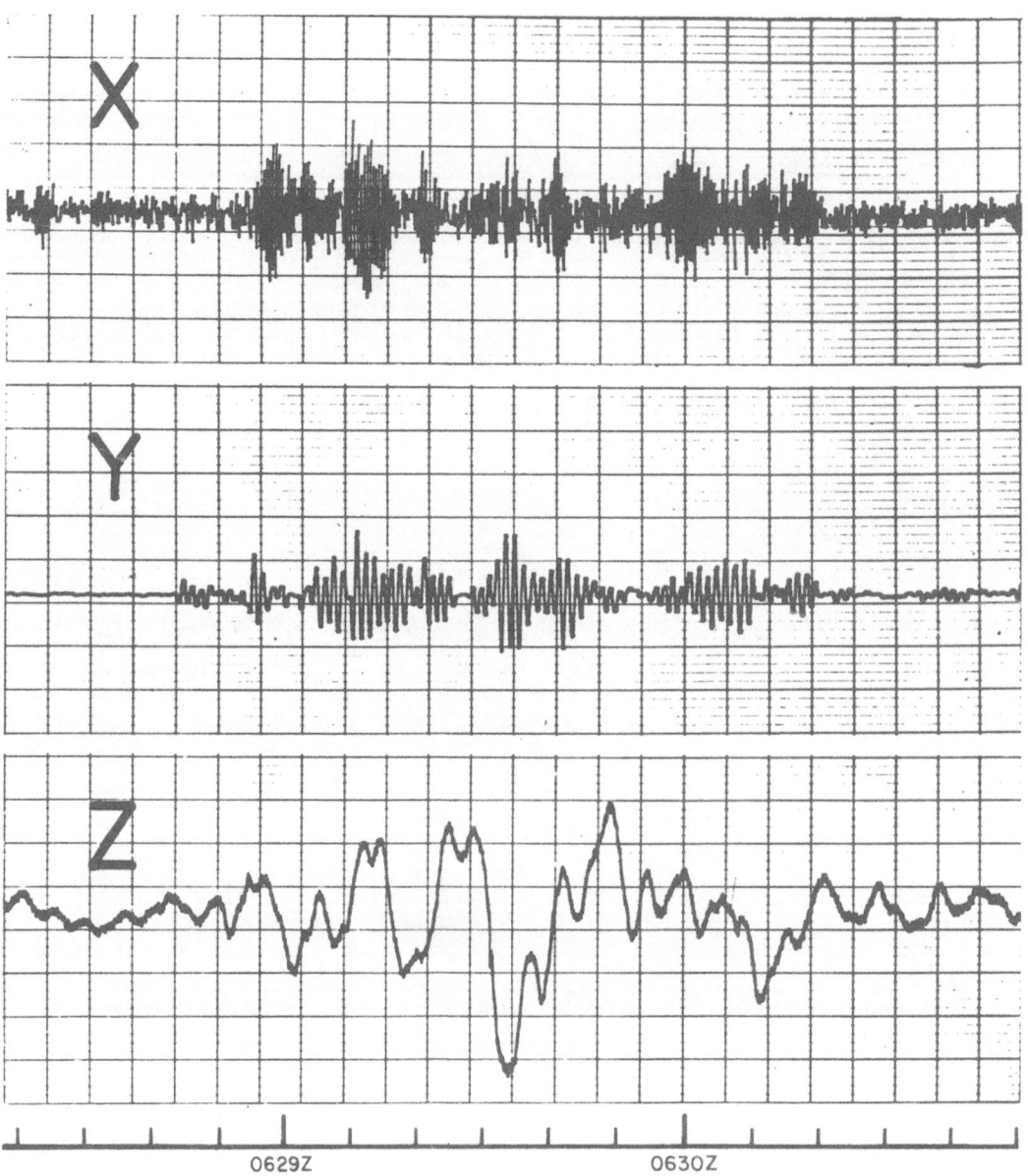

Fig. 2. A small impulsive event characteristic of those commonly encountered in the auroral zones. Component X was tuned to about 2.2 cps, Y to about 0.8 cps, while Z on broad band had a flatter response and lower sensitivity. The approximate maximum peak-to-peak flux excursions are indicated (Lokken, Shand and Wright).

In the following discussion of the observational results, some of the variances may be due to different equipment with different sensitivity and frequency response. Moreover, many investigations have been made using telluric (earth current) measurements rather than magnetic records. Since some results depend on the location of the recording station, Table 1 gives both the geographic and geomagnetic coordinates of all stations referred to in this paper.

TABLE 1
GEOGRAPHIC AND GEOMAGNETIC COORDINATES
OF STATIONS REFERRED TO IN THIS REPORT

	Geographic		Geomagnetic	
	Latitude	Longitude	Latitude	Longitude
Bangui	4°26'N	18°34'E	4.8°N	88.5°
Byrd	79°59'S	120°01'W	70.6°S	336.0°
Canton Island	2°46'S	171°43'W	5.1°S	258.0°
Chambon-La-Forêt	48°01'N	2°16'E	50.4°N	83.9°
Charcot	69°30'S	139°02'E	78.3°S	234.5°
Churchill	58°42'N	94°03'W	68.7°N	322.8°
College	64°51'N	147°50'W	64.7°N	256.5°
Dumont d'Urville	66°40'S	140°01'E	75.6°S	231.0°
Great Whale	55°17'N	77°46'W	67.8°N	350.0°
Isabella	35°40'N	118°28'W	42.6°N	303.1°
Kauai	22°09'N	159°18'W	21.7°N	265.6°
Kerguelen	49°21'S	70°12'E	56.5°S	127.0°
Legon	5°38'N	0°12'W	9.6°N	70.2°
Lovozero	68°01'N	34°01'E	63.1°N	126.1°
Macquarie Island	54°30'S	158°57'E	61.1°S	243.1°
Memambetsu	43°55'N	144°12'E	34.1°N	208.3°
Palo Alto	37°26'N	122°10'W	43.5°N	299.0°
Sodankyla	67°22'N	26°39'E	63.7°N	120.0°
Tamanrasset	22°47'N	5°32'E	26.0°N	81.5°
Tongatapu	21°14'S	175°08'W	24.0°S	258.5°
Tromsö	69°40'N	18°57'E	67.2°N	116.8°
Tucson	32°14'N	110°57'W	40.4°N	312.1°
Uppsala	59°53'N	17°39'E	58.7°N	104.5°

MICROPULSATIONS WITH PERIODS BETWEEN 3 AND 10 SECONDS

This section includes Benioff's Type B and Troitskaya's SIP, but not her PP or IPDP. Maple (1959a, 1959b) carried out an intensive analysis of the magnetic records from Tucson covering a 12-day interval. Oscillations centered around an eight-second period occurred chiefly during the night hours and were most numerous in the early morning (2300 to 0500 LMT) when sporadic E is least prevalent in that area. The oscillations in this frequency band were also closely related to the degree of magnetic disturbance, the threshold level for their appearance being at K = 3. Campbell (1960) also observed sinusoidal pulsations in southern California with periods from five to eight seconds during the night hours, and rapid oscillations with average periods around eight seconds have been reported at Eskdalemuir (Holmberg, 1953) during times of large magnetic disturbances. Maple suggests that the oscillations are excited by moving irregularities of F-region ionization, the oscillations only appearing at times of magnetic disturbance when the excitation is near the resonant frequency of the ionospheric region.

Troitskaya (1961) found her SIP on records on both quiet and disturbed days, a preliminary analysis showing that a six-second period is observed most frequently. The pulsations have their largest amplitude in the polar regions and were frequently observed simultaneously in the arctic, middle latitudes and the antarctic. SIP always appear when pt or bay disturbances are registered on standard magnetic records, i.e., they represent a micro-structure of these well-known macroscopic forms of disturbance of the earth's magnetic field. When a bay is developing after pt, SIP are more intensive, their spectral distribution is more complicated, and their duration is longer. Yanagihara (1959) noticed that at Memambetsu a train of micropulsations with periods from five to ten seconds often appeared in association with pt beginning almost simultaneously with the accompanying pt. He also found, in contrast to pt activity, that the activity of these micropulsations is greater at sunspot maximum. A very important peculiarity of SIP is that they sometimes appear when notable macroscopic disturbances of the field are absent. The diurnal variation of SIP occurrence shows a maximum near local midnight during disturbed conditions and in the early morning hours for quiet field conditions. A high degree of correlation is found between SIP and auroras - the predominant period of SIP coinciding with the main period of intensity change of auroras. Benioff (1960) also found that the occurrence of his Type B pulsations with large amplitudes coincided with times at which visible aurora were reported locally, and Alperovich (1961) found a very specific dependence between the amplitudes of these short period pulsations and the intensity of auroral activity at Lovozero.

Kato and Saito (1959) found that short period micropulsations appear mainly at night in middle latitudes (see Fig. 3). Selzer (1961) observed micropulsations with periods from three to ten seconds (usually six seconds) quite often, even under magnetically quiet conditions. They occur most frequently a little before local midnight, not only at the middle latitude stations Chambon-La-Forêt (in the northern hemisphere) and Kerguelen (in the southern hemisphere), but also at an equatorial station (Bangui). Hutton (1961) also found micropulsations with periods from 5 to 15 seconds at an equatorial station (Legon) during night hours, although such micropulsations were a relatively rare phenomenon, having occurred on only 22 days during a two-year period. According to Schlich (1961), micropulsations with periods between three and six seconds may appear at any hour of the day at Charcot in Terre Adelie, a little south of the southern auroral zone. Heacock (1963) investigated telluric current micropulsations at five stations near the auroral zone. The diurnal variation data suggest a spiral pattern of occurrence in the auroral zone for aurorally associated micropulsations (period 3 to 20 seconds). The pattern is visualized as being fixed in space,

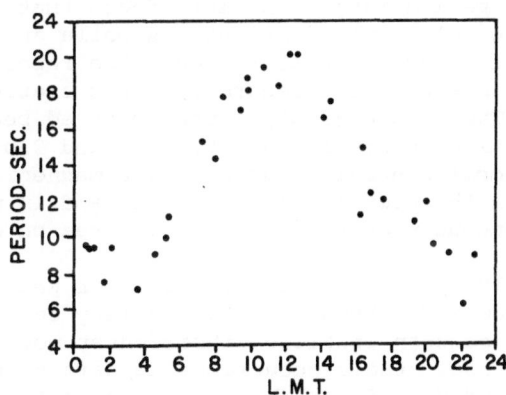

Fig. 3. Diurnal variation of short period micropulsations
in middle latitudes (Kato and Saito, 1959).

the earth rotating beneath it. Such a spiral pattern may also exist at lower
latitudes.

MICROPULSATIONS WITH PERIODS BETWEEN 0.1 AND 3 SECONDS

This section includes Benioff's Type A and Troitskaya's pulsations of pearl type (PP).
Type A oscillations are essentially nocturnal in southern California, their occurrence
increasing rapidly after sunset and falling rapidly after sunrise. The diurnal variation of occurrence frequency seems to be different in high latitudes, however. Sucksdorff (1936) found that micropulsations with periods around two to three seconds are
most likely to occur during daylight hours at Sodankyla. Harang (1936) observed that
pulsations with periods around one second or less appear most frequently at Tromsö
between 0800 and 1000 LMT with a second maximum between 2000 and 2200 LMT. Benioff
(1960) found the maximum occurrence frequency at Uppsala to be between 0700 and 1000
LMT.

Troitskaya (1961) also found that pearls in middle latitudes occurred mainly
during the night hours, although in polar regions they may appear at any time. Also
in the polar regions short separate bursts are seen whereas in middle latitudes continuous series of pearls are more often observed. The predominant period is also
different at middle and polar latitudes. Heacock and Hessler (1962) found that at
College pearl-type telluric current pulsations occurred mainly in the daytime. Some
diurnal variation in the pulse period was indicated. Thus, from 1800 to 0600 UT
(local daytime) the pulse periods average 2.4 seconds while from 0600 to 1800 UT the
periods average 1.9 seconds. This may be explained by greater daytime attenuation of
the shorter period pulses. The appearance of pearls usually coincides with times when
other disturbances within the observed period range were at a low level, i.e., pearls
as opposed to SIP do not represent simply the micro-structure of some other form of
macroscopic disturbance of the earth's electromagnetic field. There also appears to
be a rough inverse relationship to the solar sunspot cycle. Very likely the ionosphere
is an opaque shield for waves of this frequency most of the time, their occasional
appearances corresponding to intervals during which the ionosphere becomes transparent

for oscillations in this period range. Benioff found that at times the correlation between recordings at two California stations, Isabella and Palomar, some 150 miles apart, was nearly perfect. Duffus et al (1960a) also reported occasions when pulsations of frequency about one cps were recorded at two stations more than 500 miles apart (E-W) in Canada. Troitskaya (1961) often observed pearls (PP) over a very wide area, mainly during the time intervals 1600 to 2000 and 2200 to 0200 UT, although there is probably some local control. During great magnetic storms, pearls may be excited simultaneously in the arctic, the antarctic and in middle latitudes. Separate bursts may occur simultaneously over the same wide area on quiet days as well.

Jacobs and Jolley (1962) examined the records for pearls from a number of stations between October 1960 and February 1961. In many cases pearls were observed progressively earlier at eastern stations, their onset moving westward at a rate of approximately 14°/hour - i.e., their occurrence appears to be LMT dependent. On some occasions this linear trend extended right round the world, pearls reappearing at a station 24 hours after their first occurrence there. Using data taken during February 1962 at College and Macquarie Island, Dawson (1963) found a lack of agreement with regard to pearl events. College and Macquarie Island have conjugate latitudes but miss by 700 km in the east-west direction. He found a tendency for pearl events to occur on the same dates, but the beginning and terminal times were often different, and matching of individual beats was impossible.

The "pearl-beating" characteristic has been attributed to the superposition of two wave trains of slightly differing frequencies. However, the results indicate that the amplitude modulation envelope is usually associated with a far more complex superposition of waves than was previously realized. Tepley (1961a, 1961b) has shown that hm emissions often display a fine structure, characterized by wave trains of increasing frequency. The "pearl-beating" characteristic observed in amplitude-time plots results from a superposition of a number of similar wave trains of relatively short duration and rapidly increasing frequency, each wave train being delayed in time by a comparable amount from the preceding train.

Vozoff et al (1962) carried out a detailed analysis of one hour's recording of "impure" pearls. They found that most of the energy was concentrated in three bands - two, relatively narrow, centered at 0.53 cps and 0.67 cps and a much broader low frequency band. The 0.53 cps band was somewhat smaller in amplitude and most of the character was due to structure within the 0.67 cps band. When the records were re-analyzed by thirds, it became evident that there occurred drastic changes in the directional orientation and amplitudes of the 0.53 cps components. Regardless of the picture one uses to explain the fields, it would appear that the sources of the 0.53 and 0.67 cps peaks must be independent of one another, and they conclude that the changes, in time, of the relative amplitudes and orientations of the two major bands imply that the source oscillations can be excited independently.

Francis and Karplus (1960) and Francis et al (1961) considered in some detail the transmission of hydromagnetic waves through the ionosphere. Hydromagnetic wave attenuation theory indicates an inverse relationship between the amplitude transmission of hydromagnetic waves and ionospheric ion density. Thus, we may expect the signal amplitude to be relatively small during daylight hours and to become relatively large at night. This is in agreement with observation at middle latitude stations. Tepley (1961b) suggested that hm emissions are generated by the interaction during magnetically quiet conditions of solar proton streams with the geomagnetic field. [At such times there is a close correlation between the occurrence of hm emissions and polar cap absorption which is attributed to solar proton streams.] Troitskaya (1961) has reported a correlation between the occurrence of pearl oscillations (hm emissions) and solar-flare cosmic rays - the latter may also be closely correlated with polar-cap absorption events. It is interesting that the ratio of the frequencies typical of VLF emissions (near 5 kc) and hm emissions (0.5 to 5 cps) is of the order of the mass ratio of protons to electrons. This observation leads to the suggestion that VLF and hm emissions may be generated by mechanisms which are at least qualitatively similar.

As in the theory of VLF emissions, the wave frequency of the hm emission may be directly related to the geomagnetic field strength in the region where the emission is generated. Thus, it is anticipated that proton streams moving in regions of greater geomagnetic field strength would generate hm emissions at higher frequencies. This is consistent with observations (Tepley, 1961b; Duffus et al, 1958, 1960b; Troitskaya, 1961), which indicate that, although the higher oscillation frequencies (near 5 cps) occur only rarely during quiet periods, they are relatively common during periods of unusual solar activity.

It is also interesting that the hm emission frequency, 0.5 to 5 cps, corresponds to the proton cyclotron frequency range for magnetic field strengths in the range 33 to 330 gammas. Assuming an idealized geomagnetic-dipole field, the corresponding distance range for fields of this strength lies between 5 and 11 earth radii. This distance range is consistent with reasonable estimates of the location of the boundary surface of the magnetosphere along the sun-earth line.

Tepley and Wentworth (1962) found that the highest observed emission frequency decreases with increasing geomagnetic latitude (see Table 2). Table 2 also includes the maximum electron bounce frequency at the different latitudes as predicted by the theoretical model of the authors (Wentworth and Tepley, 1962). This model associates the hm-emission frequency with the bounce frequency of geomagnetically trapped electrons mirroring above or high in the ionosphere above the observation point. There is good agreement between the predicted and observed frequencies.

TABLE 2

COMPARISON OF HIGHEST OBSERVED hm-EMISSION FREQUENCY AND HIGHEST
PREDICTED ELECTRON BOUNCE FREQUENCY AS A FUNCTION OF GEOMAGNETIC LATITUDE

Station	Geomagnetic Latitude	Highest observed hm emission frequency (cps)	Maximum predicted electron bounce frequency at station (cps)
Palo Alto	45°	5	5.25
Victoria	55°	3	3.18
Uppsala	60°	3	2.46
Reykjavik	70°	2	1.34

Jacobs and Watanabe (1961) have tried to explain pearl type pulsations as hydromagnetic oscillations in the upper atmosphere as well as the exosphere below a height of about 2000 km. The height-distribution of the Alfvén velocity (see Fig. 4) selects, from the complete spectrum of hydromagnetic waves which originated in outer space, those of particular frequencies with the result that their intensity is enhanced at the earth's surface. The period of the enhanced wave is of the order of one second to several seconds, depending on solar conditions. The enhancement is also of the correct order of magnitude for the intensity of pearls observed at the earth's surface.

In addition to Tepley and Wentworth (1962) in America, Mainstone and McNicol (1962) in Australia and Gendrin and Stefant (1962) in France have used frequency time displays (sonagrams) to investigate micropulsations. They all found evidence of fine

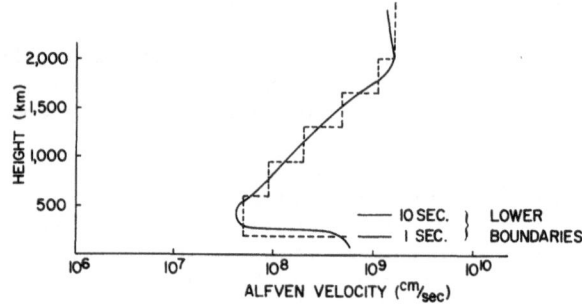

Fig. 4. Distribution of the Alfvén velocity against height - night time,
maximum sunspot activity. The chain-dotted curve is the approximation used in the calculations. The lower boundary of the
hydromagnetic region for waves of 10 and 1 second period is also
indicated (after Jacobs and Watanabe).

structure in the dynamic spectrum of short period pulsations with frequencies from
fractions of a cps to several cps, which consists of trains of rapidly increasing
frequency emissions repeated at intervals of about two minutes, the whole resembling
a fan-shaped structure (see Fig. 5). Gendrin and Stefant found that the product of
the repetition period of the rising emissions and their mean frequency was fairly
constant even though the mean frequency varied appreciably. Jacobs and Watanabe
(1963) have interpreted the fan-shaped structure in the sonagrams as caused by hydromagnetic oscillations in the lower exosphere excited by several bunches of charged
particles trapped in the earth's magnetic field. At the time of trapping a train of
geomagnetic micropulsations will be observed in those regions where the trapping
magnetic lines of force meet the earth's surface. The charged particles in each
bunch travel along the magnetic line of force toward the conjugate mirror point (in
the southern hemisphere, say) giving rise to a train of similar pulsations there half
a bouncing period later. After one bouncing period, from the beginning, the micropulsations reappear in the northern hemisphere, and after one and a half periods,
they appear in the southern hemisphere again, and so forth. Correspondence of micropulsations between a pair of conjugate areas in the two hemispheres with a time-
difference of half a bouncing period has actually been found in some cases by
Yanagihara (1963) and Lokken et al (1963). The time-difference suggests that the
velocity of the trapped particles is of the order of 10^8 cm/sec. Assuming that a
secondary disturbance is emitted from each place of origin of the micropulsations
so that pulsations observed at lower latitudes would consist of various frequencies
each of which is determined by the local conditions of the lower exosphere and
ionosphere, Jacobs and Watanabe show that the dynamic spectrum consists of several
discrete frequencies with certain repeating periods which are shorter at the lower
frequencies - which reflects the essential features of the experimental sonagrams
(see Fig. 5 and Fig. 6).

In addition to the main station at Palo Alto, Tepley (1963) has established
stations in low latitudes in the Pacific - at Kauai near Hawaii, Canton Island and
Tongatapu. Kauai and Tongatapu are approximately conjugate - the Canton station is
located near the geomagnetic equator approximately midway between them.

328

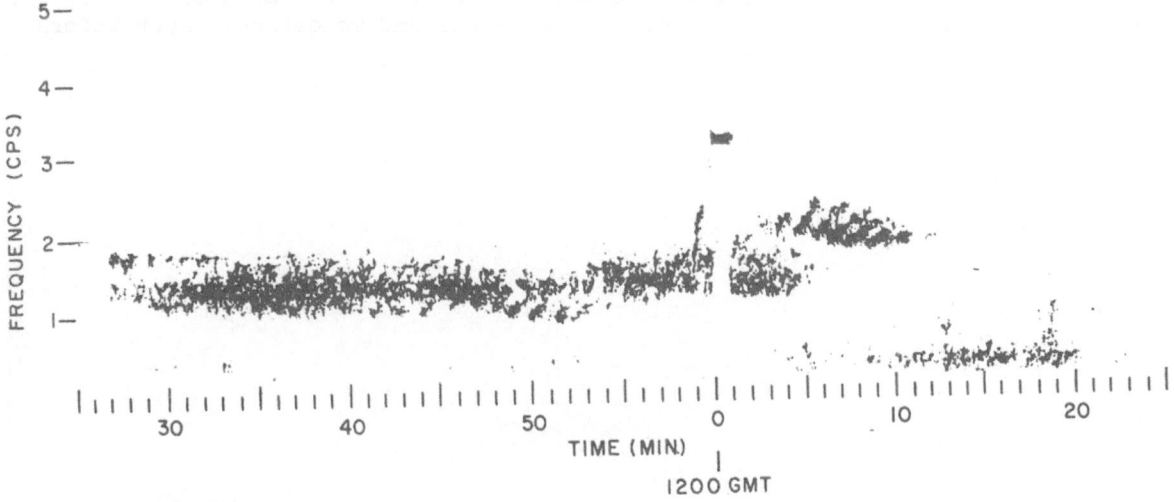

Fig. 5.   The dynamic spectrum of geomagnetic pulsations beginning about 17 minutes after the sudden commencement of a magnetic storm and showing a fan-shaped fine structure (after Tepley and Wentworth).

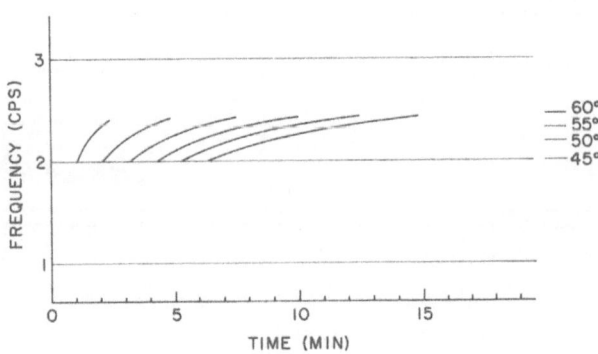

Fig. 6.   Theoretically constructed dynamic spectrum (after Jacobs and Watanabe).

Fig. 7 shows sonagrams obtained from data recorded on magnetic tape. The signals are not always observed at the same time at the four stations, but when signals are observed simultaneously at more than one station, they show a remarkable degree of similarity even to the appearance of the fine structure. At Palo Alto in the interval 1100 to 1200, five distinct bands are observed in the frequency range 0.5 to 1.0 cps. Traces of some of the same bands are observed at Kauai and Tongatapu. Just before

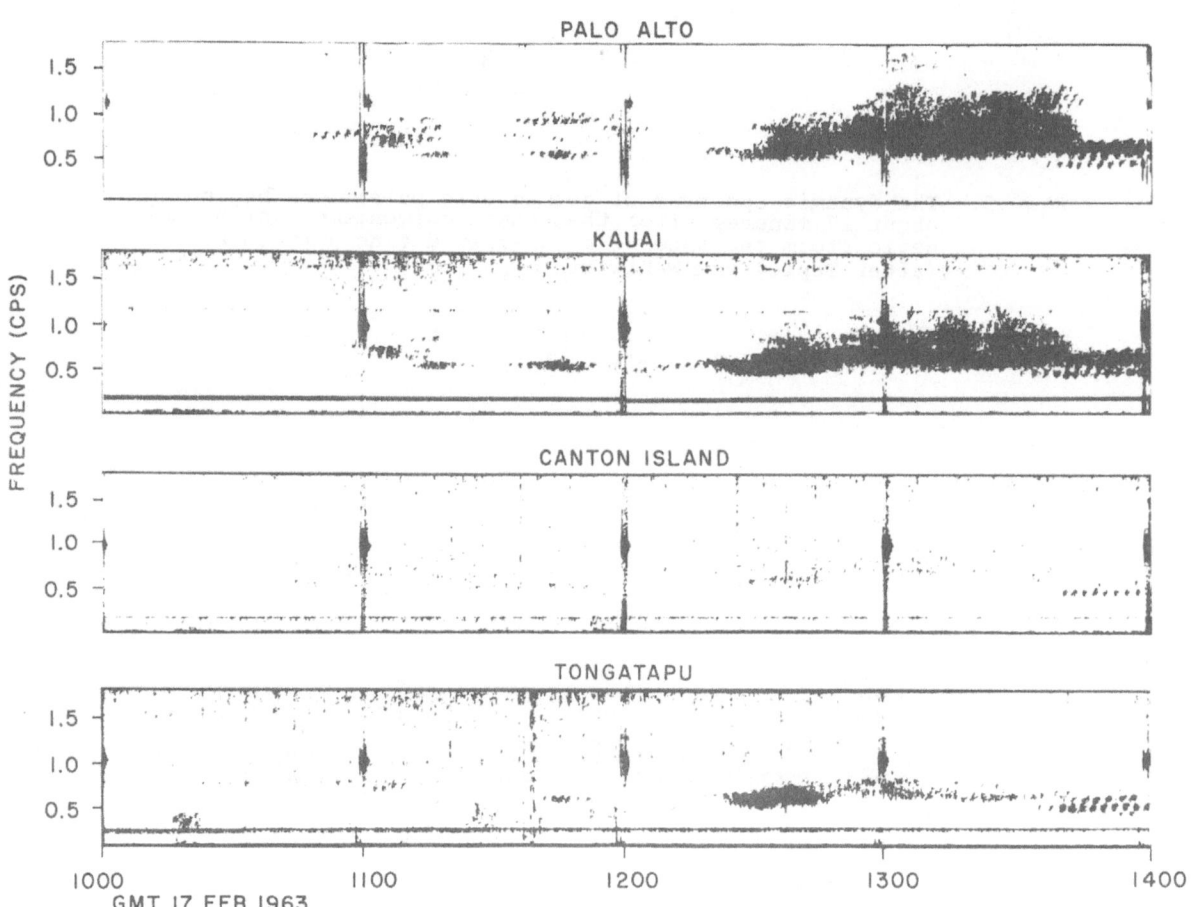

Fig. 7.   Sonagrams of hm emissions observed on February 17, 1963, 1000 to 1400 UT (after Tepley).

1230 an apparent falling-frequency fine-structure is observed at Tongatapu. The same
structure is found at Kauai and Palo Alto but appears less clearly at the latter stations. Just before 1400 a sharply defined structural pattern is observed simultaneously at all stations. Higher emission-frequencies are observed at Palo Alto than at
Kauai and Tongatapu, and at Canton Island only the lowest frequencies are observed.
This observation leads to the tentative suggestion that there may be an inverse-fre-
quency latitude relationship for hm emissions observed at middle and low latitudes.

In the four-hour time interval shown in Fig. 8, the hm emission activity is less

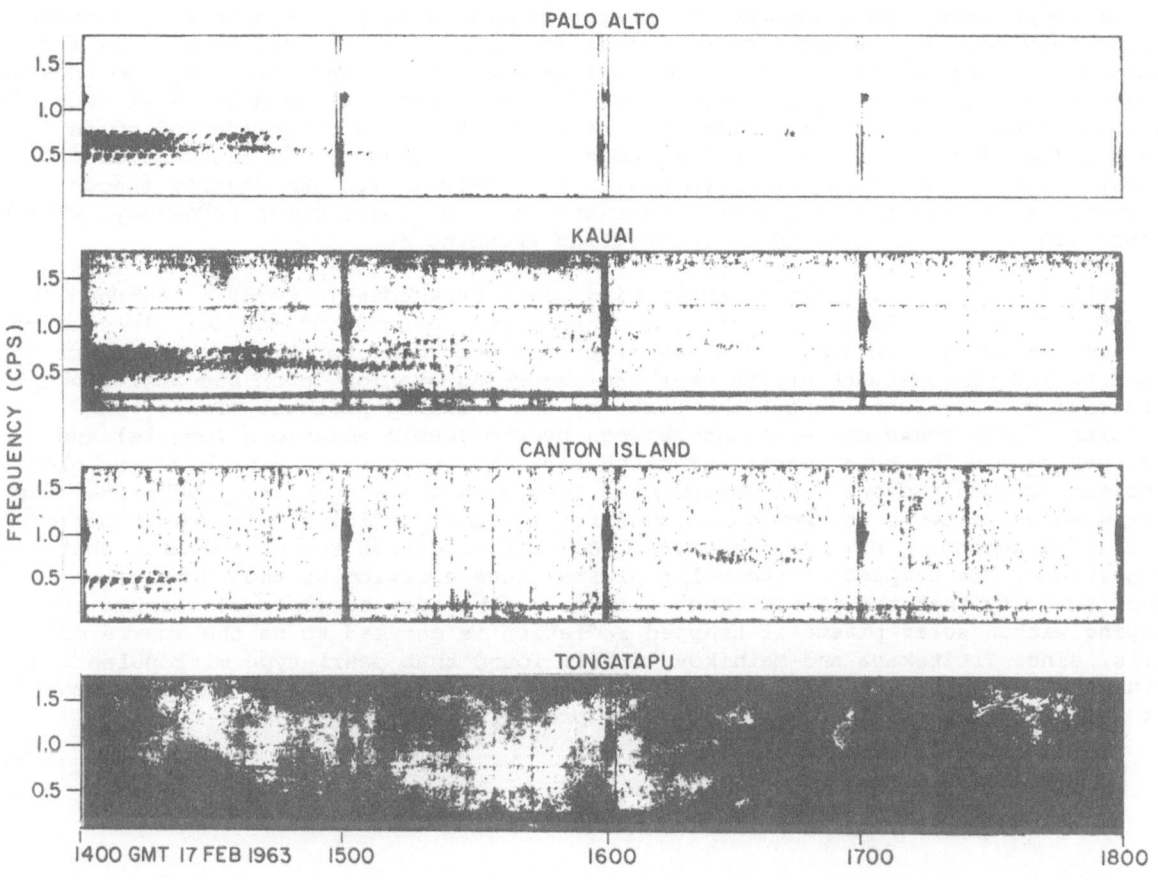

Fig. 8.   Sonagrams of hm emissions observed on February 17,
1963, 1400 to 1800 UT (after Tepley).

331

than during the earlier interval shown in Fig. 7. The sharply defined structural pattern just after 1400 is a continuation of the pattern shown at the right of Fig. 7. From a close study of this portion of the record (1335 to 1425), it appears that there is a 180° phase shift between the structural elements of the signal between Kauai and Tongatapu. Less intense emission activity is seen in the period 1620 to 1720. The signal is just barely observable at Kauai but appears clearly at the other stations. The signal intensity seems greater at Tongatapu than at Palo Alto, but from a study of the waveform records it is found that the amplitude is about the same at the two stations. In most cases the signal amplitude is greater at the high latitude station. In this case, however, the reduction in signal amplitude at Palo Alto may be attributed to propagation effects. The signal occurs considerably after local sunrise at Palo Alto so that the hydromagnetic wave is strongly attenuated in propagating downward through the ionosphere. At Tongatapu, however, the signal occurs close to local sunrise so that at least part of the propagation path may be through a night-time ionosphere where hydromagnetic wave attenuation is small.

Heacock and Hessler (1962) investigated "pearl type" telluric current micropulsations at College, Alaska. They found a number of occasions on which essentially only one pulse frequency was present in a pearl-type micropulsation sequence. On each such occasion there was a sequence of pearls nearly equally spaced several minutes apart. They call such a sequence of pearls formed by a nearly constant pulse frequency a necklace of pearls. Thus, a pearl necklace is characterized by two constants, a pulse frequency and a pearl spacing (see Fig. 9). When two or more pearl necklaces are superimposed, "beads" are formed by the interactions of the different pulse frequencies (see Fig. 10). They use the names "pearl" and "bead" to distinguish the two observed types of amplitude enlargements, i.e., "pearls" are the equally spaced intervals of amplitude enlargements associated with a single pulse frequency, whereas "beads" are the beats produced by interacting frequencies.

There is the possibility of explaining pearl necklaces in terms of bunches of trapped protons oscillating between hemispheres with periods of one to five seconds and drifting around the earth in a few minutes. According to this idea, a pearl is formed as the proton bunch drifts over the recording station, each successive pearl representing a revolution about the earth by the drifting protons. Tepley and Wentworth (1962) found correlations between hydromagnetic emissions (pearls) and X-ray events and for this reason suggested that the responsible particles are electrons rather than protons. It was pointed out by Jacobs and Jolley (1962) that the solar wind velocity is too low to account for trapped particles with bounce periods one to five seconds. However, Heacock (1963) has suggested the possibility that the particles are trapped in the solar plasma, thus arriving at the earth with a higher velocity than the plasma front. In fact, it seems necessary to assume this trapping within solar plasma if trapped radiation is assumed to be the source of pearls, since Troitskaya and Melnikova (1961) found that pearl-type micropulsations frequently appear during the few minutes following a magnetic storm sudden commencement (ssc) and/or during the several hours preceding the ssc.

MICROPULSATIONS, MAGNETIC STORMS AND OTHER UPPER ATMOSPHERIC PHENOMENA

Troitskaya et al (1962) considered the fine structure of magnetic storms with respect to micropulsations (T<20 sec.). Out of 30 storms investigated, 28 had PP excited during the interval 0 to 12 hours before the onset of the storm - usually some tens of minutes to two hours before the beginning of the storm. The duration of the series of PP is generally 10 to 15 minutes and seldom exceeds two hours. Several cases were discovered where PP pulsations occurred simultaneously with

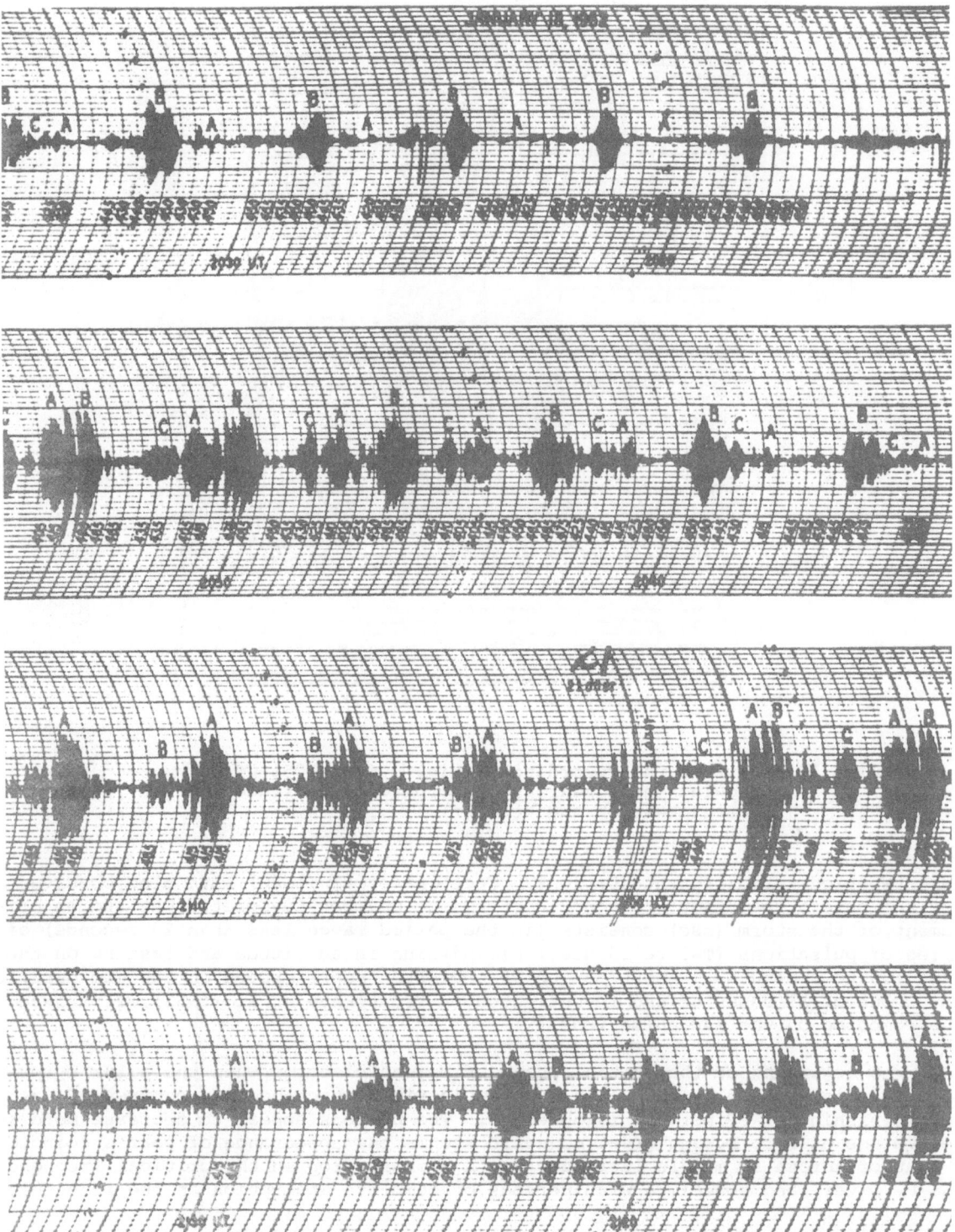

Fig. 9. Three pearl necklaces, January 18, 1962. The average pulse frequencies as determined by counting pulses over many of the 20-second intervals between successive chart time lines are given below the trace. Examination of these values establishes the existence of three pearl necklaces, A, B and C, with pulse frequencies approximately 0.41, 0.485, and 0.44 cps respectively (after Heacock and Hessler).

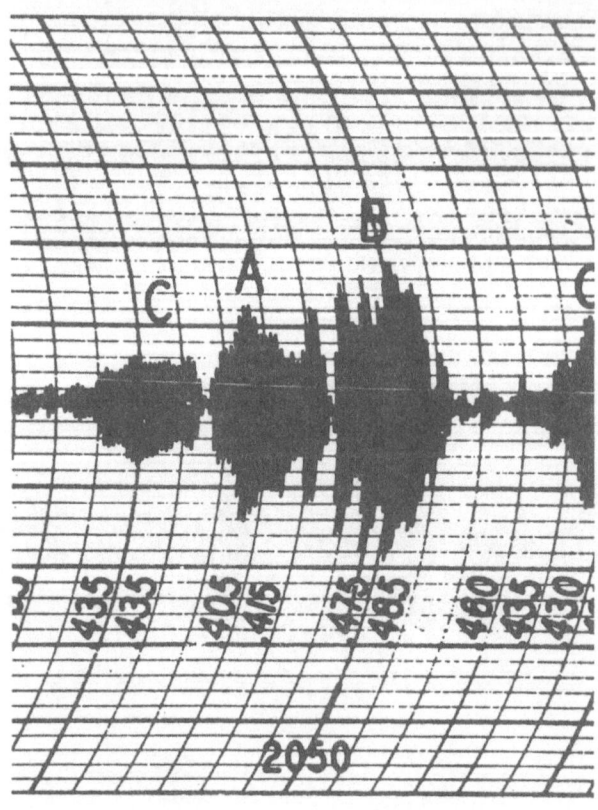

Fig. 10.   Part of the recording of Fig. 9 near 2051 UT,
showing three pearls, one from each necklace
(after Heacock and Hessler).

different periods at different stations. Cecchina et al (1960) have also observed
PP excited before the onset of magnetic storms at Chambon-La-Forêt. The sudden commencement of the storm (ssc) consists (in the period range less than 20 seconds) of
a series of pulsations (T~8 to 15 sec.) diminishing in amplitude and lasting on the
average 1.5 to 3 minutes. Short period PP are often superimposed on the ssc oscillations. [Out of 35 cases of ssc investigated by Troitskaya et al (1962), 21 showed
PP superimposed on the ssc pulsations.]

At the beginning of the initial phase of magnetic storms when the solar corpuscular stream is stopped by the earth's magnetic field which becomes compressed, a
sharp diminishing of PP periods is observed. The periods appear to decrease more at
higher values of K_p. During great magnetic storms the appearance of pearls coincides
with that of aurora in middle latitudes, with a sharp drop in the critical frequency
in the F_2 layer, and sometimes with periods of complete absorption in the ionosphere.

Investigations of a connection between PP occurrence and phenomena in the high
atmosphere have led to the discovery that very often PP are excited during periods of
sharp increases in cosmic-ray intensity in the stratosphere when the magnetic field
is quiet. When the magnetic field is disturbed, the times of such intensive cosmic-
ray bursts in the stratosphere coincide with the excitation of SIP. An analysis of
the microstructure of magnetic storms in middle latitudes by Troitskaya (1961) showed
the existence of a peculiar sequence of pulsations (IPDP) during the development of
the storm. It begins as SIP and contains pearl series which, as a rule, diminish in

334

periods (T~10 to 1 second). Troitskaya and Melnikova (1961) believe that IPDP represent an important morphological element of magnetic storms, because the beginning of several types of disturbances in the atmosphere and the ionosphere coincides with the beginning of IPDP (a sharp fall in f_0F_2, the propagation of red aurora into low latitudes, bursts of X-rays in the stratosphere, and a sharp increase in cosmic noise absorption). IPDP also coincides with times of the deepest penetration of auroras into low latitudes and the highest depression of the H component of the earth's magnetic field. Heacock and Hessler (1963) found that there was a marked tendency for micropulsations at College to occur during the first and the last hours of the main phase of a magnetic storm. Auroral evidence suggests that a narrow (in latitude) band of micropulsation occurrence existed, and that this band moved southward during the maximum part of the main phase.

Campbell and Leinbach (1961) investigated ionospheric absorption at times of auroral and magnetic pulsations. High latitude cosmic noise absorption may be classified into two categories - auroral absorption and polar cap absorption. Auroral absorption is closely related to auroral coruscations and magnetic micropulsations (Campbell and Rees, 1961), as shown in Fig. 11, indicating that the ionization

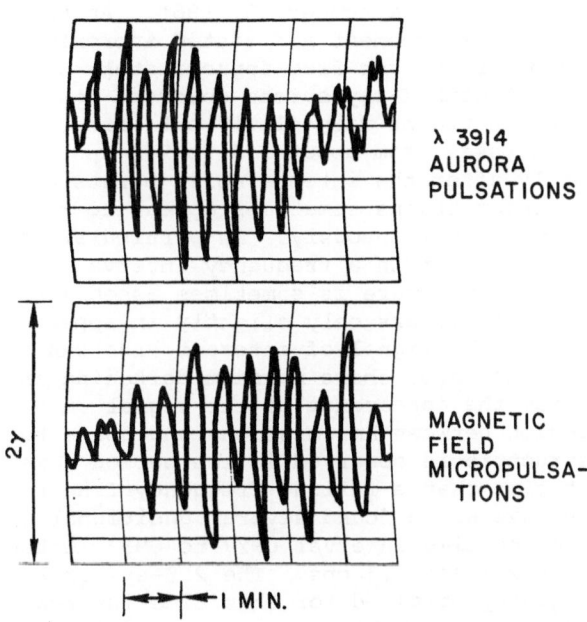

Fig. 11. Corresponding pulsations of the aurora and geomagnetic
micropulsations at stations separated by 30 miles near
College, Alaska (after Campbell).

producing this type of absorption does not effectively shield micropulsation activity and may imply that magnetic micropulsations are largest in the E region. On the other hand, decreased micropulsation activity during polar cap absorption, considered to be a result of D-region ionization by high energy protons, seems to be an example of ionospheric shielding of magnetic pulsations.

Bomke (1962) also found that at Baxter State Park (geomagnetic latitude 57°N) magnetic micropulsations coincided with visually observed light fluctuations of flaming aurora. Heacock (1963) on the other hand found no such correlation at College. In fact, a negative correlation was obtained between nighttime PP events at College and the occurrence of aurora. Bomke also concluded that the one cps oscillations are connected with the absorption of auroral particles in the lower ionosphere (~100 km). This is also the height of flaming aurora determined by Campbell and Rees (1961) by optical triangulation in Alaska.

Campbell and Matsushita (1962) carried out an investigation of auroral-zone geomagnetic micropulsations with periods from 5 to 30 seconds using a year's data from College. Micropulsation storms (sudden increases in amplitude followed by higher than normal activity) in the auroral zones are related to auroral zone electron bremsstrahlung, ionospheric currents and absorption disturbance phenomena. The data strongly imply that the observed micropulsations result from ionospheric currents set up at the onset of a storm with the arrival of bombarding electrons.

Tepley and Wentworth (1963) have examined in detail hm emissions at Palo Alto which were associated with the magnetic storm of September 30, 1961. This was the first severe magnetic disturbance from which data were obtained simultaneously from satellite, balloon and ground observations. They found that short narrow-band hm-emission bursts lasting about one-half hour occurred four hours and two hours before the sudden commencement at 2109 on September 30, 1961. An hm-emission burst occurred immediately after the sudden commencement and lasted approximately six minutes. The main narrow-band hm-emission trains run from approximately 0430 to 1330 on October 1, in approximate time agreement with X-ray bursts observed near Minneapolis. The frequencies of the above emission trains vary slowly but irregularly - over the nine-hour period the average increases from about 1.3 to 2.6 cps. The increases in frequency seem to occur shortly after the intense noise bursts at about 0600, 0900, and 1100. The narrow-band emission trains sometimes appear to merge or separate. At times many bands are observed simultaneously. In particular, between 0630 and 0730, October 1, six distinct bands occur in a frequency interval of less than one cps. A rapidly-rising-frequency fine structure is sometimes associated with burstlike signals. The composite signal differs only slightly in appearance from a series of closely spaced noise bursts. The signal of interest also exhibits a band structure which, although not sharply defined, shows maximum darkening in the frequency interval 0.5 to 1 cps. From the appearance of the signal on a sonagram, it may be placed in a transition region between hm emissions and noise bursts. A rapidly-falling-frequency fine structure is observed in the period immediately preceding 0300. This was the first time that a falling-frequency fine structure has been observed, but the authors have since found several additional examples on sonagrams of other hm emissions. In the time interval 0927 to 0933, October 1, hm-emission bands are observed at about 2.2 and 2.5 cps. The 2.5-cps emission frequency is about twice the fundamental frequency obtained for this time interval from a power-spectrum analysis of X-ray burst data from a balloon-borne scintillation counter near Minneapolis.

Micropulsations in the frequency range under discussion have also been recorded following nuclear explosions above the ionosphere. In the case of the Argus experiments (1958) micropulsations with periods from about one to three seconds were recorded nearly simultaneously at almost all the IGY French stations [Chambon-La-Forêt, Tamanrasset, Bangui, Kerguelen, Dumont d'Urville and Charcot (Selzer, 1959)] and have been described in detail by Eschenbrenner et al (1960). The characteristics of micropulsations observed on the records at Chambon-La-Forêt following the Starfish experiment (1962) have been analyzed in detail by Roquet et al (1962). They appear as a microstructure of a bay-like disturbance and may be divided into four phases: A, lasting from three to four seconds, B and C, both lasting about one minute, and the final phase D. The phase A is an impulse with pulsations initially of period 1.5 seconds, decreasing to 0.5 second. The phases B and C consist of trains of irregular pulsations with a dominant period of about two seconds for the B phase and about four to eight seconds for the C phase. Micropulsations in the D phase resemble pt's.

FINAL REMARKS

Vladimirov and Kleimenova (1962) analyzed the structure of the earth's electromagnetic field in the frequency range 0.5 to 100 cps using data between 1959 and 1961 from stations in the USSR. Most of the observed disturbances in this range are associated with lightning discharges (atmospherics), although the microvariations observed during relatively prolonged intervals of time (trains, stable oscillations, "pearls") cannot be thus explained. They concluded that the largest number of microvariations is found in the frequency range eight to ten cps. Trains of these oscillations may be observed at all hours of the day. Oscillations in the interval 0.5 to 1.5 cps are registered rather infrequently, usually before sunrise (0400 to 0600 LMT). They are generally stable oscillations with approximately constant amplitudes. Microvariations in the range three to five cps are registered least frequently - during the period of observation no trains were observed in this frequency interval. Oscillations of the "pearl" type are usually observed on quiet days, whereas stable oscillations are observed on both quiet and disturbed days. The orientation of the polarization axis is not constant but varies continuously even at the same place and at the same frequency, although it remains practically constant during a single pearl. In this respect, Dawson (1963) found that pearl type micropulsations in the auroral zones are transverse waves propagated along magnetic field lines. Linear and both clockwise and counterclockwise elliptical polarizations are observed, but counterclockwise rotation predominates. In elliptical polarization the major axis shows a tendency to precess in the same direction as the rotation.

Lokken et al (1963) investigated electromagnetic background signals in the vicinity of one cps. They found that the natural background consisted of a comparatively steady, low-level "white" continuum relieved occasionally by the intrusion of either regular or impulsive geomagnetic signals, as well as by the normal occurrence of sferics. The amplitude and the frequency of the intrusions are latitude dependent. Characteristic regularly repeating signal reinforcement patterns are sometimes observed in the auroral zones. Such patterns are displaced in time at conjugate points by one half their respective period (Yanagihara, 1963).

Yanagihara, using data obtained at Great Whale (GW) and Churchill (CH) in the Canadian Arctic and at Byrd (BY) in Antarctica by the Pacific Naval Laboratory and Stanford University during January 1961 (Lokken et al, 1961) classified auroral zone micropulsations with periods from 0.3 to 10 seconds into two main groups, viz. noise bursts and continuous type oscillations. Noise bursts include quite a wide range of frequencies and their waveforms are therefore irregular. They occur suddenly and die out after a short duration. Continuous type micropulsations on the other hand have regular waveforms in general, and Yanagihara divided them into three sub-classes. One of them is the pearl-type of oscillation and the other two are defined by their period, either shorter or longer than three seconds. A short account of this rather fundamental analysis will now be given.

Noise bursts are observed mainly during the hours 2100 to 0200 LMT at GW, their shortest period being 0.2 ~ 0.3 second (see Fig. 12). The main part of a noise burst

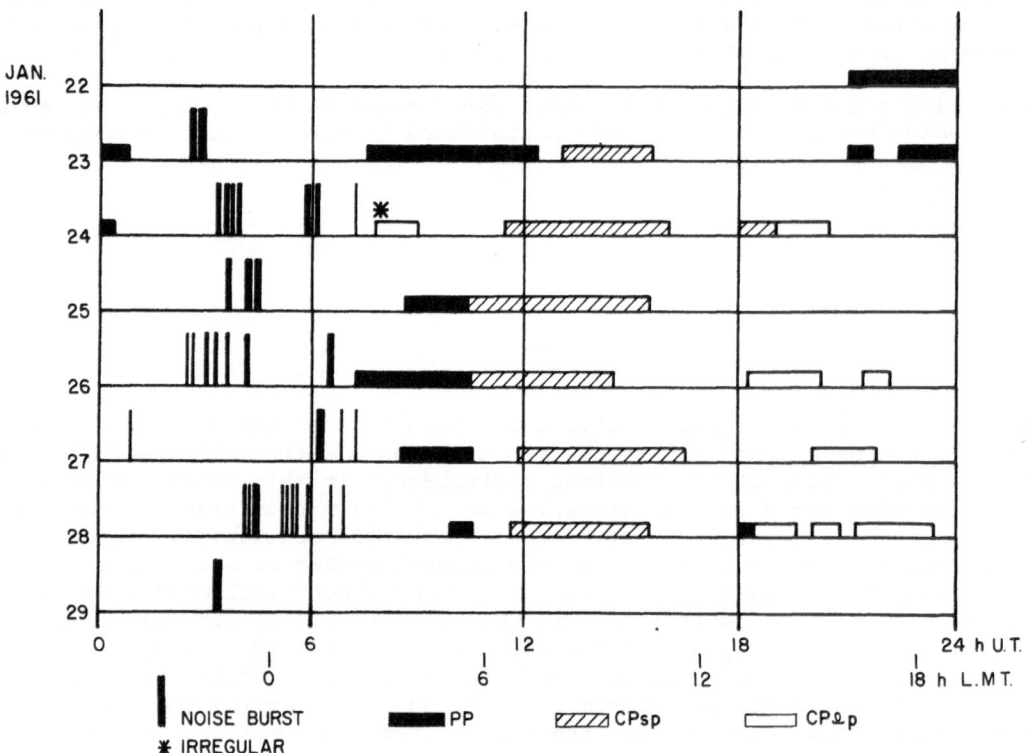

Fig. 12.
Daily occurrence of micropulsations at Great Whale (after Yanagihara).

decreases abruptly away from the central area of activity. Thus, Yanagihara found
many cases where there was no evidence of a noise burst at CH whereas the GW-record
clearly showed a noise burst, and vice versa. [The distance between CH and GW is
about 1,100 km.] Generally speaking, there is a conjugate relationship between BY
and GW (or CH) for noise bursts. A burst at BY corresponds to one at GW in some cases
and to one at CH in others. The conjugacy is thus rather loose (in this sense). The
charged particles responsible for noise bursts do not seem to be existing trapped
particles, but new ones coming from outside the magnetosphere (because of their loose
conjugacy). Their starting points are not necessarily symmetric with respect to
auroral zone conjugate points.

     The difference between PP and noise bursts is quite clear since PP has a simple
or single band of frequencies compared to the broad band of noise bursts. In the
auroral zones, the individual pearls or bunches are separated by a longer time interval than the period of the constituent pulses (see Fig. 13). The interval between
such spaced pearls is of the order of two minutes at BY, GW and CH, while the periods
of the constituent micropulsations range mainly from one to two seconds. The occurrence of auroral zone PP is more frequent than in middle latitudes, although the
diurnal variation of their occurrence frequency is similar to that found in middle
latitudes by Benioff (1960). During almost all of the active periods of PP in one
auroral zone, similar activity can be traced in the other auroral zone, although the
details of simultaneous correspondence are absent and the degree of activity on the
whole is often different. On the other hand, there are some remarkable cases when

338

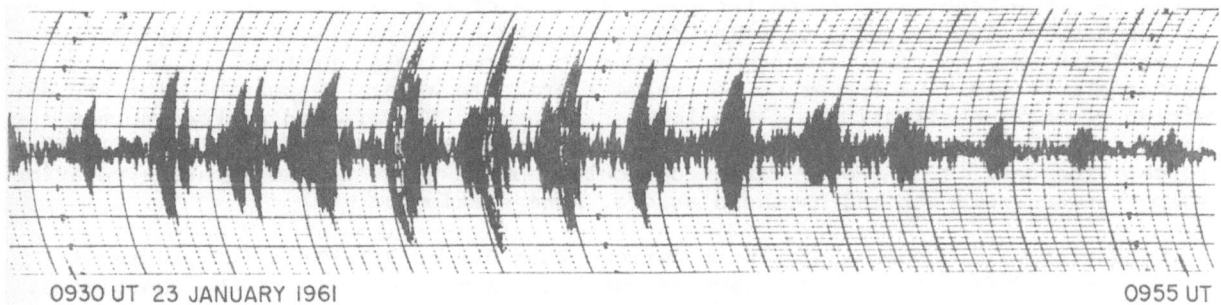

0930 UT 23 JANUARY 1961   0955 UT

Fig. 13. Pearl-type micropulsations (PP) at Churchill,
          January 23, 1961 (after Yanagihara).

pearls on a necklace occur alternately in the northern and southern auroral zones,
the maximum activity at BY occurring midway between a pair of maxima at the northern
stations (see Fig. 14). This fact suggests strongly a bouncing agent with a bounce

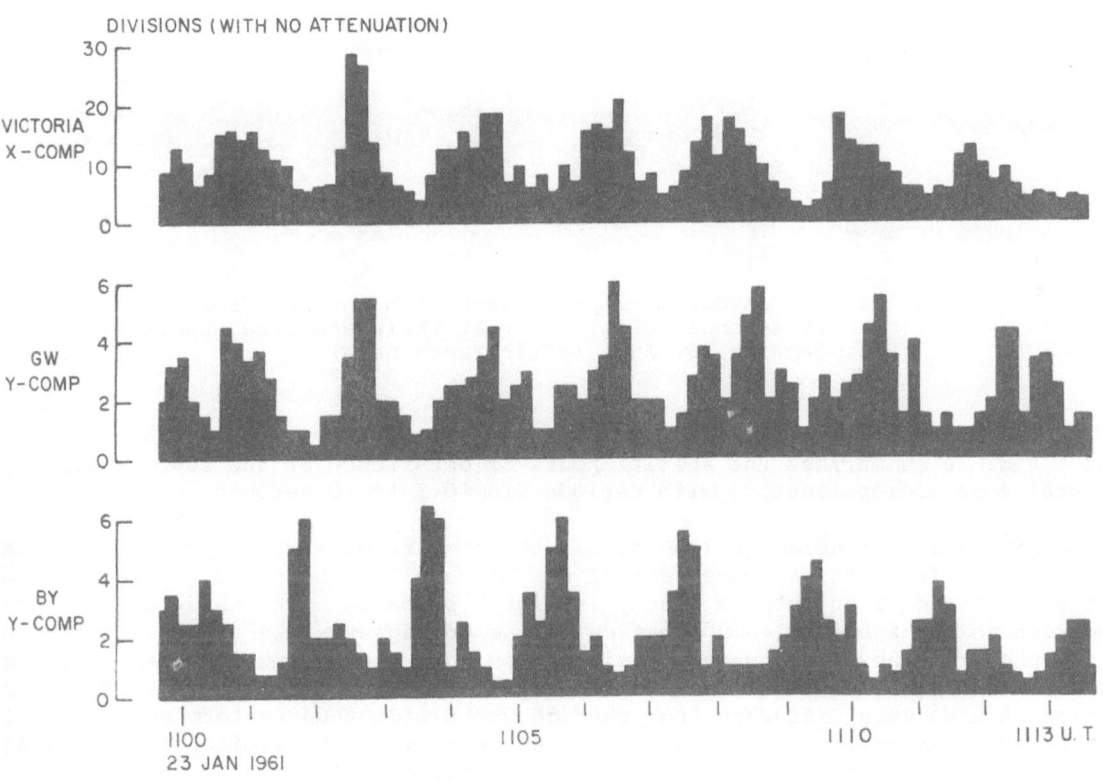

Fig. 14. Alternate occurrences of the maximum of PP at northern
          and southern auroral-zone stations (after Yanagihara).

339

time of 120 seconds - the time between the appearance of consecutive pearls on a necklace. The mean velocity of the agent is thus of the order of 10^8 cm/sec., which is the same as the velocity of solar particles deduced from the time delay between a solar flare and the commencement of a magnetic storm. Solar protons trapped in the geomagnetic field thus appear to be the most probable agents responsible for PP, exciting resonant oscillations in the lower exosphere. The trapped protons do not necessarily form a single bunch. They must be distributed at random along a line of force. The details of the PP-variation would show in this case no apparent relationships between northern and southern auroral zone stations.

Continuous micropulsations with periods from 0.3 to 3 seconds (CPsp) are similar to PP. They differ only by the lack of bunching in the activity of the constituent micropulsations. The change from PP to CPsp is generally gradual, and these two classes are sometimes mixed on the record. Micropulsations with periods from 0.3 to 3 seconds occur in PP type of activity generally during the period 0200 to 0600 LMT, and in the continuous type (CPsp) from 0600 to 1100 LMT. The generation mechanism is considered to be the same as that for PP, and similar conjugate relationships hold. Continuous micropulsations with periods from 3 to 10 seconds (CPlp) (see Fig. 15) are observed conspicuously from 1300 to 1800 LMT. Conjugate relationships are quite

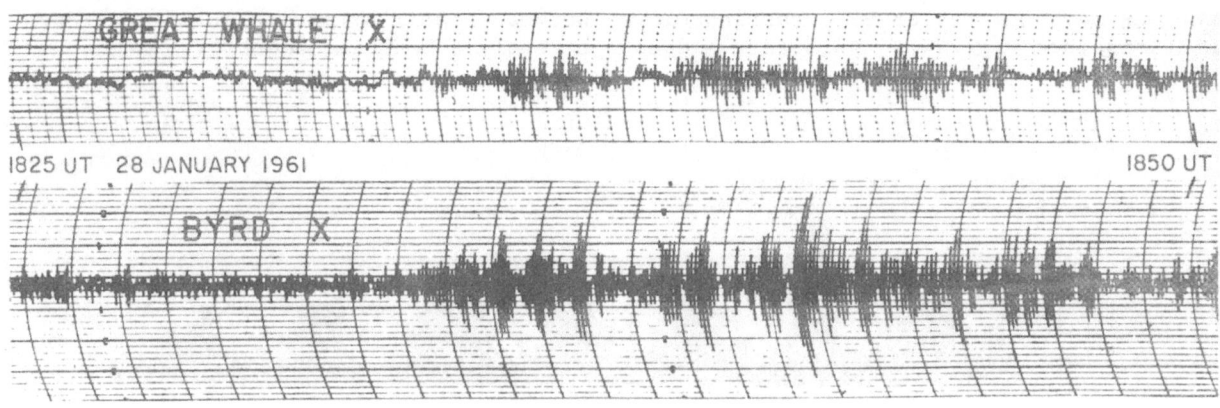

1825 UT   28 JANUARY 1961   1850 UT

Fig. 15.   Continuous micropulsations with periods from 3 to
           10 seconds (CPlp) at Great Whale and Byrd Station,
           January 28, 1961 (after Yanagihara).

similar to those for PP, although details of such simultaneous correspondence are absent. Fig. 16 summarizes the special hours of occurrence of the four groups of auroral zone micropulsations with periods from 0.3 to 10 seconds.

Yanagihara also studied in some detail the simultaneous correspondence of micropulsations with frequencies above several cps between GW, CH and BY, and concluded that the existence of an outside agency is highly probable. The whistler mode of propagation is not a suitable explanation of the conjugacy - the travel time is too long for the frequency range under consideration to explain simultaneous occurrences, which at GW and BY agree within an uncertainty of the order of 1 second. The diurnal variation is also very different from that of world-wide thunderstorm activity, although the diurnal variation of the background level is quite similar (Balser and Wagner, 1962). High energy electrons trapped in the earth's magnetic field are assumed tentatively to be the origin of these higher frequency (2~30 cps) oscillations. When they reach the end of the field lines, they may excite a resonant oscillation or emit an electromagnetic field with the frequencies under consideration.

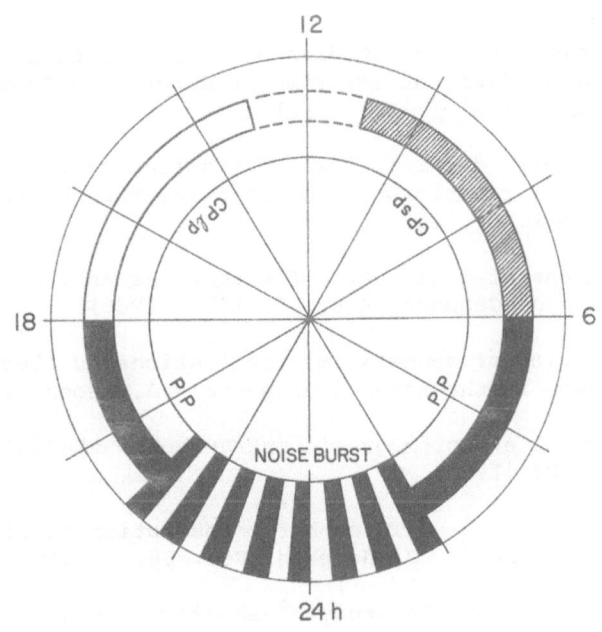

Fig. 16.
Occurrence hours of each type of auroral-zone
micropulsations (after Yanagihara).

Ness et al (1962) in a study of magnetic field fluctuations on the earth and in
space consider the observed micropulsations as the output of a system whose transfer
function (frequency characteristics) is not constant. Thus, although the input to the
system (the sources) may be identical, differences associated with system changes are
observed at the output. The resonant frequencies (eigen modes of oscillation) of the
system are independent of the source characteristics and depend only on the distribution of physical parameters within the system. However, the relative amplitudes and
phases of each eigen mode excited by a particular source is dependent on the source
characteristics. It is possible that a detailed study and analysis of the frequency
spectra of micropulsations with regard primarily to the existence and identification
of distinct modes rather than concerned with relative amplitudes will allow a determination of the model which best represents the oscillating system.

341

REFERENCES

Alperovich, L. V.: "Some results of studies of short-period pulsations of the earth's electromagnetic field at the time of auroras," Geomag. and Aeronomy (A.G.U. translation) 1, 495 (1961)

Balser, M. and C. A. Wagner: "Diurnal power variations of the earth-ionosphere cavity modes and their relationship to worldwide thunderstorm activity," J. Geophys. Res. 67, 619 (1962)

Benioff, H.: "Observations of geomagnetic fluctuations in the period range 0.3 to 120 seconds," J. Geophys. Res. 65, 1413 (1960)

Bomke, H. A.: "The relation of magnetic micropulsations to electric-current and space charge systems in the lower ionosphere," J. Geophys. Res. 67, 177 (1962)

Campbell, W. H.: "Natural electromagnetic energy below the ELF range," J. Res. N.B.S. 64D, 409 (1960)

Campbell, W. H. and H. Leinbach: "Ionospheric absorption at times of auroral and magnetic pulsations," J. Geophys. Res. 66, 25 (1961)

Campbell, W. H. and M. H. Rees: "A study of auroral coruscations," J. Geophys. Res. 66, 41 (1961)

Campbell, W. H. and S. Matsushita: "Auroral-zone geomagnetic micropulsations with periods of 5 to 30 seconds," J. Geophys. Res. 67, 555 (1962)

Cecchini, A., G. Dupouy, J. Roquet and E. Selzer: "Mise en évidence au moyen du <<concentrateur-magnétophone>> de l'Observatoire de Chambon-la-Forêt de signaux magnétiques précurseurs, lors des orages du 11 février 1958 et du 16 août 1959," C. R. Acad. Sci. 250, 4023 (1960)

Dawson, J. A.: "Pearl-type micropulsations in the auroral zones: polarization and magnetic conjugacy," Trans. Amer. Geophys. Union 44, 41 (1963)

Duffus, H. J., P. W. Nasmyth, J. A. Shand and C. S. Wright: "Sub-audible geomagnetic fluctuations," Nature 181, 1258 (1958)

Duffus, H. J., J. A. Shand and C. S. Wright: "Influence of geological features on very low frequency geomagnetic fluctuations," Nature 186, 141 (1960a)

"Geomagnetic fluctuations recorded on magnetic tape during storms," Pac. Nav. Lab. Rep. 60-6 (1960b)

Ebert, E.: "Concerning pulsations of short period in the strength of the earth's magnetic field," Terr. Mag. 12, 1 (1906)

Eschenbrenner, S., L. Ferrieux, R. Godivier, R. Lachaux, H. Larzillière, A. Lebeaux, R. Schlich and E. Selzer: "Analyse expérimentale des effets magnétiques et telluriques de <<l'expérience Argus>> enregistrés par les stations française," Ann. de Géophys. 16, 264 (1960)

Francis, W. E. and R. Karplus: "Hydromagnetic waves in the ionosphere," J. Geophys. Res. 65, 3593 (1960)

Francis, W. E., A. J. Dessler and A. J. Dragt: "Hydromagnetic wave attenuation in
          extreme model ionospheres," URSI-IRE, Austin, Texas, Oct. 1961

Gendrin, R. and R. Stefant: "Analyse de fréquence des oscillations en perle,"
          C. R. Acad. Sci. 255, 752 (1962)

Harang, L.: "Oscillations and vibrations in magnetic record at high latitude stations"
          Terr. Mag. 41, 329 (1936)

Heacock, R. R.: "Auroral-zone telluric-current micropulsations," J. Geophys. Res.
          68, 1871 (1963)

Heacock, R. R. and V. P. Hessler: "Pearl-type telluric current micropulsations at
          College," J. Geophys. Res. 67, 3985 (1962)

          "On the relation of micropulsations (T~3-20 seconds) to the main phase of
          magnetic storms and polar magnetic bays," (Abstract) Trans. Amer. Geophys.
          Union 44, 42 (1963)

Holmberg, E. R. R.: "Rapid periodic fluctuations of the geomagnetic field" I. Mon.
          Not. Roy. Astr. Soc. Geophys. Suppl. 6, 467 (1953)

Hutton, V. Rosemary S.: "Equatorial micropulsations," J. Phys. Soc. Japan 17,
          Suppl. A-II, 20 (1961)

Jacobs, J. A. and T. Watanabe: "Propagation of hydromagnetic waves in the lower
          exosphere and the origin of short period geomagnetic pulsations,"
          J. Atmos. Terr. Phys. 24, 413 (1961)

          "Trapped charged particles as the origin of short period geomagnetic
          pulsations," Planetary Space Sci., (1963) (in press)

Jacobs, J. A. and E. J. Jolley: "Geomagnetic micropulsations with periods 0.3-3
          seconds ("pearls")," Nature 194, 641 (1962)

Jacobs, J. A. and K. O. Westphal: "Geomagnetic micropulsations," Physics and Chemistry
          of the Earth Volume V, 253 Pergamon Press (1963)

Jacobs, J. A., J. E. Lokken and C. S. Wright: "Notation and classification of geo-
          magnetic micropulsations," J. Geophys. Res. (1963) (in press)

Kato, Y. and T. Saito: "Preliminary studies on the daily behaviour of rapid
          pulsation," J. Geomag. Geoelectr. 10, 221 (1959)

König, H.: "Atmospherics geringster Frequenzen," Z. angew. Phys. 11, 264 (1959)

Lokken, J. E., J. A. Shand, C. S. Wright, L. H. Martin, N. M. Brice and
          R. A. Helliwell: "Stanford-Pacific Naval Laboratory conjugate point
          experiment," Nature 192, 319 (1961)

Lokken, J. E., J. A. Shand and C. S. Wright: "A note on the classification of geo-
          magnetic signals below 30 cycles per second," Can. J. Phys. 40, 1000
          (1962)

          "Some characteristics of electromagnetic background signals in the
          vicinity of one cycle per second," J. Geophys. Res. 68, 789 (1963)

Mainstone, J. S. and R. W. G. McNicol: "Screamer activity and pearl-pulsations,"
          Int. Conf. Ionos. London, July 1962

Maple, E.: "Geomagnetic oscillations at middle latitudes. Pt. I The observational data," J. Geophys. Res. 64, 1395 (1959a)

"Geomagnetic oscillations at middle latitudes. Pt. II Sources of the oscillations," J. Geophys. Res. 64, 1405 (1959b)

Matsushita, S.: "On the notations for geomagnetic micropulsations," J. Geophys. Res. (1963) (in press)

Ness, N. F., T. L. Skillman, C. S. Scearce and J. P. Heppner: "Magnetic field fluctuations on the earth and in space," J. Phys. Soc. Japan 17, A-II, 27 (1962)

Roquet, J., R. Schlich and E. Selzer: "Perturbations transitoires mondiales du champ magnétique terrestre, observées en France lors de l'explosion nucléaire spatiale du 9 Juillet 1962," C. R. Acad. Sci. 255, 549 (1962)

Schlich, R.: "Variations rapides du champ magnétique terrestre à la station Charcot en Terre Adélie," IAGA Bull. n. 16c, 71 (1961)

Schumann, W. O.: "Über die strahlungslosen Eigenschwingungen einer leitenden Kugel, die von einer Luftschicht und einer Ionosphärenhülle umgeben ist," Z. Naturforschg. 7a, 149 (1952)

Selzer, E.: "Enregistrements simultanés en France a l'Équateur et dans l'Antartique, des effets magnétiques engendrés par l'<<Expérience Argus>>," C. R. Acad. Sci. 249, 1133 (1959)

"Resultats d'ensemble des stations françaises," IAGA Bull. n. 16c, 63 (1961)

Sucksdorff, E.: "Occurrence of rapid micropulsations at Sodankylä during 1932 to 1935," Terr. Mag. 41, 337 (1936)

Tepley, L. R.: "A study of hydromagnetic emissions," Sci. Rep. 2, Lockheed Missiles and Space Div. (LMSD-8948 14) (1961a)

"Observations of hydromagnetic emissions," J. Geophys. Res. 66, 1651 (1961b)

"Simultaneous observations of hm emissions at middle and low latitudes," Trans. Amer. Geophys. Union 44, 37 (1963)

Tepley, L. R. and R. C. Wentworth: "Hydromagnetic emissions, X-ray bursts, and electron bunches. I Experimental results," J. Geophys. Res. 67, 3317 (1962)

"Hydromagnetic emissions associated with the magnetic storm of September 30th, 1961," J. Geophys. Res. 68, 3733 (1963)

Troitskaya, V. A.: "Earth current installations at the stations of the USSR," Ann. IGY 4, 322 (1957)

"Pulsation of the Earth's electromagnetic field with periods of 1 to 15 seconds and their connection with phenomena in the high atmosphere," J. Geophys. Res. 66, 5 (1961)

Troitskaya, V. A. and M. V. Melnikova: "On characteristic intervals of pulsations diminishing by periods (IPDP) and their connection with phenomena in the high atmosphere," IAGA Bull. n. 16c, 135 (1961)

Troitskaya, V. A., L. A. Alperovich, M. V. Melnikova and G. A. Bulatova: "Fine structure of magnetic storms in respect of micropulsations (T<20 sec.)," J. Phys. Soc. Japan 17, A-II, 63 (1962)

Vladimirov, N. P. and N. G. Kleimenova: "On the structure of the Earth's natural electromagnetic field in the frequency range of 0.5 - 100 cps," Bull. Izv. Acad. Sci. USSR Geophys. Ser. (A.G.U. translation), 852 (1962)

Vozoff, K., R. M. Ellis and G. D. Garland: "Composition of pearls," Nature 194, 539 (1962)

Wentworth, R. C. and L. R. Tepley: "Hydromagnetic emissions, X-ray bursts and electron bunches. 2. Theoretical interpretation," J. Geophys. Res. 67, 3335 (1962)

Yanagihara, K.: "Some characteristics of geomagnetic pulsation pt and accompanied oscillation spt." J. Geomag. Geoelectr. 10, 172 (1959)

"Geomagnetic micropulsations with periods from 0.03 to 10 seconds in the auroral zone with special reference to conjugate point studies," J. Geophys. Res. 68, 3383 (1963)

DISCUSSION

W. H. Campbell

From a comparison of the amplitude-time and frequency-time records of the so-called "pearl" type micropulsations, the NBS group would like to call attention to two not so obvious, yet important, features.

1) From the recurrent emission fine structure, the beating appearance of the amplitude time trace arises. It can easily be shown that the apparent amplitude and "beat" periods are, to a great extent, the result of a summing (within the equipment) of two or more of the typical repeated rising frequency emissions.

2) The often random polarization sense obtained from measurements of the total pulsation field vector can result from a summation of two elliptically polarized vectors with major axes at a small angle to one another and each with the same polarization sense. In addition to the sense of rotation, the ellipticity and the inclination of the major axis of the measured resultant field vector will be time varying functions. The time at which minima occur in the different component directions will naturally vary to some extent between the components. A. Alfvén demonstrated this point with a simple pendulum experiment.

Campbell also emphasized that theories concerning possible sources of "pearl" type pulsations should be more concerned with the primary properties of the phenomena.

R. Gendrin

R. Gendrin gave an account of his theory of fine-structure micropulsations, which involves the motion of bunches of trapped particles. The correlation found between the mean frequency of emission and the repetition period of the pattern is explained by the bounce frequency and the drift period. Correct orders of magnitude are obtained only for protons with energies between 4 and 35 Mev situated on magnetic shells with McIlwain L values between 4 and 1.5. The number and distribution of particles, as a function of L, agree with recent satellite and rocket measurements on the low energy component of the inner Van Allen zone. The injection mechanism is not clear, but diffusion can explain the shape and direction of the recorded emissions. Part of the theory has been published (Gendrin, 1963a, 1963b) and a complete account will appear in Annales de Geophysique.

References

Gendrin, R.: "Sur une théorie des pulsations rapides structurées. Calcul des fréquences observées," C. R. Acad. Sci. 256, 4487 (1963a)

"Sur une théorie des pulsations rapides structurées. Calcul de l'intensité des oscillations observées," C. R. Acad. Sci. 256, 4707 (1963b)

E. Selzer

I would mainly like to clear up any misinterpretations which we may have given to some of the notations used by various authors. I will take as a good example of

346

this the "IPDP" (Irregular Pulsations of Diminishing Periods), introduced some years ago by Troitskaya, which, if well understood, refer to an important feature of most intense geomagnetic storms. But before this particular example, and in order to be a little more general, I will give very briefly some idea of the kind of classification which I have been using myself for micropulsations in the range 1/3 to 10 seconds.

My concept starts from the basic idea of considering these microphenomena as a natural extension of our previous studies in "classical geomagnetism." (This is in contrast to other investigators who consider them as an extension of their usual electromagnetic studies into the field of extreme low frequencies.) With the great progress made in the sensitivities and speeds of response of our equipment, we wish to know more and more about the microstructure of classical geomagnetic phenomena (including relatively "new" entities such as "pt's" and "pc's"). I have considered three main classes:

Class I: Micropulsations which may be considered as a microstructure of huge world-wide (magnetic) storms. I have called them "micropulsations d'orage," or "orages de micropulsations," or "orages dans l'orage," which may be compared to the "sub-storms" of Akasofu and Chapman as they later introduced this term. (J.Geophys. Res., May 1961).

Class II: Micropulsations which may be considered as a primary, or a secondary, microstructure of pt's, or of bays, respectively. I have called them "microstructure of pt's."

Class III: Micropulsations which cannot be recognized as a microstructure of any known phenomena and which take place during very quiet intervals of the "classical" geomagnetic field.

I consider that Class III is identical to the "PP's" introduced by Troitskaya. I have called them "Quiet times PP's," and I think that their study, restricted by this condition of "calmness," has the attraction of the exploration of a totally unsolved mystery.

Class II is something which is presently being extensively studied (principally in its connection with aurorae), and I will not consider them here any further. Class I takes us back to the subject of "IPDP."

These "sub-storms," or "orages de micropulsations," which may be found during the main phase of most intense storms, start, each one of them, individually, with a quite definite "commencement," which marks the beginning of the sequence of the "Irregular Pulsations of Diminishing Periods" introduced by Troitskaya. Personally, I have used the term "micropulsations d'orages" for characterizing just the most active part of each one of these sub-storms. At such a time, and for a few minutes only, the sub-storm presents its most intense and most rapid pulsations of about one second period and one gamma in amplitude.

The statement by Troitskaya (1961) that my findings seem to be different from hers (during the few minutes of maximum activity, the period of the most rapid pulsations had appeared to me as quite constant without showing any "diminishing trend") may be easily explained by the fact that, in looking for "IPDP," she was considering the whole duration of the sub-storm and not only the most intense part of it.

In conclusion, I would like to say that I am just as much convinced as Troitskaya is herself, of the importance of studying these "micropulsation sub-storms," in order to obtain a better understanding of what is happening during the main phase of geomagnetic storms. More generally speaking, I think that micropulsation studies will make more sense if they are conducted in close relation with what appears at the same time on normal magnetograms than if they are considered independently.

References

Troitskaya, V. A.: "Pulsation of the earth's electromagnetic field with periods of 1 to 15 seconds and their connection with phenomena in the high atmosphere," J. Geophys. Res. <u>66</u>, 5 (1961)

S. H. Ward

<u>Dynamics of the Magnetosphere</u>. The impedance concept, so often used in solving problems in electromagnetic theory, may be applied to the dynamics of the magneto-sphere. This concept permits description in the time or frequency domains of the transfer impedance of the magnetospheric system. The input to the system is the time series describing the impact pressure of the solar plasma while the output is the time series which constitutes a magnetogram recorded on the surface of the earth.

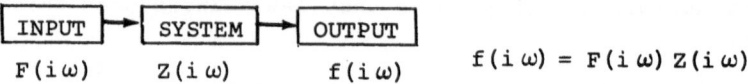

$$F(i\omega) \qquad Z(i\omega) \qquad f(i\omega) \qquad\qquad f(i\omega) = F(i\omega)\, Z(i\omega)$$

The input-output relation of a linear system in the frequency domain is illustrated in the sketch above. By Fourier transformation, we can convert to the time domain as sketched below

INPUT	SYSTEM	OUTPUT

$$x(t) \qquad h(\tau) \qquad y(t) \qquad\qquad y(t) = \int h(\tau)\, x\,(t-\tau)\, d\tau$$

Our problem then is typical of many geophysical problems; only the output is known and both the system and the input are unknown. While no mathematically unique solution may be achieved, physical reasoning and intuition may lead to a self-consistent description of the system.

The starting point for analysis is to assume an input function. Bay disturbances with a sudden commencement accompanied by pulsations of pt and spt type are, on reasonable physical grounds, thought to arise in a step function increase in the impact pressure of the solar plasma. The observed output expressed in the frequency domain, divided by the assumed input, leads to the transfer function. Repeated tests for various assumed inputs and observed outputs will determine whether or not the transfer impedance so computed is self-consistent. The advantage of using this technique lies in avoidance of guesses of the current and wave systems thought to exist in the magnetosphere. Its chief disadvantage is that the system is assumed to be linear whereas it is fairly well established that this is only true for small perturbations in the impact pressure of the solar plasma.

The transfer function is dependent upon position on the earth, and the conductivity and permeability environment of the subsurface. Such features must be evaluated if we are to utilize the transfer impedance concept in developing our understanding of the magnetosphere.

The non-linearity limitation leads us to consider random solar plasma fluctuations of minor amplitude and to attempt to correlate these with certain types of micropulsations observed at the earth's surface. Presumably these random fluctuations lead to quasi-sinusoidal magnetic perturbations as a result of acting upon an exospheric resonant system. Difficulty is encountered with this approach, however, because the quasi-sinusoidal "signals" are contaminated with "noise" considered to arise in random ionospheric currents. The application of the techniques of statistical communication theory permit separation of signal from noise to a fair degree (numerical filters are employed).

The Q of the ionospheric resonant system may be obtained from a power density spectrum computed for the time series magnetogram. Band-limited white noise or Gaussian noise, for examples, will each give a different distribution of periods in the quasi-sinusoids. By comparing the zero-crossing probability function for pure signal with that derived for various assumed random functions taken as input to a resonant system of Q equal to that observed, a description of the solar plasma fluctuations may possibly be obtained. However, much more work is required on this problem.

The "signal" and "noise" concept introduced in order to develop the method has been tested satisfactorily by filtering a pair of orthogonal E field time series about a major spectral peak. Prior to filtering, the ellipse of polarization was unstable, subsequently it was extremely stable, indicating that the power spectral peak represented a natural resonance of one of the exospheric resonant systems.

References

Ward, S. H.: "Dynamics of the magnetosphere," J. Geophys. Res. 68, 781 (1963)

O'Brien, D. P., and S. H. Ward: "Digital filtering of geoelectromagnetic disturbances," (in preparation)

A SUMMARY OF THE OBSERVED CHARACTERISTICS
OF GEOMAGNETIC MICROPULSATIONS*

J. R. Heirtzler
Lamont Geological Observatory
Columbia University
Palisades, New York

INTRODUCTION

In this paper micropulsations will be defined, following common practice, as those natural fluctuations of the ambient magnetic field strength or direction that have periods from about 0.3 to several hundred seconds. On the short period side micropulsations are bounded by the ELF region of the spectrum. The boundary on the long period side lacks a convenient definition. No special attention will be given to bays, crochets, sudden impulses, or other morphological features commonly studied from normal run magnetograms in so far as these can be separated from micropulsations.

Micropulsations of the magnetic field will be discussed. The induced earth currents (telluric field) will only be mentioned briefly. When there are so many observations and when so little can be said with certainty about the magnetic fluctuations it does not seem advantageous to introduce details of the telluric currents. If one excludes those relatively rare currents due to electrokinematic, electrochemical, or thermoelectric effects, there is no reason to doubt that all telluric currents are caused by magnetic fluctuations with sources above the earth's surface.

Within the last decade, there have been published several general reviews of micropulsations. These include articles by Holmberg (1951), Kato and Watanabe (1957a and 1957b), Coulomb (1959), Law and Boothe (1959), Aarons (1960), and Kato (1962). Within the last few years, there have been many developments, particularly in the Western Hemisphere, that have not been covered by these authors. It seems important that some of the main points of recent work be juxtapositioned.

GENERAL NATURE

Few natural phenomena have been recorded so extensively and are so little understood as the time variations of the geomagnetic field vector. Because of recent advances in the understanding of VLF and ELF phenomena, it would seem that micropulsations should next yield to some quantitative explanation. There are a number of major obstacles to extending the physical processes of the higher frequency ranges to micropulsations. These include a different possible source, different possible types of oscillation, the large wavelengths involved and the importance of the crustal structure of the earth. On the purely experimental side, uniformity of instrumentation, adequate geographical coverage, and analysis methods need to be implemented for reasonable progress in this area. In addition, there has not been a strong internationally coordinated effort and this is of vital importance as far as an observational program is concerned.

*Lamont Geological Observatory Contribution No. 741

One of the primary assumptions made by persons studying micropulsations is that they are the direct result of the motion of charged particles above the base of the ionosphere. This assumption is based on the fact that there are no conceivable electrical processes within the earth's crust that can generate activity of this frequency and possible processes deeper within the earth would have their waves severely attenuated by the overlying earth. On the other hand, as will be seen, there has been, on occasion, some correlation with micropulsation activity (generically speaking) with aurora, high altitude nuclear explosions and other perturbations of the ionosphere. The two most important questions relevant to the present subject are: where do rapid fluctuation phenomena originate, and how do they propagate? The prevalent opinion is that the origin is with solar activity, but what type of solar activity and how does this activity produce the records seen on instruments, are questions that are largely unanswered at this time.

Nearly every major study of micropulsations has utilized a different internal classification or labeling of sub-types of activity. While this makes for semantic convenience in any one paper, it is difficult to compare one paper with another. In 1957 the International Association of Geomagnetism and Aeronomy adopted names to be applied to different types of activity, but even these names are used with a great deal of flexibility and some of them are hardly used at all. The names most frequently used in the literature and the ones that will be used here are pc, pt, and pp. Pc is a rather regular sinusoidal type of micropulsation that is evident to the eye on magnetometer records during the day. Pt activity has an irregular wave shape, has about 5 to 20 minutes duration and is most evident during the night. The letters "pc" mean "pulsation, continuous" although for some of the longer period daytime pulsations there is not much continuity of phase. The letters "pt" stand for "pulsations, trains" although sometimes the "train" consists of a single disturbed period. "Pp" identify "pearly pulsations" which are pulsations with periods of less than about three seconds and which are amplitude modulated, or exhibit a beat, so that their envelope resembles a string of pearls (Troitskaya, 1959). The inconsistency of names is not an indication of the independence of the investigator, but rather, reflects the elusive nature of the phenomena as a class. The lack of basic guides, in classification and statistical measure, has led to a large number of conclusions even from extensive and careful studies later being obviously invalid. The volume of papers in this field does not reflect the volume of knowledge.

It will serve little purpose to present figures showing "typical" events since their wave shape varies considerably and depends upon the instrument used to record them. The IAGA has an album of events for persons who wish to see numerous records.

Compared to the more well known magnetic events, these fluctuations usually have small amplitudes that range from the limit of detectability up to a gamma (1 gamma = 10^{-5} oersted) and rarely, in the auroral zones, up to a few tens of gammas. The vertical component is frequently observed to have an amplitude of one-tenth that of the horizontal components. The longer periods usually have larger amplitude than the shorter periods. The amplitude spectrum commonly increases with period at a rate of 6 db/octave (20 db or a factor of 100 per decade).

The question of polarization of the incident waves is important. The polarization of observed terrestrial activity is, however, intimately connected with the anisotropy of induced earth currents through the secondary magnetic field that these currents create above the surface. No extensive and detailed study of the polarization of micropulsations has been made. Averaged over a few hours or a few days' time, the horizontal components at most stations show an elliptical polarization. On a shorter time scale of a few cycles, the polarization may be extremely variable. The direction of rotation of the ellipse as recorded at several sites simultaneously has not been critically examined. There is a general tendency for nighttime pulsations to have the major axis of the ellipse oriented North-South. For certain pulsations of several minutes' period at conjugate auroral stations, simultaneous rotations in

the opposite sense have been noticed (Wilson and Sugiura, 1961), but it is difficult
to formulate rules for this behavior (Matsushita, 1963). Lawrie (1962, 1963) has
demonstrated a spatial gradient of phase for certain magnetic storms, but this type of
analysis has not been made for micropulsations.

VARIATION OF CHARACTERISTICS OVER SMALL DISTANCES

When one considers the altitude above which micropulsations must originate and considers the likely dimensions of the source and the wavelengths, it is not probable
that the waves incident on the earth's surface will change their character any great
amount over horizontal distances of a few kilometers or a few tens of kilometers.
However, the natural magnetic activity observed on the earth's surface may be, and
frequently is, quite different for stations with small separation because of the variations in electrical characteristics of the earth at the stations. Local geological
conditions may affect the surface observations because of the earth's magnetic susceptibility or electrical resistivity structure. The former cause is usually less
important than the latter. The value of susceptibility does not differ appreciably
from that of a vacuum except over iron deposits. Resistivities for the earth's crust
range from 1 to 10^6 ohm-cm with the higher values being more common. Sea water has a
value of about 20 ohm-cm.

The currents induced in the earth reduce the vertical component of magnetic field
strength. When there is a strong horizontal discontinuity in resistivity, as a sea-
land boundary, there may be an alteration in other components as well. Although geo-
magneticians have long questioned the suitability of observatories near the sea,
recent studies have further questioned the activity at these stations. Parkinson
(1959) has demonstrated that for fluctuations of a few minutes' duration at Australian
observatories, measurements near the sea are influenced by the presence of the sea
while midcontinent ones are not. Mason (1963) has shown that on the island of Oahu,
the amplitude and phase of all components of the fluctuation are dependent upon where
the recording instruments were situated on the island. He found measurable differences for periods as great as the 24-hour period presumably due to induced currents
in the sea. These results question the utility of ocean island observatories until
further studies have been made. Duffus et al (1959) have shown the change in relative
amplitude away from the sea-land interface (Fig. 1). That these results generally
conform to what one might expect has been demonstrated by Weaver (1963). See Fig. 2.

For long-period events, Schmucker (1960) has shown that the relative spectra of
H and Z components may be used to study earth structure. Rikitake, Yabu and Yamakawa
(1962) have used the anomalous long period Z variations of Japanese observatories to
study deep earth structure. Since the earth's electrical resistivity shows more
variability near the surface than at depth, micropulsations are more likely to be
affected by the presence of the earth than are longer period variations. It should be
noted that when variations of the geomagnetic field are used to determine the earth's
electrical structure, a knowledge of the nature of that field is presupposed (Price,
1962). Similarly, if the electrical structure of the earth is known (as for rapid
variations over the oceans), some knowledge may be gained about the nature of the
rapid magnetic field variations.

That the induced currents can change character over small distances was illustrated by a series of experiments by Wescott and Hessler (1960). Fig. 3 shows how
the preferred direction and magnitude of major axis of the telluric current ellipse

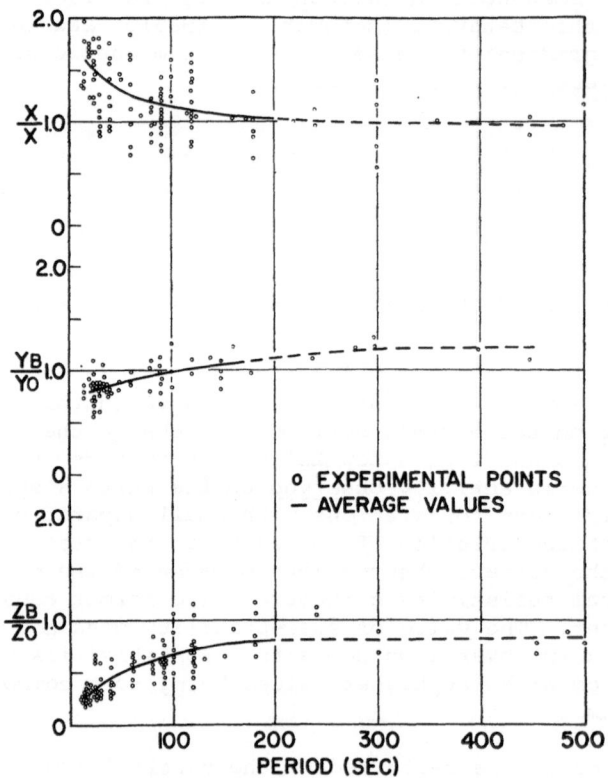

Fig. 1. The ratio of the north-south, east-west, and vertical components of micropulsation at Bear Creek to those at Albert Head, as a function of period. Albert Head is on Vancouver Island at the coast. Bear Head is approximately 12 miles inland. (Duffus et al, 1959)

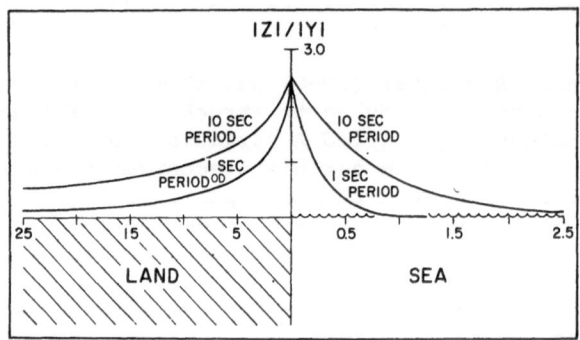

Fig. 2. Computed effect of a north-south, sea-land boundary on micropulsation amplitudes. The variations are contained in the Z component. Note horizontal scales are different on opposite sides of center of figure. East-west magnetic fluctuations. (Weaver, 1963)

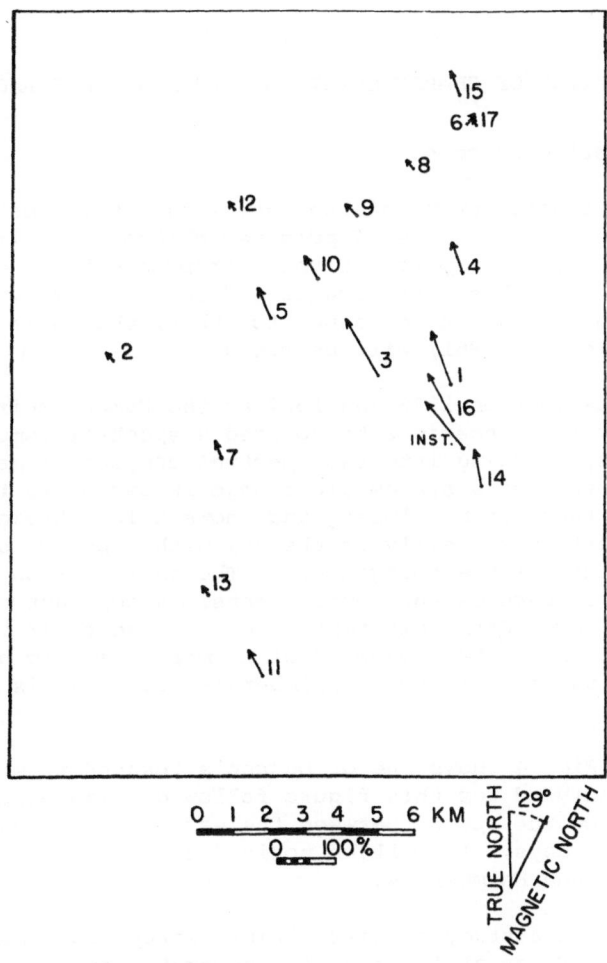

Fig. 3. Variations in direction and magnitude of the preferred axis of the
        telluric current ellipse, near College, Alaska. Amplitudes are rel-
        ative to that at the Geophysical Institute, marked INST.   (Westcott
        and Hessler, 1960)

changed over distances of a few kilometers in Alaska. On an ice island, Hessler
(personal communication) found no preferred direction for telluric currents.

    Although from place to place the change in magnetic susceptibility is usually
quite small, Ward and Ruddock (1962) have demonstrated how micropulsations differ in
amplitude over a magnetite body.

    Obviously, until more is known about the effect of the earth at the recording
site, a detailed comparison of wave shape at distant stations or a detailed study of
the polarization of the incident wave at a single station will be difficult. In an
attempt to circumvent the effects of local geology persons frequently use the characteristics of the horizontal components when investigating similarities over large
distances, although some persons study only the vertical component.

VARIATION OF CHARACTERISTICS OVER LARGE DISTANCES

Diurnal Variation of Period of pc's.

In order to compare activity of distant stations, it is necessary to come to some understanding of how the periods of pc's may change with time at a single location, so that activity may be compared at the appropriate hour, day, season, etc. for the stations. Some evidence has been accumulated for the systematic variation of pc period over the daylight hours. As evidence for this, the works of Holmberg (1951), Duncan (1961) and Ness et al (1961) will be cited.

In studying records made in 1926 and 1927 at the Eskdalemuir observatory, Holmberg states that on numerous occasions he noticed a spectral component which appeared at the late night hours, breaking into two spectral components early in the day, and that these two components have a systematic change in period as the day progresses. Fig. 4 (upper part) is taken from Holmberg and shows this behavior for selected days. The long period component starts early in the day with a period of 30 to 50 seconds and gets longer until the early evening hours. The short period component starts with a period of 20 to 30 seconds and remains sensibly constant or gets slightly shorter throughout the day. Both components seem to lose their identity early in the night. Many aspects of this extensive work of Holmberg seem to be unknown, probably because of the general unavailability of Holmberg's work (thesis at the University of London).

The lower half of Fig. 4 shows one of Duncan's results from data at Adelaide, Australia. The days analyzed for this figure followed a magnetic storm and this data is presented only to show the spread between the early morning and midday frequencies over a period of several days. It will be noticed that Duncan's long period component is appreciably shorter than Holmberg's.

Ness et al (1961), in a study of terrestrial micropulsations for a time when the U. S. satellite Explorer X was aloft, made 150 separate spectral analyses with an electronic digital computer. The spectra were determined over adjacent 20-minute intervals during the day and night for five days when there was a strong spectral component in the 5- to 20-second period region. The data analysis interval precluded any information about a long period component. Fig. 5 shows how the short period spectral peak shifted its period over the five days. Notice that the period axis increases downward so that the period gets shorter during the daylight hours. There was a sudden commencement at 1503 UT, March 27. (Depending upon one's definition, he might say that pc's occur at night as well as day!) The amplitude of Ness' peak did not show any systematic diurnal behavior, but this may be expected since his analysis interval was 20 minutes.

Ness used a digitally recording rubidium vapor (total intensity) magnetometer at Fredericksburg (large dip angle); Duncan measured the vertical component with loops buried in the ground, recorded on magnetic tape, and made frequency-time analyses; Holmberg hand-scaled records from a horizontal loop. Thus, all used the Z component. Usually the spectral peaks that result from digital electronic computer analysis are not evident to the eye on the analogue records unless they are strong spectral peaks.

Holmberg commented that the maximum frequency of occurrence seems to be in the afternoon for the long-period waves and in the forenoon for the shorter-period ones.

356

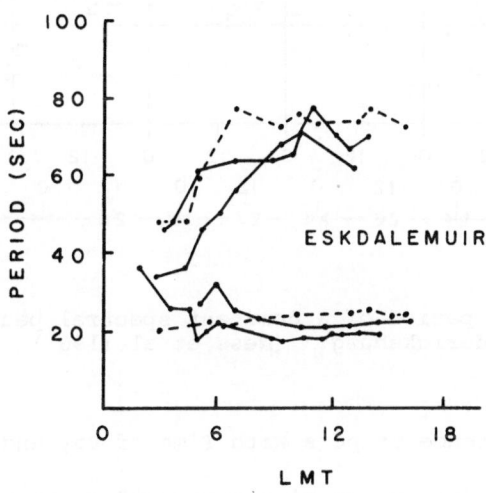

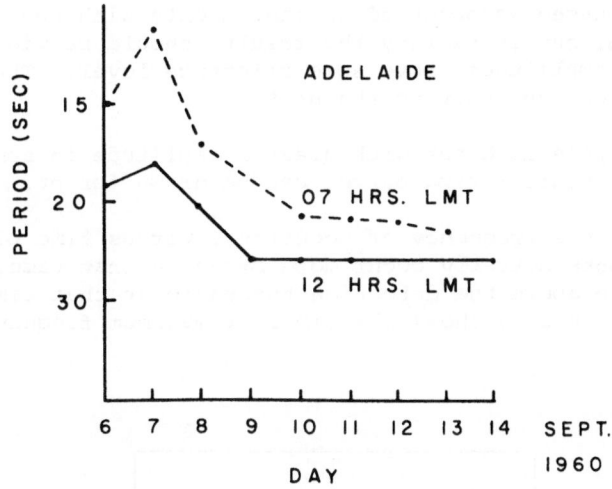

Fig. 4. (Upper) Examples of variation of pc period with
        LMT at Eskdalemuir during 1926-1927. (Holmberg,
        1951)

        (Lower) Variation of pc period with LMT over
        several days at Adelaide. (Duncan, 1961)

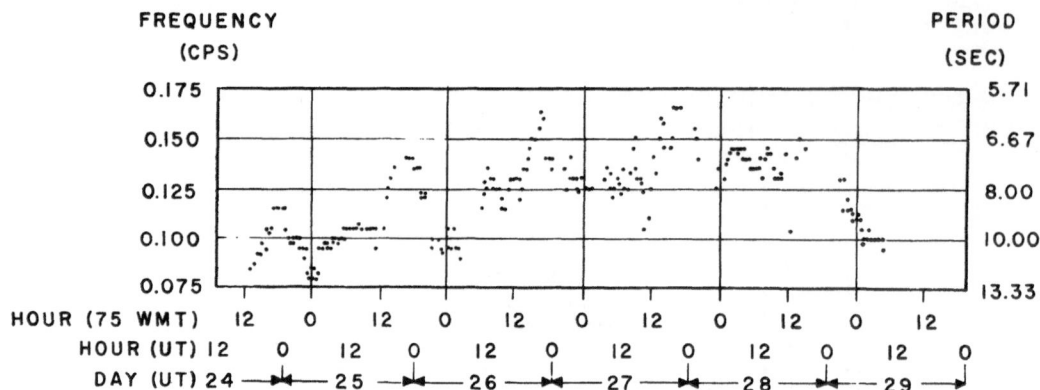

Fig. 5. The variation in period of a dominant spectral peak over
five days at Fredericksburg. (Ness et al (1961)

Variation of Frequency of Occurrence of pc's With Time of Day and Longitude.

As can be appreciated, the "frequency of occurrence" of an event is related to
the sensitivity of the instrument used to record them. This is true here since pc's
are not known to have any minimum amplitude level. In this section the frequency of
occurrence will be discussed as observed on instruments with nearly comparable
thresholds of detection, but in reality the results should be viewed as strictly
applying only to those amplitudes above some reference level. The reference level is
usually not the same from one study to the next.

That simultaneous pc's do occur with greater amplitude in some latitudes is
illustrated by Fig. 6. Similar type curves can be drawn for pt.

A curve that shows the frequency of occurrence versus time of day may have a
maximum because the events actually occur more often at that time, because a larger
percentage of events are above the detection threshold at that time, or a combination
of these circumstances. Fig. 7 shows the times of maximum frequencies of occurrence

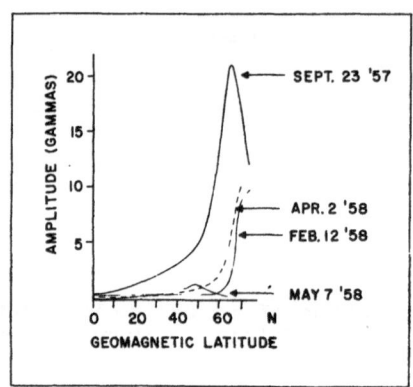

Fig. 6. Examples of the variation of pc amplitude with
geomagnetic latitude. (Jacobs and Sinno, 1959)

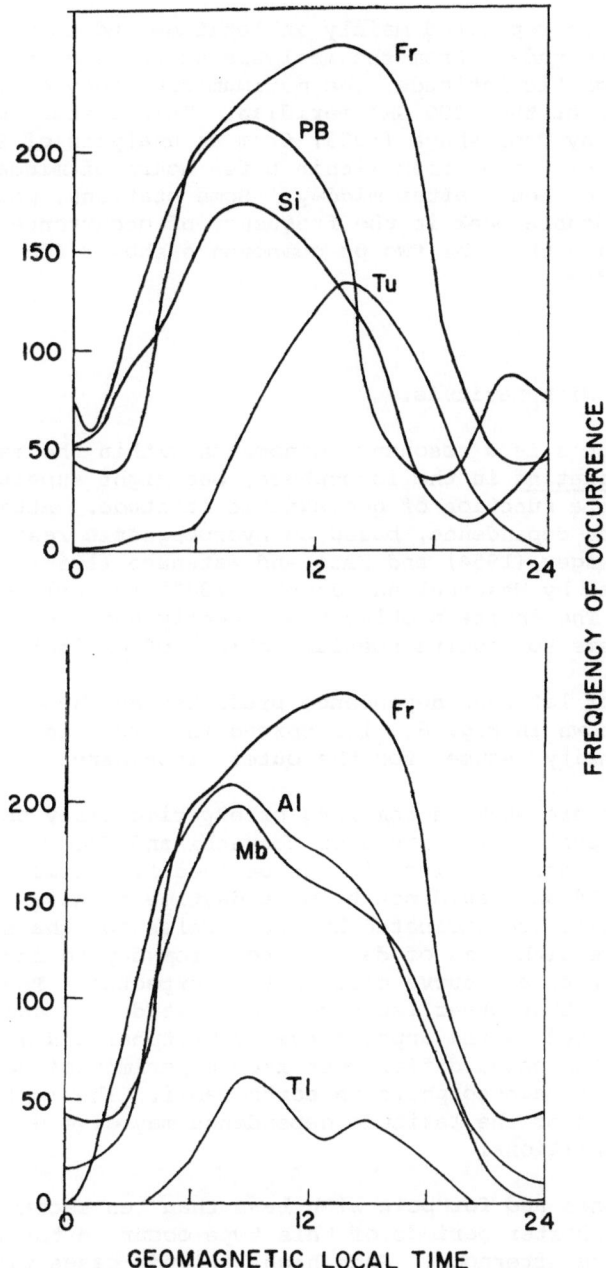

Fig. 7. Examples of the variation of frequency of occurrence of pc with
geomagnetic latitude (upper) and longitude (lower).   (Jacobs
and Sinno, 1959)

| | Geomagnetic | | | Geomagnetic |
Station	Latitude		Station	Longitude
PB Point Barrow	68.6°N		FR Fredericksburg	349.9°E
SI Sitka	60.0°N		MB Memambetsu	208.3°E
FR Fredericksburg	49.6°N		AL Alma Ata	150.5°E
TU Tucson	40.4°N		TL Toledo	74.7°E

for one group of stations separated mainly in latitude and another group of stations separated mainly in longitude. From the analysis of three stations at approximately 40 degrees north geomagnetic latitude, the maximum frequency of occurrence was found to have a highest value at the 2300 GMT meridian. This differs by several hours from the GMT meridian found by Troitskaya (1953) from an analysis of Soviet telluric stations. All stations showed a maximum within a few hours of midday; some a few hours before midday; some a few hours after midday. Some stations, particularly in the South Pacific, show a double peak in the frequency of occurrence which may correspond to Holmberg's observation that the two pc components show different amplitude maximum at different times of day.

Variation of pc Period With Latitude.

If the origin of pc's is a resonant phenomenon within the magnetosphere rather than a phenomenon originating in the ionosphere, one might expect that the period of the pulsations would be a function of geomagnetic latitude. Attempts to formulate the nature of a latitude dependence, based on hydromagnetic resonances of the magnetosphere, were made by Dungey (1954) and Kato and Watanabe (1956). Dungey's calculations have been extended by Westphal and Jacobs (1962) to include more recent knowledge of the magnetosphere. The entire problem has recently been reformulated by McDonald (1961), although he seems to require special sources of excitation.

The general type of latitude dependence predicted by these calculations, for the fundamental mode, is shown in Fig. 8. The spread in curves is due, mainly, to the different values of density assumed for the outer atmosphere.

In the same figure are shown a sampling of experimentally observed pc periods taken from published reports. The trend of Obayashi and Jacobs (1958) observations should, perhaps, not be included since it was derived from simultaneous records from stations around the world and could not be pure daytime events. Several spectra showed two peaks and these are indicated by two circles for the same observation. Since the periods change with time of day, if not from day to day, a better fit of this experimental data with any curve could not be expected. These data are presented only to emphasize that other factors than latitude are operative, and to display an effect indicated by the upper curve of Westphal and Jacobs, namely that some of the more realistic cases differ most from experimental observations. On Fig. 8, α is the radius of the magnetosphere in earth radii. The implication here is that an accurate determination of the latitude dependence may give a sensitive indication of outer atmospheric conditions.

For the auroral zones and for pc's with less than ten seconds, Yanagihara (1963) has suggested that the shorter periods of this type occur in the local forenoon and the longer periods in the afternoon. Such high latitude cases with periods less than ten seconds are apparently not of the type shown in this figure.

A thorough study of the latitude effect of period can only be undertaken with a careful selection of instruments, recording simultaneously along the same magnetic meridian and by the use of good analysis techniques. No such studies have been made, but several are scheduled at this time.

It is rarely, if ever, possible to identify the same pc wave train peak for peak over considerable distances. Over short distances, when the separation of stations increases from a few miles, it is possible to see the wave undergoing a gradual transformation in shape, but even here, it is difficult to express the alteration in a predictable and analytical way.

360

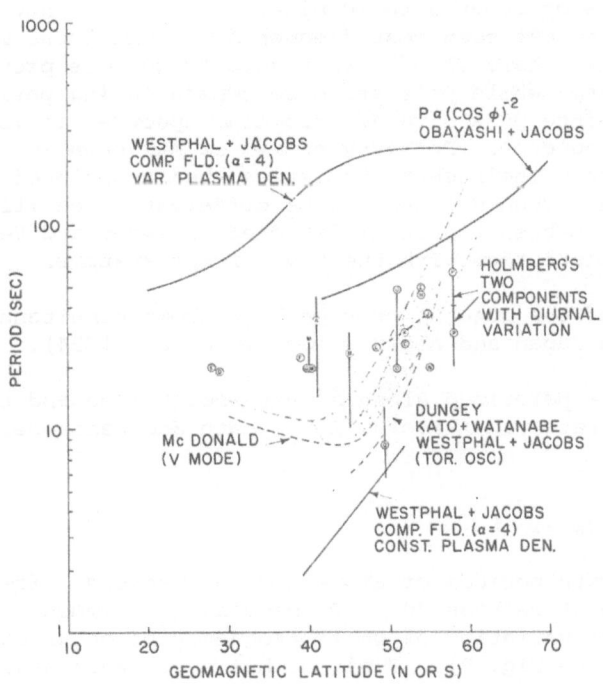

Fig. 8. Variation of pc period with latitude, predicted and observed.
Theoretical trends, (McDonald, 1961; Dungey, 1954; Kato and
Watanabe, 1956; and Westphal and Jacobs, 1962). Observations,
(Obayashi and Jacobs, 1958) and as follows:

Key	Station	Reference
A	Arizona	Berthold, Harris, Hope (1960)
B	Adelaide	Duncan (1961)
C	Hobart	Duncan (1961)
D	Townsville	Duncan (1961)
E	Onagawa	Kato and Saito (1959)
F	Los Angeles	Campbell (1959)
G	Tucson	Maple (1959)
H	Witteveen	Scholte and Veldkamp (1955)
I	Victoria	Duffus and Shand (1958)
J	Wingst	Voelker (1962)
K	Gottingen	Voelker (1962)
L	Furstenfeldbruck	Voelker (1962)
M	Austin	Bostick and Smith (1961)
N	Abbotsford	Westphal and Jacobs (1962)
O	New Jersey	Heirtzler (unpublished)
P	Eskdalemuir	Holmberg (1951)
Q	Fredericksburg	Ness et al (1961)

The Characteristics of pt's.

Less is known about the characteristics of pt's than about pc's. They consist of an irregular disturbance of about 5 to 20 minutes' duration, have been seen on records mostly at night and have been seen most frequently around local midnight (Coulomb, 1959). According to Yanagihara (1957), they seem to be more prevalent during periods of minimum solar activity, while pc's are more common during periods of sunspot maximum (Benioff, 1960). Since pt's have an irregular spectra, it is difficult to discuss them on the basis of frequency. For many years, it has been known that they tend to occur with magnetic bays. Their characteristics when associated with positive and negative bays (of the H component) seem to be different in an ill-defined way. Bays are a phenomena that have been thoroughly studied (Silsbee and Vestine, 1942). It is natural to try to associate them with the bay current systems.

Impulsive events of the general pt-type have shown simultaneity even outside the night hemisphere, as in Japan and Algeria (Grenet et al, 1954).

The pt's seem to be polarized along a north-south line and to be thought of as poloidal oscillations (Kato and Akasofu, 1956; Kato and Watanabe, 1957a, 1957b, 1958).

The Characteristics of pp's.

Pearls are waves with periods of about 0.3 to 3 seconds, frequently amplitude modulated with a period of perhaps 20 to 30 seconds. A commonly quoted characteristic of pearls is the diurnal variation of their frequency of occurrence as noted on Benioff's instruments (see Fig. 9). Their common appearance near sunrise and sunset has led to a number of independent but similar conjectures, to wit, they are associated with a transient condition near the night-day boundary in the ionosphere.

It has frequently been mentioned that pearls have lower amplitude at lower latitudes. Also, they have no known minimum so that one must exercise care in interpreting that frequency of occurrence distribution. A diurnal curve for Uppsala did not show such strong sunrise-sunset peaks, although it was based on a shorter recording period.

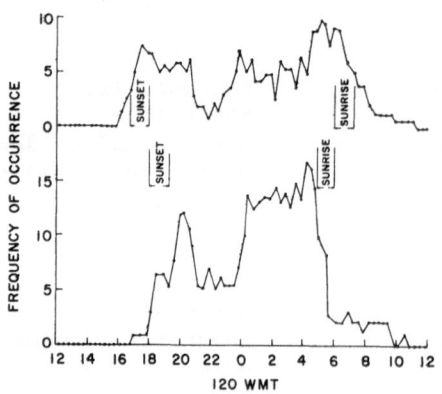

Fig. 9. Diurnal variation of frequency of occurrence of pearls in California for the years 1955-1959. Upper: summer; Lower: winter (Benioff, 1960).

362

In the same study, Benioff, using a set of identical instruments, reports that there is occasionally nearly identical pearl activity for stations 290 km apart and correspondingly less similar for stations separated by greater longitudinal distances.

Pearls are a relatively rare phenomenon. In the study, we have been quoting they occurred about once a month at midlatitude stations and did not show any strong yearly trend. Their rarity requires that considerably longer recording times must expire to get representative data (if the papers on the more common pc's and pt's offer any lesson) so that they have only been studied in recent years. Their characteristics are discussed at greater length in an accompanying article by Jacobs.

CORRELATIONS OF MICROPULSATIONS WITH OTHER PHENOMENA

Campbell and associates have made a particularly thorough study of the relation of 5 to 30-second period micropulsations to fluctuations in auroral intensity, cosmic noise absorption and electron bremsstrahlung at College, Alaska. These results have been reported in a series of letters and papers from 1959 to 1962. In particular, Campbell and Rees (1961) and Campbell and Matsushita (1962) illustrate a very near peak—to—peak correlation of the time derivative of the N-S micropulsation intensity (as measured by an induction coil) and the luminosity fluctuations of the 3914 A line of aurorae visible to them at the time. At the time of these aurorae, the electron density maximum was at 90 to 110 km in the E-region. Both phenomena are tentatively thought to be due to electron precipitation into that polar region. It is tempting to think that the magnetic activity is due to current systems at E-layer height. Presumably, this micropulsation activity falls in the pt class, since aurorae are observed at night (if the pc-pt classifications apply in polar regions). Jacobs and Sinno (1960), from IGY data, constructed pt and pc current systems which closed over the auroral zones. Others have tried but have not been able to construct current systems for micropulsations from data available. If large hemispheric current systems are responsible for pc oscillations, it would be very difficult to explain any dependence of period on geomagnetic latitude.

A number of persons have noticed the relationship of the three-hour K index, and the planetary K_p index, to shorter period micropulsation activity. The K index scale is somewhat subjective and does not give any direct information about physical processes. From Campbell's studies, one would expect to find those 5 to 30-period phenomena which are common at night to show a positive correlation with reasonably high K value, as determined by an observatory not too far away. This has been the case. However, Yanagihara (1957) finds a linear relationship between frequency of occurrence of telluric pt with planetary K_p up to $K_p = 5_o$: an observation which is hard to understand if pt's are a local auroral phenomenon. Maple (1959) suggests that there is no, or a negative, correlation between longer period micropulsations and K index so that one must exercise care in relating all pc's to auroral phenomena.

Sechrist (1962), in observing rapid phase changes in 18 kcps transmissions between Panama and Pennsylvania, noticed rapid fluctuations in phase, with periods of 10 to 15 seconds, during magnetically disturbed times. He deduces that slight motions of only a few hundred meters of the E-layer could be responsible. Chan, Kanellakos and Villard (1962) noticed small rapid shifts in frequency in 18 and 20 Mcps transmissions across the United States which correlate very closely with rapid changes in the geomagnetic field strength. Flock, Belon and Heacock (1962) have noted the relationship between telluric current activity and radio aurora.

In view of the possible relationship of micropulsations to aurora, the ionosphere or overhead current systems, it is of interest to know the characteristics of micropulsations near the magnetic equator. In a recent report on magnetic observations in the Philippines near the magnetic equator, Glover (1963) reports 50 seconds as the dominant period of pc's. This is in contrast to the theoretical trends shown in Fig. 8. He also reports maximum activity in September-October and in April-May coinciding with enhancements of the equatorial electrojet.

The only known man-made origins of micropulsations were the U. S. high altitude nuclear detonations of 1958 and 1962. The July 1962 explosion, in particular, caused transient magnetic effects, simultaneously over the world, occurring at the time of the shot and about two seconds later. (Roquet, Schlich and Selzer, 1963) The first effect may be due to energy traveling in the earth-ionosphere cavity and the second effect may be due to radiation arriving near the top of the ionosphere and, thence, taking about two seconds to propagate downward to the base of the ionosphere (Francis and Karplus, 1960) and then spreading in the earth-ionosphere cavity. A micropulsation train of about 3.5 to 4.5-second period was observed immediately following the July 1962 detonation. More details about these important effects are given in another paper in these proceedings.

Another type of experiment that is proving very valuable in the study of rapid variations of the magnetic field are those measurements that are made simultaneously at magnetically conjugate points. These effects will also be discussed in an accompanying paper.

DATA FROM SCIENTIFIC SATELLITES AND SPACE CRAFT

When one considers the number of space experiments performed by the U. S. and the U.S.S.R since 1957 and the frequency with which magnetometers are mentioned in connection with them, one would be inclined to think that this is a rich source of micropulsation data. Such has not been the case. Those satellites and space rockets which have carried magnetometers are Pioneer 1, 2, 5, Vanguard 3, Explorer 6, 10, 12, 14, and Mariner 2, launched by the U.S., and Sputnik 3 and Lunik 1 and 2 and a Venus vehicle, launched by the U.S.S.R. Many of these magnetometers were designed for specific studies of the earth's main field or the nature of the geomagnetic field boundary. Without going into particulars, the several instruments on any one satellite have allowed a relatively small output information rate to a magnetometer and yet a relatively large amount of data handling is required; both factors put restrictions on the interpretation of the data*. Additionally, the instrument is sampling a new complex environment and can confuse spatial with temporal variations. In spite of the experimental problems few would question the importance of this type endeavor.

The most complete published analysis of satellite data relevant to micropulsations is a series of four papers by Sonett and collaborators (Sonett, Sims and Abrams, 1962; Sonett, 1963a and 1963b; Sonett and Abrams, 1963). Even here only about five hours of data was available between 3.7 and 7 earth radii and between 12.3 and 14.6 earth radii. The outer region showed sharp, irregular changes in intensity and direction such as have been noticed on other occasions near the termination of the geomagnetic field and outside of it (cf. paper by Axford in this publication). The inner region between 3.7 and 7 earth radii (two hours of data) could be analyzed in a statistically

---

*However, indirect information about fluctuations of the magnetic field can be obtained from plasma probes and, to a lesser extent, from particle counters.

meaningful fashion and showed several spectral peaks. Within the limitations of the search coil magnetometer used in that experiment, these were interpreted as magneto-acoustic, transverse (extraordinary) and mixed hydromagnetic modes. The periods of the peaks varied from 1 to 16 seconds. Persons who are familiar with band limited terrestrial magnetometer records can visualize the nature of this satellite data and will appreciate that it is extremely difficult to draw general conclusions from so limited data.

Pioneer 5 also carried a search coil and acquired data on rapid changes of the magnetic vector, (Coleman et al, 1960). This data has not been extensively examined from the point of view of micropulsations or hydromagnetic modes. Mariner 2, which traveled to Venus last year with a triaxial fluxgate magnetometer system, took no measurements in the region under the influence of the earth's magnetic field (Coleman et al, 1962)

Explorer 10 was expected to yield a great deal of information on the rapid fluctuations in magnetic intensity since it carried a rubidium vapor magnetometer. Most of the telemetry channel capacity was reserved for that instrument which became inoperative shortly after launch. Two fluxgates did operate for 55 hours, but their sample rate was such that they yielded information about the earth's main field and its termination rather than about very rapid time variations (Heppner et al, 1963).

The U. S. National Aeronautics and Space Administration is planning to launch EGO in the near future and from it acquire information on ELF activity and micropulsations from its eccentric orbit. Also next year, NASA plans to launch POGO in a relatively low altitude polar orbit. The rubidium instrument on POGO will take a field reading each second. It is planned to make digital readings from 50 magnetic observatories to try to separate the time and spatial variations. Both EGO and POGO are expected to be operational for a period of at least a year.

CONCLUDING REMARKS

As Bleil mentioned in the introduction to this meeting, the last decade has seen a great increase in interest in micropulsations. During this time a few, but very few, characteristics of the phenomena have been firmly established. Many of the rules governing their behavior have been challenged. Experimentalists have applied improved recording techniques and statistical methods. Perhaps most important of all the problems have been faced more realistically.

Geographic coverage by recording instrument has been greatly enlarged although internationally coordinated programs have not been enlarged at an equal rate. In North America the number of micropulsation recording stations has probably increased ten-fold in the last ten years. Fig. 10 shows stations that record micropulsations on a fairly regular basis. This includes only those magnetic detectors that have a recording speed of 180 mm/hr or greater, although many do not make a direct paper record. Nearly all of these, and some others, are participating in a new program whereby simultaneous records, for selected times, are sent to a central location for data exchange. This program has the difficult name of Cooperative Geomagnetic Micropulsation Measurement Program (CGMMP), and is under the chairmanship of Professor H. W. Smith of the University of Texas. It is hoped that this group will serve as a nucleus for a cooperative effort by persons with common interest, willing to apply any techniques necessary to understand the phenomena. Stations with similar capabilities in other areas of the world are shown in Fig. 11, 12, 13 and 14. All of these stations plan to be recording during the IQSY.

Fig. 10.
Geomagnetic micropulsation stations in
North America.

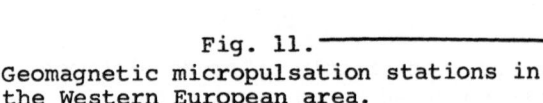

Fig. 11.
Geomagnetic micropulsation stations in
the Western European area.

Fig. 12.
Geomagnetic micropulsation stations
in Africa.

Fig. 13.
Geomagnetic micropulsation stations
in the Western Pacific area.

Fig. 14.
Geomagnetic micropulsation stations
in Antarctica.

The knowledge of the magnetosphere has enlarged at a great rate through rocket
and scientific satellite experiments and through development in the theory of VLF
phenomena.

Most persons have the feeling that it will not be long before some clue will
allow rapid progress in this area. This may be a fool-proof classification scheme,
a more adequate statistical treatment or correlative method applied to data, or a
particularly careful set of measurements with multiple instruments. At this time a
large burden rests on experimentalists. One would hope to be able to do detailed
studies on individual, short-term events, but one should be prepared to face the fact
that for a long time to come this type activity may have meaning only in a time-
average sense. In the years or decades ahead, one may discover that many different
processes, defying simple categorization, may be the source of terrestrial fluctuations.

ACKNOWLEDGEMENT

This work has been supported by Office of Naval Research Contract Nonr 266(48) and
National Science Foundation Grant GP-604.

REFERENCES

Aarons, J.: "Natural Background Noise at Very Low Frequencies," <u>The Radio Noise
        Spectrum</u>, Harvard Univ. Press, ed. D. H. Menzel, 111 (1960)

Benioff, H.: "Observations of Geomagnetic Fluctuations in the Period Range 0.3 to
        120 Seconds," J. Geophys. Res. <u>65</u>, 1413 (1960)

Berthold, W. K., A. K. Harris, and H. J. Hope: "Correlated Micropulsations at
        Magnetic Sudden Commencements," J. Geophys. Res. <u>65</u>, 613 (1960)

Bostick, F. X., Jr., and H. W. Smith: "An Analysis of the Magneto-telluric Method
        for Determining Subsurface Resistivities," Report No. 120, Elec. Eng. Res.
        Lab., Univ. of Texas (1961)

Campbell, W. H.: "Studies of Magnetic Field Micropulsations with Periods 5 to 30
        Seconds," J. Geophys. Res. <u>64</u>, 1819 (1959)

Campbell, W. H., and M. H. Rees: "A Study of Auroral Coruscations," J. Geophys. Res.
        <u>66</u>, 41 (1961)

Campbell, W. H., and S. Matsushita: "Auroral Zone Geomagnetic Micropulsations with
        Periods of 5 to 30 Seconds," J. Geophys. Res. <u>67</u>, 555 (1962)

Chan, K. L., D. P. Kanellakos, and O. G. Villard, Jr.: "Correlation of Short-Period
        Fluctuations of the Earth's Magnetic Field and Instantaneous Frequency
        Measurements," J. Geophys. Res. <u>67</u>, 2066 (1962)

Coleman, P. J., Jr., C. P. Sonett, D. L. Judge, and E. J. Smith: "Some Preliminary
        Results of Pioneer 5 Magnetometer Experiment," J. Geophys. Res. <u>65</u>, 1856
        (1960)

Coleman, P. J., Jr., Leverett Davis, Jr., E. J. Smith, and C. P. Sonett: "Interplane-
        tary Magnetic Fields," Science <u>138</u>, 1099 (1962)

Coulomb, J.: "Les Pulsations du Champ Magnetique Terrestre et des Courants
        Tellurique," Annali di Geofisica <u>12</u>, 461 (1959)

Duffus, H. J., and J. A. Shand: "Some Observations of Geomagnetic Micropulsations,"
        Can. J. Phys. <u>36</u>, 508 (1958)

Duffus, H. J., J. A. Shand, C. S. Wright, P. W. Nasmyth, and J. A. Jacobs: "Geo-
        graphical Variations in Geomagnetic Micropulsations," J. Geophys. Res.
        <u>64</u>, 581 (1959)

Duncan, R. A.: "Some Studies of Geomagnetic Micropulsations," J. Geophys. Res. <u>66</u>,
        2087 (1961)

Dungey, J. W.: "Electrodynamics of the Outer Atmosphere," Sci. Rept. No. 69,
        Ionosphere Res. Lab., Pennsylvania State Univ. (1954)

Flock, W. L., A. E. Belon and R. R. Heacock: "Geomagnetic Agitation and Overhead
        Aurora," Proc. of Internat. Conf. on the Ionosphere, July 1962, pub. by
        Inst. of Phys. and Phys. Soc., London (1963)

Francis, W. E., and R. Karplus: "Hydromagnetic Waves in the Ionosphere," J. Geophys.
        Res. <u>65</u>, 3593 (1960)

Glover, F. N.: "Recent Magnetic Observations in the Philippines," J. Geophys. Res.
        <u>68</u>, 2385 (1963)

Grenet, G., Y. Kato, J. Ossaka, and M. Okuda: "Pulsations in Terrestrial Magnetic Field at the Time of Bay Disturbance," Sci. Rept., Tôhoku Univ., Ser. 5, 6, 1, 1 (1954)

Heppner, J. P., N. F. Ness, C. S. Searce and T. L. Skillman: "Explorer 10 Magnetic Field Measurements," J. Geophys. Res. 68, 1 (1963)

Holmberg, E.R.R.: "A Discussion of the Origin of Rapid Periodic Fluctuations of the Geomagnetic Field and a New Analysis of Observational Material," Ph.D. Thesis, Univ. of London (1951)

Jacobs, J. A., and K. Sinno: "The Morphology of Geomagnetic Micropulsations, Pc," Sci. Rept. No. 1, Geophysics Laboratory, Univ of British Columbia (1959)

"World Wide Characteristics of Geomagnetic Micropulsations," Sci. Rept. No. 2, Geophysics Laboratory, Univ of British Columbia (1960)

Kato, Y.: "Geomagnetic Micropulsations," Austr. J. Phys. 5, 70 (1962)

Kato, Y., and S. Akasofu: "Outer Atmospheric Oscillation and Geomagnetic Micropulsation," Sci. Rept., Tôhoku Univ., Ser. 5, 103 (1956)

Kato, Y., and T. Saito: "Preliminary Studies on the Daily Behavior of Rapid Pulsations," J. Geomag. and Geoelec. 101, 221 (1959)

Kato, Y., and T. Watanabe: "Further Study on the Cause of Giant Pulsations," Sci. Rept., Tôhoku Univ. Ser. 5, Geophysics, 8, 19 (1956)

"A Survey of Observational Knowledge of the Geomagnetic Pulsation," Sci. Rept., Tôkohu Univ., Ser. 5, Geophysics 8, 157 (1957a)

"Studies on Geomagnetic Pulsation Pc," Sci. Rept., Tôhoku Univ., Ser. 5, 1 (1957b)

"Studies on Geomagnetic Storm in Relation to Geomagnetic Pulsation," J. Geophys. Res. 63, 741 (1958)

Law, P. F., and R. R. Boothe, Jr.: "A Study of the Characteristics and Origins of Geomagnetic Micropulsations," Rept. No. 111, Elec. Eng. Res. Lab., Univ. of Texas, ASTIA 229724 (1959)

Lawrie, J. A.: "The Spatial Distribution of Rapid Geomagnetic Fluctuations," Geophys. J. of R.A.S., 7, 102 (1962)

"The Spatial Distribution of Rapid Geomagnetic Fluctuations. Part II," Geophys. J. of R.A.S., 7, 328 (1963)

Maple, E.: "Geomagnetic Oscillations at Middle Latitudes. Part I. The Observational Data," J. Geophys. Res. 64, 1395 (1959)

Mason, R. G.: "Spatial Dependence of Time-Variations of the Geomagnetic Field on Oahu, Hawaii," (abstract) Trans. Amer. Geophys. Union 44, 40 (1963)

Matsushita, S.: "Reply (to paper of Wilson and Sugiura)," J. Geophys. Res. 68, 3320 (1963)

McDonald, G.J.F.: "Spectrum of Hydromagnetic Waves in the Exosphere," J. Geophys. Res. 66, 3639 (1961)

Ness, N. F., T. L. Skillman, C. S. Scearce, and J. P. Heppner: "Magnetic Field
Fluctuations on the Earth and in Space," Internat. Conf. on Cosmic Rays
and the Earth Storm, Kyoto (1961)

Obayashi, T., and J. A. Jacobs: "Geomagnetic Pulsations and the Earth's Outer
Atmosphere," Geophys. J. of R.A.S., 1, 53 (1958)

Parkinson, W. D.: "Directions of Rapid Geomagnetic Fluctuations," Geophys. J. of
R.A.S., 2,   (1959)

Price, A. T.: "The Theory of the Magnetotelluric Method When the Source Field is
Considered," J. Geophys. Res. 67, 1907 (1962)

Rikitake, T., T. Yabu, and K. Yamakawa: "The Anomalous Behavior of Geomagnetic
Variations of Short Period in Japan and its Relation to the Subterranean
Structure," Bull. Earthquake Res. Inst., Tokyo, 40, 693 (1962)

Roquet, J., R. Schlich, and E. Selzer: "Evidence for Two Distinct Synchronous World
Impetuses for the Magnet Effects of the Nuclear High-Altitude Detonation
of July 9, 1962," J. Geophys. Res. 68, 3731 (1963)

Schmucker, Ulrich: "Annual Progress Report: Deep Anomalies in Electrical Conductivity," Scripps Institution of Oceanography Reference 61-13 (1960)

Scholte, J. G., and L. Veldkamp: "Geomagnetic and Geoelectric Variations," J. of
Atmos. Terr. Phys. 6, 33 (1955)

Sechrist, C. F., Jr.: "Very Low Frequency Phase Perturbations Observed During
Geomagnetic Storms," J. Geophys. Res. 67, 1685 (1962)

Silsbee, H. C., and E. H. Vestine: "Geomagnetic Bays, Their Frequency and Current
Systems," Terr. Mag. 47, 195 (1942)

Sonett, C. P.: "The Distant Geomagnetic Field. 2. Modulation of a Spinning Coil
EMF by Magnetic Signals," J.Geophys. Res. 68, 1229 (1963a)

"The Distant Geomagnetic Field. 4. Microstructure of a Disordered Hydromagnetic Medium in the Collisionless Limit," J. Geophys. Res. 68, 1265
(1963b)

Sonett, C. P., A. R. Sims and I. J. Abrams: "The Distant Geomagnetic Field. 1.
Infinitesimal Hydromagnetic Waves," J. Geophys. Res. 67, 1191 (1962)

Sonett, C. P., and I. J. Abrams: "The Distant Geomagnetic Field. 3. Disorder and
Shocks in the Magnetopause," J. Geophys. Res. 68, 1233 (1963)

Troitskaya, V. A.: "Short Period (oscillatory) Disturbances in the Terrestrial
Magnetic Field," Dok. Akad. Nauk. SSSR 91, 241 (1953)

"Pulsations of the Beating Type (1-4 sec. per.); 'Pearls' in the Electromagnetic Field of the Earth," Symp. on Rapid Geomagnetic Variations,
Utrecht (1959)

Voelker, H.: "Zur Breitenabhangigkeit der Perioden Erdmagnetischer Pulsationen,"
Die Naturwissenschaften 1, 8 (1962)

Ward, S. H., and K. A. Ruddock: "A Field Experiment with a Rubidium Vapor
Magnetometer," J. Geophys. Res. 67, 1889 (1962)

Weaver, J. T.:   "The Electromagnetic Field Within a Discontinuous Conductor With Reference to Geomagnetic Micropulsations Near a Coastline," Can. J. Phys. <u>41</u>, 484 (1963)

Wescott, E. M., and V. P. Hessler: "The Effect of Topography and Geology on Telluric Currents,"   Sci. Rept. No. 3, Geophys. Institute, The Univ. of Alaska (1960)

Westphal, K. O., and J. A. Jacobs: "Oscillations of the Earth's Outer Atmosphere and Micropulsations," Geophys. J. of R.A.S. <u>6</u>,   (1962)

Wilson, C. R., and M. Sugiura: "Hydromagnetic Interpretation of Sudden Commencements of Magnetic Storms," J. Geophys. Res. <u>66</u>, 4097 (1961)

Yanagihara, K.:   Memoirs of the Kakioko Magnetic Observatory <u>8</u>, 61 (1957)

"Geomagnetic Micropulsations with Periods from 0.03 to 10 Seconds in the Auroral Zones with Special Reference to Conjugate-Point Studies," J. Geophys. Res. <u>68</u>, 3383 (1963)

INSTRUMENTATION FOR RECEIVING ELECTROMAGNETIC
NOISE BELOW 3,000 CPS

J. E. Lokken
Pacific Naval Laboratory
Defence Research Board of Canada
Esquimalt, British Columbia

INTRODUCTION

If the lower limit of frequencies, of interest to this Institute, is somewhat arbitrarily set at 0.001 cps, the central octave of the band falls between 1 and 2 cps. Our colleague, Sir Charles Wright, has often referred to this region as a "no man's land." The name is appropriate. It lies near the crossover between ionospheric and tropospheric background noise sources, and it roughly marks the practical transition from DC to AC types of receiving equipment. The lower limit has been chosen because, with some notable exceptions, instruments responding to periods much greater than 1,000 seconds are rarely sensitive enough to observe the background at a few seconds period. In any event, the background at lower frequencies is adequately covered by instrumentation now used by magnetic observatories. These measurements have been well described in the literature and, thus, are not included in this paper.

Most investigators divide the frequency range we have chosen into three bands: the micropulsation band, which extends up to approximately 5 cps and uses quasi-DC receiving systems; the low ELF or Schumann band between approximately 2 cps and the power-line frequency of 50 or 60 cps; and the band which commences at as low a frequency as is consistent with interference from the power line and its harmonics and extends up to a few tens of kilocycles. Each of these bands appears to have its own characteristics as far as spectral level, noise characteristics and siting problems are concerned, and the paper has been divided accordingly. Before instrumentation is discussed in detail, however, some general factors influencing the design and siting of equipment will be reviewed.

FACTORS AFFECTING DESIGN AND SITING OF RECEIVING SYSTEMS

Micropulsation Band.

The micropulsation band is perhaps the most difficult range for which to design equipment - if one is to cover all of the possible types of signal. As previous lecturers have pointed out, it is characterized by an exceedingly large dynamic range, a sharply rising spectrum with decreasing frequency, and an extreme variability of spectral level as a function of frequency, geomagnetic latitude, and time. Thus, even though it will be somewhat redundant, the signal characteristics are reviewed briefly, for the differences between systems used by various investigators are due in part to its variability.

On the basis of measurements made by the Pacific Naval Laboratory over the past several years, an approximation to the signal levels observed at Victoria (geomagnetic latitude 55°) is shown in Fig. 1. The cross-hatched area is intended to show the

Fig. 1. Approximate micropulsations background at Victoria, circa 1961. The shaded area is intended to show typical peak values on active days.

range of maximum values to be expected on a relatively active day. There are occasions, of course, when the amplitude exceeds that shown on the figure, and others when it is considerably lower - in fact, the lower limit is 40 to 60 db or more below the maximum values shown. Fig. 2 shows a not unusual abrupt change of amplitude

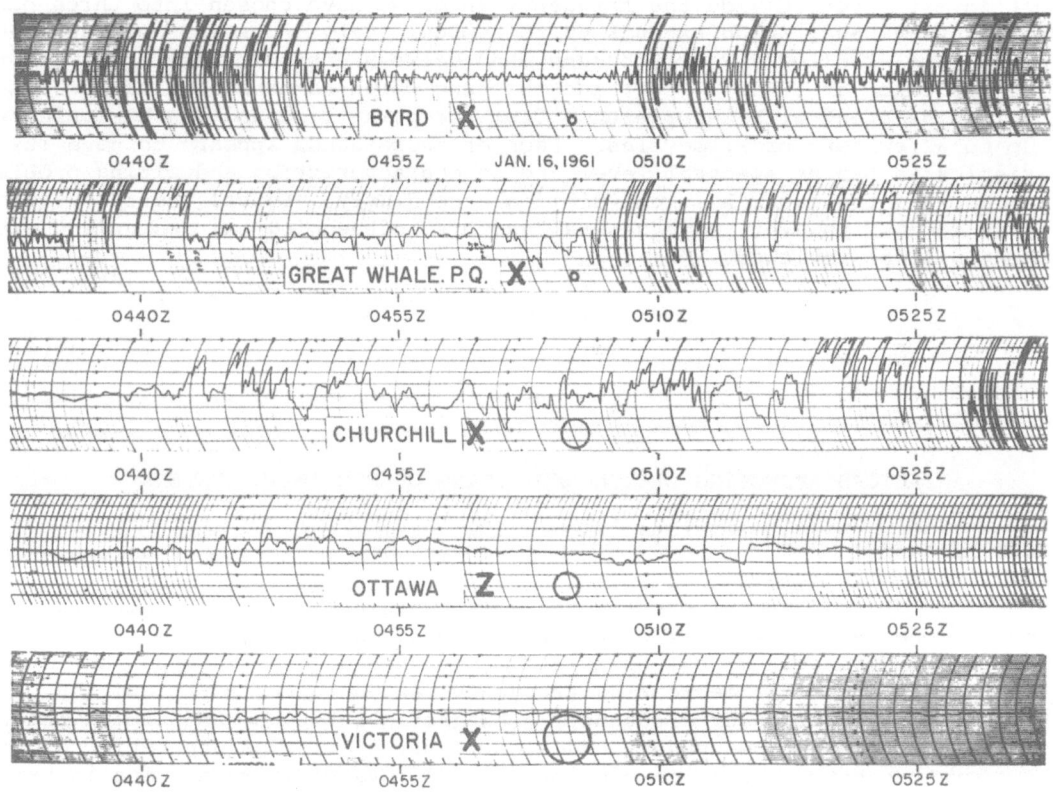

Fig. 2. Moderate impulsive activity recorded January 16, 1961. Signals below 0.02 cps were removed before recording at Byrd Station. Relative midband gains are shown by areas of circles (Lokken et al, 1962).

374

observed at several stations. Note that the range in recorded amplitude exceeds 1,000 to 1.

It can also be seen that the signals are strongly latitude dependent. Our observations of this dependence are summarized in Fig. 3, which extends from mid-latitudes

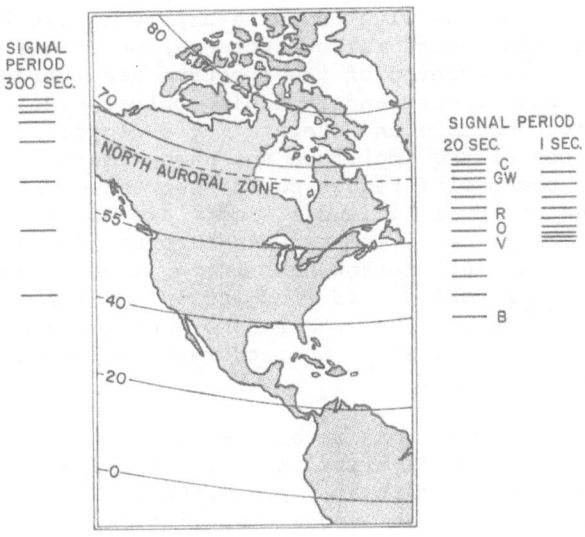

Fig. 3. Amplitude of geomagnetic micropulsations as a function of geomagnetic latitude. Density of lines indicates relative amplitudes.

to auroral latitudes, but we have insufficient observations from within the auroral zones and in equatorial regions to make meaningful comments, except that it is generally accepted that the maximum variability occurs between mid-latitudes and the auroral zones. At the lower end of the spectrum the magnitudes increase markedly with increasing geomagnetic latitude, while near and above 1 cps there is very little, if any latitude dependence.

It is evident, though we have not yet carried out a detailed analysis, that the spectral form is changing with the solar cycle. We are now leaving the period of active solar disturbances and approaching the Year of the Quiet Sun. It appears that micropulsation activity within the band 0.1 to 1 cps has increased considerably in magnitude as well as in frequency of occurrence at latitudes corresponding to southern Canada, while the longer period phenomena are distinctly lower in amplitude and probably less frequent.

It is not surprising, therefore, that the design of micropulsation equipment depends quite markedly on the location of the receiving station, on the portion of the band to be examined and perhaps even on the sun-spot cycle. For example, there are two widely held points of view with regard to the recording of magnetic field variations. One group of observers hold that the field value, $\vec{B}$, should be recorded, while the other believes that its time derivative $\frac{d\vec{B}}{dt}$ should be used.

Undoubtedly, the proponents of both methods have justifiable reasons for their choice, but they have often been influenced by the character of the background at their latitude at the time the equipment was designed, as well as by the objectives of the research program.

The history at Pacific Naval Laboratory can illustrate these points. Our early micropulsation records were obtained by integrating the signal received from loop antenna installations, we thus recorded $\vec{B}$. As we became interested in higher frequencies, however, and the equipment response was increased, the influence of the spectral levels shown in Fig. 1 became evident. It was not possible to record the activity with periods less than 20 seconds with adequate amplitude resolution if larger period activity was included on the record without attenuation. Thus, to facilitate recording, some prewhitening or flattening of the spectrum was introduced in September 1958, by removing the integrating circuit and recording only the derivatives of the field. We have found little inconvenience in interpreting a record of $d\vec{B}/dt$ rather than one of $\vec{B}$. Stable calibrations of the amplitude and phase characteristics can be obtained so that the spectra and phase differences can be corrected when required. If the observations are limited to lower latitudes - ours are not - the demands on the recorder by recording $\vec{B}$ are not as severe as when auroral zone data are included. The same is true if only a relatively narrow band is required. Under these conditions, it is entirely feasible to record $\vec{B}$. If fluxgate or total field magnetometers are used, the records are of $\vec{B}$ rather than its rate of change, and should be recorded as received. We see little need, however, to integrate the output of a loop system before making a broad-band record.

Many observers of micropulsations record telluric or earth currents as well as, or in place of, the magnetic field variations. There appear to be several advantages. Firstly, the spectrum is already partially prewhitened. Secondly, the signal amplitude can be increased by choosing an adequate spacing of the electrodes. The amplifier then need not be as sensitive as one designed for magnetic field measurements. Finally, for measurements of conductivity, it is necessary to record both the telluric currents and the magnetic field over a wide frequency range. For separation of the field into external and internal parts it appears preferable to use the magnetic field because of its relative freedom from the influence of local occlusions.

If the measurement of spatial as well as temporal coherence of micropulsations is required, it is very helpful if the frequency response and phase shift of the receiving systems are identical at each station. Visual comparison and the testing of equipment is then much simpler. Of course modern computing methods allow corrections for amplitude and phase to be made to the final calculations, but that is of little assistance when taking the measurements, or in deciding if computations are required. Furthermore, many applications require only a visual inspection or manual scaling.

Let us consider now some of the factors involved in choosing a site for micropulsation measurements. Weaver (1963) and others have shown that a conductivity discontinuity will materially affect the ratio of the vertical to horizontal components of the magnetic field. However, since even dry earth is a good reflector as far as micropulsation frequencies are concerned, the magnitude of the horizontal components should not be greatly affected by the nearness of a conductivity discontinuity. Experimentally, this has been verified to within a factor two at least; see for example Christoffel et al (1961) and Fig. 4, which shows the results of a few hours recording on Sable Island, a sand dune island on the Continental Shelf off the Nova Scotia Coast, and near the shore line at Dartmouth. The spectral levels of the horizontal components are only slightly affected by the contrast in conductivity.

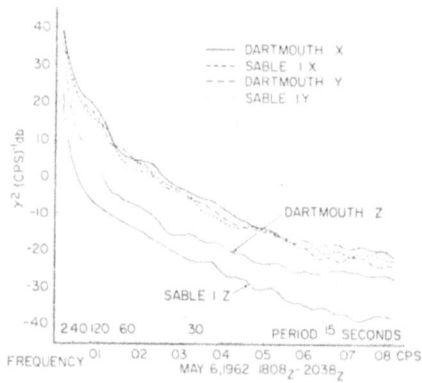

Fig. 4. Micropulsation spectrum recorded on Sable
Island and near Halifax, Canada, May 1962.

The vertical component, as expected, is very much reduced over Sable Island. Dr.
Jacobs has already mentioned that the boundary between two media of differing conductivity must be rather sharp before the vertical field anomaly becomes really pronounced. Therefore, unless one wishes expressly to investigate the behavior of
conductivity discontinuities, it is well to choose the recording site some distance
from known conductivity contrasts, or in an area where the conductivity is changing
relatively slowly. We are able to operate successfully to within a few miles of
urban areas, but would refrain from choosing sites near electric railways, teletype
lines, radio transmitters, concentrations of power lines or substations and other
equipment known to produce electrical transients.

It was originally the intention of the author to give a resume' of rocket and
satellite instrumentation, but the limitations of time preclude their inclusion.
Heppner (1963) and others have reviewed current American technology.

Lower ELF Band.

The spectrum between, say 6 cps, and the power line frequency is dominated by the
earth-ionosphere cavity resonance modes (Balser and Wagner, 1960; Lokken et al, 1962),
which are remarkably stable both in frequency and in amplitude. A change in amplitude
by a factor of 2 from the mean is large for the resonant bands, and this change occurs rather slowly and is predominantly a diurnal pattern. The detailed record in
this region, however, is characterized by bursts of a few cycles duration which may be
ten or more times the mean amplitude between bursts. Fig. 5 illustrates this bursty
character which is apparently coherent over most, if not all, of the earth. The
equipment must be designed, therefore, to cope adequately with this transient type of
signal. Until recently, the design of equipment for the low ELF frequency range was
difficult because of flicker noise in vacuum tubes. However, the development of
transistors of sufficiently low noise level and high stability - particularly field
effect transistors - has made the design and construction much easier.

Some investigators, e.g. Balser and Wagner, use a vertical electric antenna for
receiving, while others use loop antennas (often metal cored) to measure the magnetic
component (Polk and Fitchen, 1962; Evans and Horner, 1963). All significant energy
appears to be nearly vertically polarized and therefore, a vertical electric antenna
receives the activity irrespective of the direction of the point of origin. On the
other hand a measurement of the magnetic components may enable one to estimate the
direction of arrival and suitable antennas are probably somewhat easier to install.

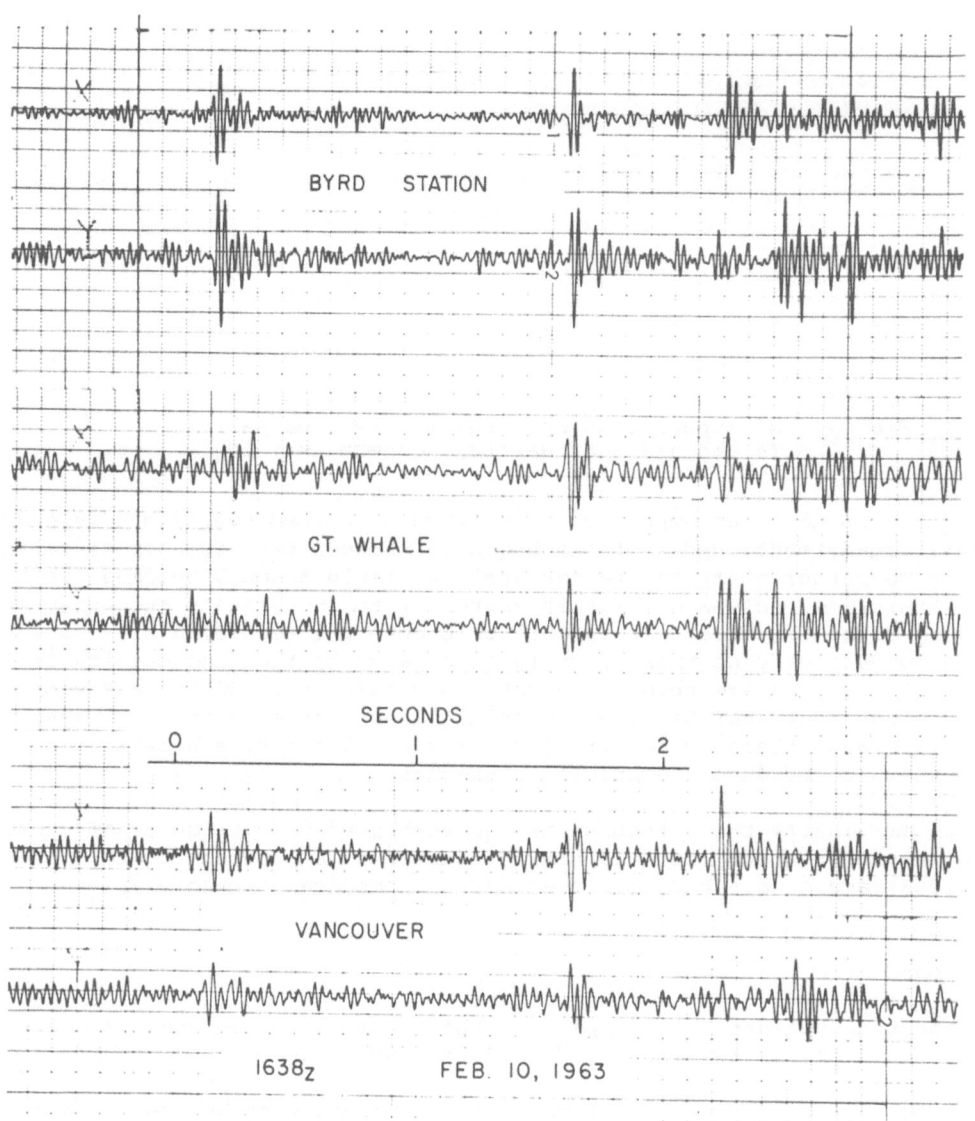

Fig. 5. ELF signals recorded at Byrd Station, Great Whale River and Vancouver.

As mentioned above, the background level both in the bursts and in the average value is nearly independent of latitude, except perhaps for a factor of 2, but during the thunderstorms season "local" electrical storm activity may seriously hamper recording. We have found that in the summertime on the Canadian Prairies the background level may be several times the value observed at stations which are more remote from thunderstorm centers. Thus, the useful recording periods may be restricted in certain localities by regional thunderstorm activity.

The same general considerations apply to the choice of a recording site as applied to stations for receiving micropulsations. The vertical magnetic component, as is the case with micropulsations, is accentuated if the receiving station is located near a conductivity discontinuity, such as at our Great Whale River station in Canada, where the vertical component is comparable in magnitude to the horizontal components, while at other stations the vertical is lower by a factor of 20 or more. The horizontal components, however, are of approximately the same magnitude. Industrial noise is more important in this part of the spectrum and the site must be chosen with more care. ELF equipment is more sensitive than micropulsation equipment to noise from industrial sources and power lines, although it can function reasonably well even in the presence of small local power lines. It may also be necessary in some localities to install radio frequency filters between the antenna and the amplifier.

ELF and VLF Band Between 60 cps and 10,000 cps.

Normally, recordings by the Whistler stations and others engaged in VLF observations extend down to frequencies as low as interference from power line harmonics will permit, and up to a few tens of kilocycles. In very favorable locations, such as some polar stations, where the man-made background is sufficiently low to prevent saturation of the amplifier and recording system, recordings may even extend below 60 cps. In urban, populated areas, however, the power line frequency and its harmonics, particularly the odd harmonics, make recording extremely difficult between 60 cps and perhaps a few kilocycles per second. The general character of the signal is similar to that in the low ELF band.

The design of amplifier and antenna in this frequency range is straightforward. Natural vibrations of the antenna due to microseisms and wind seem to contribute little to the band of interest. Either electric or magnetic antennas may be used.

Timing.

The accuracy of timing when comparing activity recorded simultaneously at separated stations is becoming increasingly important even in the micropulsation range. It has been well developed for several years in the VLF band, where low frequency radio transmitters are within the pass band and can be used to provide accurate synchronization of the recordings. No such assistance is available, however, in the low VLF and micropulsation bands and the reliance must be placed on time recorded expressly for that purpose. Fortunately, crystal clocks and receivers for VLF timing stations are now readily available at moderate cost, thus enabling one to have reasonably good time synchronization between stations. Even for micropulsations the timing should be good to within a small fraction of a second if activity near 1 cycle per second is being examined in detail. It must be borne in mind, however, that the time of onset of the different spatial components may differ at a given station, by as much as one cycle of the dominant frequency. Thus, timing depends not only on the availability of a good time on the record, but also on an interpretation of the commencement and the possible effects of local conductivity structure and polarization of the incoming field.

MICROPULSATION BAND

Receiving Equipment.

Magnetic field.

The instrumentation in use at Magnetic Observatories has been discussed in detail by several authors, e.g. Whitham (1960) and several authors in the Annals of the International Geophysical Year, Vol. IV, Part V. It would be repetitious and outside the scope of this institute to discuss standard observatory measuring techniques, except to review briefly those suitable for measuring micropulsations. The techniques and precautions required for absolute field measurements, secular variations and other phenomena below 0.001 cps will not be considered.

The equipment used by investigators of micropulsations varies widely both in sensitivity and physical size. Loop antennas at fixed stations can be made as large as copper, real estate, and manpower will allow, and can therefore be extremely sensitive. Similarly, carefully constructed Hall-effect and fluxgate type magnetometers in a well-regulated constant temperature environment can be made very sensitive - perhaps of the order of one milligamma. Portable versions of micropulsation receiving equipment must be convenient to handle and require a minimum of set-up time, and often must be designed to use a minimum of power. Thus, a compromise is required between convenience, cost and sensitivity in any installation. Since the specific research objectives, available sites, manpower and funds govern the final choice, it is doubtful that any one optimum station exists or can be designed. Rather than advocate a particular set of equipment, then, we will review those in current use, or some which show promise, and attempt to point out advantages and limitations of each type.

Total field magnetometers.

There are several methods available for measuring "total field," i.e. the magnitude of the total field vector, $\vec{F}$, but in this paper we will restrict the discussion to those utilizing a property of the atom or its nucleus. We note, however, that "total field" can be measured by a single-component magnetometer aligned parallel to the total field vector. Since we are concerned with fluctuations whose magnitude are exceedingly small compared to the total field, only the component parallel to $\vec{F}$ and not the total fluctuating vector is measured by "total field" magnetometers.

Proton-precession magnetometer. The study of nuclear resonance and nuclear induction has led to several new magnetometers, including the proton magnetometer. It can be shown (for example, Whitham, 1960) that, for protons, absorption phenomena occur at the frequency

$$f = \frac{\beta H}{2\pi I} = \frac{\gamma H}{2\pi} ,$$

where   f = frequency of applied electromagnetic field,

   β = nuclear spin moment,

   H = magnetic field,

   I = nuclear spin angular momentum,

and γ = gyromagnetic ratio

   = 2.67513 $\pm$ 0.00002x10^{-4}sec^{-1}oersted^{-1} for distilled water, uncorrected for the diamagnetic effect of the water molecule (Driscoll and Bender, 1958).

(Value officially adopted at the 13th General Assembly of the IUGG, August 1963.)

The duration of the absorption is governed by the spin-lattice interaction time, which can vary widely, but is of the order of one second for proton-rich liquids. Packard and Varian (1954) reported the observation of free audio-frequency induction signals in water after a polarizing field $\overrightarrow{H_p}$ perpendicular to the earth's field $\overrightarrow{F}$ was quickly removed. The signal, lasting for more than one second, was observed by a receiving coil perpendicular to both $\overrightarrow{H_p}$ and $\overrightarrow{F}$, with a signal-to-noise ratio of 20 for $\overrightarrow{H_p}$ = 100 oersted. The current practice is to use the same coil for producing $\overrightarrow{H_p}$ and receiving the precession frequency, f. The latter is independent of the direction of $\overrightarrow{F}$, although the signal-to-noise ratio does depend on the orientation.

For a single-coil magnetometer with solenoid axis at an angle θ with respect to $\overrightarrow{F}$, the component of the magnetic moment which precesses is proportional to $\sin \theta$, as is the flux cutting the receiving coil. Thus, the signal amplitude is proportional to $\sin^2 \theta$. Three mutually perpendicular coils can be used to reduce the orientation dependence to zero (Whitham, 1960). The signal amplitude depends also on the volume of the sample and on the magnetic field gradient across it. If the same coil is used for polarizing and receiving, the signal-to-noise ratio can be written (Whitham, 1960)

$$\frac{Vs}{Vn} = Qav\chi 4\pi \left[\frac{W_i}{4kT\Delta f}\right]^{\frac{1}{2}}$$

where Q is the figure of merit of the tuned receiving coil, considered as a solenoid, without end correction, Δf the receiving bandwidth, a the filling factor between the sample and coil, v the volume, χ the nuclear susceptibility of the sample, and W_i the input power on polarizing. For a sample of a few hundred cubic centimeters and a coil of about 1000 turns, the induced signal is a few microvolts.

The relaxation time can be varied over a considerable range by the choice of suitable proton-rich liquids, and, if necessary, by certain additives. Water is the most common liquid used and, like most liquids, has a relaxation time of the order of one second - 2 to 3 seconds for distilled water. As expected, if the magnetic field gradient across the sample is large, the relaxation time is reduced because of the resulting mixture of the precession frequencies.

Since the operation of proton magnetometers is well known and adequately described by a number of authors (Whitham, 1960, Serson, 1962 and several pamphlets by Varian Associates), only the outline of typical systems will be discussed here. Fig. 6 is a block diagram of a typical proton magnetometer. The same coil is used for

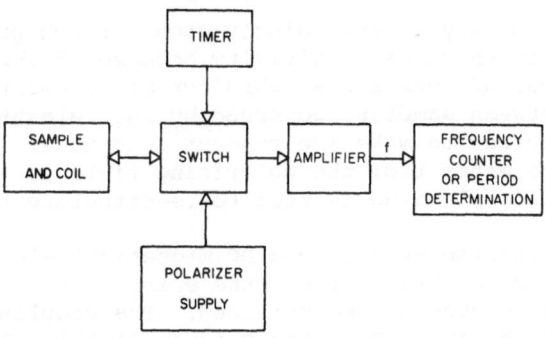

Fig. 6. Block diagram of proton precession magnetometer.

polarizing the sample and receiving the precession frequency. Since the precession signal is available for only about one second, simple counting of the cycles for a fixed period leads to an error of the order of 25 gammas. Improved accuracy can be obtained by measuring the period. Many electronic counters are now available for accurate determination of the period of a signal of, say 1000 cycles. By this method a recording accuracy of one gamma is possible, but since $|F| = \dfrac{2\pi f}{\gamma p}$, one is measuring directly in inverse of the field F.

A recent development by Serson (1961; 1962), however, has been the introduction of a frequency multiplier and a phase-locked filter for determination of the frequency. Fig. 7 shows a block diagram of such a magnetometer and Fig. 8 the frequency

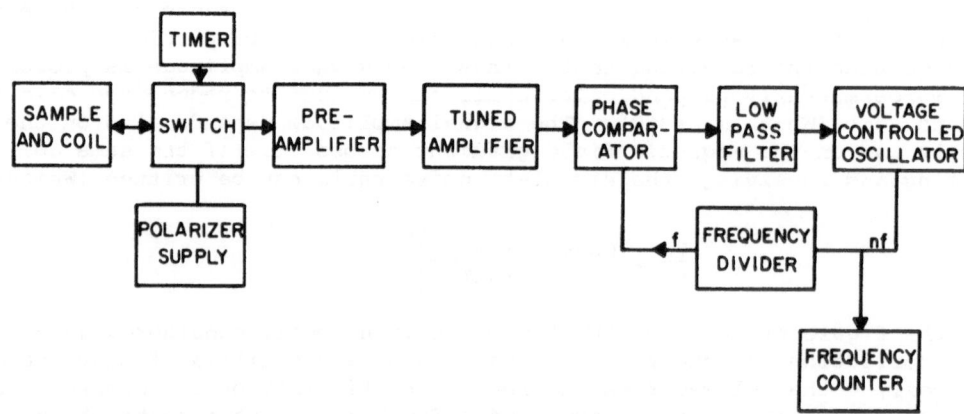

Fig. 7. Block diagram of improved proton precession magnetometer. The frequency multiplying factor n and the counting interval are adjusted to give direct reading in gammas or tenths of gammas.

multiplier used by Serson. The voltage controlled oscillator is a multivibrator, typically locked at 160 times the precession frequency. The output to the counter is at twice this frequency giving a total multiplication of 320 times. The counting time is chosen to give a direct readout in gammas or tenths of gammas. Not only does this method provide a readout directly in gammas, but the noise content of the proton signal is averaged by the low pass filter. Serson reports that as noise was artificially added to the proton signal, no deterioration in the consistency of the readings to ± 1 gamma was observed until the signal-to-noise ratio had been reduced to 2 to 1. Since the counting time was about 0.15 second a signal-to-noise ratio of 50 to 1 would be necessary to obtain comparable consistency using conventional methods.

While the absolute accuracy is exceedingly good, proton precession magnetometers are limited to about one-tenth gamma sensitivity because of the low precession frequency and the small number of cycles available for measurement. A further disadvantage is the time taken between samples, governed by the relaxation time. Attempts are under way at some laboratories to make the readout continuous, through the physical transport of the polarized water from the polarizing field to a receiving coil (Benoit, 1958). Appropriate plumbing is used to re-circulate the fluid.

In addition to polarization by a strong DC magnetic field perpendicular to the unknown field, the magnetic coupling between the spins of the electron of a paramagnetic free radical and the proton may be utilized. The coupling, or Overhouser effect, is discussed in detail by Abragam (1955) and Solomon (1955). They showed that if an electronic resonance line is saturated (the number with positive and negative spins made equal) by the application of a high frequency field, f_e, perpendicular to the

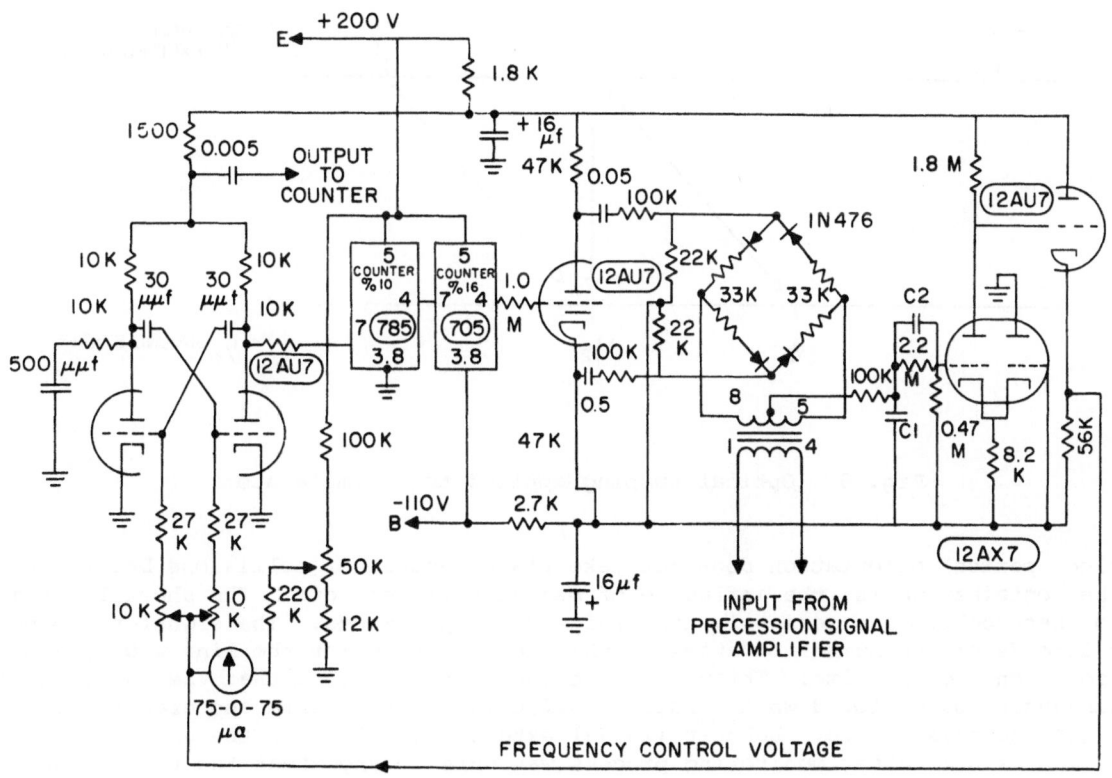

Fig. 8. Frequency multiplier and phase-locked filter
used by Serson to give direct reading in gammas.

unknown magnetic field, then the polarization is augmented by an amount proportional
to f_e/f, where f is the precession frequency.

    Magnetometers based on the Overhouser effect have been described in detail by
Freycenon and Solomon (1960), Freycenon (1960) and by Bonnet (1962a; 1962b). They
use nitrous disulphate of potassium as the radical and a frequency of 56 megacycles.
One of the early disadvantages, aside from the RF oscillator, was the short lifetime
of the radical. Selzer (1963), however, reports that except for temperatures such as
in the Sahara Desert, lifetimes of months are being achieved at present.

    Optically pumped magnetometers. Kastler (1954) and later Hawkins (1955) discuss
optical pumping, which was proposed by Kastler in 1950. Optical pumping consists of
orienting an ensemble of atoms through appropriately polarized light such that the
average magnetic moment is not zero. Consider a simple atomic system whose ground
state has a maximum total angular momentum F = ½, and a first excited state having
F = 3/2. If an atom in the ground state with m = -½ absorbs a photon of angular
momentum +1 it may, on return to the ground state, occupy either the m = +½ or
m = -½ substate. An atom initially in the m = +½ substate, however, on absorbing
such a photon will go into the excited substate m = 3/2 and must on return to the
ground state re-enter the m = +½ substate. (Electric dipole selection rules have
been assumed.) The possible transitions are shown in Fig. 9. Thus, after irradia-
tion with completely circularly polarized light, all atoms will end up in the m = +½
ground state. Such an ensemble would be "perfectly" oriented. If the polarization
is reversed the ensemble will be oriented in the m = -½ ground state. In actual

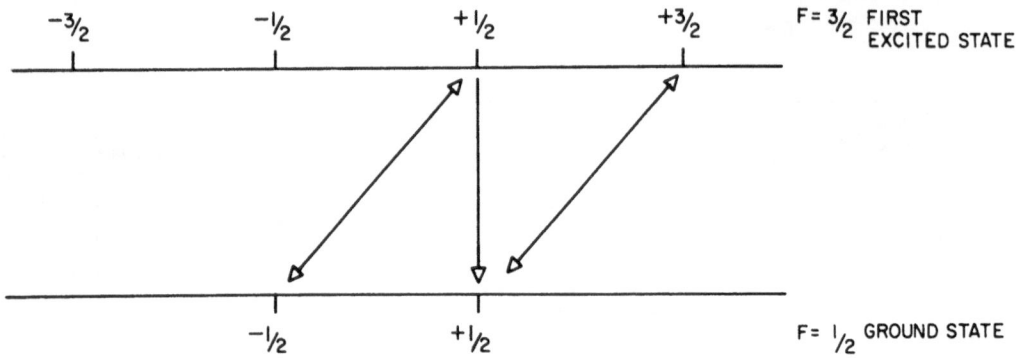

Fig. 9. Optical pumping applied to a simple atom.

practice, perfect orientation does not take place because of collisions between atoms
and the container walls, the influence of magnetic fields, etc. The short lifetime
due to these collisions seriously hampered early experiments. The signal-to-noise
ratio depends on the number of atoms in the excited state (or the line width) and,
therefore, on the lifetime. Through an accident Kastler found that the addition of
certain gases would slow down the rate of diffusion to the walls and greatly improve
the signal-to-noise ratio. Dehmelt (1957a) extended the work on the use of buffer
gases, and with a better monitoring process was able to show that the relaxation time
from collisions with the wall could be increased to 0.2 seconds by the addition of
Argon gas as a buffer, up to a pressure of 33 mm of mercury. The computed relaxation
time due to diffusion of the sodium atoms to the wall was about 0.4 second, the reduction being caused by collisions with the buffer gas. He later showed that coating the
cell wall with a material having a similar electron structure to Argon may be even
more effective than buffer gases in increasing the relaxation time. Commercial magnetometers, however, do not use coated cells because of the lack of stability and
difficulty of reproduction.

Dehmelt's most important contribution was not in increasing the relaxation time,
but in the method of observing the degree of pumping. He used changes in the transmitted intensity of the pumping light as a monitor of the absorption of photons. In
a later paper the same year Dehmelt (1957b) pointed out that light of the pumping
frequency when transmitted through the absorption cell perpendicular to the pumping
and magnetic field axis, Fig. 10, would be modulated at the Larmor frequency. The
precessing angular momentum component is created by the application of an RF field at
the Larmor frequency, about 700 (kc/sec)/gauss for sodium. Experimental verification
was soon published by Bell and Bloom (1957). If the RF oscillator in Fig. 10 is set
to a frequency near that corresponding to the field H_O and ΔH is swept, a dip in
transmitted intensity is observed on 'scope 1, and an increase on 'scope 2 as the
resonance of the RF oscillator is passed. Dehmelt and Bell and Bloom also showed that
an atomic oscillator whose frequency is proportional to H_O can be constructed by coupling the output of the cross-beam photocell to the RF oscillator. Dehmelt further
pointed out that the two beams could be replaced by a single beam oriented at about
45° to the magnetic field.

Commercially available magnetometers are now made using rubidium vapor and metastable helium. Magnetometers of this type, often referred to as optically pumped
magnetometers, will be discussed in order of historical development. A cesium magnetometer is under development but will not be reported here.

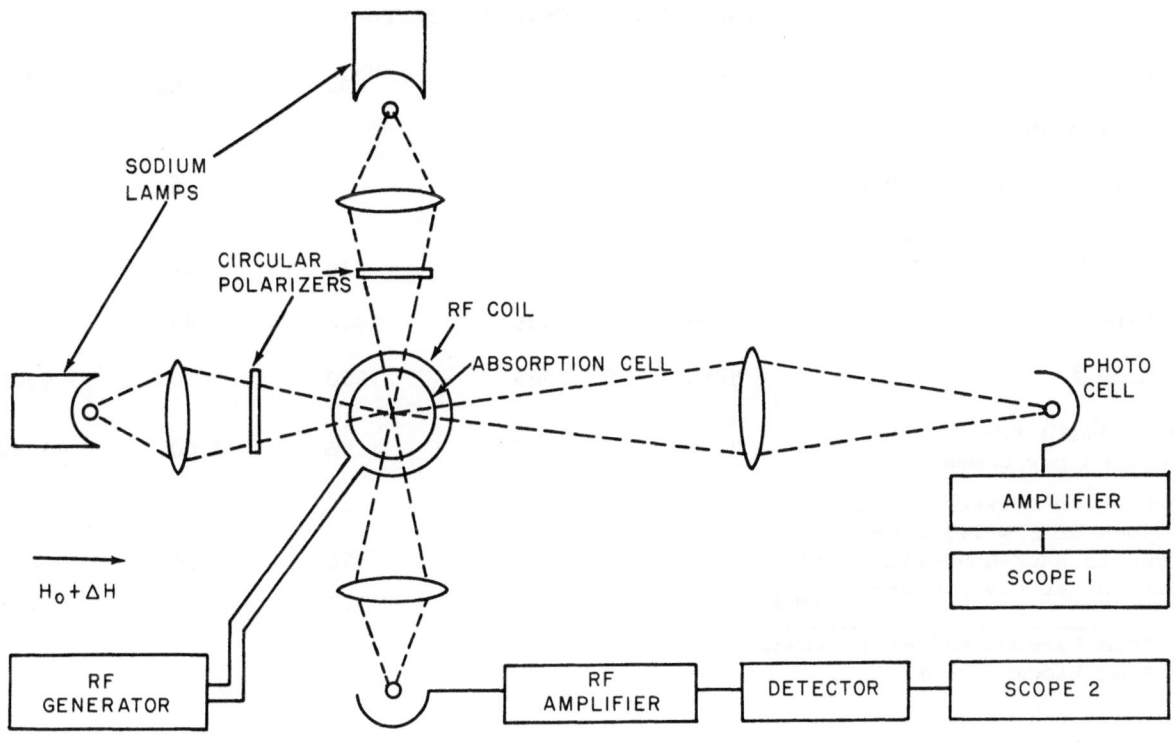

Fig. 10. Diagram of magnetic resonance experiment in alkali metal
vapor. H_O is the earth's magnetic field and ΔH a variable swept field parallel to H_O.

Rubidium vapor magnetometer. Following the suggestion of Dehmelt and Bell
and Bloom that optically pumped alkali vapor could be used as a magnetometer, many
groups have produced such instruments. For several reasons rubidium has received
most attention. Among the principal ones, evident on examining Table 1, are lower
operating temperature and easier isolation of the D_1 line. Furthermore, the rubidium
Larmor frequency is higher, and there are fewer spectral lines than for cesium.

The complex line structure, Fig. 11, presents problems, especially to the de-
signer of equipment for a moving platform. There is a difference in energy between
pairs of Zeeman sub-levels proportional to H^2, owing to the Back-Goudsmit effect,
which is appreciable for fields of the order of 0.5 gauss, but negligible for fields
of 0.1 gauss (Bloom, 1962). As the angle between the pumping axis and the magnetic
field is changed the relative population of the various sub-levels changes, giving
rise to an appreciable asymmetry in the lobal pattern when the average line is used
for detection. Methods of reducing the orientation error will be discussed later.
For stationary application the Back-Goudsmit effect broadens the line width, but does
not appear to add appreciably to the noise level of the instrument.

The design of practical magnetometers is discussed in some detail by Bloom (1962)
and by Parsons and Wiatr (1962). The former describes the techniques and design considerations used by Varian Associates in designing magnetometers for several applications, while the latter reviews the theoretical aspects in more detail and describes

TABLE 1
COMPARISON OF ATOMIC CONSTANTS *

Atom	^{23}Na	^{39}K	^{85}Rb	^{87}Rb	^{133}Cs
Nuclear Spin I	3/2	3/2	5/2	3/2	7/2
Number of Spectral Lines	6	6	10	6	14
Temperature for vapor pressure of 10^{-6}°C	126	63	34	34	23
D_1 Line Å	5896	7699	7948	7948	7944
D_2 Line Å	5890	7665	7800	7800	8521
Approximate f/H cycles per gamma	7.00	7.00	4.66	7.00	3.50
Separation between adjacent lines of F = I+J level due to Back-Goudsmit effect at 0.5 gauss** (cps	138	531	36	36	6.7
(gamma	19.8	76	7.7	5.2	1.9

* From Parsons and Wiatr (1962)
** From Bloom (1962)

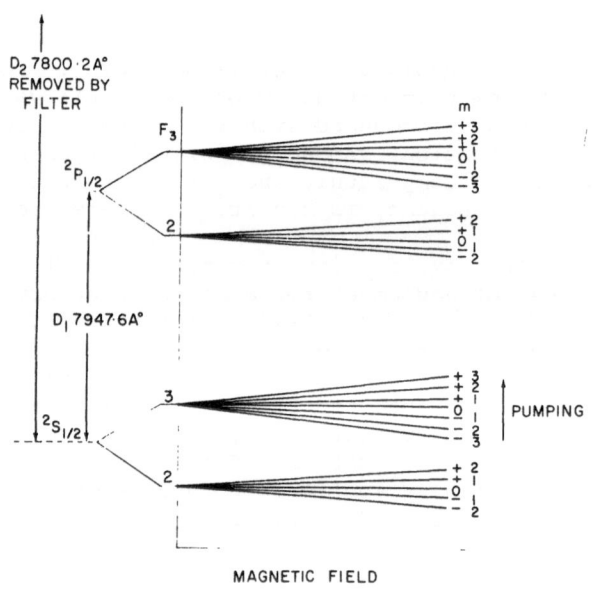

Fig. 11. Schematic energy level diagram for ^{85}Rb.

386

an experimental magnetometer built at the Signals Research & Development Establishment at Christchurch. Most of the following brief account of the design of typical magnetometers is taken from the above papers. Skillman and Bender (1958) describe a rubidium magnetometer built at the Fredericksburg Magnetic Observatory for absolute measurements of the earth's magnetic field. They used both the 87_{Rb} and 85_{Rb} isotopes, which gave complete agreement to 0.1 gamma, but were 6 gammas lower than the value in terms of the International Magnetic Standard obtained from the Fredericksburg magnetographs. The uncertainty in the Rubidium constants correspond to about 2 gammas. The field was determined by careful observation of the dip in pumping intensity as the radio frequency was varied, corresponding to 'scope 1 in Fig. 10. They reported observing the six components of the Zeeman transition.

Self-tracking magnetometers can be designed following two basic methods. In the first, the Larmor-frequency modulated cross beam is coupled to the RF oscillator, while in the second the absorption frequency is found by sweeping either the magnetic field or the RF oscillator frequency and using a servo-loop to correct the RF oscillator frequency. Parsons and Wiatr (1962) report on a magnetometer using the latter method, Fig. 12. A small magnetic field supplied by the low frequency oscillator is

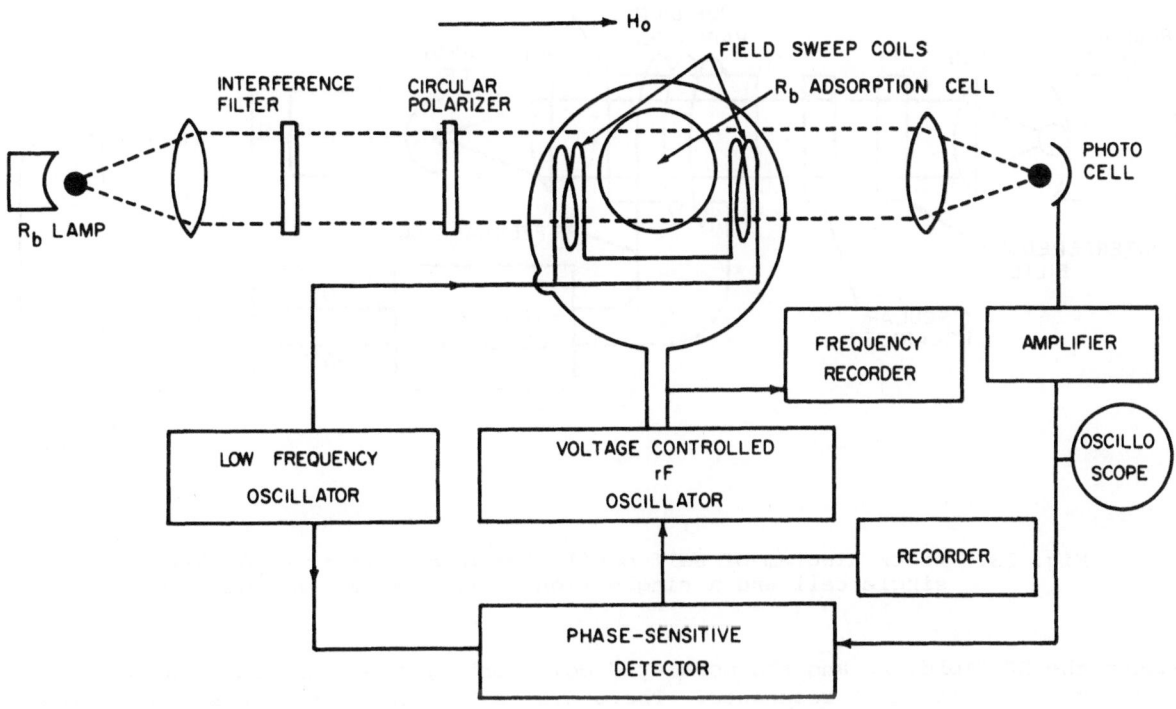

Fig. 12. Block diagram of rubidium magnetometer using swept-field frequency locking.

used to sweep the field through the absorption line. An alternative method is to superimpose a small frequency modulation on the RF oscillator. The frequency control voltage is proportional to the frequency and can be used as the output. Alternatively, the frequency of the oscillator may be recorded. Parsons and Wiatr report the

387

F = 3 resonance to be much stronger than the F = 2, the respective signal-to-noise ratios being 100 and 20. Using a field augmented to 2.5 oersted the individual lines could be observed on the oscilloscope. The lines were not observed at 0.47 oersted. This could be due to an inhomogeneous field across the absorption cell, interference from mains frequency fields, lamp characteristics or photocell noise. A sensitivity of 0.5 gamma and frequency response to 0.1 cps were achieved in the experimental model.

Varian Associates have examined self-oscillating magnetometers in detail (Bloom, 1962, and others). A typical magnetometer is outlined in Fig. 13. A 90° phase change

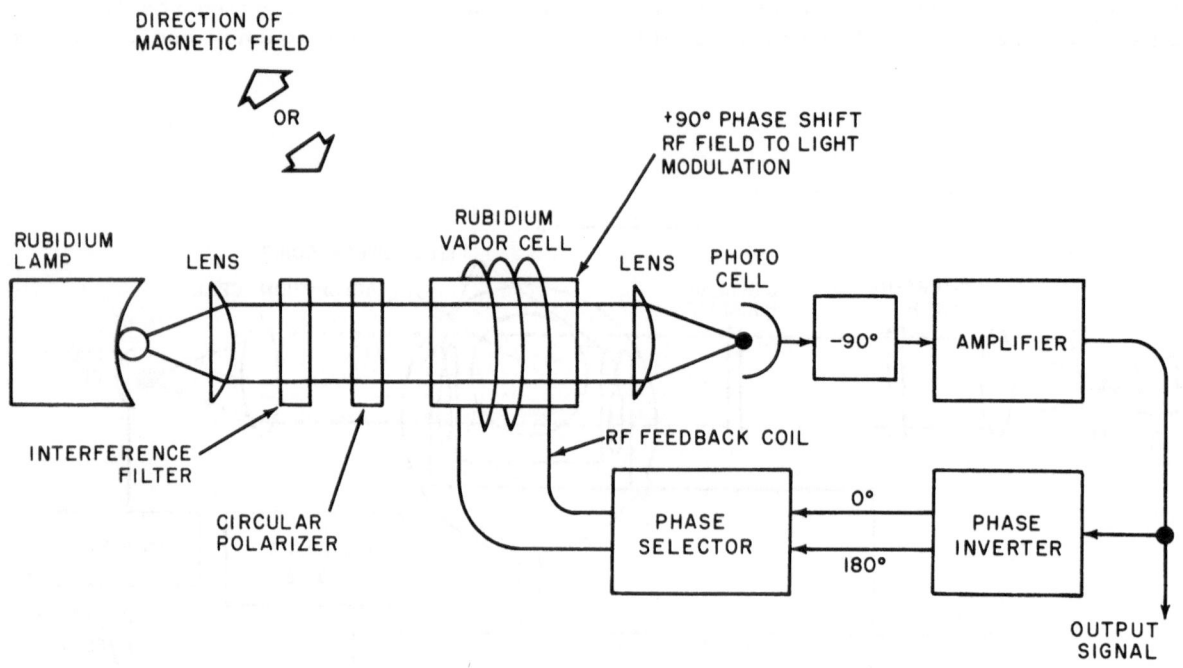

Fig. 13.   Block diagram of self-oscillator magnetometer employing a
            single cell and a single light path (Varian Associates).

between the RF field, H, and the modulated component of the light perpendicular to H_O, M_X, is predicted at exact resonance. Therefore, a 90° phase shift must be added by the external circuit. If the phase shift has the correct sign for angles between the optic axis and H_O lying between 0° and 90°, it has the wrong sign for angles between 90° and 180°. This problem may be overcome either by automatic searching for the correct sign of phase shift, or by constructing a magnetometer symmetric with respect to a reversal of H_O, Fig. 14. The dual cell design of Fig. 14 has the further advantage that the line asymmetry due to the Back-Goudsmit effect can be made opposite in the two cells (corresponding to exchanging +m and -m levels) resulting in an average line which is unchanged on rotation by 180°. In practice, however, irregularities in manufacture contribute some asymmetry. As will be discussed later, the configurations of Fig. 13 and Fig. 14 have dead zones about the optic axis and at

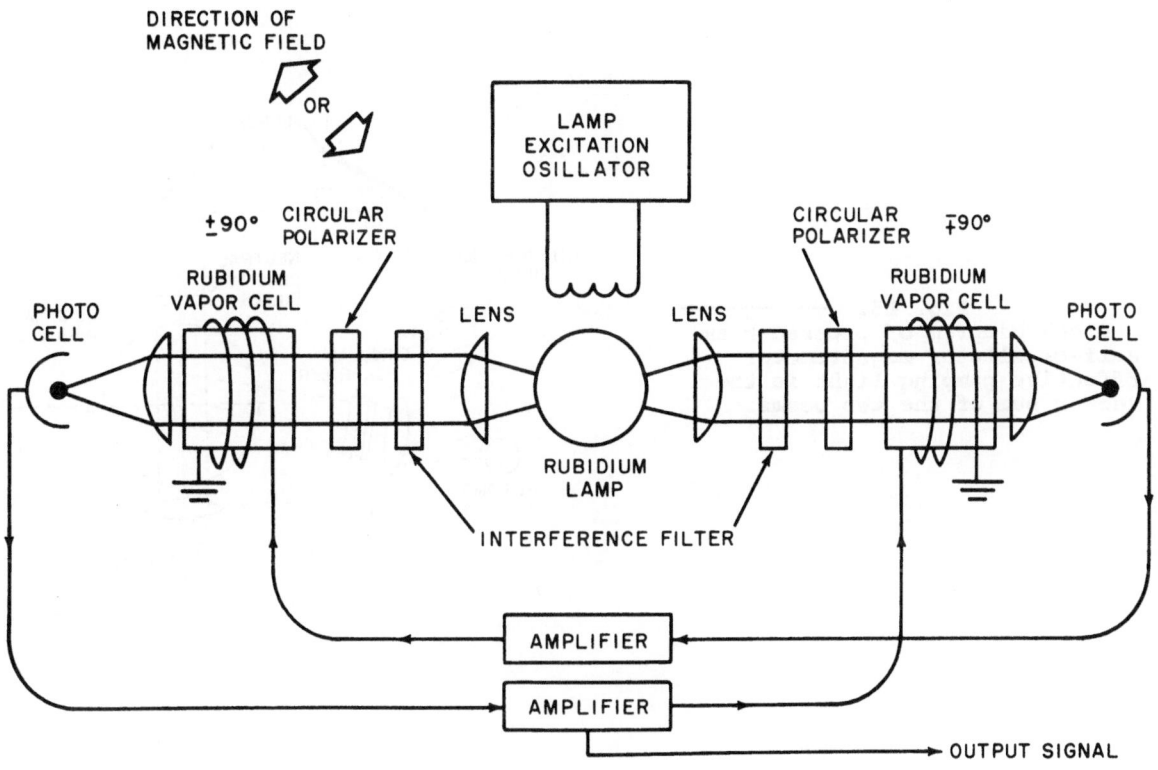

Fig. 14. Block diagram of dual cell self-oscillation
magnetometer (Varian Associates).

right angles to it. A third configuration, using crossed beams and corresponding to
the original Bell and Bloom experiment, Fig. 10, has no dead zone along the optic
axis. Fig. 15 is a block diagram of the optical system of a self-oscillating magnetometer of this design recently delivered to the Pacific Naval Laboratory.

While Zeeman splitting is nearly independent of the orientation, signal-to-noise
ratio is a function of the magnetic field direction. With the angle between the optic
axis and the magnetic field denoted by θ the following angular dependence applies for
a single-beam magnetometer using circularly polarized light: a) the pumping process
varies as $\cos\theta$, b) the detection of the absorption line varies as $\cos\theta$, c) the
generation of free spin precession varies as $\sin\theta$, d) the detection of the Larmor
modulation varies as $\sin\theta$ (Bloom, 1962). The signal-to-noise ratio of the magnetometer outlined in Fig. 12 varies as $\cos^2\theta$ while those in Fig. 13 and Fig. 14 vary as
$\sin\theta \cos\theta$. When the effect of the variation of the perpendicular component of the RF
field is taken into account, the curves in Fig. 16 apply to a single-axis self-
oscillator magnetometer. In practice, the dead zones are about 15° at the poles and
10° at the equator. In the crossed-beam design, however, the axis of the detecting
light is rotated nearly 90°, resulting in an approximately $\cos^2\theta$ dependence, thus
eliminating the polar dead zone. However, the lobal pattern for rotation about an
axis parallel to the effective pumping axis deteriorates somewhat, but still falls
within ±0.5 gamma. Some of the deviation may be due to residual current loops or to
small amounts of magnetic material.

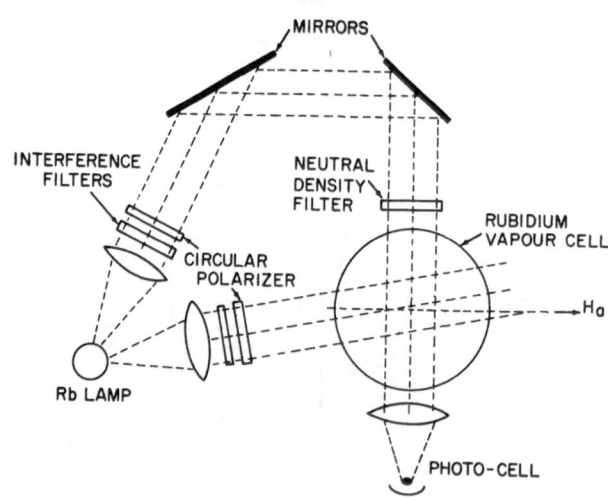
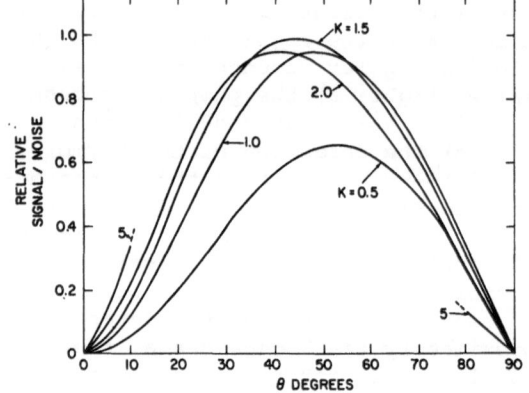

⟵ Fig. 16.
Signal amplitude or signal-to-noise
ratio of the self-oscillator as a function of θ. These curves differ from
the simple $\sin\theta \times \cos\theta$ dependence in
that they include the effect of varia-
tion of the perpendicular component of
the rf field. $k = \gamma H_a (T_1 T_2)^{\frac{1}{2}}$ where $2H_a$
is the peak value of the alternating rf
field along the coil axis, T_1 and T_2 are
relaxation times and γ is the ratio of
the Larmor frequency to the applied
field (Bloom, 1962).

390

Bloom (1962) claims that the self-oscillating magnetometer is lighter in weight, has lower power consumption and wider dynamic range than the swept-field type. The minimum detectable field is limited by the line width to about 2 gammas. Practical noise levels of less than 0.02 gamma have been achieved. In an analog recording system considerable noise is contributed by the frequency-to-voltage converter. The inherent noise level, with appropriate filtering of the self-oscillator output frequency, heterodyned with a well-stabilized oscillator would appear to be of the order of 0.01 gamma or less in low-gradient areas. Gradients across the absorption cell may be troublesome unless the sensing head is mounted well above the ground, and the usual precautions are taken with respect to ferromagnetic objects (nails, etc.) and current-carrying cables, especially power-lines. For a cell diameter of 5 cm a gradient of 200 gammas per kilometer will produce a variation across the cell of 0.01 gamma. Thus, in some locations significant line broadening and degradation of sensitivity may occur. It is obvious that motion in even a small gradient will introduce noise. This problem has been discussed in more detail by Ward and Ruddock (1961).

Metastable helium magnetometer. Franken and Colegrove (1958) first reported the alignment of metastable helium atoms, using unpolarized resonant radiation. A more extensive report followed two years later (Colegrove and Franken, 1960). The relevant energy diagram is given in Fig. 17 and the experimental layout in Fig. 18. A fraction of the helium is kept in orthohelium form by weak RF excitation. The lifetime (10^{-4} second) is sufficient to consider the metastable $^3S_1$ level as the groundstate of a new atom, orthohelium, contained in a buffer gas of groundstate helium, parahelium.

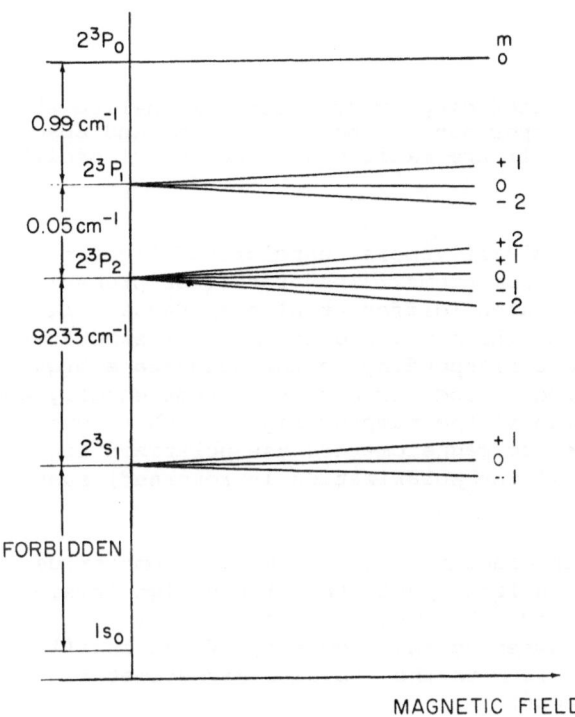

Fig. 17.   Schematic energy diagram of
metastable helium.

391

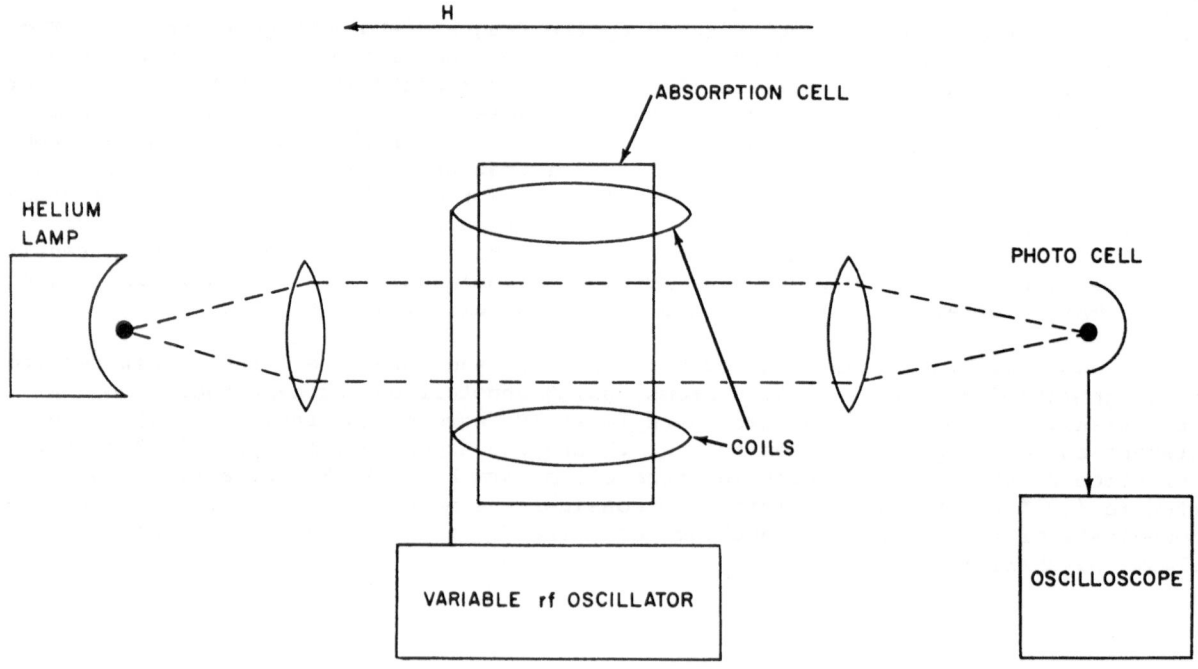

Fig. 18.   Block diagram of metastable helium absorption
           experiment. The helium lamp and absorption
           cell are excited by a common rf oscillator.

    In contrast with the alkali vapors, unpolarized light may be used to achieve sub-
stantial orientation of helium atoms. The transition probabilities between the ^3P
substates and the $^3S_1$ substates (different n) are unequal, leading to different
steady-state populations of the m = 0 and the m = $\pm$ 1 substates. As with pumped
alkali vapor, an RF field corresponding to the substate energy separation of about
28 cps per gamma is applied to redistribute the atoms equally amongst all substates.
This leads to an absorption of the pumping light, with as much as 0.8% change in
intensity. Commercial helium magnetometers use polarized light to enhance the
population of +m (or -m, if the polarization is reversed) substates by pumping
action.

    Fig. 19 shows a block diagram of a helium magnetometer using swept-frequency
detection of the absorption line. Note that the design is similar to that of the
rubidium magnetometer in Fig. 12. The operation is the same except that the absorption line is detected by sweeping the frequency of the oscillator rather than by
sweeping the field. But, as with the rubidium magnetometer, the field rather than
the frequency could be swept.

    The angular dependence of the signal-to-noise ratio is a function of several var-
iables, including the polarization of the pumping light, the mixing of the ^3P substates (hence the density of the buffer gas) and the RF field. For unpolarized light

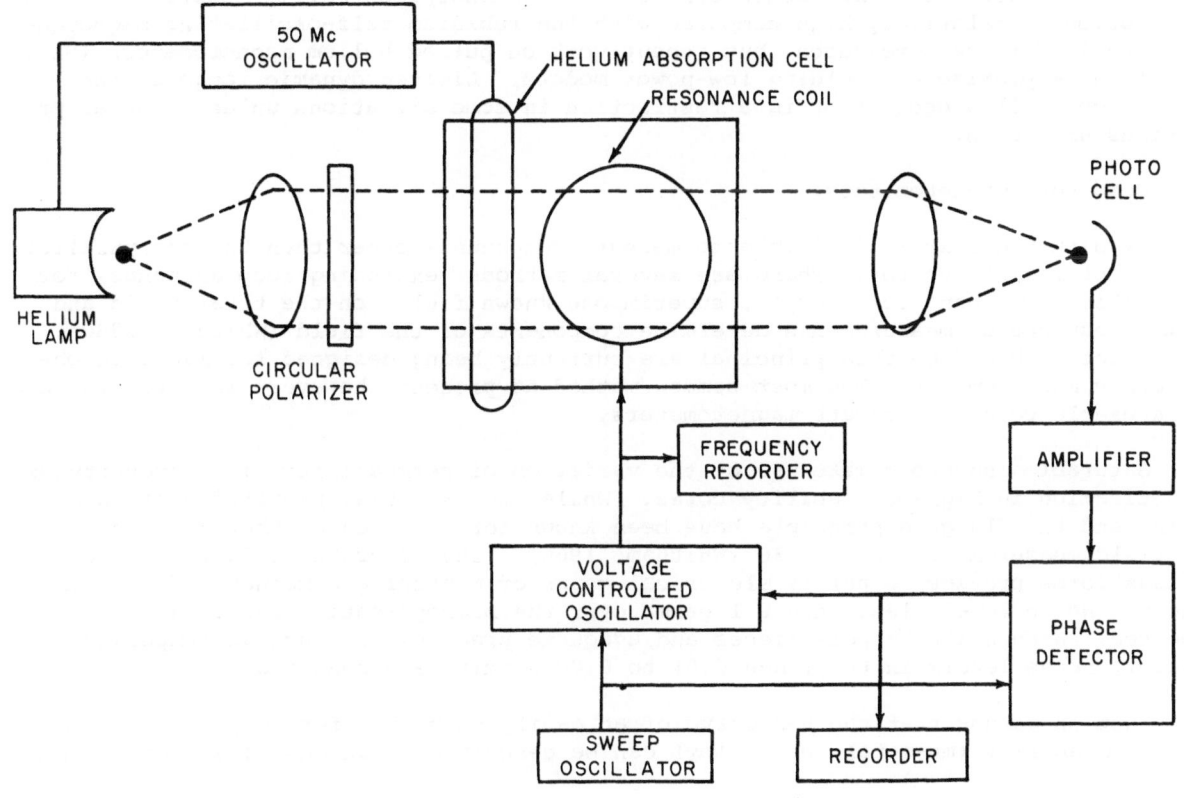

Fig. 19. Block diagram of metastable helium magnetometer.

the angular dependence is approximately $S(\theta) = (3\cos^2\theta-1)^2$, where θ is the angle between the optic axis and the magnetic field. Thus, there is a null at $\theta = 55°$. With circularly polarized light, however, the angular dependence under the same conditions is given by $S(\theta) = \cos^4\theta$ and the null occurs at $90°$ rather than $55°$. Something less than fourth power dependence of $\cos\theta$ is observed in practice.

The resonant frequency for constant H should be nearly independent of the angle between the ambient magnetic field and the optical axis of the magnetometer. It is reported, however, that rotation through $180°$ leads to a shift in frequency corresponding to nearly one gamma. Very little has been published on the effect as yet and it does not appear to be well understood. However, as with rubidium magnetometers, it can be reduced by using two oppositely directed optical paths.

In both the rubidium vapor and metastable helium magnetometers, current loops, glass-to-metal seals, etc. are a potential source of perturbing fields, which will alter the lobal patterns. The noise level of the helium magnetometer seems to be about the same as that of rubidium magnetometers - about 0.01 gamma. The size of the absorption cell in demonstration models is larger than some rubidium cells, and, therefore, may be more sensitive to field gradients. The usual requirements for a stable platform in areas of severe gradients, and for removal from local inhomogeneioties, apply.

Advantages claimed for the metastable helium magnetometer (Rice, 1961; Keyser, et al, 1961) include: a) a single resonant line (all $\Delta m = \pm 1$ transitions involve the same energy); b) the strictly linear relationship between the magnetic field and the output; c) no need for temperature control of the absorption cell. However, the power consumption is relatively high compared with the rubidium self-oscillating magnetometers built for space research, but recent work on pulsed helium magnetometers (Bleil, 1963a) shows promise of yielding low-power models. Limited dynamic range of the voltage-controlled oscillator is a restriction in some situations unless special precautions are taken.

Component magnetometers.

In many situations it is desirable to measure components other than the one parallel to the total field vector. There are several methods, excluding loop antennas, for doing this. One can, for example, superimpose known fields on the total field and from a sequence of measurements determine components of the field (Whitham, 1960). Magnetometers based on this principal are currently being designed for magnetic observatory measurements. The most common method at present, however, involves the use of saturable-core or fluxgate magnetometers.

Saturable inductors make use of the variation of permeability with intensity of magnetization in high-permeability cores. While the essential properties of such cores, and the fluxgate principle have been known for many years, they were not used for field measurements until 1936 (Whitham, 1960). The saturable inductor in its various forms provides a purely electrical means of measuring a magnetic field component. Noise levels less than 0.1 gamma over the micropulsation range have been reported. With suitable pole pieces and adequate precautions, such as temperature control, noise levels in the range 0.01 to 0.001 gamma have been realized.

Let us assume that the B-H curve of an easily saturable ferromagnetic core of high permeability (mu-metal, permalloy) can be described by an odd power series in H,

$$B = \sum_{n=o}^{\infty} b_{2n+1} H^{2n+1}$$

Hystresis is neglected in this qualitative discussion. The coefficients b are fixed by the geometry and core material while the field H is given by $H = H_o + \cos \omega t$, where H_o, less than the saturation level, is the unknown field and $h \cos \omega t$ is an impressed field of constant amplitude. Expanding in harmonics of the impressed field we obtain

$$b = \sum_{n=o}^{\infty} B_n (H_o, h) \cos n \omega t.$$

The coefficients of the even harmonics, including the constant term, involve odd powers of H_o but even powers of h. Similarly, the coefficients of the odd harmonics are even functions of H_o but odd functions of h. Thus, the odd harmonics do not vanish for $H_o = o$, but the even ones do. Furthermore, the phase of the even harmonics is sensitive to the sign of H_o. This property is exploited in various ways, the most common of which are discussed below.

The peak voltage method is one of the simplest, a block diagram, one configuration being shown in Fig. 20. Two parallel inductors with the primaries wound in opposite senses are used. Alternatively, a single inductor with a center-tapped primary may be used. The secondary voltage is a function of the sum of the induction, B_1 and B_2, of the two inductors, which add in-phase opposition, Fig. 21.

It can be seen from the figure that the peak value of the resultant field, $B_1 + B_2$, is proportional to the unknown field, H_o. In the ideal case the voltage induced in the common secondary is symmetrical but when hysteresis is taken into account it is

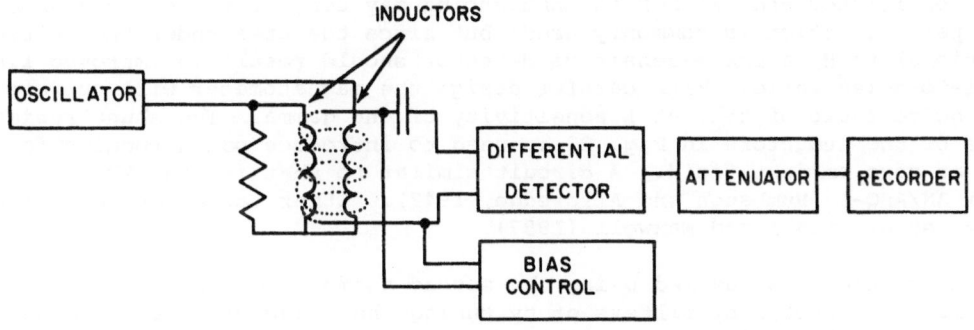

Fig. 20.   Block diagram of peak voltage magnetometer
with bias control. The resistor reduces the
noise level at zero ambient field.

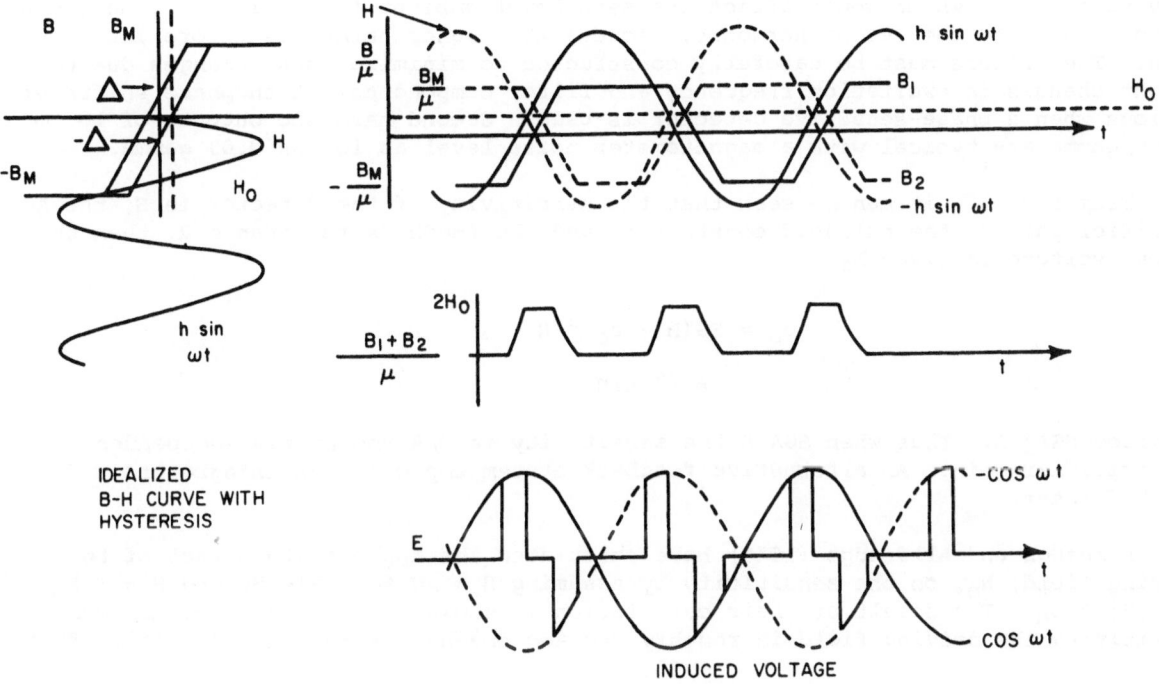

Fig. 21.   Wave forms in an idealized two-element
fluxgate magnetometer with hysteresis.

not, and the peak value changes sign when the sign of H_O reverses. Furthermore, loading of the oscillator and deliberate unbalancing may contribute to the asymmetry. In practice, peak detection is commonly used, but since the area under the voltage curve is proportional to H_O a phase-sensitive detector should result in improved linearity and signal-to-noise ratio. With careful design the magnetometer will operate with a signal-to-noise ratio of ten, at a sensitivity of one gamma. The shunt resistor across one of the inductors in Fig. 20 is used to introduce odd harmonics for minimizing noise at zero ambient field. A circuit similar to that in Fig. 20 was used in the MAD system AN/ASQ-1 (Rumbaugh and Alldredge, 1949). Other variations are discussed by Vacquier et al (1947) and Maxwell (1957).

Recent practice has favored using the second harmonic of the unbalance voltage only. It may be selected by filters or by tuning the inductor to the second harmonic. Great advantages in sensitivity and simplicity (Whitham, 1960) are claimed for the tuned inductor method.

In the untuned system either a single element or two matched, oppositely excited, inductors may be used. If two inductors are used the odd harmonics are approximately cancelled, thus relaxing the requirement on the filtering circuits. It is common practice to remove most of the component that is to be measured by adding a known constant field of appropriate sign - usually through a winding - and recording deviations from this base directly, or with the use of negative feedback. Feedback systems can be designed to make the output almost independent of amplifier gain and to have improved linearity, since changes in the ambient field at the detector are greatly reduced. In some designs only an error field remains.

A typical single-element system has been described by Schonstedt and Irons (1955) and by Whitham (1960), and is outlined in Fig. 22. The filter tuned to the fundamental of the excitation frequency minimizes the second harmonic content in the driving current, which would affect the zero level ambient field signal. The second filter, tuned to the second harmonic, reduces the higher harmonics before amplification. The filters must be carefully constructed to minimize phase changes due to slight changes in excitation frequency and filter components. Such phase shifts are serious when a phase-sensitive detector is used. Second harmonic outputs of 10 to 35 μv/gamma are typical with a magnetometer noise level as low as 0.03 gamma.

From Fig. 22, it can be seen that the sensitivity if the detector is S, the AC amplifier gain G, the solonoid constant A, and the feedback resistance R, then the output voltage is given by

$$e_O = SG(H - e_O \; ^A/R,$$

$$\doteq (^R/A)H$$

provided SGA>>R. Thus when SGA R the sensitivity is $^R/A$ and nearly independent of the amplifier gain. An alternative feedback system employing an integrator is described later.

Rumbaugh and Alldredge (1949) have calculated the approximate effect of the driving field, h_O, on the sensitivity by assuming $B = \mu H$ for $|B| < B_M$ and $B = \pm B_M$ for $|B| > B_M$. The result of their calculation is shown in Fig. 23. For maximum sensitivity the driving field is roughly one and one-half times the saturation field.

A suitably tuned inductor results in the second harmonic component of the signal being so greatly enhanced that band-pass filters are unnecessary (Serson and Hannaford 1956; Serson et al, 1957; Serson, 1957).

The effect is unlike that of a tuned inductance elsewhere in the circuit, for the circulating current of the resonant system directly affects the saturable inductor

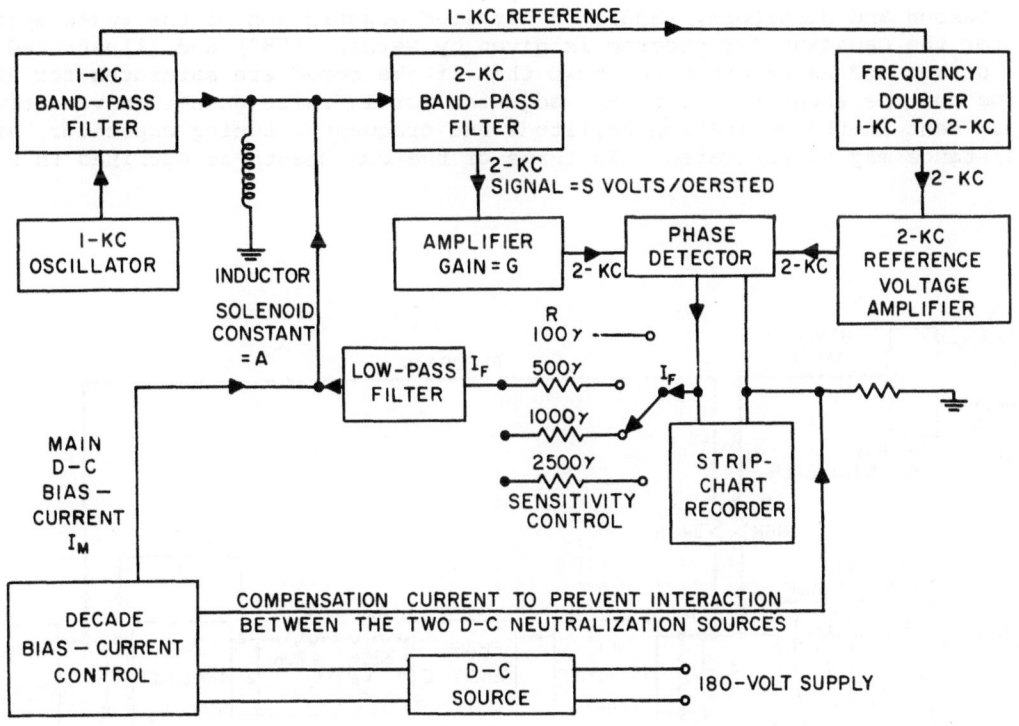

I-KC REFERENCE

I-KC BAND-PASS FILTER

I-KC OSCILLATOR

INDUCTOR
SOLENOID CONSTANT = A

2-KC BAND-PASS FILTER

2-KC SIGNAL = S VOLTS/OERSTED

AMPLIFIER GAIN = G

2-KC

PHASE DETECTOR

2-KC

2-KC

FREQUENCY DOUBLER I-KC TO 2-KC

2-KC

2-KC REFERENCE VOLTAGE AMPLIFIER

LOW-PASS FILTER I_F

R
100γ
500γ
1000γ
2500γ
SENSITIVITY CONTROL

I_F

STRIP-CHART RECORDER

MAIN D-C BIAS- CURRENT I_M

DECADE BIAS — CURRENT CONTROL

COMPENSATION CURRENT TO PREVENT INTERACTION BETWEEN THE TWO D-C NEUTRALIZATION SOURCES

D-C SOURCE

180-VOLT SUPPLY

Fig. 22. Block diagram of an untuned second harmonic magnetometer with feedback and bias (redrawn from Schonstedt and Irons, 1955).

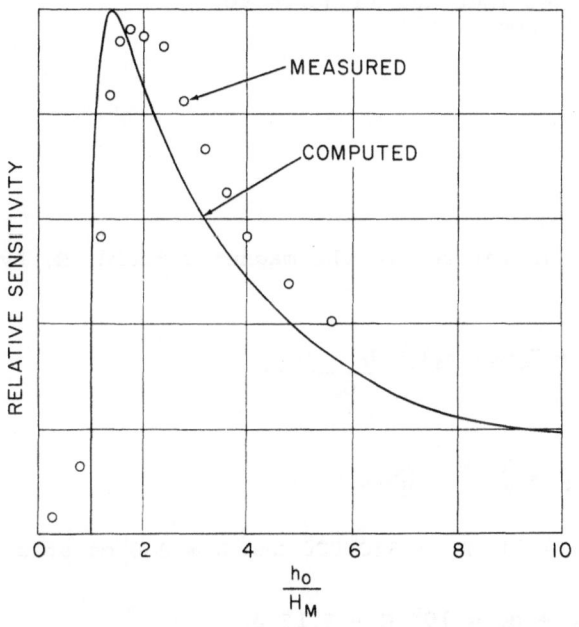

MEASURED

COMPUTED

RELATIVE SENSITIVITY

$\frac{h_0}{H_M}$

Fig. 23. Variation of sensitivity with the amplitude of the driving field h_0. (Rumbaugh and Alldredge, 1949).

397

resulting in a large increase in sensitivity. In fact, "infinite" sensitivity may be
achieved (Serson and Hannaford, 1956). A detailed description of the tuned system
designed for the Canadian IGY program is given by Serson (1957) and illustrated in
Fig. 24. Serson and Hannaford have shown that if the cores are saturated for 28.5%
of the time and the shunt resistor, R_1, adjusted for infinite gain then variations of
several per cent in the excitation amplitude and frequency, tuning capacitor, and
shunt resistance may be tolerated. In terms of the components as outlined in Fig. 24,

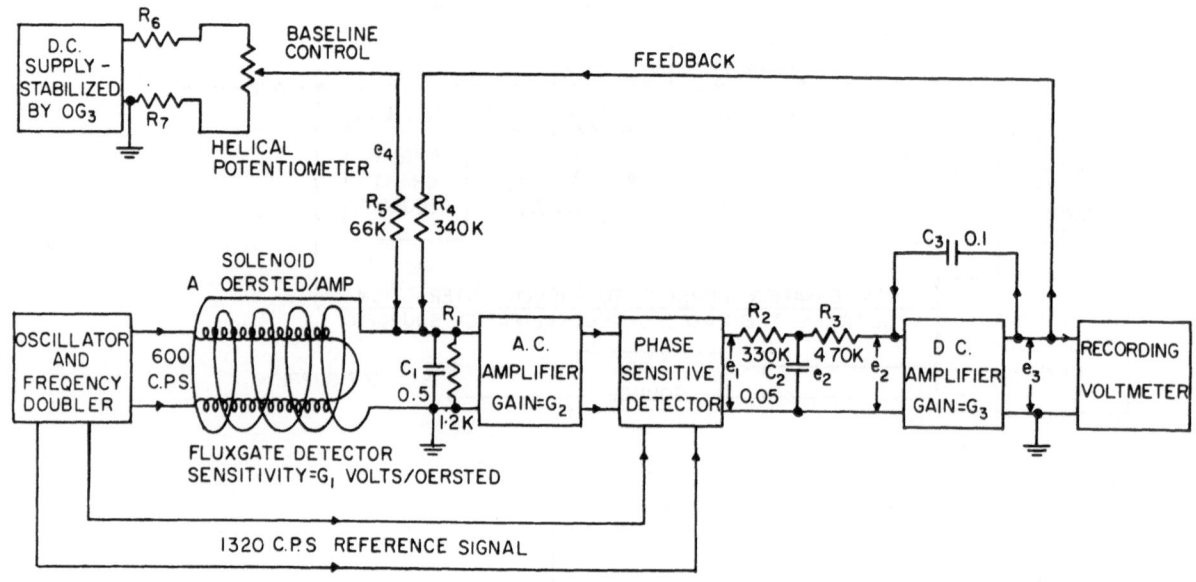

Fig. 24. Block diagram of a tuned inductor magnetometer
employing feedback (Serson, 1957).

Serson has shown that the output voltage, e_3, is related to the magnetic field, H, by
the differential equation.

$$\frac{R_2 C_2 R_3 C_3 R_4}{G_1 G_2 A} \ddot{e}_3 + \left(\frac{R_2 C_2}{G_3} + R_2 C_3 + R_3 C_3 \right) \frac{R_4}{G_1 G_2 A} \dot{e}_3$$

$$+ \left(1 + \frac{R_4}{G_1 G_2 G_3 A}\right) e_3 = \frac{R_4}{A} H - \frac{R_4}{R_5} e_4$$

Substituting G_1 = 100 v/oersted, G_2 = 200, G_3 = 50, R_4 = 340,000 and A = 340 oersted/
ampere, we obtain:

$$3.9 \times 10^{-5} \ddot{e}_3 + 4.0 \times 10^{-3} \dot{e}_3 + e_3 = 10^3 H - 5.15 e_4$$

where e_3, e_4 are in volts, H in oersteds and time in seconds. The response is similar to that of a servo mechanism with inertia and viscous damping and a natural frequency of 36 cycles per second. The high natural frequency allows the integrator to
respond to field changes of 4000 gammas per second even though the phase sensitive
detector saturates at $\pm$ 2 volts. Sensitivity measured between different instruments
was identical with 1%, limited by the resistor tolerance of 1%. Temperature effects
are small in terms of percentage - 0.0055% per °C for the solenoid but when measuring
a large component they become appreciable. The recorded value of the vertical component in Canada increases 3 gammas per °C as the temperature of the head rises. Other
components may be more easily thermostatted and temperature effects minimized.

Snavely (1963) has described a magnetometer utilizing the difference in time of
saturation when sawtooth excitation is applied as a measure of the unknown field.
The principle of operation is illustrated in Fig. 25 and Fig. 26.

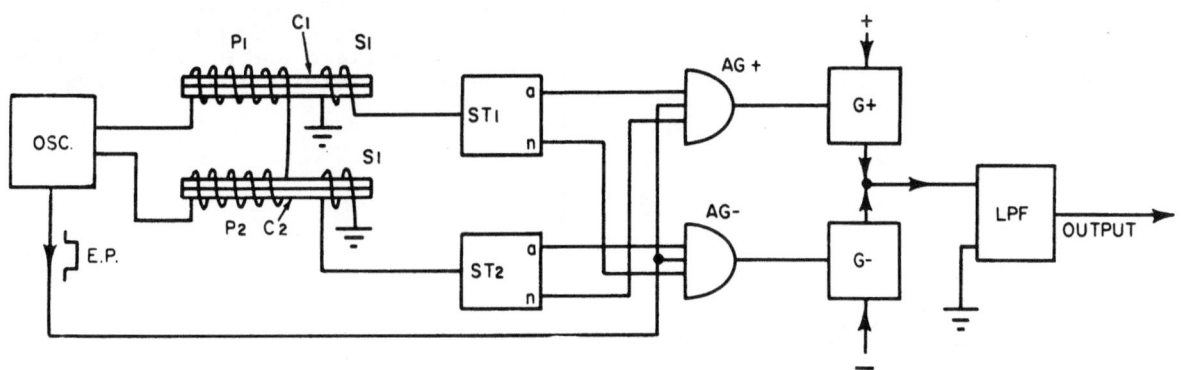

OSC.= SAWTOOTH OSCILLATOR
C1, C2 = SATURABLE CORES (PARALLEL)
ST1, ST2 SCHMITT TRIGGERS

G+,G- = GATES CONTROLLING + AND-VOLTAGES TO LPF
EP = ENABLE PULSES
AG+, AG-=AND GATES
LPF = LOW PASS FILTER

Fig. 25.   Pulse fluxgate magnetometer (Snavely, 1963).

A recent development in fluxgate magnetometers has been the introduction of
toroidal cores as the sensitive element in a manner similar to second harmonic
magnetic modulations (Geyger, 1962a, b, c). Since there is no air gap the driving
power is low and a simple switching transistor multivibrator may be used such as in
the circuit of Fig. 27 (a). The primary and secondary windings are accurately wound
on two halves of the core. The core and windings are analogous to a conventional

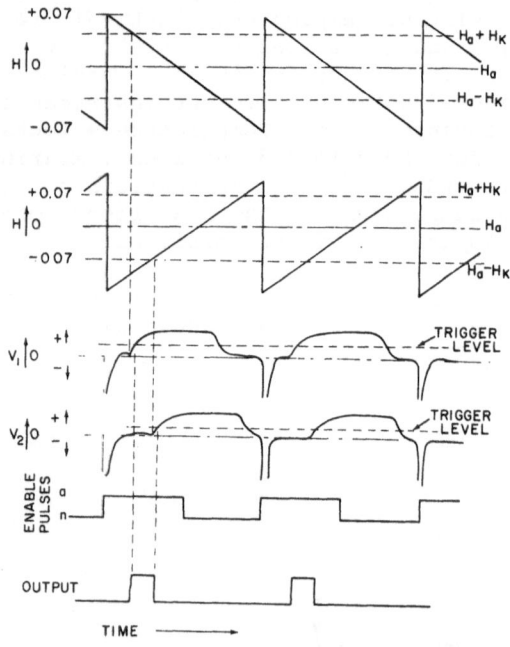

Fig. 26. Wave forms in pulse fluxgate magnetometer (Snavely, 1963).

Fig. 27.
   (a) Toroidal core fluxgate magnetometer driven by a switching transistor multivibrator. FG is the fluxgate core, TM unsaturated coupling core and DS the detector.

   (b) Principle of operation. H_{AC} is the alternating driving field, H_X the unknown field and H_1 and H_2 the resultant field in the two halves.

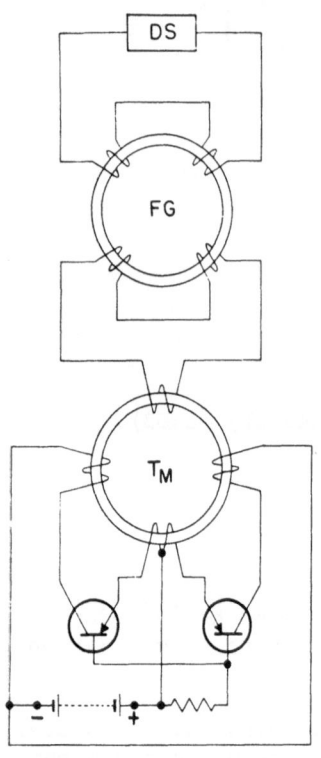

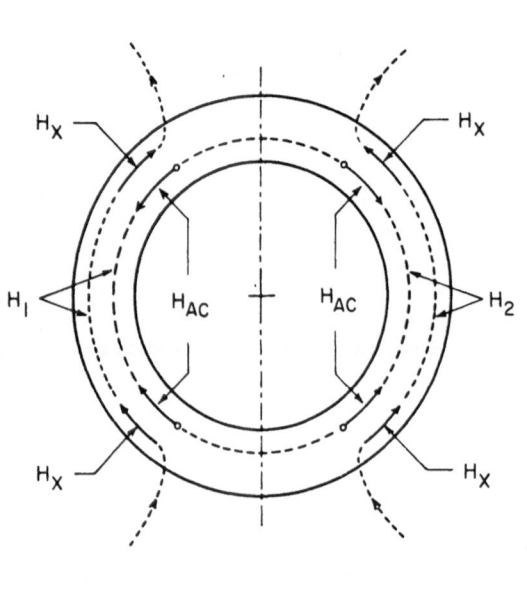

two element inductor, Fig. 27 (b), but due to the thin laminar windings of the core material or the use of a ferrite core, a higher frequency may be used. Since the same core is used for both elements of the inductor physical balancing and treatment of the inductor is simplified. A novel feature is the ability to use several windings and, therefore, obtain a nearly circular directional characteristic. Noise levels of less than 0.01 gamma have been reported for the ring core magnetometer (Bleil, 1963b).

In addition to the conventional saturable-core reactor one can use the variable permeability directly. Gregg (1947) has described a probe for measurements of magnetic fields using the variation of inductance of a mumetal core with applied field. A transformer was wound on a 0.1" x 0.01" diameter permalloy wire, and with a constant input voltage the output was used as a measure of the applied field. With this method it is necessary to use an exceedingly stable input source since amplitude variations, unless corrected, appear as spurious field measurements.

The Electro-Mechanics Company (1959) discuss a variable permeability magnetometer in which the variable inductance forms the inductive element of an oscillator, the frequency of the oscillator then being a function of the applied magnetic field. In their final design Ceramag 27 with annealed soft iron pole-pieces formed the variable inductor, Fig. 28. They report sensitivities of the order of 1 cycle per

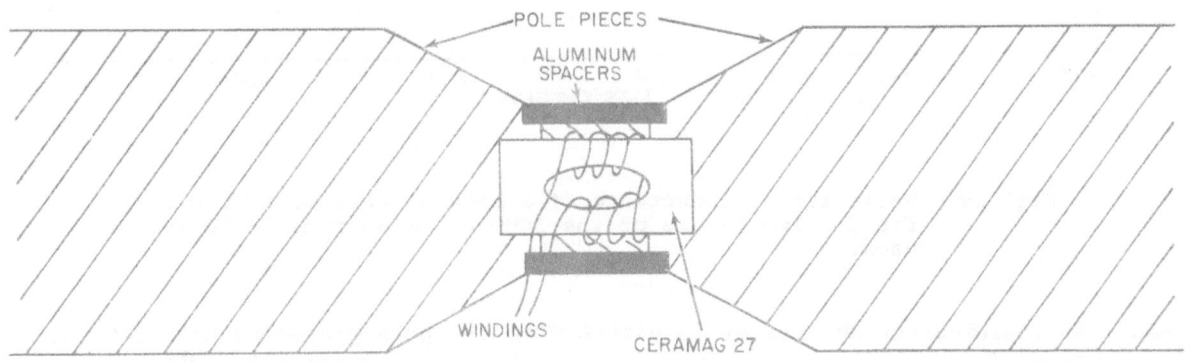

Fig. 28. Variable inductor used by Electro-Mechanics Company.

gamma, but unfortunately, the design is very sensitive to temperature changes. The temperature effects may be reduced by operating two inductors either magnetically biased in opposite directions or with the pole-pieces removed from one, and recording the beat frequency between the oscillators.

Lieberman (1963) has investigated μ-metal as core material for a variable inductance magnetometer and found that it is most sensitive when well saturated. The reason is simple, for $\Delta L/\Delta H$ is a maximum when μ is changing most rapidly, Fig. 29. He has found that sensitivities of 0,001 gamma may be achieved if care is taken to minimize noise and drift and an auxiliary magnet used to obtain the most sensitive operating condition.

Hall effect magnetometers also show appreciable promise. Bleil (1963a) describes Hall effect magnetometers of about 1 gamma sensitivity. Using large permalloy flux collectors Lieberman (1963) has reported achieving sensitivities as high as 0.001

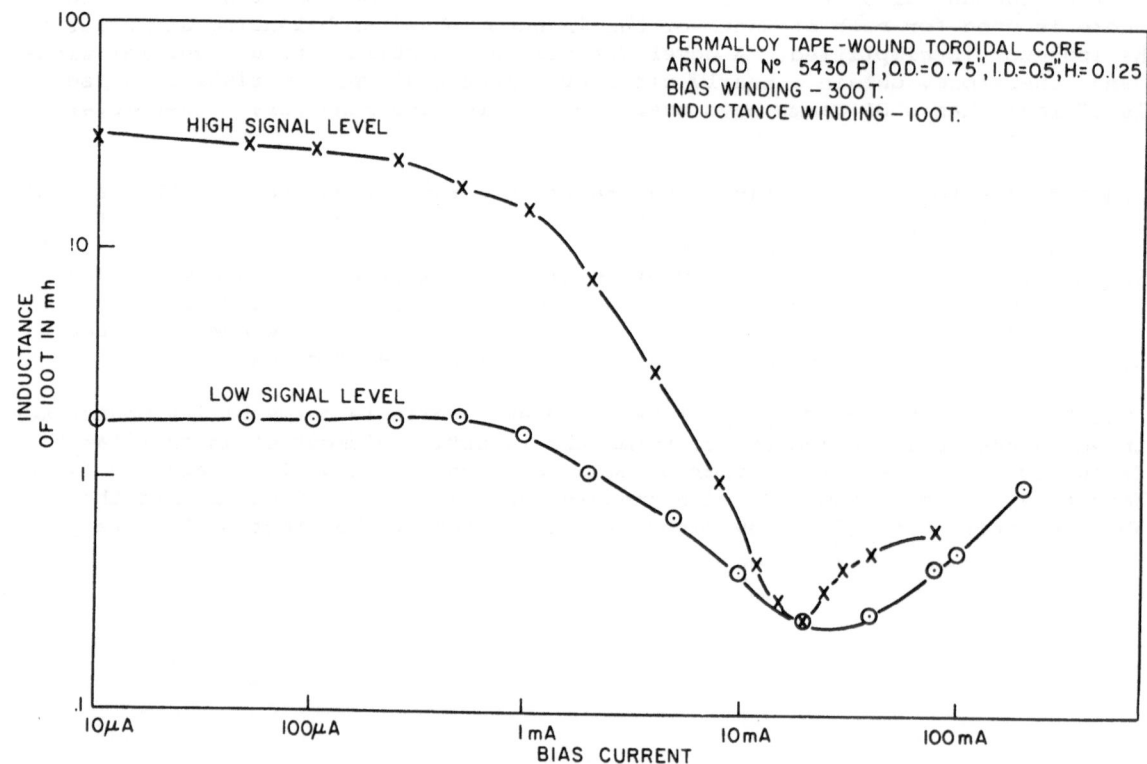

Fig. 29. Variation of inductance of a μ-metal core with DC bias
field. Two levels of the 1000 cps measuring field were
used.

gamma. An indium-arsenide crystal insulated from the pole-pieces by thin ferrite
slices was used as the detector. In order to achieve the sensitivity claimed the
temperature of the pole-pieces was maintained constant by circulating water.

Thellier (1957) and Whitham (1960) discuss in detail variometers which use suspended magnets whose rotation is then a function of the magnetic field and is detected optically. Auxiliary magnets may be used to increase the sensitivity to any
desired degree, but in practice it is limited by considerations of stability and
linearity to about 1 gamma per millimeter. It can be shown (Thellier, 1957) that
the frequency amplitude response is given by

$$A = \frac{1}{\{(1 - r^2)^2 + 4\alpha^2 r^2\}^{\frac{1}{2}}}$$

and the system phase lag by

$$\tan\beta = 2\alpha \frac{r}{(1 - r^2)}$$

where $r = \frac{\omega}{\omega_0} = \frac{T_0}{T}$, with T_0 the undamped period of the variometer, and α is the ratio of the damping to the critical damping. The responses for various damping ratios are plotted on Fig. 30. Thellier recommends that half critical damping be used.

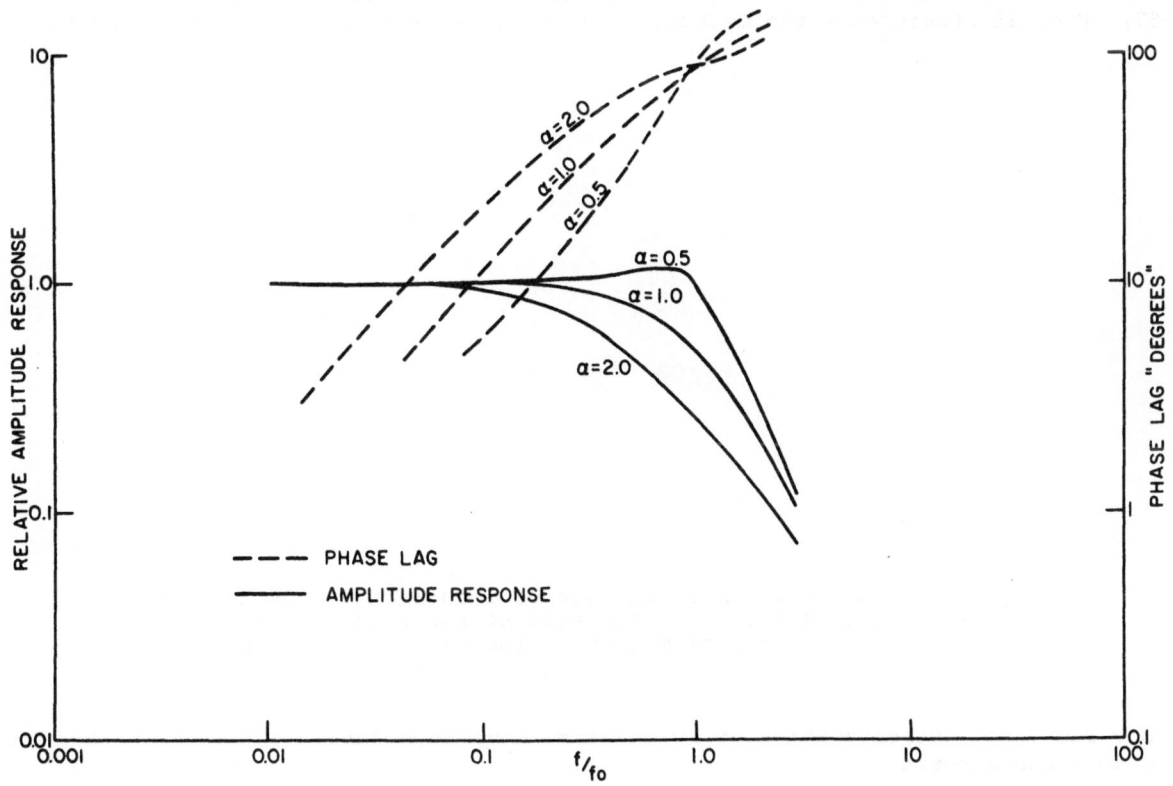

Fig. 30.   Frequency response and phase characteristics of a variometer. α is the ratio of damping to critical damping.

There are several other methods of measuring magnetic fields, for example, electron beam methods, but these are not in common use and do not appear to offer advantages over systems that have been discussed. Whitham (1960) and others discuss some of these.

Measurements of $\frac{dB}{dt}$.

As is well known, micropulsations are so small compared to the steady state value of the field, that unless special precautions are taken, devices such as fluxgate magnetometers introduce noise from the biasing field. Furthermore, the magnitude of micropulsations increases sharply with decreasing frequency. For these reasons, systems for measuring $\frac{dB}{dt}$ directly are attractive.

A further advantage when using coils as detectors is the ability to increase the sensitivity of the instrument by increasing the effective area of the coil. If a response proportional to B instead of $\frac{dB}{dt}$ is required over a limited frequency range, the output may be integrated with a suitable time constant. Although alternative ways for obtaining $\frac{dB}{dt}$ are available, such as differentiation of the output of a fluxgate magnetometer, we shall discuss only the use of coils in this paper.

The coil may be used to detect the motions of a suspended magnet which in turn is a measure of fluctuations in the magnetic field (Grenet, 1949; 1957 and Thellier, 1957). Fig. 31 illustrates the method. A large magnetometer is used limiting the fre-

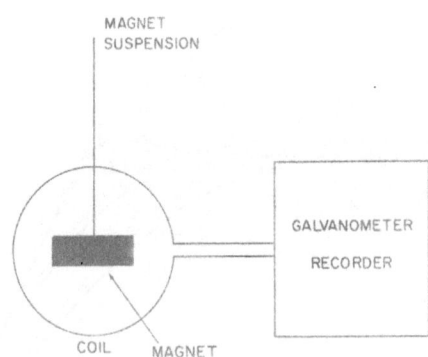

Fig. 31. Block diagram of the Grenet induction variometer, the axis of the coil, the axis of rotation of the magnet, and the axis of magnetization are mutually perpendicular.

quency response. Sensitivities of 10 mm/gamma have been easily obtained in the band 4 to 40 seconds period.

Alternatively, the coil may be used to detect the fluctuations directly through the induced voltage. There are two common methods of detecting the induced voltages; galvanometers and electronic amplifiers, each of which leads to distinct design considerations.

In general, the detecting coil together with the galvanometer or electronic amplifier should have an acceptable signal-to-noise ratio and be of a size and weight compatible with requirements of installation and of portability. The coil should be some distance from the recording building and from vehicular traffic, and preferably buried to minimize wind motion. The impedance should, therefore, be such that the induced signal can be transmitted over several hundred feet of cable to the galvanometer or amplifier. Furthermore, if more than a few sets of equipment are required, construction problems should be minimized. Obviously, these requirements are conflicting and a compromise must be reached.

Thellier (1957), Selzer (1957) and Whitham (1960) discuss various coil/galvanometer combinations. In view of the complete discussions available only an outline of the design considerations will be given. The design of the coil must take into account the characteristics of the galvanometer. If A is the area of one turn in square centimeters, the number of turns in the coil n, the flux density in gammas, then the induced voltage will be

$$= -nA10^{-13}\frac{dB}{dt} \text{ volts}, \tag{1}$$

404

and the equation of motion of the galvanometer, neglecting inductance of the coil,

$$\frac{d^2S}{dt^2} + 2\alpha\omega_0 \frac{dS}{dt} + \omega_0^2 S = -S \frac{\omega_0^2 nA}{R} 10^{-13} \frac{dB}{dt} \qquad (2)$$

where S = galvanometer response in millimeters at one meter,

Sg = galvanometer sensitivity in millimeters per ampere at one meter,

ω_0 = natural undamped frequency of the galvanometer,

α = ratio of damping to critical damping,

R = total resistance of the circuit, and

B = the magnetic field in gammas.

The response to a sinusoidal signal, B sinωt, is given by

$$S = \frac{-r}{\{(1-r^2)^2 + 4\,\alpha^2 r^2\}^{\frac{1}{2}}} -\omega_0 \frac{S_{g}nAB}{R} 10^{-13} \cos(\omega t -\psi)\,mm \quad \text{at one meter,}$$

where

$$\psi = \arctan \frac{2\,\alpha\,r}{1-r^2}$$

and r is the ratio of the frequency to the natural frequency of the galvanometer. The frequency response is a function of the coil resistance through the damping factor α. Fig. 32 shows the frequency response for various values of α.

If the inductance of the coil is important, the response to a sinusoidal signal becomes:

$$\frac{d^2S}{dt^2} + 2\alpha\omega_0 \frac{dS}{dt} + \omega_0{}^2 S = - \frac{S_{g}\omega_0^3 nA10^{-13}}{Z} \, B \cos(\omega t - \psi_c),$$

where $Z = \sqrt{R^2 + \omega^2 L^2}$

is the impedance of the coil-galvanometer combination and $\psi_c = \arctan \frac{\omega L}{R}$. The response at high frequencies is decreased and an additional phase lag introduced. To illustrate the design of a detector let us assume an air core coil in which other considerations (for example, portability) have fixed the coil diameter and weight. The coil resistance and inductance are given by

$$R_c = \frac{\rho \pi^2 d^2 n^2}{\sigma W} = \frac{k_1}{W} \, n^2, \qquad (3)$$

$$L = 0.00629dn^2 \left[\log \left(\frac{8d}{d_1}\right) -1.75\right] \times 10^{-6} \text{ henries}$$

where

σ = conductivity of the conductor,

d = diameter of coil,

d_1 = diameter of the cross section of the winding,

$$W = \text{weight of the conductor, and}$$

$$\rho = \text{density of the conductor.}$$

Since we assume $\frac{d}{d_1}$ is large one may approximate the inductance by:

$$L = 6.29 \; d \; \ln\left(\frac{8d}{d_1}\right) = 10^{-9} n^2$$

$$= k_2 n^2 \text{ henries.} \qquad (4)$$

Note that for a given mass of conductor the ratio of resistance to inductance is independent of the number of turns. If we neglect frictional losses in the galvanometer the damping factor α is given by

$$\alpha = \frac{R_O + R_g}{R_C + R_\ell + R_g}$$

where R_O is the critical damping resistance, R_C the resistance of the coil, R_ℓ the resistance of the lead-in wire, and R_g the resistance of the galvanometer. Substituting for n, α and L in terms of R_C gives

$$S = \frac{\sqrt{R_C W} \; \omega_O \; S_g \; A \; B \; X \; 10^{-13} \text{ mm}}{\sqrt{K_1 \left[(R_C+R_\ell+R_g)^2 + r^2 \; \omega_O^2 \; \left(\frac{k_2 W}{k_1}\right)^2 R_C^2\right]^{\frac{1}{2}} \left[(1-r^2)^2 + 4\left(\frac{R_C+R_g}{R_C+R_\ell+R_g}\right)^2 r^2\right]^{\frac{1}{2}}}}$$

Usually R_ℓ, R_g may be neglected by comparison with R_C and R_O. With this approximation S simplifies to:

$$S = \frac{r \sqrt{RW} \; S_g \; \omega_O \; A \; B \; X \; 10^{-13} \text{ mm}}{\sqrt{k_1 \left[1 + r^2 \; \omega_O^2 \; \left(\frac{k_2 W}{k_1}\right)^2\right]^{\frac{1}{2}} \left[(1-r^2)^2 \; R^2 + 4R_O^2 r^2\right]^{\frac{1}{2}}}}$$

where we have written R for R_C.

If R is varied to give maximum response

$$R = 2 \; R_O \; \frac{r}{1-r^2} \; .$$

Note that it is independent of the ratio of resistance to inductance and is infinite when $\omega = \omega_O$. If we wish to maximize the response at frequencies low compared to the resonance of the galvanometer the coil resistance must be decreased (damping increased). Usually, however, improved response at higher frequencies is required and $R = 2R_O$ is a suitable choice. Since the sensitivity, neglecting inductance, is then given by

$$S = S_g \; d \left(\frac{W\sigma}{8\rho R_O}\right)^{\frac{1}{2}} \frac{r \; x \; 10^{-13}}{\sqrt{1 - r^2 + r^4}} \text{ mm/gamma}$$

it has been pointed out that a galvanometer with a maximum value of $S_g / \sqrt{R_O}$ should be chosen.

As a sample calculation, assume a coil of 2 meters diameter, 50 kg weight and a good one-second galvanometer with R_g = 500 ohms, R_O = 12,000 ohms and S_g = 1.5 x 10^9 mm/ampere. The coil consists of 12,500 turns of No. 29 wire and the theoretical response is given in Fig. 32 by the curve α = 0.5, neglecting inductance. The change in response for other values of resistance, but retaining the same mass of copper, is evident from the figure.

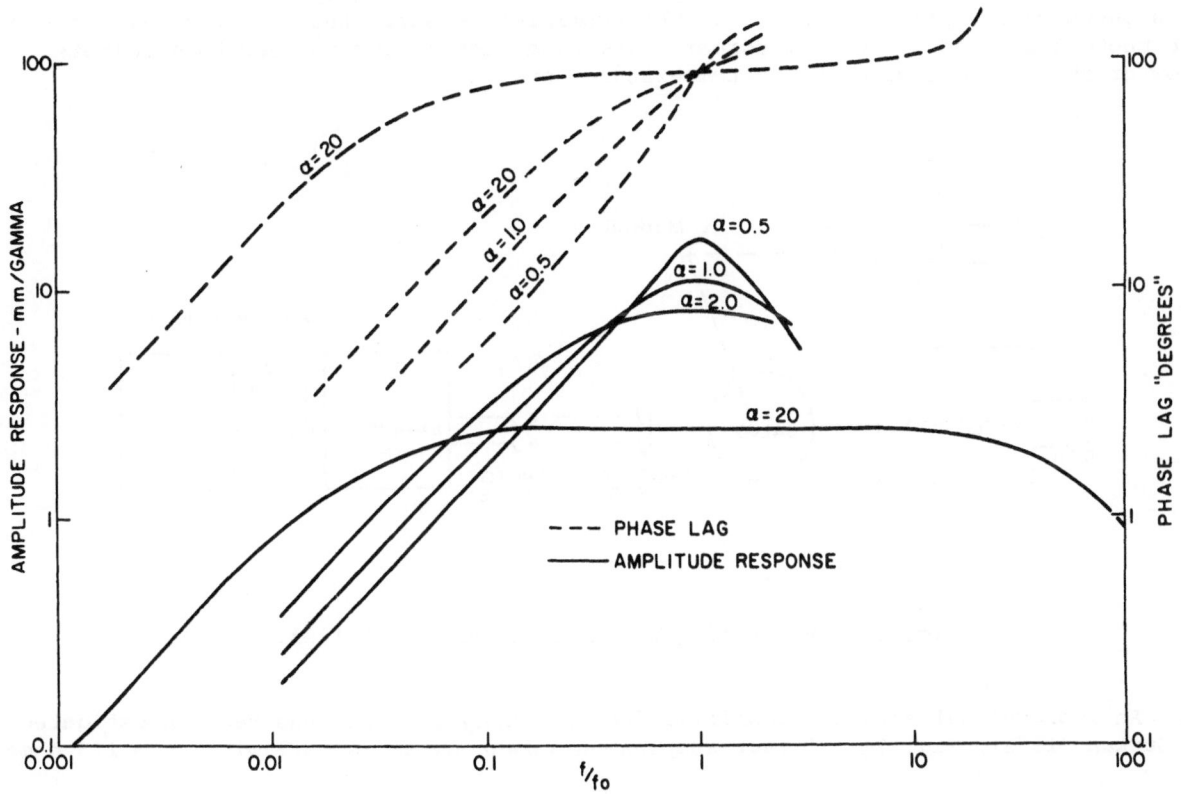

Fig. 32.   Theoretical response of constant diameter and weight coil-galvanometer system discussed above. α is the ratio of critical damping resistance to resistance of the coil.

Thellier (1957) and Selzer (1957; 1963) have pointed out that if in Eq. (2), $\alpha \gg 1$ (highly overdamped) then over a considerable frequency range the equation reduces to

$$2\,\alpha\,\omega_O\,\frac{dS}{dt} \;=\; -S_g\,\frac{\omega_O\,^2 nA}{R}\,10^{-13}\,\frac{dB}{dt}$$

By integration, one obtains

$$S - S_O = -\;\frac{S_g\omega_O\,nA10^{-13}}{2\alpha}\;(B - B_O).$$

Therefore, the response is proportional to the field rather than its derivative. Such an overdamped galvanometer system has been called a "fluxmeter," If a high permeability core is used in the coil, then the term "barre-fluxmeter" is used (Selzer, 1957). The response of a fluxmeter with α = 20 is included in Fig. 32.

Selzer (1963) also discussed the use of multiple galvanometer systems (photocell coupled) increasing the response in a particular band. For example, a combination of two galvanometers with natural periods of 0.2 and 1.8 seconds may be used to increase the response at "pearl" frequencies.

The galvanometer displacement may be recorded on photographic paper or detected by a photo-cell amplifier, Fig. 33. The amplifier measures the differential light on the photo-tubes and drives a recorder. The band-pass filter is used to eliminate slow drifts of the system due to temperature, etc.

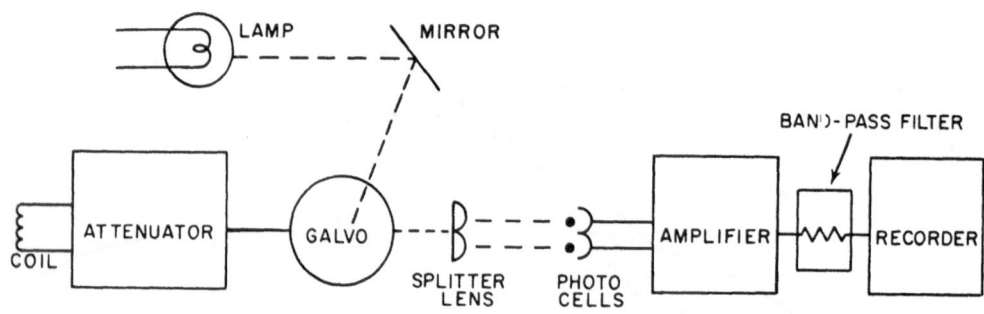

Fig. 33. Block diagram of photo-cell amplifier.

An improved galvanometer amplifier incorporating feedback has been investigated by Surkan (1963), Fig. 34. A taut-suspension galvanometer is used with R_o = 125 ohms.

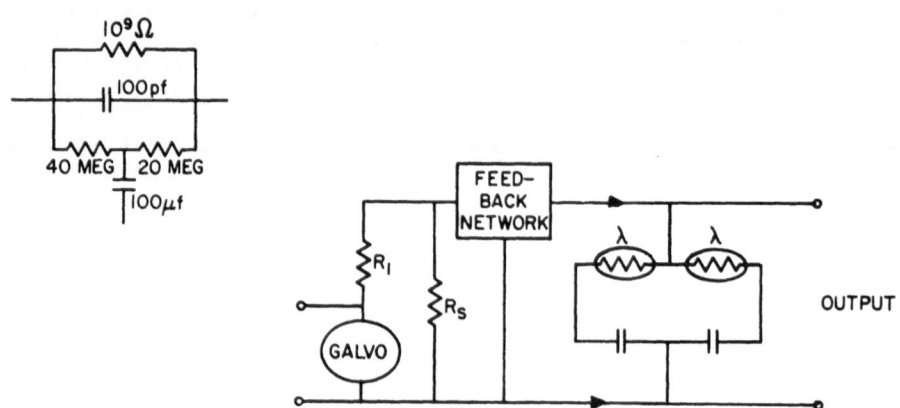

Fig. 34. Diagram of photo-cell amplifier with feedback. The optics now shown are photo-resistors.

The gain is varied by changing R_S, typically R_1 = 100,000 ohms and the balancing potentiometer in the photo-resistor circuit 10,000. The feedback network is used to provide drift correction (attenuate low frequency response). A capability of measuring 10^{-16} watts is claimed by the author.

As an alternative to galvanometers, one may use low noise quasi-dc electronic amplifiers to detect the induced voltage. Evans (1960) has stressed that the characteristics of the amplifier must be taken into account when designing the coils or loop antennas. It is necessary to fix some of the parameters such as the minimum signal-to-noise ratio, mass of the coil, and perhaps its diameter. The dependence of amplifier noise on source impedance must also be known. The discussion on coil design below is based largely on the work of Evans.

Let us consider first the design of air core antennas or coils. Fig. 35 shows the equivalent noise factor as a function of source resistance for the preamplifier used by the Pacific Naval Laboratory. It is evident that the minimum noise factor occurs with a source resistance between 40 and 150 ohms.

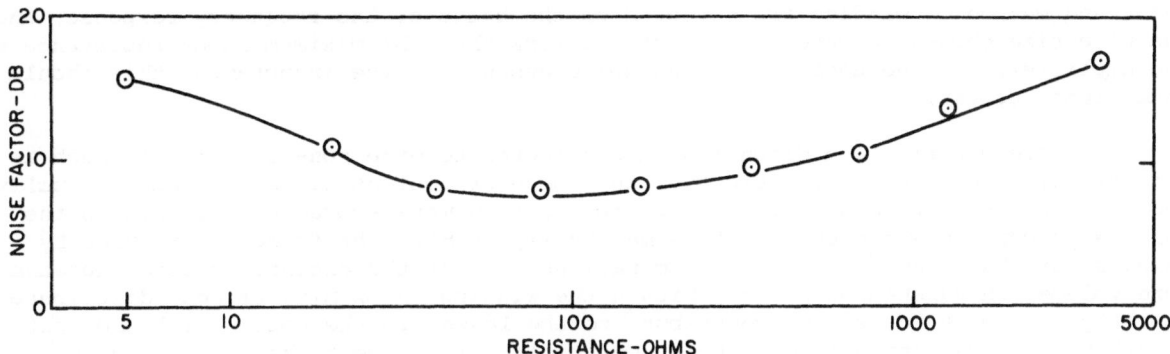

Fig. 35. Noise Factor as a function of source resistance for PNL micropulsation amplifier.

It follows from Eq. (1), (3) and (4) that for a fixed diameter coil the ratio of resistance to inductance is given by

$$\frac{R}{L} = \frac{k_1}{k_2 W}$$

and the power into a load resistance R_{in} equal to the resistance of the coil is given by

$$I^2 R_{in} = \frac{k_1 \; \omega^2 W \; A^2 B^2 \times 10^{-26}}{4k_1^2 + W2k_2^2 \omega^2} \text{ watts,}$$

where ω is the angular frequency and B the rms field in gammas. Therefore, both the ratio of resistance to inductance of the coil and the power to the load depend only on the mass of the conductor and not on the number of turns. Thus, the coil resistance can be varied to match the input resistance of the amplifier. If the mass of the conductor is now varied to give maximum power transfer, it can be shown that this occurs when $\omega L = 2R$. If the mass of the conductor is increased above the optimum value at $f = f_2$ then the power delivered at lower frequencies will be increased at the expense of the signal near f_2. For example, if a coil of 2 meter diameter is constructed

409

weighing four times the optimum value for a frequency of three cycles per second and having 100 ohms resistance, it will consist of 3100 turns of No. 12 wire and weigh about 600 kilograms. Clearly, this is prohibitively heavy for most applications. Thus, the weight of the coil will usually be fixed by other considerations at less than the "optimum value" and the inductance can be neglected.

If the amplifier must have a fixed source resistance, the induced voltage is proportional to the radius of the coil and this should be increased to the largest possible value. On the other hand, if, for example, the equivalent noise voltage at the input of the amplifier is proportional to the square root of the source resistance, the signal-to-noise ratio is proportional to the coil radius and it should be increased at least until the coil resistance is equal to the maximum value for which the noise voltage is proportional to $\sqrt{R}$.

If both the weight and diameter of the coil are fixed by other considerations and the equivalent noise voltage at the input of the amplifier is proportional to the square root of the resistance, it can be shown that the signal-to-noise ratio is independent of the resistance. The wire size may, therefore, be determined by the available stock, ease of making the coil, etc.

The design of metal cored coils is more complex, for unlike air cored coils, inductance is the limiting factor. Evans (1960) and Hill and Bostick (1962), among others, have designed coils of this type for use with DC amplifiers. The number of turns and method of winding are selected on the basis of the frequency response, and the wire size chosen to make the resistance less than the maximum input resistance of the amplifier. If the amplifier input noise depends on the inductance, this should be taken into account.

The core material and dimensions are selected to make non-linearities, such as variable permeability, negligible over the range of ambient fields present - usually about 0.6 oersted or less. Losses are important in metal-cored coils. Within the core, eddy-current and hystresis loss may be important. The former is reduced by using a core built up of insulated laminations, but at the expense of some increase in core volume and Barkhausen noise. Within the winding, proximity effect, distributed capacity, and wire resistance contribute to the losses in the coil. Hill and Bostick (1962) have shown that skin effect is negligible. The same authors measured the effective resistance at 8.8 cps of a coil of the type described in Table 2, and found it to be about 750 ohms in excess of the DC resistance. Two hundred and fifty ohms was attributed to eddy-current losses in the core and the remainder to proximity effect, eddy currents in the winding, and hystresis. It was assumed that the frequency dependence of the losses was such as to make them much less than the resistance of the winding at one cycle per second.

Mumetal is the most commonly used core material but due to the demagnetizing factor of long slender rods the effective permeability μ^1, is not strongly dependent on the permeability provided it is a few thousand (Bozorth, 1951). μ^1 is more dependent on the length to diameter ratio, Fig. 36. In order to avoid saturation in the earth's ambient field, the field in the bar should be less than 1000 gauss. For an ambient field of about 0.5 oersteds this leads to μ^1 = 1000 to 2000. A length to diameter ratio between 70 and 100 will produce the desired permeability. The induced field decreases toward the ends nearly parabolically, resulting in the optimum winding cross section also being parabolic. Details of typical coils described by Evans and by Hill and Bostick are summarized in Table 2. Electro-static shielding is provided in all cases.

The inductance of the cored detector coils in Table 2 seriously limits the sensitivity at frequencies near 1 cps where the natural background is extremely low. For some purposes it can be reduced by series tuning (Lokken, Shand and Wright, 1963). Response curves with and without tuning are shown in Fig. 37.

TABLE 2

	Evans	Hill and Bostick
Core material strip size (in)	.75 x .015	1.2 x .032
Number of strips	35	36
Core area (in)^2	.39	1.38
Core length (in)	72	72
Winding section	Parabolic	Cylindrical, 4 section
Average number of turns	20,732	30,000
Wire size	18	22
Resistance (ohms)	63	360
Inductance (henries)	210	660
Calibration winding (turn)	10	30
Sensitivity $\frac{microvolts}{(gamma)(cps)}$		150
Resonant frequency (cps)	82	210
Weight of complete coil (lbs)	92	94

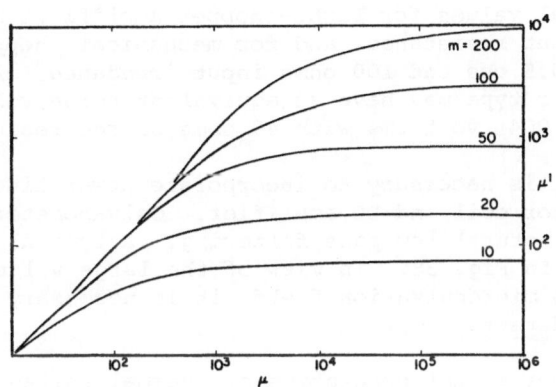

Fig. 36. Relation of apparent to true permeability of a
cylinder of given length to diameter ratio, m.
(Drawn from Bozorth, 1951).

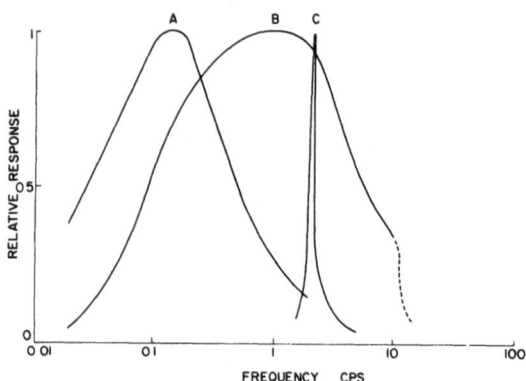

Fig. 37.   Response of Pacific Naval Laboratory metal-cored detector coils and
           amplifier, with and without series tuning. Curve A is with an Esterline-
           Angus chart recorder, curve B with magnetic tape or high frequency paper
           recorder, and curve C tuned to 2.2 cps recorded on magnetic tape.

    Calibration of air cored detector systems can be conveniently achieved by injec-
ting a voltage in series with the antenna. In the case of metal cored antennas,
however, the preferred method is to apply a known field by means of auxiliary "calibration" windings. The relation between current in the calibration winding and
magnetic field can be obtained by comparison with a known field from a large diameter
coil or from a solenoid. A multi-turn coil 200 feet in diameter has an axial flux
density that is uniform to within 2 per cent over a region extending to 20 feet from
the coil center. Vozoff (1961) has shown that for frequencies below 100 cps the
influence of the ground may be neglected for a coil less than 100 meters in diameter.
This type of coil, while requiring a large flat area, is easier to construct than a
good solenoid.

    Commercial DC amplifiers are available or specially designed amplifiers may be
built. High input impedance amplifiers may use light-choppers, while good low input
impedance amplifiers use mechanical choppers. Parametric amplifiers look promising
and one has recently been announced for frequencies down to about 0.01 cps in which
it is claimed the input noise voltage is equivalent to 10,000 ohms, but problems with
drift still exist. Typical values for light-chopper amplifiers below 1 cps are 0.2μ
volts rms and 1 megohm input resistance, and for mechanical chopper amplifiers $.005\mu$
volts peak to peak below 0.5 cps and 100 ohms input impedance. A carefully constructed amplifier of the latter type may have an equivalent noise voltage in the band
0.02 to 3 cps as low as 0.004μ volt rms with 40 ohms source resistance, Fig. 35.

    In most localities it is necessary to incorporate power line frequency rejection
filters between the detector coil and DC amplifier. Galvanometer amplifiers of the
non-feedback type provide natural low pass filtering. A typical 60 cycle per second
rejection filter is shown in Fig. 38. In view of the large value of the ratio of
permanent ambient field to micropulsation field, it is necessary to carefully bury or
otherwise immobilize the detector coils.

    Another source of error is the generation of thermal voltages at joints in the
wires, or due to small temperature differences along the wire. Since it is not practical to thermally insulate the entire system, negative feedback can be applied to
reduce drifts and keep the recorder on scale. The long time constants required can
be obtained either electronically or by adding a thermocouple pair in series with the
lead-in wire and applying heat to one of them. The thermal time constant of the

412

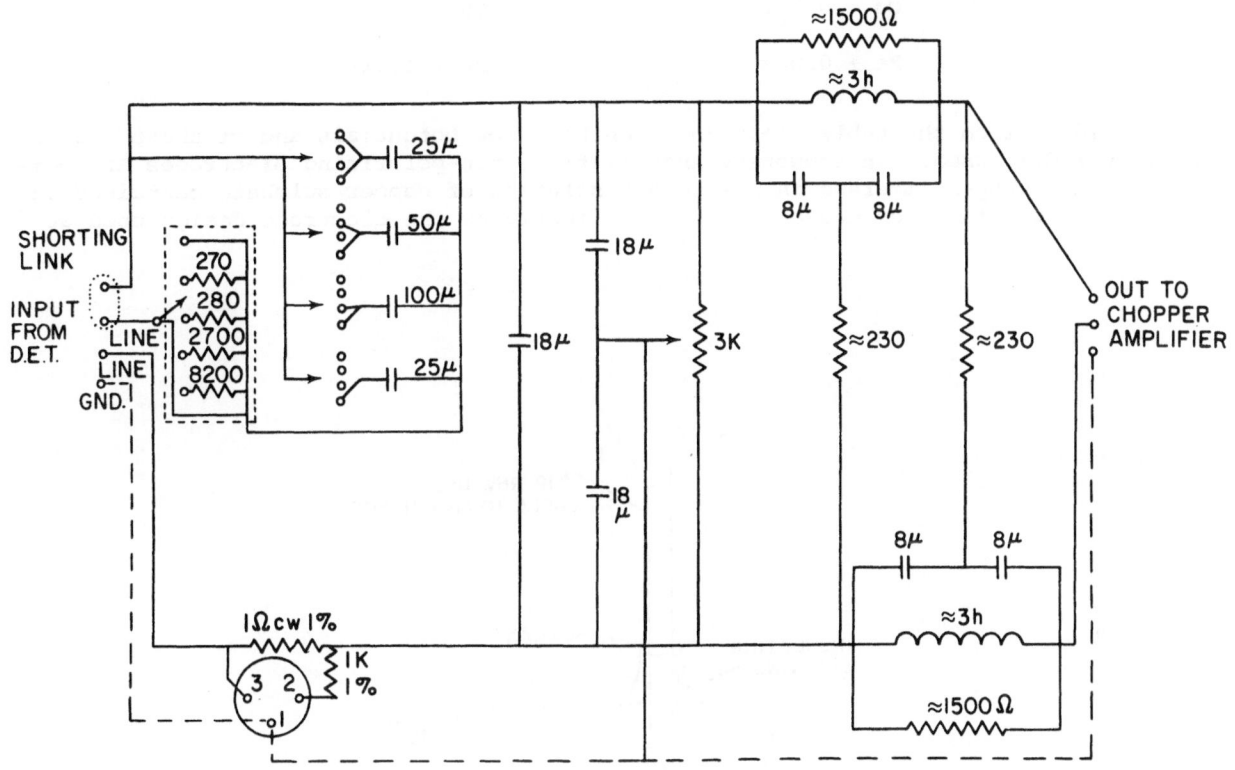

Fig. 38. 60 cps rejection input filter. Provision is made for calibration by voltage injection and for series tuning the detector coil.

heated thermocouple and its associated heat sink then provide the time constant.

Telluric currents.

Garland (1960) discusses telluric currents in considerable detail. Earth currents were observed in the ground return signals of telegraphy before 1850. Near the end of the Nineteenth Century measuring systems for them became part of the equipment at certain magnetic observatories. The currents are due both to natural and artificial causes, for example, oxidation of metallic minerals. However, the larger part of the earth currents are due to electromagnetic induction. Troitskaya (1957) discusses the system used by the Russian observers and installed for the IGY. In practice, the potential observed between pairs of buried electrodes is measured in terms of millivolts per kilometers. Therefore, improved sensitivity can be obtained by increasing the separation of the electrodes until a suitable signal-to-noise ratio is obtained for the particular amplifier under consideration.

It is usual to use a saturated solution of a salt of the metal for electrodes, or, for long term stability, a metallic grid buried in the ground. Since the potential between a metal and its solution is proportional to the temperature, it is desirable that a metal for electrodes be chosen for which the potential is small. Typical potentials between a metal and a solution of its salt are given below:

Al + 0.22v		Pb − 0.10v
Cd + 0.19v		Cu − 0.60v
Fe + 0.06v		Ag − 1.01v

As is evident from the table, lead and iron have low potentials and of these lead is the most widely used. In temporary installations non-polarizing electrodes are commonly used. Copper immersed in a saturate solution of copper sulphate contained in an unglazed pot is often chosen. Fig. 39 illustrates the electrode design used at

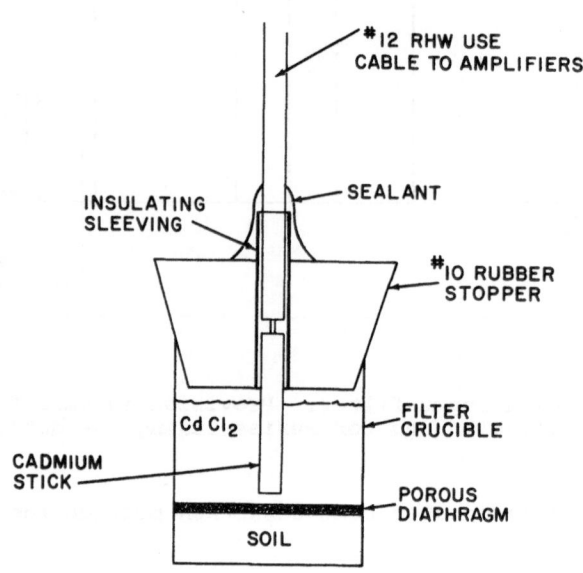

Fig. 39. Electrode for ground potential measurements suitable for mobile use (Williams and Hopkins, 1961).

the University of Texas (Williams and Hopkins, 1961). Electrical contact with the earth is made through the solution. Temperature effects are still present, however, and the difference in potential between the metal and the earth also varies with the rate of diffusion of the solution. Garland suggests that for long term stability and freedom from temperature effects, electrodes of a suitable metal such as lead buried several feet below the surface are probably most effective. Care should be taken to maintain similar earth characteristics at each electrode site. Hessler and Heacock (1962) report that in frozen ground they have found a network of lead to be effective.

Since potentials of the order of millivolts are to be recorded, amplifier characteristics are governed by considerations such as long term stability. Often the

414

amplifier on the recording instrument is used directly.

Recording Equipment.

Paper charts.

Magnetograms are a form of paper chart in wide use at magnetic observatories. At micropulsation frequencies, however, the recording speed must be such that the desired information can be resolved. Often the individual cycles are required and a high paper speed is necessary. However, if only a measure of the quantity in a particular frequency band selected by a filter will suffice a slow chart speed may be used. Drum recorders can provide a fast time base on a small area of paper chart. The dynamic range, however, is limited, and it is sometimes difficult to take quantitative measurements from the record. Strip chart recorders offer an improved dynamic range, but at the expense of an increased volume of recording. Typical recording speeds range from 3" per hour to 2" per minute. Speeds in terms of inches per hour are used for continuous recording, while the high speed recordings are used for limited time intervals which may extend to weeks. An interesting method of recording is reported by Schlicht (1962). A continuous-fold paper chart is passed through the recorder, which has its dynamic range restricted, at a sufficiently high speed that the detail of interest (frequencies of the order of 1 cps) can be resolved. A day later the paper is again fed through the recorder with the base line displaced, thus producing parallel records for each day. This allows rapid observation of diurnal trends. Professor Selzer showed examples of such recordings. Hessler and Heacock (1962) have installed a clutch arrangement on their recorder which records reversals of pen travel and the count of such reversals is recorded. This provides some information on frequency content even though individual cycles cannot be resolved on the record.

Magnetic tape.

Taking quantitative data from paper chart recordings is slow and laborious. The development of automatic data-processing techniques has led to an increase in the use of magnetic tape recording. In view of the low frequencies involved in micropulsations, extremely low speed magnetic tape recording is feasible - speeds as low as 0.015" per second. The limit is set by the tape recording mechanism which introduces noise at such low speeds due to the tape sticking, and to the difficulty of providing mechanical smoothing of the tape-drive mechanism. Nevertheless, the cost of taking magnetic tape records at conventional speeds makes ultra-slow-speed recording attractive for some purposes. Campbell (1963) reports a dynamic range of about 90 db by recording the signal on three channels having a range of 35 db each.

With conventional recording it is necessary to use frequency or amplitude modulation of a carrier in order to resolve micropulsation frequencies. The most common method in use in North America is a multi-channel recorder, usually 7 channels, at a recording speed of 1-7/8 inches per second. At this speed a four-hour record may be obtained on one 2500 foot spool of tape. Obviously, the cost and storage problem would be prohibitive if continuous recording were made at this speed. Therefore, only selected intervals for specific purposes are recorded.

Multi-channel recording may also be obtained by taking advantage of the excess bandwidth available at conventional speeds and using a frequency division multiplex system. A system for recording 8 channels of information on one two-track ½-inch tape at 3-3/4 is described by Weir (1963). Fig. 40 is a block diagram of the multiplex system with six channels for micropulsation data and one channel for radio and timing. Thirty per cent modulation rather than the conventional forty per cent modulation is used. The dynamic range is reduced somewhat to approximately 35 db

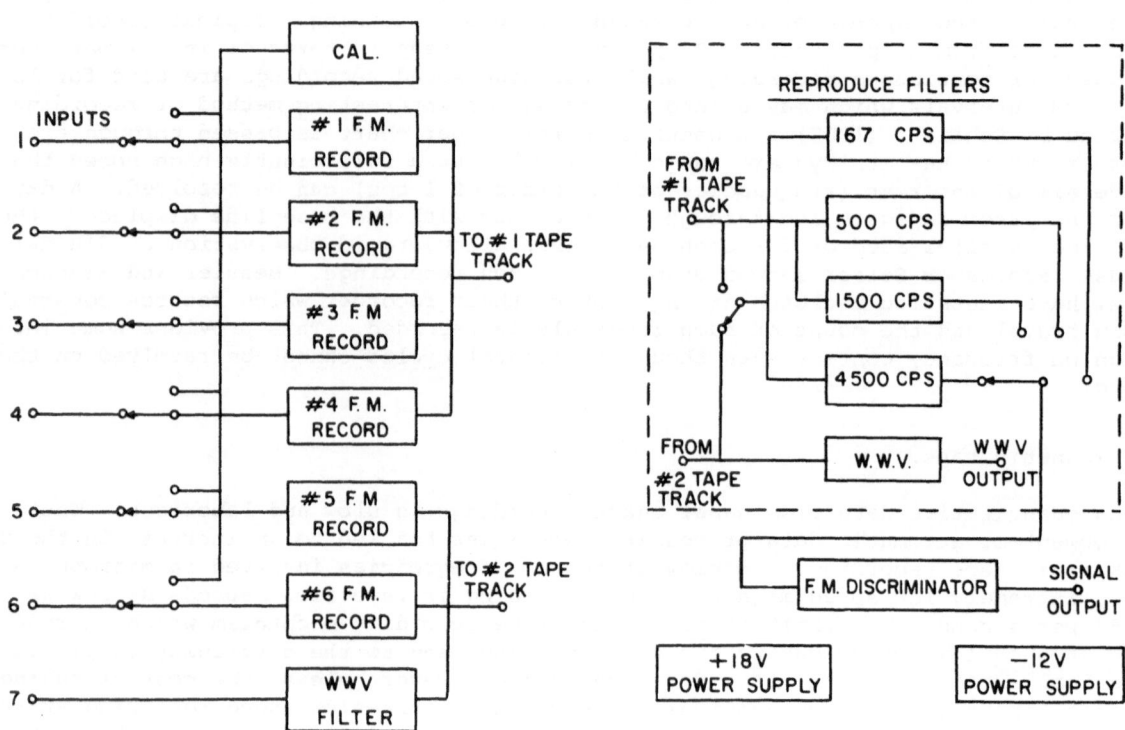

Fig. 40. Block diagram of frequency-multiplex tape recording
system (Weir, 1963).

over the frequency band of interest. This compares with 40 db for a good 7-channel
½-inch tape recorder at 1-7/8 inch per second.

Receiving Equipment.

   Magnetic field.

Similar design considerations apply to equipment for the frequency band between 2 cps and the power line frequency as to the micropulsation receiving equipment. For example, the characteristics of the amplifier must be taken into account in the antenna design. The signal strength of the horizontal components is sufficiently high that metal-cored detector coils may be used, but in order to obtain the same signal-to-noise ratio, air coils are necessary for the vertical component in most localities. Since it is not usually difficult to achieve large areas for coils receiving the vertical component, the antenna may be made of large diameter and involve a small number of turns. As pointed out in the section on Micropulsation Band, this leads to the more efficient design. For convenience in handling, metal-cored antennas are typically used for the horizontal components.

   Polk and Fitchen (1962) use an amplifier with high input impedance and detector coils having thousands of henries inductance, while for convenience in winding the coils the Pacific Naval Laboratory uses a specially designed transistor amplifier with a low input impedance, about 50 ohms, and a noise figure less than 6 db.

   Evans and Horner (1963) describe a low impedance metal-cored detector. The core material and size is the same as that given in Table 2 for micropulsation use. Tests on the amplifier showed that the ratio of induced voltage to amplifier noise (2 to 30 cps) is approximately constant over the range 2 to 9 henries. Frequency response characteristics, therefore, can be used to fix the inductance. In this case 2 henries was chosen. The resistance is not critical but should be small compared to the reactance at 30 cps. In the final design the windings consist of 10 turns of No. 26 gauge wire for calibration spaced $\frac{1}{4}$ inch apart at the center of the core and embedded in the PVC tube used as coil form, and for the signal 1530 turns of No. 15 gauge wire wound in 9 layers 12 inches long. Tests on a series of detector coils showed that the average inductance was 1.88 henries, the DC resistance 2.1 ohms and Q = 29 at 30 cps.

   The air cored coil for the vertical component was designed to give approximately the same output voltage from the amplifier, taking into account the lower signal level. A six-conductor shielded cable 1000 feet long formed into a circular coil with the conductors connected in series to provide a total of 6 turns and 4.41×10^8 sq. cm. effective area was chosen. The resistance of the detector and 500-foot, 2-conductor shielded cable is about 50-ohms and the calculated inductance 21 millihenries.

   Fig. 41 is a block diagram of the Schumann ELF system currently in use by the Pacific Naval Laboratory, and Fig. 42 gives the response of the amplifier system with a 50-ohm resistive source impedance with the detector coils described above. The preamplifier is normally housed in the recording hut, but a heater for temperature control is provided if it should be necessary to place the amplifier outside the building. The detector coils for both vertical and horizontal components must be buried as usual to reduce wind vibration. The insulation resistance to ground of each detector should be greater than 20 megohms in order to minimize ground loops. The shield must also be well insulated from ground. Care should be taken, therefore, to avoid lead-in cables with outer jackets of "lossy" insulating material.

   Other observers, such as Balser and Wagner (1962), record the vertical electrical component using a high mast. Since almost all the signal received in the Schumann ELF band is vertically polarized the electrical antenna provides an effective measure of the signals irrespective of the direction of arrival. Wind vibration of the mast has not been a problem.

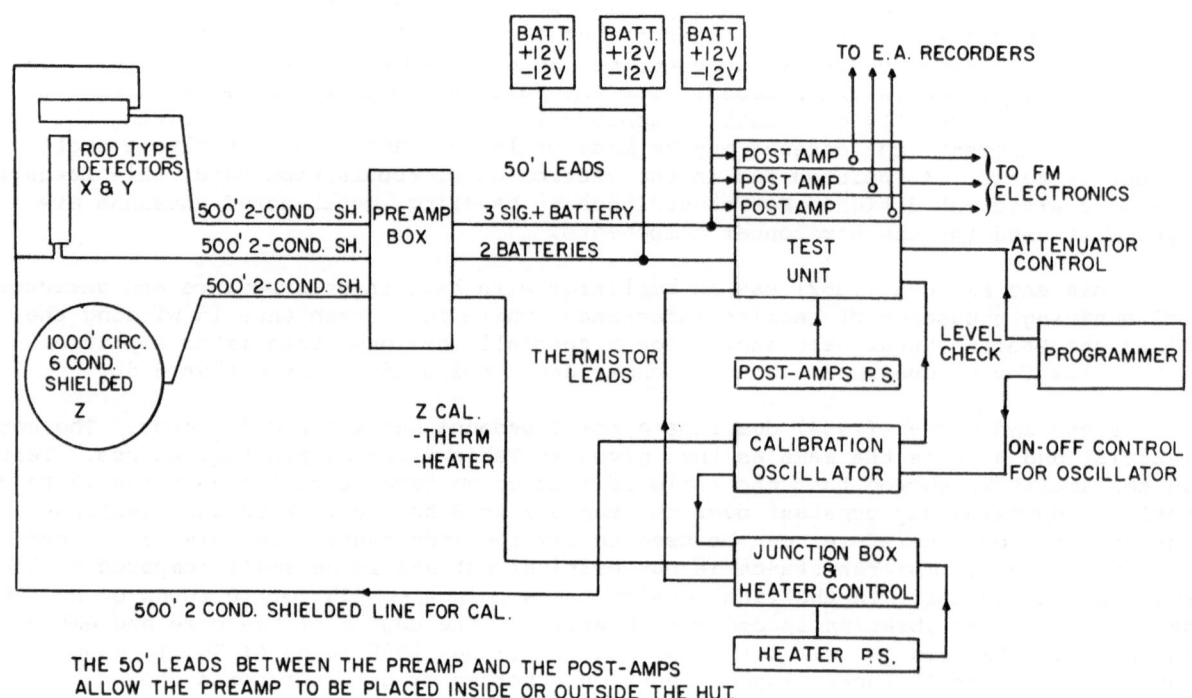

Fig. 41.   Block diagram of Pacific Naval Laboratory
            ELF system (Evans and Horner, 1963).

Recording Equipment.

Polk and Fitchen (1962) use a high-speed paper chart recorder to record the individual cycles in the band with sufficient resolution to determine the frequency content. Selected intervals each day are recorded by this method and provide diurnal and other statistical information. Some of the general features of the Schumann ELF band can be recorded on slow speed paper chart records by rectifying and smoothing with a suitable time constant, Fig. 43. The average value follows closely the diurnal pattern of the prominent modes and individual bursts can be recognized on recordings taken at several stations. Polk uses band-pass filters centered on the principal mode frequencies before rectifying and smoothing. By this method, it is possible to maintain a constant monitor of the average amount of activity in the different modes.

Magnetic tape records are commonly used for subsequent spectral analysis by machine methods, although as mentioned above, Polk has used paper records. FM recording can be used for later digitizing or reproduction on high speed paper charts, or digital recording may be used directly. Balser and Wagner (1960) have used the latter

418

method which provides the dynamic range required to accommodate both the low level quiet signal and the much larger bursts which sometimes tend to overload a conventional analogue magnetic tape record in the ELF band.

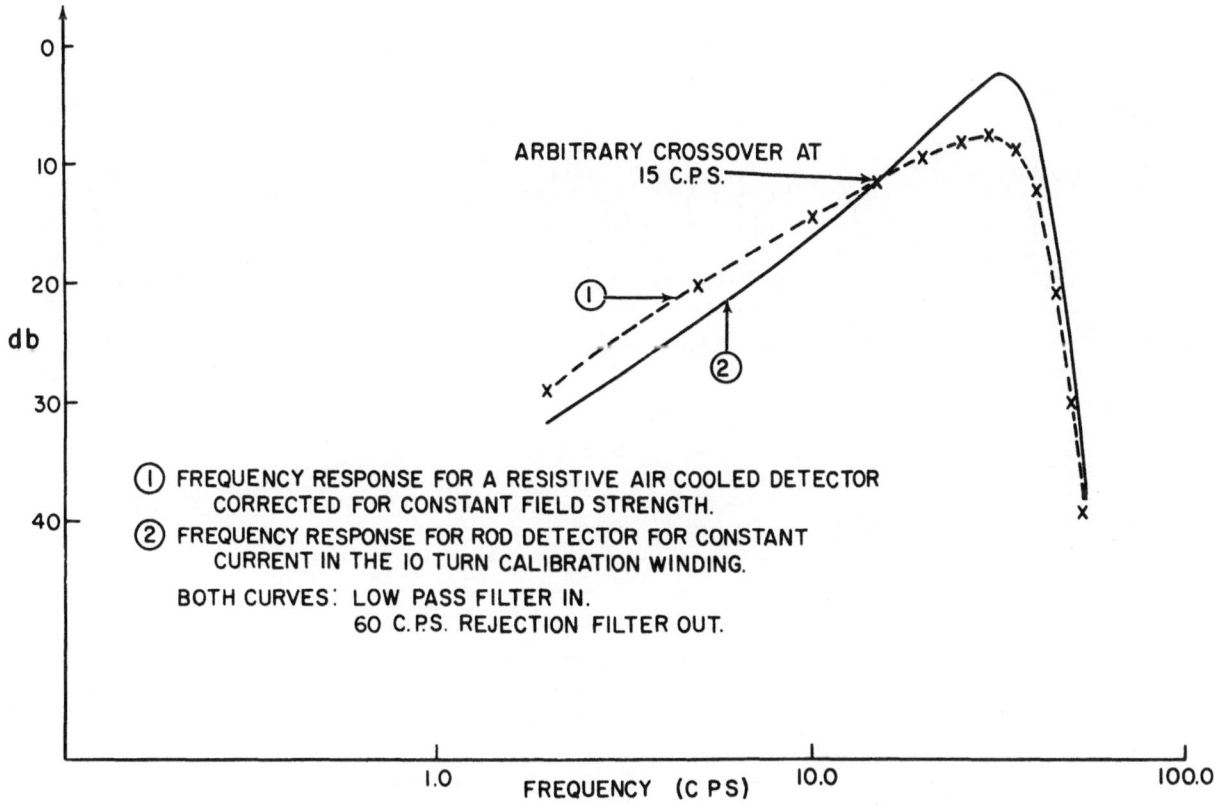

Fig. 42. Response of Pacific Naval Laboratory ELF System (Evans and Horner, 1963).

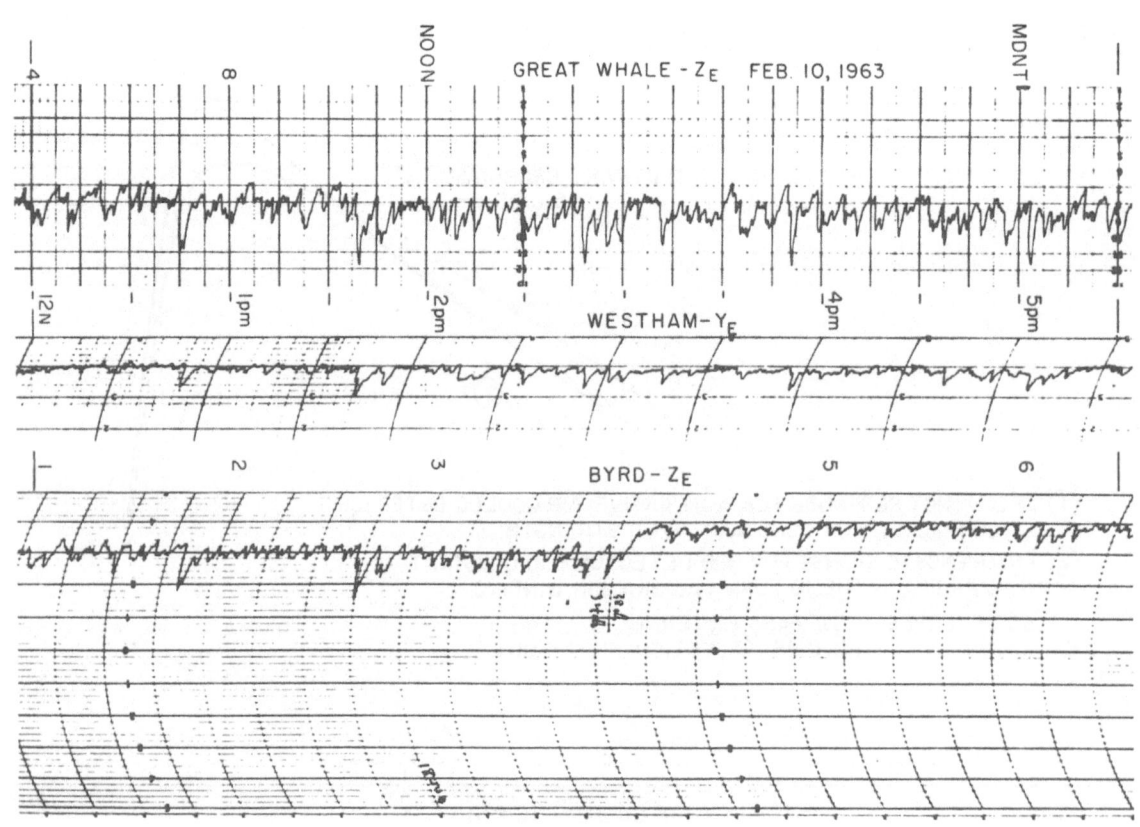

Fig. 43.   Rectified and integrated ELF records,
February 1963.

420

ELF and VLF ABOVE THE POWER LINE FREQUENCIES

Widespread VLF recordings started with the introduction of whistler recorders in connection with the IGY. The discussion given below is largely abstracted from Marks (1962) and discusses whistler and hiss recorders in use by Stanford University in Antarctica. The antenna consists of a single-turn loop of No. 6 soft-drawn copper wire in the form of an isosceles triangle with a 59-foot base and a 29.5-foot altitude. The loop antenna is matched to the preamplifier by means of a special transformer.

Marks recommends that the antenna and preamplifier be installed about 2000 feet from the recording building and from any AC power lines, generators or radio transmitting antennas. The tower should be well guyed. For proper damping of wind vibration at least 6 guy wires are recommended - three at the top and three in the middle of the tower. [Harang (1963) has used a multicore wire for the loop in order to damp out wind vibration.] Loose bolts and other connections in the antenna system may result in vibration noise being picked up in the signals. At Byrd Station an antenna buried beneath the snow surface has been installed. Fig. 44 illustrates the equipment in use at Byrd and Pole stations.

The preamplifier circuit is of the cascode type, with all tube heaters supplied with direct current. The preamplifier is mounted in an aluminum weatherproof box and is connected to the main amplifier rack by means of separate transmission lines and cables for the output, the calibration signals and the preamplifier power.

A "mixer monitor-unit" mixes the preamplifier signals and the time marks, and provides for the automatic substitution of WWV for the preamplifier signals at the beginning of each recording.

The signals are recorded on a single channel halftrack $\frac{1}{4}$ inch tape recorder at $7\frac{1}{2}$ or 15 ips using constant current recording. In addition to magnetic tape records, statistical information about hiss background can be obtained by recording the envelope of the signal received through various pass-bands. [Short pulses from the rectified signal may be rejected by a modified "Laffineur and Whitehead" (1955) circuit, Fig. 45, before recording on film or paper chart.] Gallet (1963) has found that recording the negative of the signal on film can also be used to reject short noise bursts. Such recordings have been found to be particularly useful when attempting correlations with other geophysical phenomena such as geomagnetic micropulsations, auroral intensity fluctuations or riometer absorption. It has also been pointed out that narrow band records may lead to erroneous interpretation unless special care is used in analysis. Arrons et al (1960) use a swept frequency spectrum analyzer whose output is recorded on a paper chart recorder requiring several minutes for each sweep. Gallet and his group at NBS, as well as the Stanford group, have recently introduced a similar technique but with a much faster sweep by recording on film from an oscilloscope display.

The following account by Dr. Katsufrakis (1963) describes two special recording techniques recently adopted by Stanford University.

"Stanford University is presently making three hour-long continuous recordings of the VLF spectrum from 50 cps to 10,000 cps.

These are in addition to the regular synoptic recordings adopted during IGY. The response of the latter are from 50 cps to 30,000 cps.

Commencing 1 January 1964 recordings will be made on a 24 hour continuous basis. The 4-channel tape recorder will be driven at 1-7/8 ips. The tape to be used is $\frac{1}{4}$" x 3600'. This will give one tape per day from each station.

We have modified the tape recorder to give us response up to 10 kc. The record-er is down 5 db at 10 kc.

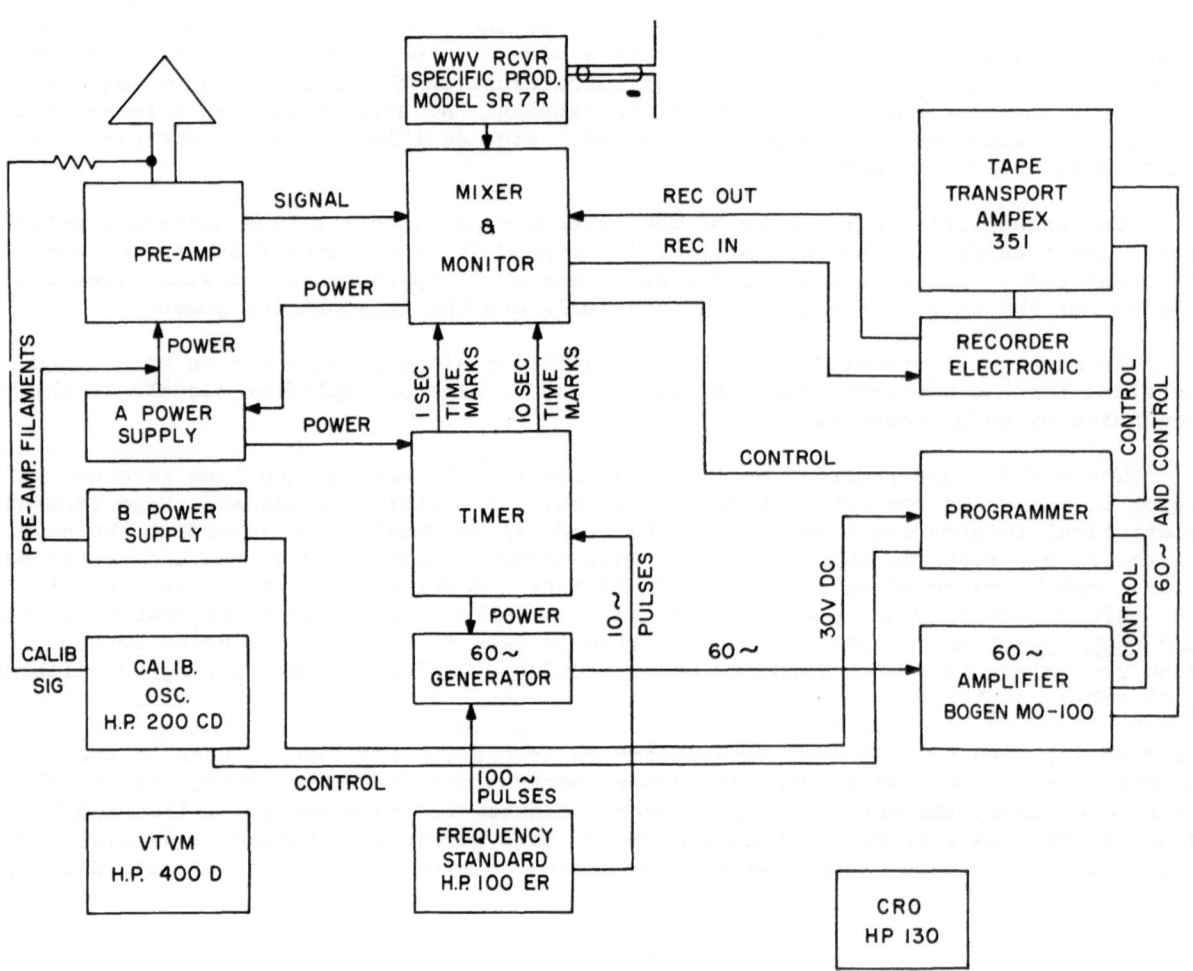

Fig. 44. Block diagram of Byrd and Pole Station
          Whistler Recorders (Marks, 1962).

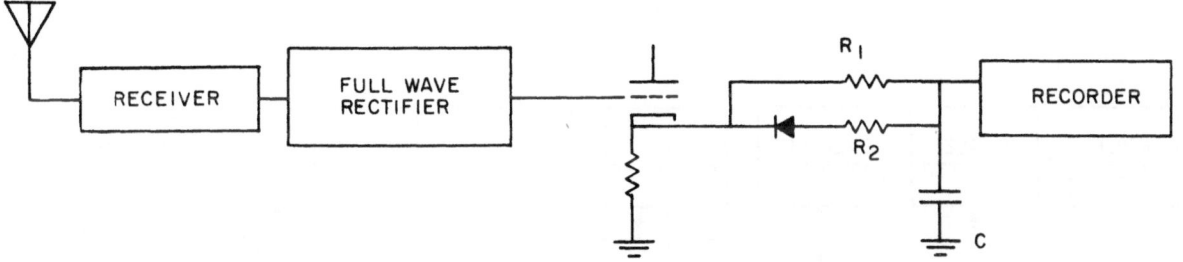

Fig. 45. Circuit for suppressing short pulses in the rectified VLF
signal. The discharge time constant is much shorter than
the charge time constant.

VLF Spectrum Measurement.

The spectral density of the VLF band has been recorded synoptically for the complete year of 1962 at Byrd Station, Antarctica. This is the first quantitative
measurement of the VLF spectrum in the virtual absence of interference.

The instrumentation used is called the Panoramic System and is a sweeping filter
type spectrum analysis system which covers the audio frequency range from 40 c/s to
20 kc/s. The filter is swept across the spectrum once per second and the resulting
information is displayed in two forms. One is a log amplitude, log frequency plot
which is photographed each minute averaging sixty sweeps. The other is a plot of
frequency vs. time with the amplitude information displayed as intensity variation.
This is accomplished by slowly moving a film past the intensity modulated spot on the
cathode ray tube screen as it is swept in frequency. The amplitude frequency display
defines the power spectral density while the frequency time display shows the spectral variation with time of the noise bands.

The system block diagram is shown in Fig. 46. [A similar system is used by
Gallet (1963).] The operation of the system is as follows: The signal enters and is
attenuated to the proper value, then is mixed with the output of the swept oscillator
and run through the intermediate frequency amplifier. It is then detected, amplified
and displayed on a cathode ray tube.

Since the frequency scale is logarithmic, it is necessary to vary the bandwidth
of the I.F. amplifier to obtain maximum frequency definition. The method used to
vary this bandwidth also varies the gain, thus compensation is necessary to retain
flat frequency response. The variation of the bandwidth with frequency has a fundamental effect upon the response of the system to random type signals. This effect
has been taken into account in the calibration of the system and has the fortunate
effect of decreasing the sensitivity at low frequencies where the natural noise power
is greatest.

This instrument has proved very useful in the analysis of VLF emission signals
because the logarithmic frequency scale expands the low frequency portion of the spectrum which contains the bulk of the interesting emissions, and coverage of 20 kc/s is
also retained. The log amplitude feature in conjunction with the bandwidth variation
gives the required dynamic range.

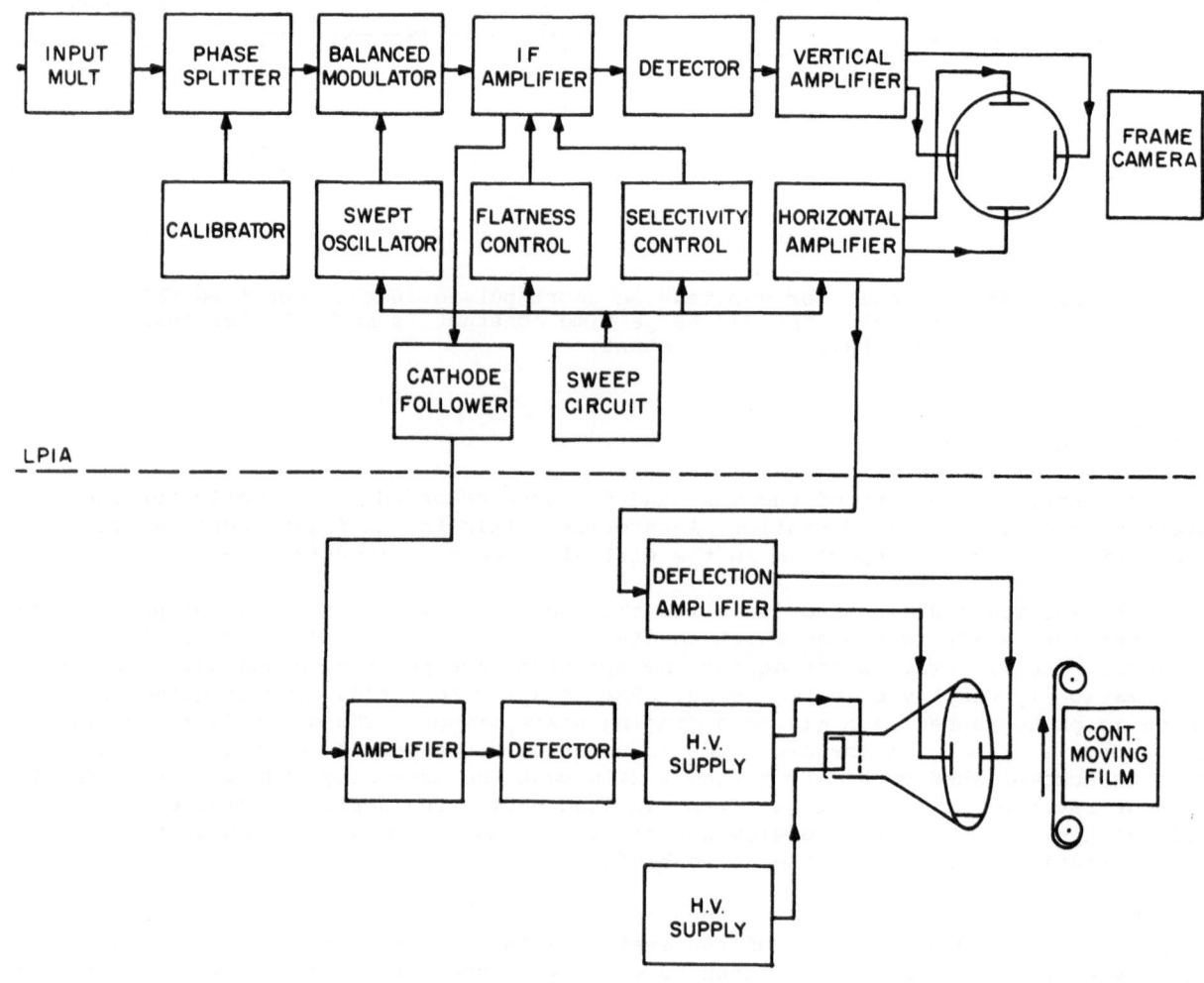

Fig. 46. Block diagram of VLF spectrum measurement
equipment at Byrd Station.

The data analyzed thus far show spectral densities in the order of 10^{-10} watts
per square meter for VLF emissions. Very interesting band structure and movement
has also been noted. "

ACKNOWLEDGEMENTS

I am indebted to my colleagues at the Pacific Naval Laboratory for stimulating discussions and kind assistance in checking the calculations and in reading the manuscript. I should especially like to thank Mr. F. A. Tregear and Mrs. C. A. Powell for the preparation of the illustrations, and Mrs. M. A. Coombes for typing the manuscript.

A debt of gratitude is also due to the many investigators who kindly supplied me with information on the equipment they are using. A special thanks is due to Dr. J. Katsufrakis for assistance in preparing the VLF section. Any errors or omissions, however, are the sole responsibility of the author.

REFERENCES

Abragam, A.: Phys. Rev. 98, 1729 (1955).

Arrons, J., G. Gustafsson and A. Egeland: Nature 185, 148 (1960).

Balser, M. and C. A. Wagner: Nature 188, 638 (1960).

J. Geophys. Res. 67, 619 (1962).

Bell, W. E. and A. L. Bloom: Phys. Rev. 107, 1559 (1957).

Benoit, H.: Compt. Rend. 246, 3053 (1958).

Bleil, D. F.: Private communication, (1963a).

Discussion at the Institute, (1963b).

Bloom, A. L.: Applied Optics 1, 61, (1962).

Bonnet, G.: Ann. Phys. (Paris), 18, 62 (1962a).

Ann. Phys. (Paris), 18, 160 (1962b).

Bozorth, R. M.: Ferromagnetism, D. Van Nostrand, New York (1951).

Campbell, W.: Private communication, (1963).

Christoffel, D. A., J. A. Jacobs, E. J. Jolley, J. K. Kinnear and J. A. Shand: Pacific Naval Laboratory, Report 61-5, (August 1961).

Colegrove, F. D. and P. A. Franken: Phys. Rev. 119, 680 (1960).

Dehmelt, H. G.: Phys. Rev. 105, 1487 (1957a).

Phys. Rev. 105, 1924 (1957b).

Driscoll, R. L. and P. L. Bender: Phys. Rev. Letters 1, 11, 413 (1958).

Electro-Mechanics Co., Austin, Texas: Variable mu magnetometer, Final Report, (January 1959).

Evans, D. J.: Pacific Naval Laboratory, Unpublished Manuscript (1960).

Evans, D. J. and S. Horner: Pacific Naval Laboratory, Lab. Note 63-1, Unpublished Manuscript (1963).

Franken, P. A. and F. D. Colegrove: Phys. Rev. Letters 1, 316 (1958).
    Phys. Rev. 119, 680 (1960)

Freycenon, J. and I. Solomon: Onde Elect. 40, 590 (1960a).

Freycenon, J.: Onde Elect. 40, 596 (1960b).

Gallet, R.: Discussion at the Institute, (1963).

Garland, G. D.: Methods and Techniques in Geophysics: Earth Currents, p. 277, Interscience Publishers, New York (1960).

Geyger, W. A.: Commun. and Electronics, No. 59, 65, (1962a)

    J. Appl. Phys. 33, Supplement, 1280 (1962b).

    Electronics, 35, 48 (1962c).

Gregg, E. C.: Rev. Sci. Instrum. 18, 77 (1957).

Grenet, G.: Ann. Geophys. 5, 188 (1949).

    Ann. Intern. Geophys. Yr. IV, 302 (1957).

Harang, L.: Discussion at the Institute (1963).

Hawkins, W. B.: Phys. Rev. 98, 478 (1955).

Heppner, J. P.: Goddard Space Flight Center, Publication X-611-63-107, May 1963. [To be published in Space Science Reviews, Dordrecht, Holland.]

Hessler, V. and R. R. Heacock: Earth Currents in the Auroral Zone, Geophys. Inst. University of Alaska, Final Report (December 1962).

Hill, L. K. and F. X. Bostick: Electrical Engineering Research Laboratory, University of Texas Report 126 (May 1962).

Kastler, A.: Proc. Phys. Soc. A67, 853 (1954).

Katsufrakis, J.: Private communication (1963).

Keyser, A. R., J. A. Rice and L. D. Shearer: J. Geophys. Res. 66, 4163 (1961).

Laffineur, M. and J. D. Whitehead: J. Atmos. Terrest. Phys. 9, 347 (1956).

Liebermann, L.: Private communication (1963).

Lokken, J. E., J. A. Shand, and C. S. Wright: Can. J. Phys. $\underline{40}$, 1000 (1962).

        J. Geophys. Res. $\underline{68}$, 3, 789 (1963).

Marks, K. E.: Stanford University Internal Memorandum No. 1108-1 (August 1962).

Maxwell, A.: Ann. Intern. Geophys. Yr. $\underline{IV}$, 281 (1957).

Packard, M. and R. Varian: Phys. Rev. $\underline{93}$, 941 (1954).

Parsons, L. W. and Z. M. Wiatr, J. Sci. Instrum. $\underline{39}$, 292 (1962).

Polk, C. and F. Fitchen: J. Res. Nat. Bur. Stand. $\underline{66D}$, 313 (1962).

Rice, J. A.: IRE Intern. Conv. Record $\underline{9}$, 244 (1961).

Rumbaugh, L. H. and L. R. Alldredge: Trans. Amer. Geophys. Union $\underline{30}$, 836 (1949).

Schlich, R.: Bulletin des T.A.A.F. No. 18, (1962).

Schonstedt, E. O. and H. R. Irons: Trans. Amer. Geophys. Union $\underline{36}$, 25 (1955).

Selzer, E.: Ann. Intern. Geophys. Yr. $\underline{IV}$, 287 (1957).

        Discussion at the Institute, (1963).

Serson, P. H.: Can. J. Phys. $\underline{35}$, 1387 (1957).

        "Proton Precession Magnetometer" Canadian Patent Office, Pat. No. 618,762 (1961).

        "A Simple Proton Precession Magnetometer" Dominion Observatory, Ottawa (May 1962).

Serson, P. H. and W. L. W. Hannaford: Can. J. Technol. $\underline{34}$, 232 (1956).

Serson, P. H., S. Z. Mack and K. Whitham: Publ. Dominion Observatory, Ottawa $\underline{19}$, 15 (1957).

Skillman, T. L. and P. L. Bender: J. Geophys. Res. $\underline{63}$, 513 (1958).

Snavely, B. L.: Pulse Fluxgate Magnetometer, Naval Ordnance Laboratory, Disclosure of an Invention (May 1963).

Solomon, I.: Phys. Ref. $\underline{99}$, 559 (1955).

Surkan, A. J.: Paper in preparation (1963).

Thellier, E.: Ann. Intern. Geophys. Yr. $\underline{IV}$, 255, (1957).

Troitskaya, V. A.: Ann. Intern. Geophys. Yr. $\underline{IV}$, 322 (1957).

Vacquier, V., R. F. Simons, and A. W. Hull: Rev. Sci. Instrum. $\underline{18}$, 483 (1947).

Vozoff, K.: J. Geophys. Res. $\underline{66}$, 1983 (1961).

Ward, S. H. and K. A. Ruddock: Paper presented at the Conference on Telluric and Geomagnetic Field Variations, University of Texas, Austin, Texas (October 1961).

Weaver, J. T.: Can. J. Phys. <u>41</u>, 484 (1963).

Weir, R. C.: Pacific Naval Laboratory Tech. Memorandum 63-1 (1963).

Whitham, K.: Methods and Techniques in Geophysics: Measurement of the Geomagnetic
    Elements, p. 104, Interscience Publishers Incorporated, New Yo k (1960).

Williams, F. U. and G. H. Hopkins: Electrical Engineering Research Laboratory,
    University of Texas, Report No. 123 (1961).

SPECTRAL, CROSS-SPECTRAL, AND BISPECTRAL ANALYSIS OF LOW FREQUENCY ELECTROMAGNETIC DATA

T. Madden

Geology and Geophysics Department
Massachusetts Institute of Technology
Cambridge, Massachusetts

ABSTRACT

The techniques and the significance of various spectral and higher order spectral analyses are examined, and examples of the application of these analyses to geophysical electromagnetic data are given.

INTRODUCTION

In all branches of geophysics the use of frequency analysis to represent time varying phenomena is so deeply ingrained it is usually taken for granted. Frequency analysis is essentially the implementation of the Fourier transform

$$f(t) = \frac{1}{2\pi} \int_{-\infty}^{\infty} F(\omega) e^{i\omega t} d\omega \qquad (1)$$

$$F(\omega) = \int_{-\infty}^{\infty} f(t) e^{-i\omega t} dt \qquad (2)$$

The value of this representation scheme in geophysics stems from the simplification of analytic developments in the frequency domain, the ease of approximating the transformation by means of electronic and mechanical filters, and the great efficiency of the trigonometric functions in describing many geophysical phenomena because of their cyclical or resonance properties.

The transforms defined in Eq. (1) and (2) are only valid for functions with finite energy, and a different treatment is necessary for functions (or data) which exist for all time so that their total energy ($\int_{-\infty}^{\infty} f^2(t) dt$) is not finite. In such cases one must consider the spectral properties of the energy density rather than the spectral properties of the amplitude fluctuations. Further complications stem from the noise or random behavior of geophysical phenomenon, and it is only fairly recently that the mathematicians have advanced the techniques for handling such data to the point of practical implementation.

In this chapter we wish to examine in a non-rigorous manner the various operations involved in spectral analysis in order to develop a feeling for what the operations accomplish, and to illustrate some typical applications with examples of such analyses on low frequency electromagnetic data. The operations we will be concerned with are assumed to involve digital computations, but many of the principles apply to operations using analogue devices.

NOISE REPRESENTATION

The continuous nature of noise data causes difficulties with normal Fourier analysis, as we have mentioned above, and one must resort to a power density analysis. An obvious extension of the Fourier analysis would be the following operation

$$\text{Power density estimate} = \frac{1}{2T} \left| \int_{-T}^{T} f(t) e^{-i\omega t} dt \right|^2 \lim \text{ as } T \longrightarrow \infty \qquad (3)$$

This operation is known as a periodogram analysis, and was widely used in the past, but it does not converge to any constant value at $T \longrightarrow \infty$. As a matter of fact the variance of the periodogram estimate is independent of the length of time (2T) used in making the estimate. The reason for this difficulty is not hard to understand, and the difficulty is automatically avoided when analogue methods are used to simulate the operation. If the noise is passed through a narrow band filter, and the mean squared output of the filter is used as the power density estimate, then the estimate will converge for stationary noise as the averaging time increases. As we shall show below, however, the periodogram operation involves a narrower and narrower filter as T increases, and the increased amount of data is just balanced by the narrowing of the filter and no improvement in the statistical fluctuations of the estimate results.

An estimate that does converge, however, and from which one can compute power density estimates is the autocovariance function estimate. Before examining this approach it is probably worthwhile to illustrate a noise representation scheme that is very useful in helping one to understand the results of various operations on noise data.

We must first define what is meant by noise or noise data. A geophysicist usually uses the word "noise" to describe those signals which are interfering with his observations of signals of a different origin which he happens to be studying. In electromagnetic measurements the power line signals, and radio station signals may often be referred to as noise. This is not the definition we wish to use here, however. We define noise as signals which contain an essential element of randomness. In principle therefore the power line signals are not noise, but in practice the amplitude and phase of the power line signals undergo unpredictable changes and they too are thus noise, but with a very narrow frequency structure. A constant d.c. level in noise data is removed since it represents a predictable zero frequency component. Any frequency components in the data with periods comparable to or longer than the data length cannot be tested statistically, and it is often useful to filter them out also.

It is usually assumed that noise is continuous. When the statistical properties of the noise are time invariant the noise is referred to as stationary noise. In practice, since one never tests all the statistics of any noise data, a much looser definition of stationarity is used. Most of the mathematical developments concerned with noise have centered on stationary noise, which represents the simplest case. In geophysics the noise data is probably never stationary, but if the properties are slowly varying in comparison to the length of data sections being analyzed the assumption of stationarity does not cause any great errors.

A very simple scheme to represent stationary noise processes was proposed by Wold (Wold, 1938) and Robinson (Robinson, 1954) and is sometimes called the Wold decomposition, or the random wavelet representation (Robinson, 1954; 1962). This representation scheme makes a very clean separation between the predictable elements of a noise process and the random elements. The predictable elements are represented by a transient waveform which is called a wavelet. The random elements are represented by a random sequence of impulses. The noise process is represented by a sequence of these wavelets arriving at the time and with the amplitude given by the random impulse sequence. In more mathematical terms we say the noise is represented by the convolution of the wavelet and the random impulse sequence.

$$h(t) = \int_{-\infty}^{\infty} f(t - \tau_1) \, g(\tau_1) \, d\tau_1 = f*g = g*f \qquad (4)$$

noise = wavelet * random impulse sequence

The wavelet contains all the frequency information of the noise. The random impulse sequence is a white noise sequence and thus does not have any frequency information, but it contains other statistical properties such as the amplitude distribution.

The convolution operator plays an important role in power spectrum analysis because of its simple Fourier transform properties. These are indicated below.

Time Domain	Frequency Domain	
$f(t)$	$F(\omega)$	
$g(t)$	$G(\omega)$	
		(5)
$f*g$	FG	
fg	$F*G$	

We see from Eq. (5) that multiplication in the time domain is equivalent to convolution in the frequency domain, and that likewise convolution in the time domain is equivalent to multiplication in the frequency domain. Thus, since white noise has a flat power density spectrum, the convolution of the white light random sequence with a wavelet produces a noise process with frequency properties determined by those of the wavelet.

One method of extracting this frequency information is to examine the autocorrelation function of the noise. The autocovariance is simply the average lagged product. Each wavelet acting on itself gives a contribution to this product, but because of the random relationship of the different wavelets to each other, no net contribution results from a wavelet acting on other wavelets. Thus, the autocovariance of the noise data is equal to the autocorrelation of the wavelet that composes the noise.

$$\frac{1}{2T} \int_{-T}^{T} h(t) h(t + \tau_1) \, dt = \langle h(t) h(t + \tau_1) \rangle =$$

$$\lim \tau \longrightarrow \infty$$

$$\int_{-\infty}^{\infty} f(t) f(t + \tau_1) \, dt \qquad (6)$$

The autocorrelation is a backwards convolution, so that its Fourier transform gives us $F(\omega)\overline{F(\omega)}$, where $\overline{F(\omega)}$ stands for the complex conjugate of $F(\omega)$. Thus, the Fourier analysis of the autocorrelation estimate when properly normalized gives us an estimate of the power density spectrum.

SPECTRAL ANALYSIS

The essence of power density spectral analysis by the use of the autocorrelation was given in the previous section. There are several important points that arise concerning the details of the methods of computation, however, which we wish to discuss a bit more. All the steps in such an analysis and their consequences in the frequency domain are illustrated in Fig. 1.

431

Operation	Time Domain	Freq. Domain
1. digitigation	multiplication by  (impulse train) ΔT	convolution by 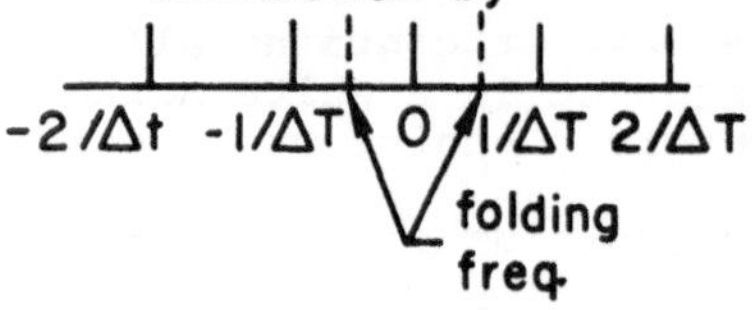 (impulse train) $-2/\Delta t$ $-1/\Delta T$ 0 $1/\Delta T$ $2/\Delta T$ folding freq.
2. selecting a finite interval	multiplication by (box)	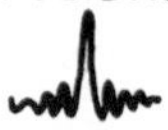 convolution by (sinc-like function)
3. autocorrelation	backwards self convolutions	squaring, ie $F(w)\,\overline{F(w)}$
4. window shaping	multiplication by  (wavy function)	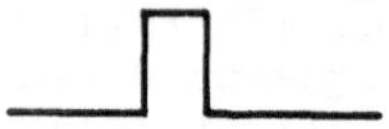 convolution by (box)
5. cosine transform	going from time domain to freq. domain	

Fig. 1. Steps in Power Density Spectrum Estimate

The first step which is necessary for subsequent computer analysis of the data is the digitization of the data. This can be represented as a multiplication of the data by a function which is equal to 1 at discreet intervals of time separated by the sampling interval ΔT, and zero at all other times. This operation is equivalent to convolution in the frequency domain by a sequence of impulses centered at zero and separated by 1/ΔT. Since convolution is essentially the process of looking at a function through a window shaped by the other function, we see that if all the frequency information lay between -1/2ΔT and +1/2ΔT the digitization process would lose no information. This is a remarkable fact since most of the data appears to be thrown away during digitization, and it gives us a very efficient method for storing low frequency data. If no energy of any consequence exists above 1 cps, one can store 100 days of data at a dynamic range of 45 db on a standard high density binary tape. If energy exists at frequencies outside of the range ± 1/2ΔT, those frequencies are folded into the range and one cannot reconstruct the original frequency content of the data. This is the aliasing problem which is an important limitation to the digitizing process and necessitates the use of low pass filtering before digitization.

The second step involves the selection of a finite interval of data. This is equivalent to multiplication in the time domain by a box of unit height and of width T. In the frequency domain this is equivalent to convolution by the function $\frac{\sin\ (\omega T/2)}{\omega}$, so that each frequency component represents an average of the frequencies in its immediate vicinity. The longer T is made, the narrower $\frac{\sin\ (\omega T/2)}{\omega}$ becomes, and the finer the frequency resolution possible. The averaging process in the frequency domain is an important element in reducing the statistical fluctuations of the estimates, and it is because the $\frac{\sin\ (\omega T/2)}{\omega}$ window decreases in width in direct proportion to T that the periodogram estimate does not converge as $T \longrightarrow \infty$.

The third step is the autocorrelation of the data interval. This operation is the same as conjugation in the frequency domain or the squaring of the absolute value.

The fourth step is a window shaping step. In order to reduce the variance of the power density estimates, and also in order to reduce the number of frequency points needed to describe the power density it is desirable to further average the power density function. The ideal way to do this is to convolve the power density function with a box function, and the resulting power density estimates are known as Daniell estimates. This is simply accomplished by multiplying the autocorrelation function by a $\frac{\sin (\Delta\omega t/2)}{t}$ function where $\Delta\omega$ is the width of the spectral window desired. When power density estimates are made every $\Delta\omega/2$ radians only neighboring estimates are correlated, and the variance of these estimates is given by $\frac{M}{N}$ x (Power)^2 where M is the number of estimates and N is the number of data points. With the Daniell window it is really only necessary to make estimates every $\Delta\omega$ radians, however.

Other windows are also used that save computational time by not needing all the lags in the autocorrelation computation, but sacrifice some clarity in the frequency resolution as side lobes beyond $\Delta\omega/2$ are present in the convolution operator. The most common ones used are the Hamming and Hanning windows (Blackman and Tukey, 1958).

The final step is simply the cosine transform of the weighted autocorrelation function which gives us the final power density spectral estimate. For details of all these computational steps and the proper normalizing factors we refer you to Simpson (Simpson, 1961), Blackman and Tukey,(Blackman and Tukey, 1958) and Hannan (Hannan, 1960).

Using special high speed computational programs developed for seismic analysis (Simpson, 1962), it is possible to perform very high resolution spectral estimates (500 - 1000 frequency points) of the Daniell type on data with frequencies of up to 100 cps in real time on an IBM 7090. Speed is not so much a factor in computational work as is cost, however. At the present commercial rates it costs about $5 to perform a high resolution analysis on 6000 data points.

When continuous frequency information is desired it becomes too expensive to use computational techniques unless the frequencies are very low (0.1 cps or less) and one must resort to analogue techniques. The principles of an analogue analysis can also be briefly summarized with reference to Fig. 1. The basis of the analogue methods is the use of narrow band filters to separate out the various frequency components. Narrow band filtering is very similar to step 2, and thus, the output of a narrow band filter involves the past inputs for a length of time dependent on the filter width. If the output is then squared (it is usually only rectified) we are essentially performing step 3. The resulting output may then be time averaged. This step is not quite the same as step 4, as the spectral window was solely determined by the original filtering, but it does give us an output dependent on a time interval longer than the time resolution of the filter, and therefore also reduces the statistical fluctuations of the output.

In recent times some very useful instruments have been designed that allow a multifrequency analysis to be performed by analogue techniques. A widely used, commercially available unit is the Kay Sonograph, but the list of commercially available equipment is continuously growing. Many laboratories, including the Bell Laboratories and the Stanford Electronics Laboratories, have designed their own instruments. The Stanford University Real-Time Spectrum Analyzer can perform continuous analyses in real time with fine frequency resolution between 500 and 5000 cps, the time frequency plots being put on a continuous film strip (Helliwell et al, 1961). This instrument has proven to be an important tool in studying the whistler and VLF emission phenomenon. The flexibility of magnetic tape recording systems allows one to vary the apparent time scale of recorded data, so that instruments such as these are not limited to any particular frequency range.

The application of power density spectral analysis to geophysical problems is so varied we will make no attempt to cover this topic, but we can illustrate a few typical examples.

One common application is that of picking out certain special signals which have a narrow frequency structure. In Fig. 2 is shown an analysis of magnetic fluctuations at Tuscon. The diurnal variations and many of their harmonics are clearly distinguishable, and in this example almost any technique to find these signals would have given good results.

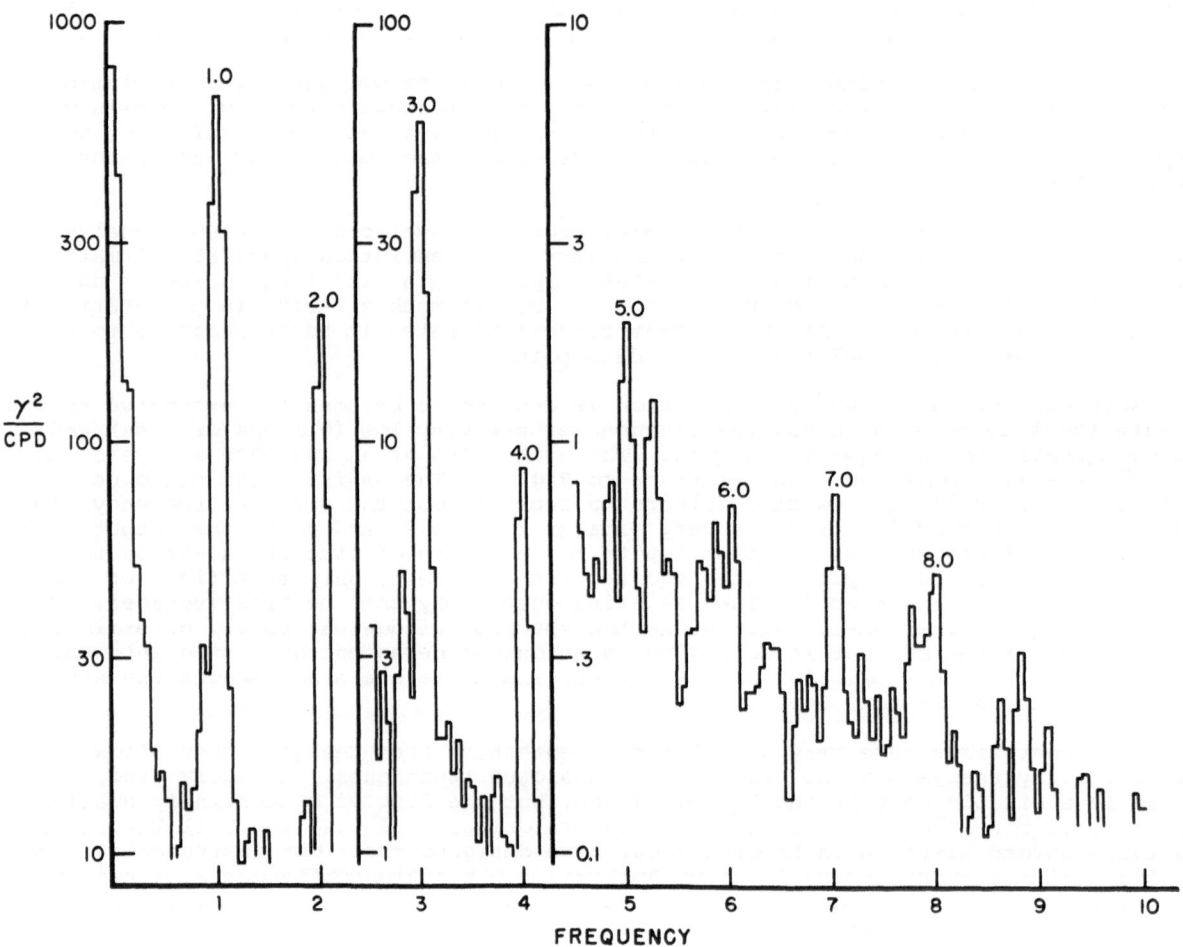

Fig. 2.   Frequency in Cycles Per Day

Another example of the same sort of data is shown in Fig. 3. This example illustrates the importance of computing power density estimates at many frequencies in order to determine whether a power density estimate truly involves the special signal or is only due to a general background of fluctuations. The higher harmonics of the diurnal variations at Weston are completely masked by the background, and if the power estimates at 2 or 3 cycles per day had been equated to the diurnal variation the conclusions drawn would have been completely false.

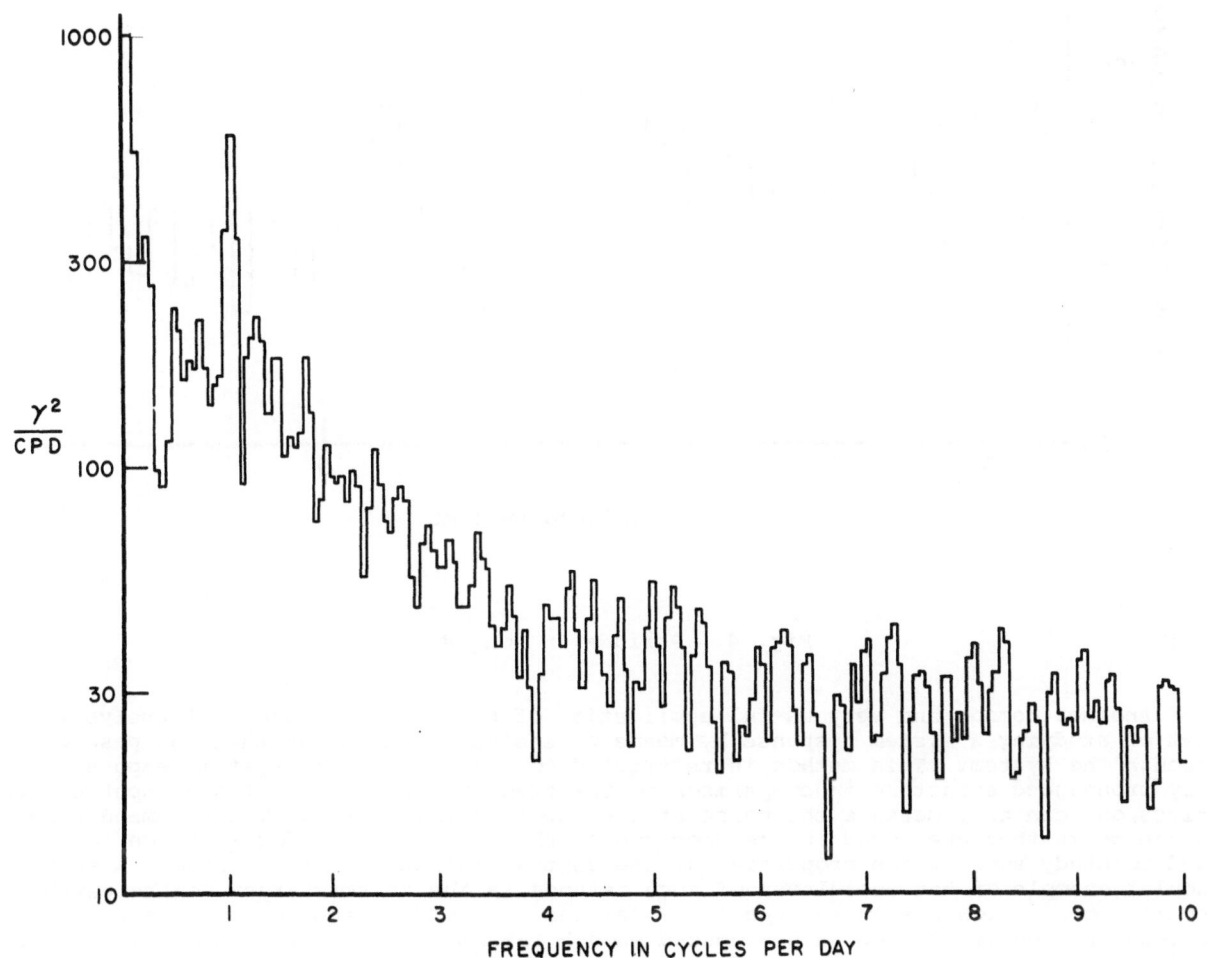

Fig. 3. Frequency in Cycles Per Day

435

Fig. 4 is another such example involving much lower frequencies. This illustrates also the fact that high resolution frequency analyses are not restricted to data handled only with modern recording systems, but tabulated data collected years before can also be treated, provided the aliasing problem had been properly taken care of. The search for six month magnetic fluctuations necessitates a high resolution because the internally generated magnetic fluctuations create a very high level background.

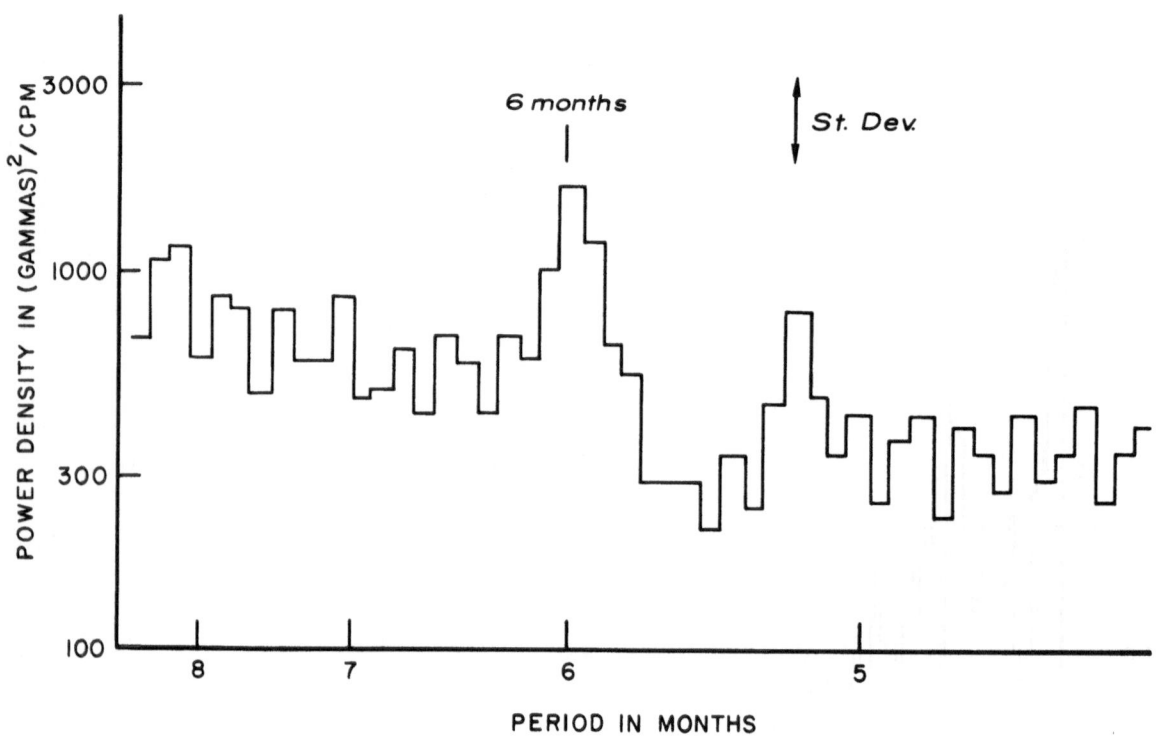

Fig. 4.   Periods in Months

Another common and very useful application of power density spectral analysis is that of studying a system response by means of analyzing the noise that has passed through the system. This method is restricted to cases where the system response has very pronounced structure in comparison to the frequency structure of the input noise, unless one can also measure the noise at the input to the system. The Schumann resonance peaks that are found in the spectrum of the ELF noise, for instance, can be used to study some of the properties of the earth-ionosphere cavity system. A spectacular example of the usefulness of such methods is the seismic study of the earth's oscillations. High resolution spectral analyses of records taken of the great Chilean earthquake of 1960 have produced enough information to allow checks to be performed on the seismic velocity profiles determined from all the past history of earthquake recordings.

Other special features of noise data are sometimes revealed in spectral analyses which might otherwise go undetected. In Fig. 5 is shown a low resolution power spectrum analysis of pearl oscillations. These very curious signals at around 1 cps often show a regular repetition rate which can at times be simply picked by eye from a pen recording, or more easily even from a time frequency sonograph plot. When the

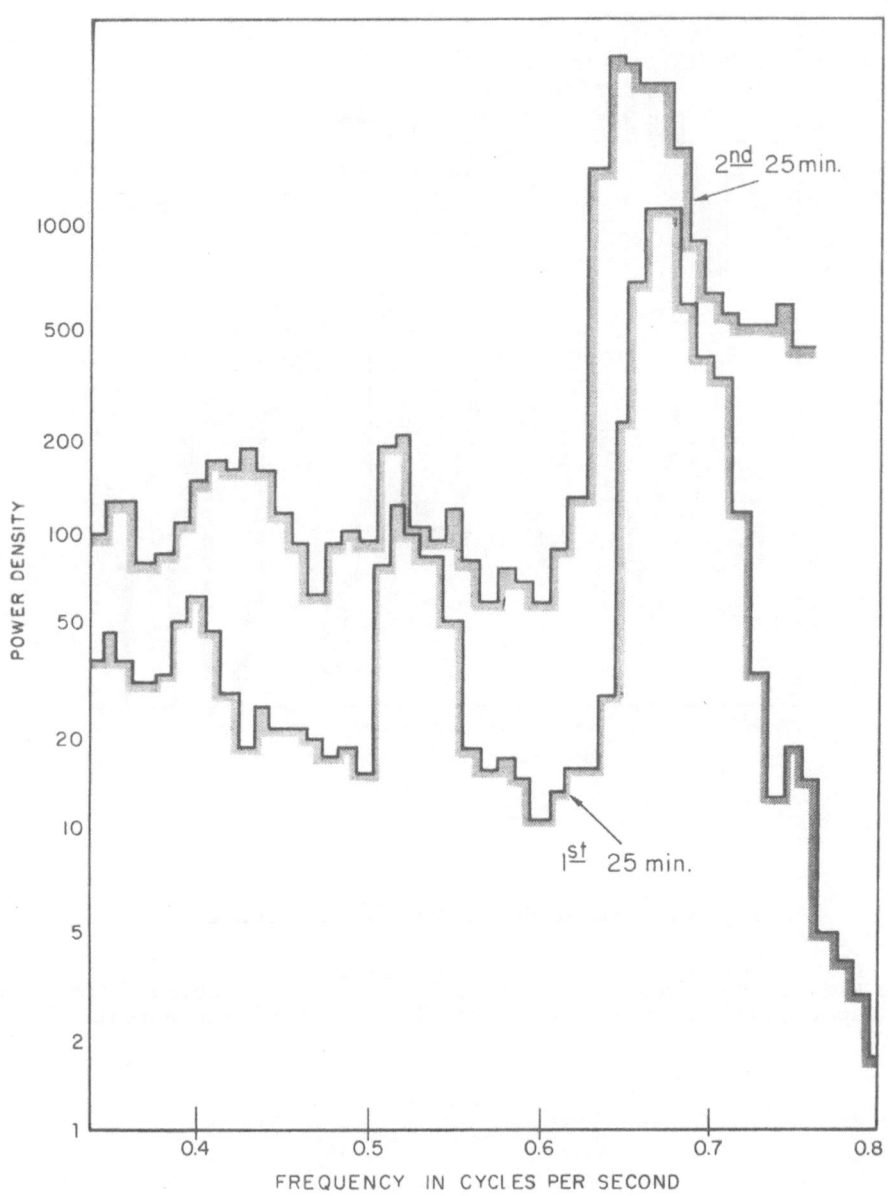

POWER DENSITY

FREQUENCY IN CYCLES PER SECOND

Fig. 5.   Frequency in Cycles Per Second

repeating signals overlap too much, however, this information tends to be obscured. Repeating a signal is equivalent to convolving that signal with a series of equally spaced impulses, and therefore in the frequency domain this is equivalent to multiplying the signal spectrum by a series of impulses equally spaced in frequency. The spacing of these impulses is equal to the inverse of the period of the repetition rate, and thus, the repetition rate is revealed in the spectral analysis. In Fig. 6, an expanded spectrum of these same oscillations is shown, and a comb-like frequency structure is clearly revealed. The separation between the peaks was .008 cps indicating that a 125 second repetition rate existed. Dr. Vozoff, who collected this data, reported that the data appeared to repeat with a period of 145 seconds, which illustrates some of the difficulties of spotting such things by eye. The narrowness

437

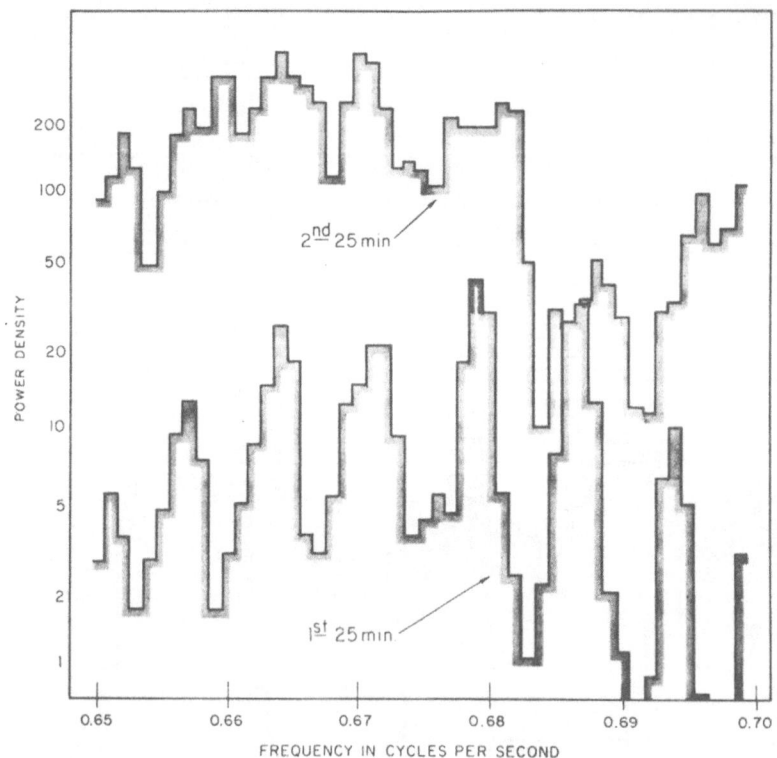

Fig. 6. Frequency in Cycles Per Second

of the peaks in Fig. 6 are a measure of the faithfulness of reproduction of the waveforms in their repetition, and this is one of the most remarkable features of these
records.

COHERENCY ANALYSIS

Another important extension of the principles of spectral analysis of noise data is
that of coherency analysis. Coherency is a measure of the degree of linear relationship between two groups. When we are dealing with noise data we express the same
thing by saying the coherency measures the degree to which one noise data set can be
predicted from another noise data set (by a linear operation). In the random wavelet
representation of noise data, the wavelet represents a predictable element, while the
white light impulse sequence represents the random element. In order for two noise
series to be predictable one from the other, it is only necessary that they have the
same impulse sequence. This implies that both sets of data resulted from the same
source, and therefore there are many uses for the coherency measure in geophysical
work. The environment may have affected the two data sets in different ways so that
coherent noise data do not necessarily look alike, and if some unpredictable noise is
added it may become very difficult to notice any coherency at all without carrying
out a detailed analysis. The heart of the measure of coherency comes from the consistency of the phase relationships between the two data sets. When the data are

completely coherent the arrival of a new wavelet in one record is always matched by the arrival of another wavelet in the other record which has a constant amplitude and time relationship to the first one. Thus, the phase relationships between the records remain constant.

This can be used as the basis of an analogue measure of coherency. If the signals are narrow band filtered and a phase sensitive detection of one signal with respect to the other of the same frequency is made, allowing for an appropriate phase shift, the average output of incoherent signals will be zero, while the output of coherent signals will simulate rectification of the signal being detected.

The computational approach to coherency analysis follows very closely the techniques of spectral analysis. The cross-covariance of two time series produces a net output that results only from the cross-correlation of the associated wavelets in the two series. Thus, if

$$h_1 = f_{1a} * g_a + f_{1b} * g_b \qquad\qquad f \equiv wavelet$$

$$h_2 = f_{2a} * g_a + f_{2c} * g_c \qquad\qquad g \equiv random\ impulse\ sequence$$

$$\frac{1}{2T}\int_{-T}^{T} h_1(t)h_2(t+\tau_1)\,dt = \int_{-\infty}^{\infty} f_{1a}(t)f_{2a}(t+\tau_1)\,dt \qquad (7)$$

$$\lim T \longrightarrow \infty$$

The estimation of this quantity and its Fourier representation from finite data follows steps 1, 2 and 4 of Fig. 1. The autocorrelation of step 3 is replaced by a cross-correlation, and the cosine transform of step 5 is augmented by a sine transform which gives the quadrature component of the cross-correlation.

The frequency domain components of the cross-correlation are known as the crosspower, and the coherency is defined as the ratio of the crosspower to the root mean autopowers

$$coherency = \frac{\Phi ab(\omega)}{(|\Phi aa||\Phi bb|)^{\frac{1}{2}}} \qquad (8)$$

Φab = crosspower, Φaa, Φbb = autopower

The coherency here is defined as a complex quantity, although many people refer to the modulus of this quantity as the coherency.

The inphase and quadrature components of the crosspower are often referred to as the co-spectrum and the quadrature spectrum.

There is a subtle difficulty that appears when we work with data of finite length. The cross-correlation of two transients does not lose any of the power of the transients, and this would appear to guarantee a coherency of 1. This difficulty is overcome by making the spectral window average together the crosspower of a group of neighboring frequencies. When the two noise signals are random to each other, the phases of each frequency component are random, and a vector addition of the crosspowers of a group of neighboring frequencies tends to cancel. When the two noise signals are coherent, the phases of each frequency in the crosspower are those of the phase spectrum of the linear operator which generates one signal from the other. If this spectrum is slowly varying as a function of frequency, the addition of neighboring crosspowers do not cancel out. This necessitates some care in evaluating coherency estimates if time delays are involved, or if very sharp frequency structure exists in the relationship between the data.

The variance of the crosspower estimates follows much the same rule as that for the autopower, but an important modification appears in the variance of the coherency estimate. (End effects and differences in computational techniques will make small modifications to these variances.)

$$\text{variance of crosspower estimate} = \frac{M}{N} \, (\text{Power}(a) \; \text{Power}(b))$$

$$\text{variance of coherency estimate} = \frac{M}{N} \, (1 - \left| \text{coherency} \right|^2) \tag{9}$$

This modification arises because the variance in the crosspower estimate is balanced to a certain extent by the variance of the autopowers. When the two noise series are coherent the balancing is exact, and a consistent coherency estimate results. When there is no coherency between two noise series, it takes a large N/M ratio to estimate accurately that the coherency is zero. Because of this one finds, if one attempts to analyze coherency by a visual inspection of records, that coherent records are easily spotted, but slightly or somewhat coherent records appear indistinguishable from incoherent ones. In Fig. 7 are shown sections of synthetic noise data which can be used to demonstrate this effect.

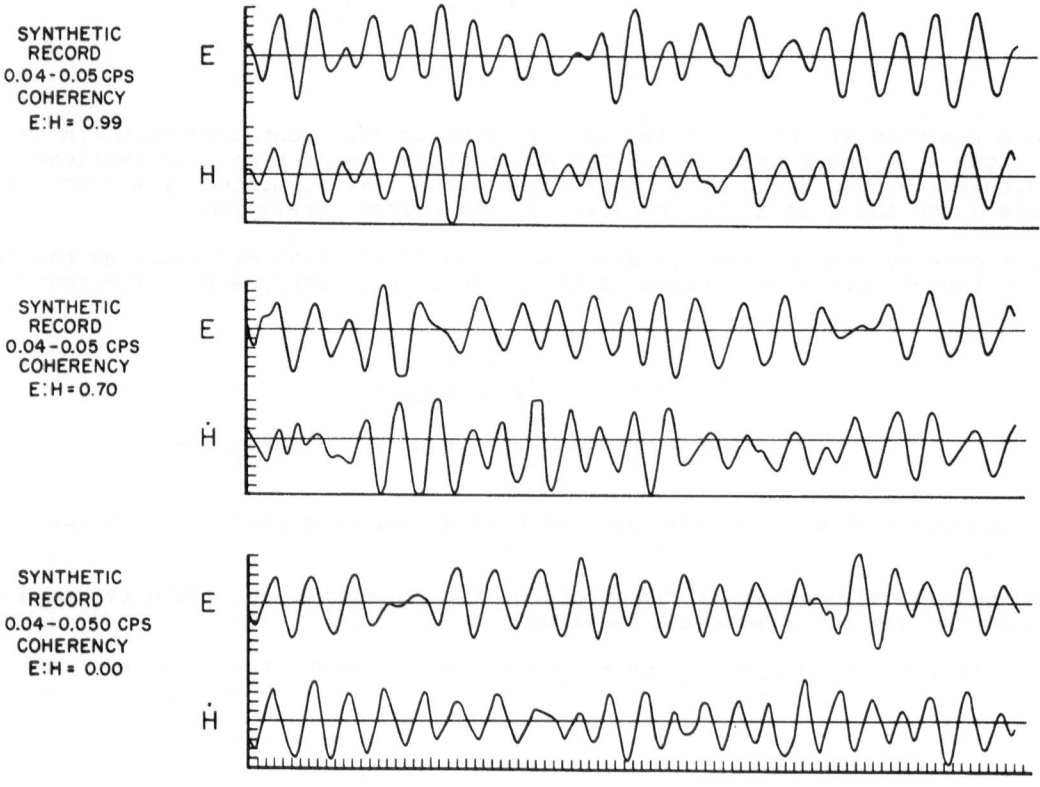

SYNTHETIC RECORD 0.04-0.05 CPS COHERENCY E:H = 0.99   E H

SYNTHETIC RECORD 0.04-0.05 CPS COHERENCY E:H = 0.70   E Ḣ

SYNTHETIC RECORD 0.04-0.050 CPS COHERENCY E:H = 0.00   E Ḣ

Fig. 7. Synthetic Electric-Magnetic Records

Since the coherency is a measure of the amount of relationship between two records, it is very useful in problems where one is interested in establishing that relationships exist. It also establishes some of the characteristics of the relationship. If the coherent signals in the two records are related to each other through an operator or filter which has a certain gain and phase spectrum, the

coherency phase will give us the phase spectrum (if there is any significant coherency). The gain is indeterminant unless the coherency has a magnitude of one, or
unless one knows a priori how much of the power of the incoherent signals is in one
record.

Fig. 8 is an example of some geophysical data, the analysis of which was helped
by the use of coherency computations. This data shows the power density spectrum of
the north-south horizontal magnetic component fluctuations for periods of days and

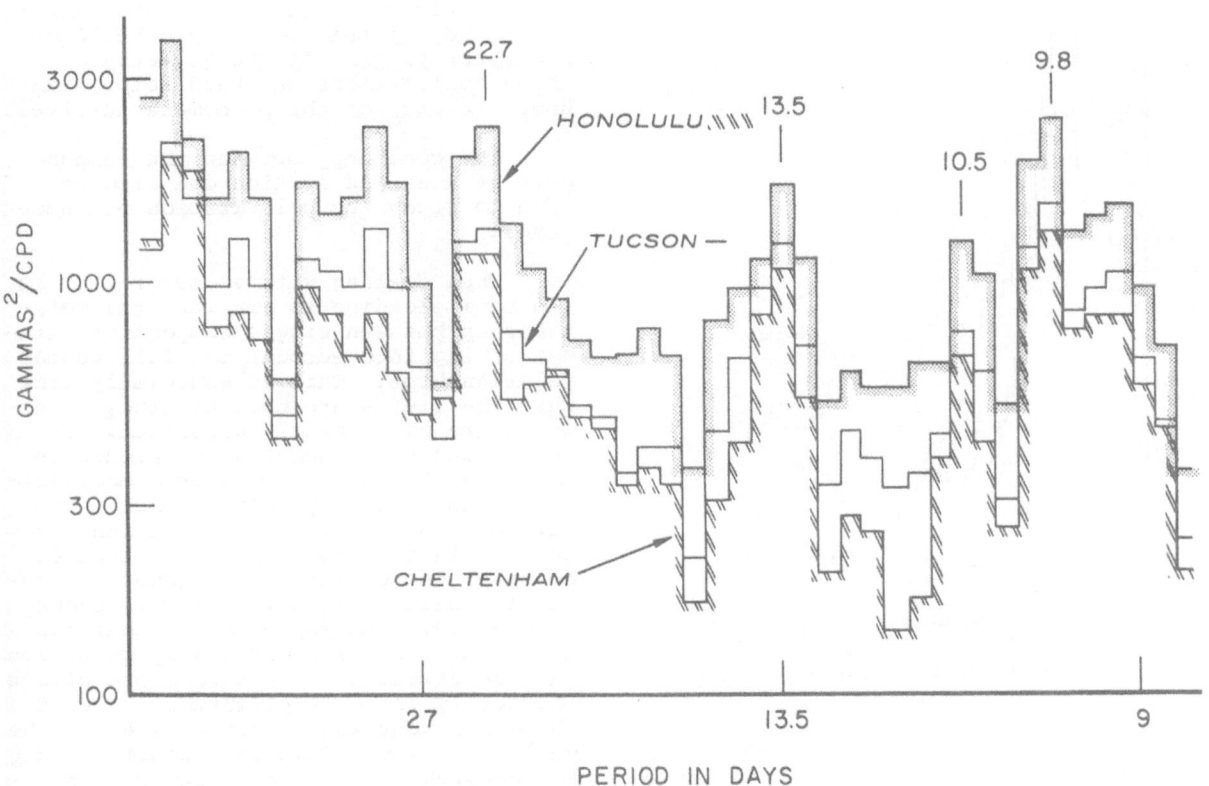

Fig. 8. Period in Days

weeks at several stations. The similarity of the spectra is an indication that the
various stations were observing the same fluctuations, but it is not a proof. The
coherencies between stations, however, were around .95 throughout much of the frequency range with virtually no phase shifts. This is the proof that the fluctuations
were world wide and simultaneous. The source of these fluctuations could also be
checked by studying the coherency of these fluctuations with respect to the magnetic
K index. The variations of the K index had a coherency of 0.9 and better with these
low frequency magnetic fluctuations, and the phase shift was $180°$. This is a strong
indication that almost all of these magnetic fluctuations were due to the negative
main phase of magnetic storms. The latitude and longitude dependence of these fluctuations was almost entirely described for these mid latitude stations by a P_1^0 spherical harmonic, which indicates that the source system had a simple geometry, which is
often referred to as the ring current.

Fig. 9 shows another example of magnetic fluctuations. These power spectra
cover the frequency range associated with oscillations called PC's. (The two power
spectra are displaced vertically in order to be able to observe them separately on

441

the same plot.) It has been suggested that such oscillations might be resonance oscillations of a vorticity mode of magnetohydrodynamic waves traveling back and forth along a small group of magnetic field lines (MacDonald, 1961). The data shown in Fig. 9 were obtained from stations set up by Prof. Smith and his group at the University of Texas and the Air Force Cambridge Research people, and the data can be used to test the vorticity mode hypothesis. The spectral peaks shown in the frequency analysis had the same frequency of about .045 cps at Austin and at Trinidad, and the coherencies between these stations were over .95 at these frequencies. This indicates that the oscillations involved very large horizontal wave lengths and were not restricted to a small group of magnetic field lines. A more extensive study of this sort could build up a clearer picture of the geometry of these oscillations and thus help in an understanding of the phenomena involved.

The coherency between the components at a single station can also be used to study the polarization of these waves.

When dealing with vector fields it can be misleading to consider the relationship between single components without taking into account the full vector relationships. This is especially true when the fields are used to study waves which are not linearly polarized. In such cases the concept of coherency as expressed in Eq. (8) needs some modification. In fact, any multichannel noise data need a more involved treatment when their relationships are being studied. Consider the electric and magnetic fluctuation data from Weston, Mass., shown in Fig. 10. The magnetic fluctuations appear more or less circularly polarized, but the electric fluctuations are almost completely linearly polarized. The NE-SW measurement was almost aligned in the null direction. This is undoubtedly due to the proximity of the ocean-continent boundary, as the continental shelf runs approximately NE-SW in this area. In general, one expects a horizontal electric fluctuation to be related to the horizontal magnetic fluctuation at right

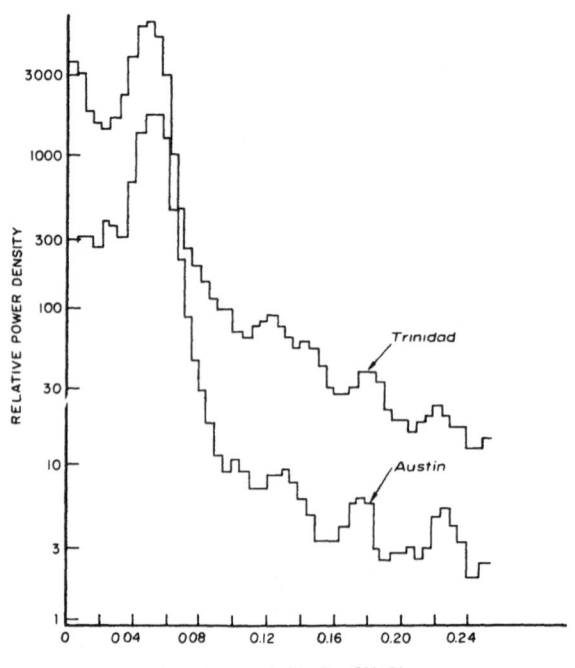

Fig. 9. Frequency in Cycles Per Second

angles to it. At Weston, however, in the frequency range depicted, any magnetic variation produces a NW-SE electric field, and thus, the component of this field in any particular direction is not simply related to the magnetic component in any one other direction. As a matter of fact the coherency between the electric north and the magnetic east variations was very close to zero, and no coherencies between any electric and magnetic components were higher than 0.60. These figures are very misleading by themselves, however, and a more complete analysis showed that the electric fluctuations were almost completely predictable from the magnetic fluctuations, although the converse was not true.

In order to properly study the electric and magnetic field relationships it is necessary to use a matrix treatment

$$E_i(\omega) = A_{ij}(\omega) H_j(\omega) \tag{10}$$

The matrix A_{ij} is the linear operator which predicts the electric field given the magnetic field variations. One can determine this matrix from the crosspower between E and H. If we define

$$\langle E_i H_k \rangle = \text{crosspower between } E_i \text{ and } H_k \text{ (at frequency } \omega)$$

$$\langle\!\langle E_i H_k \rangle\!\rangle = \text{crosspower matrix}$$

$$\langle\!\langle H_i H_k \rangle\!\rangle = \text{magnetic autopower matrix}$$

then

$$\langle\!\langle E_i H_k \rangle\!\rangle = A_{ij} \langle\!\langle H_j H_k \rangle\!\rangle \tag{11}$$

$$A_{ij} = \langle\!\langle E_i H_k \rangle\!\rangle \langle\!\langle H_j H_k \rangle\!\rangle^{-1} \tag{12}$$

Eq. (11) shows that the crosspower between E and H can depend on the crosspower between the H components, and it is possible for the individual coherencies to be much less than one, even though the E field is entirely predictable from Eq. (10). Eq. (12) shows, however, that certain input fields, H_i, will not allow the solution for A_{ij} to be found, as the autopower matrix $\langle\!\langle H_j H_k \rangle\!\rangle$ may be singular. As a matter of fact, it is often data which seems highly coherent which produces a singular condition. The reason for this is that such data does not sample properly all the relationships that A_{ij} expresses.

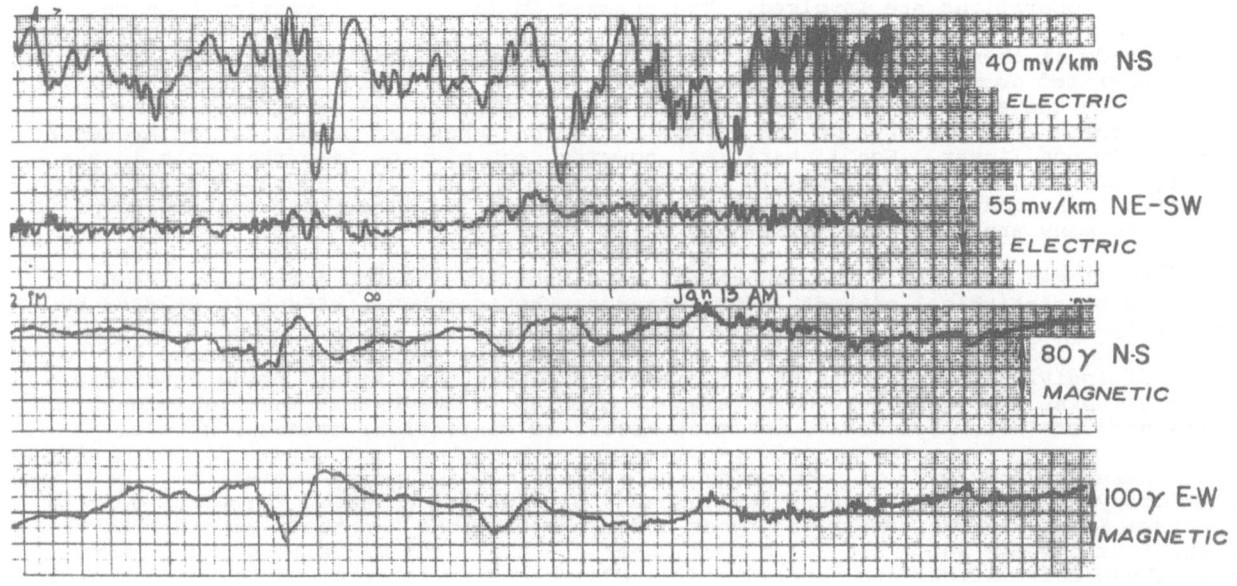

Fig. 10. Magneto Telluric Signals at Weston, Mass.

A similar treatment could have been used to find the operator which predicts H from E. In the case depicted in Fig. 10, this operator, which is the inverse of the A_{ij} matrix, could not be found as the electric autopower matrix was essentially singular.

In place of the concept of the coherency of two components, it is probably most useful to talk of the coherency of a component in relation to a multicomponent field. This can be computed as the coherency of the predicted component to the measured component, when the predicted component is given by Eq. (10) and A_{ij} is computed from Eq. (12). For the data referred to above the electric field components were found to be from .90 to .95 coherent with respect to the magnetic fluctuations.

BISPECTRAL ANALYSIS

In the representation scheme referred to in the first section, it was assumed that the stationary random impulse sequence was a Gaussian white noise process. This assumption not only sets the amplitude distribution of the noise to be a Gaussian distribution, but it also implies that the impulse sequence is completely random, and no further statistics exist that add any information about the series. When this white process is convolved with a wavelet to produce the noise process, the noise is no longer completely random, as the wavelet represents a kind of memory for the immediate past of the noise. The noise is still Gaussian, however, but now the covariance is also needed to describe the statistical properties of the series. All further statistics, such as higher order moments, can be determined simply from just the knowledge of the amplitude distribution and the covariance. Knowledge of the power density spectrum is of course equivalent to knowledge of the covariance. Any subsequent <u>linear</u> operations performed on such a Gaussian noise, such as filtering or the addition of different Gaussian noise processes, will not change the basic Gaussian character of the noise (Middleton, 1960; Wiener, 1958). This is no longer true if <u>non linear</u> operations are involved. The changes in the noise statistics from the Gaussian statistics are related to the nature of the non linearities of the operations and a study of these statistics can therefore give us information about the system that generated or modified the noise. This is a very large and difficult field which is still in the stage of development. The largest impetus for carrying out these studies comes from the area of the communication sciences, as communication processes are highly non linear. Non linearities, however, can play an important role in electromagnetic phenomenon occurring in the ionosphere and in outer space, and the statistical techniques that are developed to investigate non linear effects should prove useful in many areas of geophysics.

Non linear effects in the ionosphere have been studied for some time without recourse to statistical methods. For instance, the Luxembourg effect is studied by measuring the interaction of two transmitted radio signals on each other. Some very interesting ideas on other non linear interactions have been presented based on the study of individual transients of VLF emissions. When the signals being studied are noise signals, and when these noise signals overlap to the extent that individual transients are not discernible, one must resort to statistical methods, however.

There are many ways of describing the statistics of a noise series, but for stationary noise a valid representation and a natural extension of the power density spectrum ideas are the family of higher order correlations and their multidimensional Fourier transforms which are the higher order spectra. For Gaussian noise of zero mean these correlations are expressible in a simple form in terms of the second order covariance function (Wiener, 1958).

$$\langle h(t)h(t + \tau_1)\rangle = \varphi(\tau_1) \quad \text{autocovariance}$$

$$\langle h(t)h(t + \tau_1)h(t + \tau_2)\rangle = 0 \tag{13}$$

$$\langle h(t)h(t + \tau_1)h(t + \tau_2)h(t + \tau_3)\rangle = \varphi(\tau_1)\varphi(\tau_3 - \tau_2) +$$

$$\varphi(\tau_2)\varphi(\tau_3 - \tau_1) + \varphi(\tau_3)\varphi(\tau_2 - \tau_1)$$

etc.

The higher the order, the more prodigious is the task of computing these correlations and presenting the results. We shall only examine the next higher term which is the mean third order product and its two-dimensional Fourier transform which is called the bispectrum (Hasselman et al, 1963).

A simple way of visualizing what the bispectra means is to jump directly into the computational steps and apply the concepts that were discussed in earlier sections. When dealing with finite data lengths we have no problems with limits or the order of integration, and thus we sidestep certain mathematical technicalities. Considering f(t) as our chopped off data set of finite length, the bispectral estimate when properly normalized would be

$$B(\omega_1, \omega_2) \cong \int_{-\infty}^{\infty} \int_{-\infty}^{\infty} \int_{-\infty}^{\infty} f(t)f(t + \tau_1)f(t + \tau_2)e^{-i\omega_1\tau_1}e^{-i\omega_2\tau_2} dt_1 dt_2 dt \tag{14}$$

If we now perform the integrations with respect to τ_1 and τ_2 first we have

$$B(\omega_1,\omega_2) \cong F(\omega_1)F(\omega_2) \int_{-\infty}^{\infty} e^{i\omega_1 t} e^{i\omega_2 t} f(t)dt \tag{15}$$

$$B(\omega_1,\omega_2) = F(\omega_1)F(\omega_2)F\overline{(\omega_1 + \omega_2)} \tag{16}$$

In order for the computations to have a statistical significance, we know that we must average this estimate over a group of neighboring frequencies. This can be done in a similar fashion as with the power spectra by modifying the triple product function before taking its Fourier transform so that the resulting transform is an average over an area of the $\omega_1\omega_2$ plane. It will probably prove more efficient to compute $B(\omega_1,\omega_2)$ directly from Eq. (16) rather than to work with the triple product time function $\langle f(t)f(t + \tau_1)f(t + \tau_2)\rangle$, however. In general as we mentioned in discussing coherency analyses, the frequencies in noise data have independent phase relations with respect to other frequencies, so that an average of the triple product $F(\omega_1)F(\omega_2)F\overline{(\omega_1 + \omega_2)}$ over a group of frequencies should tend towards zero. The variance of this estimate is given as

$$\text{variance of } B(\omega_1\omega_2) \text{ estimate} = \frac{M^2}{N^2} \text{ Power } (\omega_1) \text{ Power } (\omega_2)$$
$$\text{Power } (\omega_1 + \omega_2) \tag{17}$$

where M and N have the same meaning as in the power density estimates.

When non linear interactions take place, it is possible for two frequencies to interact on each other and produce a sum or difference frequency. A simple example is the modulation effect

$$\cos(\omega_1 t + \Phi_1)\cos(\omega_2 t + \Phi_2) = \tfrac{1}{2}\Big[\cos[(\omega_1 + \omega_2)t + \Phi_1 + \Phi_2]$$

$$+ \cos[(\omega_1 - \omega_2)t + \Phi_1 - \Phi_2]\Big] \tag{18}$$

The phase of the generated sum and difference frequencies have a consistent relationship to the sum and differences of the phases of the generating frequencies, and this consistency is measured by the bispectra.

Because of the symmetry of the Fourier components

$$F(-\omega) = \overline{F(\omega)} \tag{19}$$

Eq. (16) has a three fold symmetry and only needs to be determined in an octant of the $\omega_1 \omega_2$ plane.

$$B(\omega_1,\omega_2) = B(\omega_2,\omega_1) = \overline{B(-\omega_2,\omega_1 + \omega_2)} = B(-\omega_1 -\omega_2,\omega_1)$$

$$= \overline{B(-\omega_1,-\omega_2)} = B(\omega_2,-\omega_1 -\omega_2)$$

$$= B(\omega_1,-\omega_1 -\omega_2) = \overline{B(\omega_1 + \omega_2,-\omega_1)} \tag{20}$$

Just as the significance of the amplitude of the crosspower is given by the coherency, so the significance of the amplitude of the bispectra is given by a complex partial skewness coefficient.

$$\text{coefficient of skewness } (\omega_1,\omega_2) = \Big[\Phi_{\omega_1} \frac{B(\omega_1,\omega_2)}{\Phi_{\omega_2}\, \Phi_{\omega_1 + \omega_2}}\Big]^{\frac{1}{2}} \tag{21}$$

Φ = autopower density   B = bispectral density

It is important to remember that both the numerator and denominator of Eq. (21) represent sums of products of terms and not products of sums of terms.

The variance of this skewness coefficient estimate behaves in much the same fashion as the variance of the coherency estimate

$$\text{variance skewness estimate} = \frac{M^2}{N^2}(1 - |\text{skewness}|^2) \tag{22}$$

All the difficulties that were mentioned about the determination of coherency also apply to the determination of the skewness. In fact the skewness coefficient is simply the coherency at the frequency $(\omega_1 + \omega_2)$ between the signal and a signal formed by multiplying the ω_1 and ω_2 band signals together. Thus, if the non linear effects occurred a long distance away, any differences in the time delay of different frequency components can cause a lowering of the bispectral estimate. This can be offset by taking longer data sections and finer frequency intervals. The criteria is that MΔT, where M is the number of independent frequency intervals and ΔT the sampling time must be large compared to the differential delays.

The concepts we have discussed here are not limited to a single noise data set, and in fact, some of the most interesting non linear coupling effects take place between different phenomena. For instance, we can envision gravity waves in the ionosphere modifying magnetohydrodynamic waves that are propagating through the ionosphere. If the gravity waves do not produce a measurable electromagnetic signal, a bispectral analysis of the electromagnetic signal will not show this interaction. In order to observe such cross relations, we must deal with a cross-bispectrum

$$B_{abc}(\omega_1,\omega_2) = \int_{-\infty}^{\infty}\int_{-\infty}^{\infty} \langle b_a(t)b_b(t+\tau_1)b_c(t+\tau_2)\rangle e^{-i\omega_1\tau_1 - i\omega_2\tau_2}\,dt_1\,dt_2 \qquad (23)$$

If we are looking for a gravity wave effect as mentioned above, one would use a signal which was a measure of these waves as well as a signal which was a measure of the magnetohydrodynamic waves and compute $B(\omega_1,\omega_2)$.
$$MH,G,MH$$

Conversely, gravity waves modified by magnetohydrodynamic waves could be studied from $B(\omega_1,\omega_2)$.
$$G,MH,G$$

These same ideas can be further extended to the interactions of more than two phenomena, and to higher order correlations.

Bispectral calculations have not been of common use up to now, and we do not have any good examples to illustrate their use in low frequency electromagnetic measurements. A simpler calculation which is related to the bispectrum can be performed by analyzing the modulation envelope and its relations to the low frequencies in the noise data. The crosspower of the modulation envelope, determined by squaring the data, and the low frequencies, is in fact just one part of the bispectral estimate calculation.

$$\text{crosspower of modulation and signal} = \int_{-\infty}^{\infty}\langle b(t)b(t)b(t+\tau_1)\rangle e^{-i\omega\tau_1}\,dt \qquad (24)$$

For many simple interactions this measure will detect the presence of these interactions, but with a saving of computational effort. At first glance it appeared that the pearl oscillation records previously mentioned might have provided such an example. In Fig. 11 is shown the modulation spectrum of these oscillations displaced

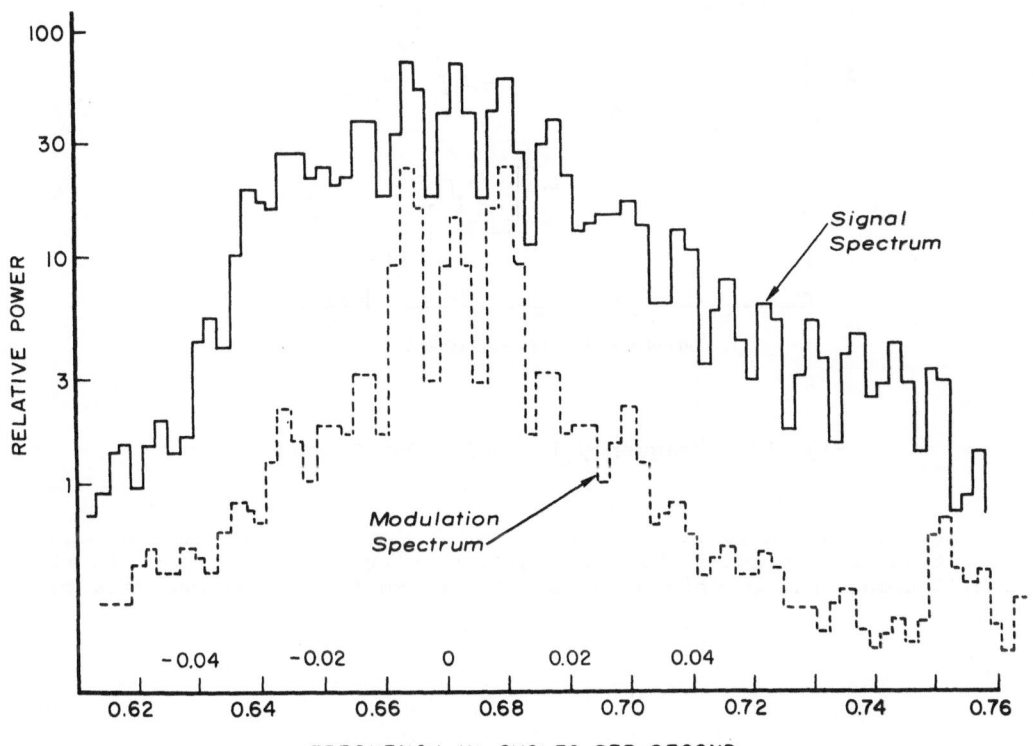

Fig. 11. Frequency in Cycles Per Second

so as to lie alongside the frequency spectrum of the oscillations. If one does not look at the details too closely, the two spectra appear very similar. It is tempting to suggest, therefore, that the oscillation spectrum is the result of convolving the center frequency with the modulation spectrum. Since convolution in frequency is multiplication in time, this would imply that the observed spectrum was due to the modulation of a single narrow band of frequencies. If the modulating phenomenon also produced an electromagnetic signal, then one should observe a significant crosspower between the modulation and the low frequency signals. Fig. 12 shows the low frequency spectrum and the modulation spectrum superimposed. It is seen here that the spectra do not look alike, and as we might suspect, the coherencies were not significant. From our previous discussion of these oscillations, however, we realize that a much better explanation of the oscillation spectra was that of a repeating set of signals.

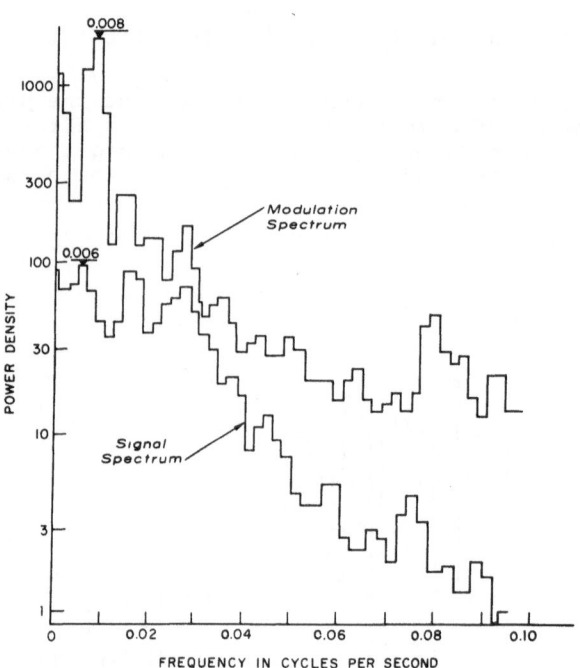

Fig. 12.   Frequency in Cycles Per Second

We have seen in our telluric data examples of micropulsation modulations with period of about 10 to 15 minutes which appeared to correlate with 10 to 15 minute telluric oscillations, but to date we have not carried through an analysis to test this effect.

CONCLUSIONS

This chapter has been concerned with reviewing the concepts that apply to ordinary and higher order spectral analyses of natural signals. The treatment here has been intuitive rather than rigorous, and if any readers unfamiliar with these ideas are interested in applying them in their work, it is important that they refer themselves to the work of the experts in this field. By now the ideas concerning ordinary spectral analysis are well developed and widely known. The application of these ideas to multichannel systems is still undergoing development with a good deal of the stimulus stemming from seismology. The non linear treatments are, however, quite a new field and one that promises to be of considerable importance to geophysicists involved with space studies. In the author's opinion, based on his own experiences, the intelligent application of all these techniques can greatly increase the insight that an investigator develops about natural phenomena from his studies of the noise signals generated.

ACKNOWLEDGEMENTS

This author is not an innovator in this field, and most of his knowledge about the subject has come directly from his associations with scientists who are. In this connection, he is fortunate to have been in close contact with Prof. Simpson of M.I.T. and Prof. Robinson, formerly of the University of Wisconsin and now visiting at the University of Uppsala, who are concerned with many aspects of the theory and practice of the analysis of noise data. Some of the ideas developed by Prof. Wiener and expanded upon by Prof. Lee and his associates at M.I.T. on the representation of non Gaussian noise and non linear operators has filtered down to the author through graduate students. Prof. Munk of the University of California, San Diego, first interested the author in the application of bispectral analyses to geophysical phenomena.

The author wishes to acknowledge the support he has received from the Office of Naval Research for his researches on low frequency electromagnetic phenomena. He also wishes to acknowledge the cooperation of Prof. Vozoff of the University of Alberta and of Mr. Orange of Air Force Cambridge Research, who provided data for some of the analyses referred to. All of the author's researches in the field of low frequency electromagnetic phenomenon have been carried out in close cooperation with Prof. Cantwell of M.I.T., and his indebtedness to Prof. Cantwell covers every phase of this work. Among the graduate students and former graduate students whose ideas and researches and help the author has leaned upon should be mentioned Mr. Ken Larner, Mr. James Everett, Mr. John Claerbot, Mr. Ralph Wiggins, Dr. Donald Eckhardt, Dr. James Galbraith, and Dr. Freeman Gilbert. Prof. Richard Haubrich and Prof. Freeman Gilbert of the University of California, San Diego, read the manuscript and pointed out certain corrections that were incorporated.

Mrs. Luella Brozewich, Mrs. Barbara Gold, Mrs. Mary Mahar, and Miss Elizabeth Frank of the staff of the Institute of Geophysics and Planetary Physics have been very helpful in preparing this manuscript and I wish to thank them and all the people mentioned above.

REFERENCES

Blackman, R. B., and J. W. Tukey: "The Measurement of Power Spectra from the Point of View of Communications Engineering," Dover Publications, New York (1958)

Goodman, N. R.: "On the Joint Estimation of the Spectra, Cospectrum, and Quadrature Spectrum of a Two-Dimensional Stationary Gaussian Process," Scientific Paper No. 10, Engineering Statistics Laboratory, New York University (1957)

Hannan, E. J.: "Time Series Analysis," Methuen, London (1960)

Hasselmann, K., W. Munk, G. MacDonald: "Bispectra of Ocean Waves," Chap. 8 in Proceedings of the Symposium on Time Series Analysis, M. Rosenblatt, editor, John Wiley and Sons, New York (1963)

Helliwell, R. A., J. H. Crary, J. P. Katsufrakis, M. L. Trimpi: "The Stanford University Real-Time Spectrum Analyzer," Technical Report No. 10, Radioscience Laboratory, Stanford Electronics Laboratories, Stanford University (1961)

MacDonald, G.: "Spectrum of Hydromagnetic Waves in the Exosphere," J. of Geophys. Res., 66, 3639-3671 (1961)

Middleton, D.: "Statistical Communication Theory," McGraw-Hill, New York (1960)

Robinson, E. A.: "Predictive Decomposition of Time Series with Applications to Seismic Exploration," Ph.D. Thesis, M.I.T. (1954)

"Random Wavelets and Cybernetic Systems," Griffin's Statistical Monographs and Courses 6, Charles Griffin, London (1962)

Simpson, S. M.: "Time Series Techniques Applied to Underground Nuclear Detection and Further Digitized Seismic Data," Scientific Report No. 2 of contract AF 19(604)7378 Project Vela Uniform, M.I.T. (1961)

"Magnetic Tape Copies of M.I.T. Geophysics Program Set I (Time Series Programs for the IBM 709, 7090)," Scientific Report No. 4 of contract AF 19(604)7378 Project Vela Uniform, M.I.T. (1962)

Wiener, N.: "Non linear Problems in Random Theory," John Wiley, New York (1958)

Wold, H.: "A Study in the Analysis of Stationary Time-Series," Thesis, University of Stockholm, Almqvist and Wiksells, Uppsala, Second Edit. (1954)

SUMMARY REPORT
of
NATO Advanced Study Institute on
Low Frequency Electromagnetic Radiation
Bad Homburg, Germany

S. H. Ward

INTRODUCTION

This conference has been concerned chiefly with natural electromagnetic radiation in the frequency band from 10^{-4} cps to 3×10^4 cps. While this _eight_ decade range of frequencies may seem to be an excessive range to cover in a two-week conference, it has been brought out in formal presentations and discussions that we cannot hope to understand any one phenomenon within this range without knowledge of all other phenomena. In fact, it has become obvious that our progress in understanding our environment has been hindered by a "pigeon-hole" approach to geomagnetism and aeronomy. The correlation between the sudden commencement of polar elementary storms and the onset of such phenomena as micropulsation activity, ionospheric absorption, VLF hiss, aurora, bremsstrahlung X-rays, and attenuation of r.f. propagation in the earth ionosphere wave guide is evidence for the desirability of correlation studies of all electromagnetic phenomena.

The ultimate source of energy of all natural low frequency electromagnetic radiation is the sun, and our main problem in this field is in studying the mechanism for converting the sun's particulate and radiative energy to the electromagnetic fields which we observe at the surface of the earth. Of course, we may aid our studies by utilizing artificial sources such as radio transmitters and atmospheric nuclear detonations.

The sun, according to recent theory, is expanding continuously. Since the sun is rotating, the gaseous matter of the sun, a plasma, moves radially into space. The flow is irregular since the sun is a turbulent regime. The turbulence leads to many local magnetic fields, which, because of the velocity in the solar corona, will be "frozen in" the plasma.

The earth, in its orbit about the sun, will intercept this irregular plasma. An interaction between the solar plasma and the earth's main magnetic field results. The nature of the reaction is controversial. Spreiter and co-authors and Beard treat the problem from a purely dynamical viewpoint, although Beard lately allows for the frozen in field of the solar wind. The theoretical result is that the earth's dipolar magnetic field is confined to a cavity - the magnetosphere - elongated opposite to the direction of the advancing solar wind. A sharp discontinuity in magnetic field direction, and often in intensity, has been established experimentally at the sunward boundary of the magnetosphere. No experimental evidence is available on the opposite side or "tail" of the magnetosphere. Energy may be transferred from the solar plasma across the sunward side of the magnetospheric boundary by generation of hydromagnetic waves. The transfer of particulate matter from the solar wind is not accounted for by this theory.

Axford and Hines have modified the theory to allow the particulate matter to enter the tail of the magnetosphere by convection. Turbulent energy can be transferred to the night side of the magnetosphere in this manner.

Alfvén has tested the Birkeland-Störmer theory of precipitation of energy and

451

particulate matter into the magnetosphere by laboratory experiments on terella subjected to plasma streams. Alfvén concludes from these experiments and physical reasoning that a "hard-shelled" magnetosphere does not exist. Particle precipitation and energy transfer is readily accounted for in this manner. Alfvén does obtain a sharp field discontinuity for a limited region on the sunward side, but no such discontinuity elsewhere.

While these theories are interesting, no one theory can be conclusive; experimental evidence is required. This evidence must come from magnetic field and particle measurements made from satellites traversing the region from 5 to 50 earth radii about the earth. Such experimental evidence should come from the forthcoming Ego and Pogo satellites. While these satellite excursions should settle the controversy concerning the nature of the magnetosphere, many more questions will remain unanswered. Perhaps the most important unanswered question today and quite probably for some time, is a description of the time dependency and constitution of the solar wind. We have attempted to monitor the solar wind with a plasma probe aboard the successful Mariner space flight to Venus. This has confirmed our intuitive guess that the plasma is irregular. Succeeding space probes are not apt to learn significantly more, for each such experimental probe disturbs the environment in which it is placed. A shock wave seems to be formed in advance of the space vehicle so that plasma probes in the vehicle are unlikely to be able to monitor the undisturbed solar plasma. For a general description of the plasma, this is no great hindrance, but we are now in need of detailed knowledge of the time dependency, ionization, and departure from neutrality of the plasma. We must then contemplate the design of an experiment which will provide us this information. Perhaps a complete description of the solar plasma is the most important accomplishment we could make in the ensuing years.

WITHIN THE MAGNETOSPHERE

With a description of the solar plasma and of the boundary of the magnetosphere, we then presumably would have knowledge of the mechanisms whereby energy is transferred from the plasma to the magnetosphere. With this knowledge, and with full cognizance of the purely radiative contributions of solar energy to the magnetosphere, we would be in position to start to unravel the intricacies of the more detailed field and particle behavior in the magnetosphere. Even today, working solely on the output of the system without detailed knowledge of the input, we have explained many facets of natural electromagnetic radiation. It is evident that whistler and VLF emission theory developed by Gallet, Helliwell and others permits some remarkable predictions and deductions concerning the exosphere. Perhaps no other aspect of geomagnetism and aeronomy has met with such success. Yet one hesitates to accept all of the theory for the origin and propagation of VLF emissions and whistlers since local irregularities in the particle density is not taken into account by the theory. In this regard, we are reminded of the dynamic irregular ionosphere which is required to explain sporadic E ionization. Surely, similar inhomogeneities exist in the exosphere. We should not become permanently enamored of simple theories originally designed to aid our understanding. Nature is complex!

Much emphasis has been placed on geomagnetic storms and the auroral zones in attempting to obtain an understanding of solar-terrestrial relations. While this approach has yielded much information, it is evident that both geomagnetic storms and auroral zones are anomalous features.

We should, perhaps, concentrate now on average geomagnetic disturbance and especially average geomagnetic disturbance in middle and low latitudes. This approach is logical when one considers that the magnetospheric system in general is non-linear but may be treated as linear for small perturbations. The auroral zone is far too active, magnetically, to permit a linear analysis. Thus, low-latitude conjugate point studies of micropulsations, ionospheric absorption, etc. need to be conducted. In fact we need conjugate point experiments for all L values because we know that the particle density in the exosphere is an irregular function of L.

The conjugate point experiments completed to date indicate the need for a widely spaced array of stations about one terminus of the field line since transverse electric fields cause a wandering of the conjugacy. Multicomponent recording of the various electromagnetic radiations would permit definition of this "loose" conjugacy. Mode conversion and ionospheric absorption can lead to an apparent loose conjugacy if only one component is measured at one location at the terminus of a field line. Of course only one station with three components is required at the other terminus. Considering the manner in which conjugate point experiments have been conducted to date, it is rather surprising that any degree of conjugacy has been recognized in the micropulsstion and Schumann resonance bands.

Of course, we have very little knowledge yet of the origin of micropulsations, and particularly so in the 3 to 300 second band. Definitive experiments which should be conducted include:

(1) Correlation of events between dark and light hemispheres without regard to classical subdivisions, but with a full statistical evaluation. This should be conducted for low and middle latitudes.

(2) The attenuation of the ionosphere for micropulsations.

(3) The direct correlation of particle precipitation events with micropulsations in all frequency bands. This can perhaps be carried out best with both counters and magnetometers carried aloft in balloons.

(4) A very complete study of the latitudinal dependency of the periods of micropulsations. Past studies of this feature have yet to produce entirely conclusive results.

Wherever possible, these experiments should be conducted utilizing three magnetic and two electric components. The effect of ground conductivity and permeability should be evaluated at each site so that we might, for example, determine whether high coherency between any orthogonal pair is a function of the interior or exterior of the earth.

In the "pearl" band, 0.3 to 3 second period, remarkable progress has been made in studying the probable sources of the various types of signals. Care must be exercised to recognize that several types of signals and several sources do exist in this band.

WITHIN THE EARTH-IONOSPHERE CAVITY

The prediction by Schumann of resonant modes of propagation within the earth-ionosphere cavity has been verified experimentally under many ionospheric conditions. Diurnal variations, seasonal variations, and variations induced by nuclear detonations have all been considered, but no evidence has yet been presented of a study of the solar cycle variation of resonant modes.

Dr. T. R. Madden has pointed out that the determination of the Q of the cavity has been made routinely but is erroneous when line splitting occurs.

The effect of an anisotropic ionosphere, of a sharp and an exponential lower boundary of the ionosphere have all been allowed for in theoretical analyses of the resonant modes.

The excitation of the cavity seems to be mostly due to lightning strokes, but VLF emissions are generated frequently and can in some latitudes contribute a substantial proportion of the energy in the earth-ionosphere cavity. A study of the contributions of VLF emissions and of sferics to the energy supplied to the cavity is indicated.

Of course, cavity resonances are but one aspect of natural electromagnetic propagation in the earth-ionosphere wave guide. We should not all "climb aboard the Schumann bandwagon," but carefully consider other problems of VLF propagation in this natural wave guide.

THE DESIGN OF EXPERIMENTS

One cannot help but feel that more emphasis needs to be placed on experimentation and less on speculation in studying our environment. However, many of our past experiments have been imperfectly designed. There appears to be need for all of us to contemplate experiment design with at least as much vigor as we contemplate rushing into print with our latest results. One cannot overemphasize the importance of good design in experimentation.

Perhaps I should close this summary report with a request to all to consider the design of the ultimate experiment - a controlled nuclear explosion outside the magnetosphere. With this artificial sun of known characteristics, we could expose answers to many of our most vexing problems. Yet, I doubt that any man today is capable of designing such an experiment. Certainly many of our past atmospheric nuclear detonations have been poorly designed and potentially dangerous experiments.

It is time to look before we leap.

ATTENDEES AT NATO ADVANCED STUDY INSTITUTE ON LOW-FREQUENCY ELECTROMAGNETIC RADIATION

Bad Homburg, Germany
July 22 - August 2, 1963

Mr. Hans Joachim Albrecht
Fluggeraetewerk Bodensee G.M.B.H.
Ueberlingen, West Germany

Mrs. H. J. Albrecht
P. O. Box 62
Unteruhldingen, West Germany

Prof. H. Alfvén *
The Royal Institute of Technology
Department of Electronics
Stockholm 70, Sweden

Dr. W. I. Axford *
Department of Astronomy
Cornell University
Ithaca, New York

Mr. Arnold B. Bailey
Mitre Corporation
Bedford, Massachusetts

Mr. A. J. B. C. Bal
Physisch Laboratorium R.V.O.-T.N.O.
Vlakte van Waalsdorp
The Hague, The Netherlands

Dr. David F. Bleil * *
U. S. Naval Ordnance Laboratory
White Oak
Silver Spring, Maryland

Dr. Derrill Joseph Bordelon
NATO Scientific Affairs Division
NATO Building
Place du Marechal
De Lattre de Tassigny
Paris, France

Mr. Lars Brock-Nannestad
NATO Saclant ASW Research Center
Viale San Bartolomeo, 92
La Spezia, Italy

Dr. Wallace Hall Campbell
National Bureau of Standards
Boulder, Colorado

Dr. Edwin Denis Dracott
Boundary Hall, Tadley
Basingstoke, Hants.
England

Dr. H. J. Duffus
Head of Physics Department
Canadian Services College
Royal Roads, Victoria
British Columbia, Canada

Dr. Alv Egeland
The Royal Swedish Academy of Science
Kiruna Geophysical Observatory
Kiruna C, Sweden

Dr. I. Estermann * *
Office of Naval Research
Branch Office, 429 Oxford St.
London, England

Dr. Warren L. Flock
Geophysical Institute
University of Alaska
College, Alaska

Mr. Maurise Fournet
Sud-Aviation
40 rue de l'Industrie
Courbevoie, Seine
P. O. Box 106
Courbevoie, France

Mr. Hugo Fournier
Garchy
Nievre, France

Mr. D. C. Fraser
Admiralty Research Laboratory
Teddington
Middlesex, England

Dr. Janis Galejs *
Applied Research Laboratory
Sylvania Electronics Systems
40 Sylvan Road
Waltham, Massachusetts

Dr. Roger M. Gallet *
National Bureau of Standards
Boulder, Colorado

Dr. Roger Gendrin
Groupe de Recherches Ionospherique
Centre National d'Etudes des
    Telecommunications
38 Avenue du General Ledere
Issy-les-Moulineaux (Seine)
France

Dr. John J. Gibbons *
Ionosphere Research Laboratory
Penn State University
University Park, Pennsylvania

Mr. Perikles Giouleas
National Observatory of Athens
Ionospheric Institute
Theseum, Athens (3), Greece

Prof. Leiv Harang *
Norwegian Defence Research Establishment
Kjeller, Norway

Mr. John G. Heacock
Office of Naval Research
Washington 25, D. C.

Dr. James R. Heirtzler *
Lamont Geological Observatory
Columbia University
Palisades, New York

Dr. Ernest A. Hogge
U. S. Navy Mine Defense Laboratory
Panama City, Florida

Dr. E. H. Hurlburt
Office of Naval Research
Washington 25, D. C.

Dr. J. A. Jacobs *
University of British Columbia
Vancouver 8, British Columbia, Canada

Dr. John Katsufrakis *
Radio Science Laboratory
Stanford University
Stanford, California

Dr. Ing. Herbert L. Konig
Elektrophysikalisches
Institut of the Technische
Hochschule München
Munchen, Arcisstrasse 27, Germany

Mr. Martin B. Kraichman
U. S. Naval Ordnance Laboratory
White Oak
Silver Spring, Maryland

Dr. G. Lange-Hesse
Max-Planck-Institute for Aeronomy
3411 Lindau ueber Northeim/Hann
Germany

Dr. Edward A. Lewis
Air Force Cambridge Research Laboratory
Hanscom Field
Bedford, Massachusetts

Dr. J. E. Lokken *
Department of National Defence
Defence Research Board
Pacific Naval Laboratory
Esquimalt, British Columbia, Canada

Dr. Theodore R. Madden *
Department of Geology and Geophysics
Massachusetts Institute of Technology
Cambridge, Massachusetts

Dr. Robert W. E. McNicol
Boeing Scientific Research Laboratories
Geo-Astrophysics Section
P. O. Box 398
Seattle 24, Washington

Mr. Alvin McNish *
National Bureau of Standards
Washington 25, D. C.

Prof. Carl E. Menneken * *
Office of Naval Research
Branch Office, 429 Oxford St.
London, England

Mr. Georges Moraitis
National Observatory of Athens
Ionospheric Institute
Theseum, Athens (3), Greece

Dr. Minoru Paul Nakada
National Aeronautics and Space Administration
Goddard Space Flight Center
Greenbelt, Maryland

Dr. Alastair W. Nichol
University of Exeter
Department of Physics
Exeter, Devon, England

Dr. Charles Polk
Department of Electrical Engineering
University of Rhode Island
Kingston, Rhode Island

Prof. Ivo Ranzi
Istituto Superiore P. T.
Viale Trastevers 189
Rome, Italy

Mr. Bodo Reinisch
Ionospharen Institut
Breisach/Rhein, Germany

Dr. Eduoard Selzer *
Institut de Physique du Globe
191 rue Saint-Jacques
Paris V, France

Dr. Arnold Shostak
Office of Naval Research
Washington 25, D. C.

Dr. Harold W. Smith
Electrical Engineering Research Laboratory
University of Texas
P. O. Box 7789
Austin 12, Texas

Dr. Hans M. Strack
Kiel-Western Germany
Fa. HAGENUK

Dr. Masahisa Sugiura *
National Aeronautics and Space
Administration
Goddard Space Flight Center
Greenbelt, Maryland

Dr. Stanley H. Ward
Space Sciences Laboratory
University of California
Berkeley, California

Mr. Erwin G. Weber
Control Data
Bad Homburg, Germany

Mr. Wallace D. Westfall
U. S. Navy Electronics Laboratory
San Diego 52, California

Mr. John B. Wilcox
U. S. Naval Ordnance Laboratory
White Oak
Silver Spring, Maryland

Sir Charles Wright *
Pacific Naval Laboratory
H.M.C. Dockyard
Esquimalt, Victoria
British Columbia, Canada

Mr. Isidore Zietz
U. S. Department of the Interior
Geological Survey
Washington 25, D. C.

* Faculty Members
** Directors

AUTHOR INDEX

A

Aarons, 158-159, 165, 351, 421
Abe, 59
Abragam, 382
Abraham, 237
Abrams, 26, 364
Ahluwalia, 5
Aikin, 95, 207-208
Akasofu, 51, 62, 65-66, 291, 311, 314
      347, 362
Albrecht, 206
Alfven, 33-34, 61, 67, 327-328, 451-452
Allcock, 263, 265
Alldredge, 396-397
Alperovich, 324
Amazeen, 11, 51
Angenheister, 288
Anger, 292, 296
Apel, 62
Appleton, 103, 179
Arendt, 288
Axford, 5-6, 11, 16, 18, 20, 23, 25-26,
      52, 67, 293, 312, 364, 451

B

Back, 385, 388
Baker, 59
Balser, 205, 213-214, 216, 219, 227, 234-
      235, 237, 240, 252, 340, 377, 417
      418
Balton, 132
Bandyopodhyay, 19
Barcus, 292, 295
Barkhausen, 262
Barrington, 207, 266
Barron, 165
Bartels, 5, 19-20, 51, 59-60, 62, 65
Bath, 116
Beard, 10, 51, 451
Bell, 384-385, 389
Bellas, 98
Belon, 363
Belrose, 206, 266
Bender, 380, 387
Benioff, 57, 288, 320, 324, 326, 338,
      362-363
Bennett, 77
Benoit, 216, 382
Berthold, 57, 113, 361
Bessel, 220, 242
Biermann, 5, 51
Birkeland, 66, 77, 82, 287, 290-291, 300
      311, 451

Blackman, 433
Bleil, 1, 365, 394, 401
Block, 41
Bloom, 384-386, 388-391
Boardman, 262
Bomke, 119, 129, 132, 336
Bonetti, 12
Bonnet, 383
Booth, 351
Bostick, 361, 410-411
Bourdeau, 408, 210, 231, 233, 235
Bowles, 103
Boyd, 293, 298
Bozorth, 410-411
Brackmann, 64
Bremmer, 221
Brice, 267, 315
Briggs, 10, 51
Brock-Nannestad, 205
Brown, 208, 292, 295
Bryant, 298
Budden, 103, 205, 226, 264
Burton, 262

C

Cahill, 11, 13, 43, 51
Cain, 62
Campbell, 92, 129-130, 267, 292-295, 319,
      324, 335-336, 346, 361, 363, 415
Cantwell, 449
Carpenter, 14, 175, 265-266
Casaverde, 59
Cecchina, 334
Cerenkov, 167
Chamberlain, 25, 66
Chan, 363
Chapman, 5, 7, 9, 11, 19-20, 25-26, 51,
      59-62, 65-67, 84, 205, 215, 226,
      233, 291, 311, 314, 347
Christoffel, 376
Chrzanowski, 89
Claerbot, 449
Clark, 208, 210, 231, 233, 235
Cole, 6, 20, 67
Colegrove, 391
Coleman, 3, 5, 365
Cook, 89
Corcuff, 265, 266
Coulomb, 351, 362
Cowling, 67
Crain, 114, 125-127

T

Tamarkin, 114, 125-127
Tanner, 214, 215, 229
Taylor, 222
Tepley, 213, 320, 326-332, 336
Thellier, 402-404, 407
Thomas, 27, 266
Thompson, 205, 221, 225, 247, 252
Thrane, 207
Triom, 20
Troitskaya, 2, 113, 288, 314-315, 320,
        324-327, 332, 334-335, 347
        352, 360, 413
Tukey, 433
Turtle, 307-308, 310, 316

U

Ulwick, 207
Ungstrup, 143
Unthank, 59
Unwin, 114

V

Vacquier, 396
Van Allen, 3, 11-14, 77, 87, 111, 169,207
van Bremmelen, 288
Vancour, 207
Veldkamp, 361
Venkatesan, 62
Vestine, 25, 59-60, 65-67
Vilbig, 206
Villard, 363
Vitousek, 131
Vladimirov, 319, 337
Voelker, 361
Vogan, 25, 27, 29
Vozoff, 326, 412, 437, 449

W

Wagner, 205, 213-216, 219, 227, 234-235,
     237, 240, 252, 340, 377, 417-418
Wait, 205-206, 213, 215, 219, 221, 225-
     227, 229, 234, 239, 245, 247

Waitr, 385-387
Ward, 2, 216, 259, 348, 355, 391, 451
Warren, 27
Watanabe, 58, 327-329, 351, 360-362
Watson, 205
Watt, 206, 213
Weaver, 354, 376
Weeks, 59
Weir, 415-416
Wentworth, 62, 320, 327, 329, 332, 336
Wentzel, 3
Westcott, 293, 297, 353, 355
Westphal, 319, 360-361
Whipple, 208, 210, 231, 233, 235
Whitehead, 421
Whitham, 380-381, 394, 396, 402-404
Wiener, 444, 449
Wiggins, 449
Williams, 213, 239, 414
Wilson, 26, 51-55, 58, 353
Wold, 430
Wright, 216, 287, 291, 321-322, 373, 410

Y

Yabu, 353
Yamakawa, 353
Yanagihara, 2, 324, 328, 337-341, 360,
      362-363
Young, 89, 92
Yu, 216

Z

Zeeman, 385, 387, 389
Zhigulev, 51
Zmuda, 216
Zschörner, 266

Receiving equipment
  for micropulsation band, 417
  low ELF band, 417
  total field magnetometer, 390
Recording equipment
  frequency-multiplex tape, 416
  magnetic tape, 415, 418
  paper charts, 415, 418
Recording galviometer, 404
Recovery phase
  storms, 49, 62
Recovery rate,
  change, 63
Reflection
  group height, 98
  ratio waves from ionosphere, 82
Rejection input filter, 413
Ring current, 15, 25
Riometer events, 305
Riometer recordings, 148
Russian tests
  characteristics of, 109
Satellites
  data from, 364
  Explorer X, 12
  Explorer XII, magnetosphere surface
    transition, 11, 13
  Explorer XIV, omnidirectional intensity
    of electrons, 11
  INJUN I, 12, 29
  Pioneer I, magnetic record, 13
  Pioneer III, counting rate, 12
  Pioneer IV, counting rate, 12
Schumann resonance, 436
  diurnal variations, 217
Shock wave, 7
Solar corona, 5
  tide, 19
  particles, 26
Solar magnetic field, 5
Solar wind on magnetosphere, 23
Sound waves
  longitudinal, 26
Space charge waves, 180
Space probes, 5-6, 11-12
Spectra
  formulas for computing frequency-time,
    195
  VLF emission, 194
Spectral analysis, 431
  power density, 432
  Schumann resonance, 436
  pearl oscillations, 437
Spectrogram
  chorus bursts, 281
  hiss, 283
  polar chorus, 282
SSC's
  elliptically polarized, 54

SSC's (continued)
  Honolulu, 53
  polarization of horizontal perturba-
    tion, 56
  polarization rules, 55
  Tucson, 50
Starfish characteristics, 109
Stationary noise, 430
Stations
  Byrd (BY), 276
  Carde, 276
  Dunedin (DU), 276
  Eights (EI), 276
  Great Whale River (GWR), 276
  in Africa, 367
  in Antarctica, 368
  in North America, 366
  in Western Europe, 366
  in Western Pacific, 367
  map of conjugate stations, 292
  sky-Hi, 279
  Unalaska (UN), 276
Storms
  antiparallel, 46
  geomagnetic, 49
  induced electromotive force, 62
  initial phase, 60
  main phase, 61
  noise, 154
  parallel, 46
  recovery phase, 49
  sudden commencement, 49
Subsonic wind, 7
Sudden commencement
  storm, 49
Supersonic wind speed, 6-7
Surface impedance measurements, 218
Teak characteristics, 109
Telluric currents, 413
  at Garchy, 118
  at Kerguelen, 119
  electrodes for, 414
  instrumentation for, 413
Telluric current ellipse
  variations in, 355
Terrella, 1, 33
  antiparallel fields, 43
  model, 41
  parallel fields, 43
  quantities, 42
Terrestrial propagation
  ELF, 205
Transformation factors, 42
Trapped particles, 3
Trapped radiation, 11, 20, 24
  electrons, 11, 25, 29
Traveling wave tube mechanism
  generalized, 187
  whistler frequencies, 187

ERRATA